Springer Proceedings in Mathematics & Statistics

Volume 325

Springer Proceedings in Mathematics & Statistics

This book series features volumes composed of selected contributions from workshops and conferences in all areas of current research in mathematics and statistics, including operation research and optimization. In addition to an overall evaluation of the interest, scientific quality, and timeliness of each proposal at the hands of the publisher, individual contributions are all refereed to the high quality standards of leading journals in the field. Thus, this series provides the research community with well-edited, authoritative reports on developments in the most exciting areas of mathematical and statistical research today.

More information about this series at http://www.springer.com/series/10533

Jacek Banasiak · Adam Bobrowski ·
Mirosław Lachowicz · Yuri Tomilov
Editors

Semigroups of Operators – Theory and Applications

SOTA, Kazimierz Dolny, Poland, September/October 2018

In Honour of Jan Kisyński's 85th Birthday

Springer

Editors
Jacek Banasiak
Department of Mathematics
and Applied Mathematics
University of Pretoria
Hatfield, South Africa

Institute of Mathematics
Łódź University of Technology
Łódź, Poland

Mirosław Lachowicz
Institute of Applied Mathematics
and Mechanics
University of Warsaw
Warsaw, Poland

Adam Bobrowski
Department of Mathematics
Lublin University of Technology
Lublin, Poland

Yuri Tomilov
Institute of Mathematics
Polish Academy of Sciences
Warsaw, Poland

ISSN 2194-1009　　　　　ISSN 2194-1017　(electronic)
Springer Proceedings in Mathematics & Statistics
ISBN 978-3-030-46081-5　　　ISBN 978-3-030-46079-2　(eBook)
https://doi.org/10.1007/978-3-030-46079-2

Mathematics Subject Classification: 34-XX, 45-XX, 46-XX, 47-XX, 60-XX, 92-XX

© Springer Nature Switzerland AG 2020

This work is subject to copyright. All rights are reserved by the Publisher, whether the whole or part of the material is concerned, specifically the rights of translation, reprinting, reuse of illustrations, recitation, broadcasting, reproduction on microfilms or in any other physical way, and transmission or information storage and retrieval, electronic adaptation, computer software, or by similar or dissimilar methodology now known or hereafter developed.

The use of general descriptive names, registered names, trademarks, service marks, etc. in this publication does not imply, even in the absence of a specific statement, that such names are exempt from the relevant protective laws and regulations and therefore free for general use.

The publisher, the authors and the editors are safe to assume that the advice and information in this book are believed to be true and accurate at the date of publication. Neither the publisher nor the authors or the editors give a warranty, express or implied, with respect to the material contained herein or for any errors or omissions that may have been made. The publisher remains neutral with regard to jurisdictional claims in published maps and institutional affiliations.

This Springer imprint is published by the registered company Springer Nature Switzerland AG
The registered company address is: Gewerbestrasse 11, 6330 Cham, Switzerland

Preface

The presented volume is, to some extent, a sequel to the book of proceedings *Semigroups of Operators: Theory and Applications* published in 2015. They both contain selected, peer-reviewed papers related to conferences that have been organized under the same title and held every five years since 2008 to bring together the semigroup community and to celebrate consecutive birthdays (75th, 80th, and 85th, respectively) of Prof. Jan Kisyński. The previous volume reflects, to some degree, the second edition of the conference which took place in Będlewo, Poland, in October 2013. The present volume is devoted to its third edition in Kazimierz Dolny, Poland, in October 2018.

Since the semigroup history does not change as quickly as the history of some other events, most of what we wrote in the preface to the previous volume is still correct today, and thus we will take the freedom of incorporating parts of it here. The foundations of the theory of semigroups of operators can be traced back to the formulation of the Cauchy equation,

$$f(s+t) = f(s)f(t),$$

and the work of G. Peano and his student M. Gramegna who, at the turn of the nineteenth and twentieth century, derived formulae for the solutions of, respectively, finite and infinite dimensional linear systems of differential equations in terms of exponential functions of a bounded linear operator defining the right-hand side of the system. Later, in the first half of the twentieth century, a formalization of the view, expressed by A. Hadamard, that an autonomous deterministic system should be described by a one-parameter semigroup of transformations, provided a more general background for the development of the theory of semigroups designed as a functional analytic language of partial differential equations. However, the concept soon also proved to be an important tool in stochastic processes, where mathematicians like W. Feller were instrumental in developing the foundations of the theory.

It can be safely said that the theory of semigroups, as a part of functional analysis, reached its maturity in the fourth and fifth decades of the previous century when the major generation theorems were established thanks to the work of K. Yosida, E. Hille, R. S. Phillips, I. Miyadera, and W. Feller. The state of the theory was summarised in the famous book *Functional Analysis and Semi-Groups* by E. Hille and R. S. Phillips, which firmly placed semigroups in the domain of functional analysis. It is interesting to note that the first book under that title (from which the famous quote: "I hail a semi-group when I see one and I seem to see them everywhere! Friends have observed, however, that there are mathematical objects which are not semi-groups" originated), published in 1948, was written only by E. Hille and was just 285 pages long. Such rapid was the advance of the theory in the next few years that a revised version, written together with R. S. Philips in 1955, grew to 808 pages. It has continued to be a basic reference in the field till now and possibly anybody who dares to read all 800 pages of it there will still find many challenging and generally unknown results.

Since then, semigroup theory not only has been rapidly developing as an independent field of research, posing new and fascinating questions but it also has become an indispensable tool and language for applications ranging from the classical ones such as partial differential equations and stochastic processes to less standard such as integro-differential and functional-differential equations, quantum mechanics, population biology, or control theory. Furthermore, though inherently linear and autonomous, semigroups became a cornerstone in the analysis of both nonlinear and nonautonomous evolutionary phenomena.

The achievements of that period were reflected in a series of excellent books by E. B. Davies, A. Belleni-Morante, A. Pazy, J. A. Goldstein, A. McBride, H. Fattorini, or Ph. Clément et al., that presented state of the art in the field of the late 70s and early 80s. These books extended the fundamental treatise of E. Hille and R. S. Phillips by not only presenting new theoretical results but also adding a strong PDE oriented component to the theory. They were complemented by monographs by S. N. Ethier and T. G. Kurtz, W. Feller, and E. B. Dynkin, which were instrumental in understanding the role of semigroups in Markov processes.

Subsequent years witnessed the emphasis on special classes of semigroups, such as positive semigroups and their applications in both natural sciences and stochastic processes, as well as an intensive investigation of the long term behavior of semigroups. Fundamentals of these developments can be found, e.g., in monographs by R. Nagel and K. J. Engel, W. Arendt, Ch. Batty, M. Hieber and F. Neubrander, A. Lunardi, A. Lasota and M. C. Mackey, or A. Bátkai, M. Fijavž and A. Rhandi, published in the last three decades.

The organisers of the series of conferences *Semigroups of Operators: Theory and Applications* and, at the same time, the editors of the present volume feel nevertheless that the exchange between different communities working on semigroups of operators and their applications has been too slow, partly because of isolation and partly due to the hermetic languages used in particular applications. The conferences are, therefore, meant to be a forum for the exchange of ideas between various groups of mathematicians using and developing the theory of semigroups so that both theoreticians and practitioners could present their ideas and bounce them against colleagues working on other topics so that both sides can benefit by possibly acquiring new knowledge.

It is a pleasure and honour that at the same time we could celebrate consecutive birthdays of Prof. Jan Kisyński, who has played an essential role in the development of the theory and bringing together its probabilistic and analytic aspects.

Our most recent conference in Kazimierz Dolny, Poland, commemorated his 85th birthday. The conference brought together 103 participants from 16 countries, who presented 5 plenary and 80 contributed talks. The present volume contains 24 papers based on the talks presented in 2018 that have undergone a strict refereeing process before being accepted. Clearly, 24 papers cannot cover any significant part of the field, but the editors believe that the presented selection, that includes topics ranging from hydro and gas dynamics, pedestrian flow, hyperbolic heat equation, degenerate semigroups, population dynamics, through Markov processes, approximate methods in evolution equations to control problems and long term behaviour, will give the reader a glimpse of this fascinating area of research.

The editors are grateful for the support without which the conference would not have been possible. We received financial support from the International Banach Centre, the Mathematical Institute of the Polish Academy of Sciences, Lublin University of Technology, the Faculty of Mathematics, Informatics and Mechanics, University of Warsaw, Juliusz P. Schauder Center for Nonlinear Studies, the European Regional Development Fund within the Economic Marketing of the Lubelskie Voivodeship II, EY (former Ernst and Young), the Zamoyski Museum in Kozłówka and the National Research Foundation of South Africa. The event was held under the honorary patronage of the Mayor of Lublin and Gazeta Wyborcza.

Great thanks go to the members of the Local Organizing Committee: Małgorzata Murat, Katarzyna Steliga, Adam Gregosiewicz, and Zbigniew Łagodowski, and to all those who in various ways contributed to the event, including Magdalena Karczmarczuk, Iwona Malinowska, and Łukasz Stępień of the Department of Mathematics at Lublin University of Technology. Finally, we thank the organisers of special sessions, plenary speakers, and all participants.

Do not let population dynamics, boundary conditions, singular perturbations, stochastic or deterministic control prevent you from seeing semigroups everywhere, and especially in this volume!

Pretoria, South Africa/Łódź, Poland	Jacek Banasiak
Lublin, Poland	Adam Bobrowski
Warsaw, Poland	Mirosław Lachowicz
Toruń, Poland	Yuri Tomilov
February 2020	

Contents

85th Birthday Lecture

Topologies in the Set of Rapidly Decreasing Distributions 3
Jan Kisyński

Theory

The Method of Chernoff Approximation 19
Yana A. Butko

Laplacians with Point Interactions—Expected and Unexpected
Spectral Properties 47
Amru Hussein and Delio Mugnolo

Remarks on a Characterization of Generators of Bounded
C_0-Semigroups 69
Sylwia Kosowicz

Semigroups Associated with Differential-Algebraic Equations 79
Sascha Trostorff

Positive Degenerate Holomorphic Groups of the Operators
and Their Applications 95
Sophiya A. Zagrebina and Natalya N. Solovyova

Applications

Microscopic Selection of Solutions to Scalar Conservation Laws
with Discontinuous Flux in the Context of Vehicular Traffic 113
Boris Andreianov and Massimiliano D. Rosini

Newton's Method for the McKendrick-von Foerster Equation 137
Agnieszka Bartłomiejczyk and Monika Wrzosek

Singular Thermal Relaxation Limit for the Moore-Gibson-Thompson Equation Arising in Propagation of Acoustic Waves 147
Marcelo Bongarti, Sutthirut Charoenphon, and Irena Lasiecka

Applications of the Kantorovich–Rubinstein Maximum Principle in the Theory of Boltzmann Equations 183
Henryk Gacki and Roksana Brodnicka

Propagators of the Sobolev Equations 205
Evgeniy V. Bychkov

The Fourth Order Wentzell Heat Equation 215
Gisèle Ruiz Goldstein, Jerome A. Goldstein, Davide Guidetti, and Silvia Romanelli

Nonlinear Semigroups and Their Perturbations in Hydrodynamics. Three Examples 227
Piotr Kalita, Grzegorz Łukaszewicz, and Jakub Siemianowski

Method of Lines for a Kinetic Equation of Swarm Formation 251
Adrian Karpowicz and Henryk Leszczyński

Degenerate Matrix Groups and Degenerate Matrix Flows in Solving the Optimal Control Problem for Dynamic Balance Models of the Economy 263
Alevtina V. Keller and Minzilia A. Sagadeeva

Degenerate Holomorphic Semigroups of Operators in Spaces of K-"Noises" on Riemannian manifolds 279
Olga G. Kitaeva, Dmitriy E. Shafranov, and Georgy A. Sviridyuk

Optimal Energy Decay in a One-Dimensional Wave-Heat-Wave System 293
Abraham C. S. Ng

Implicit Convolution Fokker-Planck Equations: Extended Feller Convolution 315
Wha-Suck Lee and Christiaan Le Roux

Asymptotic Properties of Stochastic Semigroups with Applications to Piecewise Deterministic Markov Processes 329
Katarzyna Pichór and Ryszard Rudnicki

On Polynomial Stability of Coupled Partial Differential Equations in 1D 349
Lassi Paunonen

Degenerate Nonlinear Semigroups of Operators and Their Applications 363
Ksenia V. Vasiuchkova, Natalia A. Manakova, and Georgy A. Sviridyuk

Sharp Interior and Boundary Regularity of the SMGTJ-Equation with Dirichlet or Neumann Boundary Control 379
Roberto Triggiani

Inverse Problem for the Boussinesq – Love Mathematical Model 427
Alyona A. Zamyshlyaeva and Aleksandr V. Lut

Optimal Control of Solutions to Showalter–Sidorov Problem for a High Order Sobolev Type Equation with Additive "Noise" 435
Alyona A. Zamyshlyaeva and Olga N. Tsyplenkova

85th Birthday Lecture

Topologies in the Set of Rapidly Decreasing Distributions

Jan Kisyński

To the memory of Professor Janusz Mika

Abstract Two topologies are introduced in the set of rapidly decreasing distributions on Euclidean space. One of these turns the standard convergence structure carried by this set into a topological convergence structure. The other allows the set in question to be topologically identified with the multiplier space for the Schwartz space of rapidly decreasing functions.

Keywords Fourier transformation · Rapidly decreasing distributions · Convolution · Locally convex space

1 Introduction

The set of rapidly decreasing distributions on Euclidean space was conceived by its originator, L. Schwartz, as a convergence space—a set equipped with a convergence structure. Informally, a convergence structure on a set is merely a rule indicating which nets converge to which points, and may not be derived from a topology. As it turns out, alongside Schwartz's convergence structure, the set of rapidly decreasing distributions carries natural topological structures. In this paper we reveal two topologies for this set that we denote by $\widetilde{b}$ and op. The $\widetilde{b}$-topology serves to show that Schwartz's original convergence structure is in fact topological: convergence relative to $\widetilde{b}$-topology is the same as convergence relative to the Schwartz convergence structure. The other topology has to do with the fact that each rapidly decreasing distribution acts continuously, by convolution, on the Schwartz space of rapidly decreasing functions. The corresponding convolution operators are continuous automorphisms of the Schwartz space. The topology op is derived, as the so-called

J. Kisyński (✉)
ul. Tymiankowa 56 m. 4, 20-542 Lublin, Poland
e-mail: jan.kisynski@gmail.com

initial topology relative to a specific mapping, from the topology of bounded convergence on the space of continuous linear automorphisms of the Schwartz space. What matters next is that the Fourier image of any convolution operator of the type just described is a multiplication operator, by a slowly increasing smooth function, on the Schwartz space. Furthermore, the multiplication operators can be topologized using again the topology of bounded convergence. Now, and this is one of the main points of our discussion, the passage from a rapidly decreasing distribution all the way to a multiplication operator, with a convolution operator in between, defines a linear topological automorphism. We close the paper with an addition to the lore on convolution of distributions: we propose a convolution of a rapidly decreasing distribution with a tempered distribution that obeys the principle of Fourier exchange.

2 The Fréchet Spaces $S_\mu^1(\mathbb{R}^n)$, $\mu \in [0, \infty[$, and the J. Horváth Space $\mathcal{O}_C(\mathbb{R}^n) = \lim \mathrm{ind}_{\mu \to \infty} S_\mu^1(\mathbb{R}^n)$

For each $\mu \in [0, \infty[$, let $S_\mu^1(\mathbb{R}^n)$ be the space of infinitely differentiable complex functions ϕ on $\mathbb{R}^n$ such that

$$\pi_{\mu,\alpha}^1(\phi) < \infty \quad \text{for every multiindex } \alpha \in \mathbb{N}_0^n,$$

where

$$\pi_{\mu,\alpha}^1(\phi) = \int_{\mathbb{R}^n} (1 + |x|^2)^{-\mu/2} |(\partial^\alpha \varphi)(x)| \, dx.$$

Every $S_\mu^1(\mathbb{R}^n)$ is a Fréchet space whose topology is determined by the countable system of seminorms $\{\pi_{\mu,\alpha}^1 : \alpha \in \mathbb{N}_0^n\}$. If $\mu, \nu \in [0, \infty[$ are such that $\mu < \nu$, then $S_\mu^1(\mathbb{R}^n) \hookrightarrow S_\nu^1(\mathbb{R}^n)$.

Define the space $\mathcal{O}_C(\mathbb{R}^n)$ of very slowly increasing functions on $\mathbb{R}^n$ by

$$\mathcal{O}_C(\mathbb{R}^n) := \lim \mathrm{ind}_{\mu \to \infty} S_\mu^1(\mathbb{R}^n).$$

The idea of using the inductive limit in this definition goes back to J. Horváth [10, Sect. 2.12, Example 9]. Originally Horváth defined $\mathcal{O}_C(\mathbb{R}^n)$ as $\lim \mathrm{ind}_{\mu \to \infty} S_\mu(\mathbb{R}^n)$, where the spaces $S_\mu(\mathbb{R}^n)$ are different from, but similar to, the spaces $S_\mu^1(\mathbb{R}^n)$. The fact that replacing the $S_\mu(\mathbb{R}^n)$ by the $S_\mu^1(\mathbb{R}^n)$ does not affect $\mathcal{O}_C(\mathbb{R}^n)$ is a consequence of [14, Sects. I–III].

In what follows it will be important that the Schwartz space of rapidly decreasing functions $\mathscr{S}(\mathbb{R}^n)$ is sequentially dense in each $S_\mu^1(\mathbb{R}^n)$ and in $\mathcal{O}_C(\mathbb{R}^n)$. For $S_\mu^1(\mathbb{R}^n)$ this can be proved by employing routine analytic tools, while the denseness of $\mathscr{S}(\mathbb{R}^n)$ in $\mathcal{O}_C(\mathbb{R}^n)$ can be proved as follows. If $p \in \mathcal{O}_C(\mathbb{R}^n)$, then $p \in S_{\mu_0}^1(\mathbb{R}^n)$ for some μ_0. By the sequential denseness of $\mathscr{S}(\mathbb{R}^n)$ in $S_{\mu_0}^1(\mathbb{R}^n)$, there is a sequence $(p_k)_{k \in \mathbb{N}} \subset$

$\mathscr{S}(\mathbb{R}^n)$ converging to p in the topology of $S^1_{\mu_0}(\mathbb{R}^n)$. A fortiori $(p_k)_{k\in\mathbb{N}}$ converges to p in the topology of $\mathscr{O}_C(\mathbb{R}^n)$.

The J. Horváth space $\mathscr{O}_C(\mathbb{R}^n) = \lim\text{ind}_{[0,\infty[\,\ni\mu\to\infty} S^1_\mu(\mathbb{R}^n)$ was intended to be the predual space of $\lim\text{proj}_{\infty\leftarrow\mu\in[0,\infty[}(S^1_\mu(\mathbb{R}^n))'$, but for a proper realization of this idea the dual spaces involved have to be endowed with a topology. For the weak dual topology in the spaces $(S^1_\mu(\mathbb{R}^n))'$, we have

$$\lim\text{proj}_{\infty\leftarrow\mu\in[0,\infty[}(S^1_\mu(\mathbb{R}^n))'_w = (\mathscr{O}_C(\mathbb{R}^n))'_w;$$

see [20, Sect. IV.4, Theorem 4.5]. Some earlier results of the present author suggest that

$$\lim\text{proj}_{\infty\leftarrow\mu\in[0,\infty[}(S^1_\mu(\mathbb{R}^n))'_b = ((\mathscr{O}_C(\mathbb{R}^n))', \tau_b),$$

where $(S^1_\mu(\mathbb{R}^n))'_b$ stands for the strong dual space of $S^1_\mu(\mathbb{R}^n)$ and τ_b is the $\mathfrak{S}$-topology in $(\mathscr{O}_C(\mathbb{R}^n))'$ corresponding to the covering $\bigcup_{\mu\in[0,\infty[} B_\mu$ of $\mathscr{O}_C(\mathbb{R}^n)$, where, for every $\mu\in[0,\infty[$, B_μ is the family of all bounded subsets of $S^1_\mu(\mathbb{R}^n)$.

Let us mention that in [18] N. Ortner and P. Wagner investigated the J. Horváth spaces $\mathscr{O}_C^m(\mathbb{R}^n)$, $m\in\mathbb{N}_0\cup\{\infty\}$. Various formulas for representing the space $\mathscr{O}_M(\mathbb{R}^n)$ of slowly increasing functions on $\mathbb{R}^n$, closely related to $\mathscr{O}_C(\mathbb{R}^n)$, in terms of projective and inductive limits can be found in the paper [16] by J. Larcher. These formulas permit obtaining $\mathscr{O}_M(\mathbb{R}^n)$, by Fourier transformation, directly from L. Schwartz's definition of the convergence space $\mathscr{O}'_C(\mathbb{R}^n)$. The methods used by J. Larcher in [16] and [17] are well adapted to relating $\mathscr{O}'_C(\mathbb{R}^n)$ to the notion of continuity of a bilinear mapping with respect to both variables simultaneously. In abstract terms a similar enrichment of the structure of a convergence space is considered in [2].

3 The Isomorphisms of Horváth

The following theorem, which is a variation on the result given in [9, Sect. 2.5, Example 8], reveals the apparently obvious isomorphisms between the spaces of the form $S^1_\mu(\mathbb{R}^n)$.

Theorem 3.1 *Let $\mu, \lambda \in \mathbb{R}$ and $\psi \in C^\infty(\mathbb{R}^n; \mathbb{C})$. Then $\psi \in S^1_\lambda(\mathbb{R}^n)$ if and only if there is a $\phi \in S^1_\mu(\mathbb{R}^n)$ such that*

$$\psi(x) = (1+|x|^2)^{(\mu-\lambda)/2}\phi(x) \quad \text{for every } x \in \mathbb{R}^n. \tag{1}$$

Moreover, equality (1) defines an isomorphism $I_{\mu,\lambda}$ of the locally convex space $S^1_\mu(\mathbb{R}^n)$ onto the locally convex space $S^1_\lambda(\mathbb{R}^n)$.

If, for $\mu, \lambda \in \mathbb{R}$, $I'_{\lambda,\mu}$ denotes the mapping adjoint to $I_{\lambda,\mu}$, then [4, Sect. IV.4.2, Proposition 6] implies

Corollary 3.2 *Given* $\mu, \lambda \in \mathbb{R}$, $I'_{\lambda,\mu}: (S^1_\lambda(\mathbb{R}^n))'_b \to (S^1_\mu(\mathbb{R}^n))'_b$ *is an isomorphism of the locally convex space* $(S^1_\lambda(\mathbb{R}^n))'_b$ *onto the locally convex space* $(S^1_\mu(\mathbb{R}^n))'_b$.

A specific application of Corollary 3.2 will be of relevance in the next section.

4 Translation of the Language of L. Schwartz's Convergence Space $\mathscr{O}'_C(\mathbb{R}^n)$ into the Language of Locally Convex Spaces

It will sometimes be useful to distinguish clearly between a topological space or a convergence space and the set of elements of this space, without any topology. So, we shall denote by $[E]$ the set of all elements of a topological space E or a convergence space E. L. Schwartz in [22, Sect. VII.5, p. 244] defines his convergence space $\mathscr{O}'_C(\mathbb{R}^n)$ in terms of bounded distributions (that is, elements of the set $\mathscr{S}'(\mathbb{R}^n)$ of tempered distributions on $\mathbb{R}^n$ whose translations constitute bounded sets in the strong dual $\mathscr{S}'(\mathbb{R}^n)_b$ of $\mathscr{S}(\mathbb{R}^n)$). L. Schwartz denotes the locally convex space $S^1_0(\mathbb{R}^n)$ by $\mathscr{B}$ and in [22, Sect. VI.8] he proves that the elements of $\mathscr{B}'$, the strong dual space of $\mathscr{B}$, are bounded distributions. It follows from Corollary 3.2 above that without using $\mathscr{B}$ and $\mathscr{B}'$, the convergence space $\mathscr{O}'_C(\mathbb{R}^n)$ can be equivalently defined by the following two conditions:

(A) $[\mathscr{O}'_C(\mathbb{R}^n)] = \bigcap_{\mu \in [0,\infty[}[(S^1_\mu(\mathbb{R}^n))']$,
(B) a net $(T_\iota)_{\iota \in J}$ of elements of $[\mathscr{O}'_C(\mathbb{R}^n)]$ converges to zero if and only if, for every $\mu \in [0, \infty[$, this net converges to zero in the sense of the topology of $(S^1_\mu(\mathbb{R}^n))'_b$.

To speak meaningfully about the intersection in condition (A), we treat each $[(S^1_\mu(\mathbb{R}^n))']$ as a subset of the common set $\mathscr{S}'(\mathbb{R}^n)$. That this makes sense follows from the fact that, given that $\mathscr{S}(\mathbb{R}^n)$ is sequentially dense in $S^1_\mu(\mathbb{R}^n)$ and that the topology of $S^1_\mu(\mathbb{R}^n)$ restricted to $\mathscr{S}(\mathbb{R}^n)$ is coarser than the topology of $\mathscr{S}(\mathbb{R}^n)$, any continuous linear functional on $S^1_\mu(\mathbb{R}^n)$ is uniquely determined by its restriction to $\mathscr{S}(\mathbb{R}^n)$ and any such restriction is continuous in the topology of $\mathscr{S}(\mathbb{R}^n)$. Note that with these identifications, $[\mathscr{O}'_C(\mathbb{R}^n)]$ emerges as a subset of $\mathscr{S}'(\mathbb{R}^n)$.

Consider the locally convex space $E := \varprojlim_{\infty \leftarrow \mu \in [0,\infty[} (S^1_\mu(\mathbb{R}^n))'_b$. Then $[E] = [\mathscr{O}'_C(\mathbb{R}^n)]$ by condition (A). Define $\tilde{b}$ as the topology induced in $[\mathscr{O}'_C(\mathbb{R}^n)]$ from E. Then $\tilde{b}$ coincides with the initial topology in $[\mathscr{O}'_C(\mathbb{R}^n)]$ with respect to the family of projections $\pi_\mu: [\mathscr{O}'_C(\mathbb{R}^n)] \to (S^1_\mu(\mathbb{R}^n))'_b$, $\mu \in [0, \infty[$, where π_μ is the dual map of the canonical embedding of $S^1_\mu(\mathbb{R}^n)$ into $\mathscr{O}_C(\mathbb{R}^n)$. Consequently, the definition (B) of the convergence of nets in $[\mathscr{O}'_C(\mathbb{R}^n)]$ can be formulated in equivalent form as follows:

(B') a net $(T_\iota)_{\iota \in J}$ of elements of $[\mathscr{O}'_C(\mathbb{R}^n)]$ converges to zero if and only it converges to zero in the sense of the topology of $\varprojlim_{\infty \leftarrow \mu \in [0,\infty[} (S^1_\mu(\mathbb{R}^n))'_b$.

In this way the convergence space $\mathscr{O}'_C(\mathbb{R}^n)$ is shown to be a convergence space induced by a locally convex topology. We note in passing that, conversely, some important relations of the theory of locally convex spaces can be clearly expressed in terms of convergence spaces—see Sects. 9.9 and 10.9 of the book [12] by H. Jarchow. The article [2] by R. Beattie and H.-P. Butzmann is an example of an advanced study of relations between locally convex spaces and convergence spaces.

Remark The proof of the coincidence of $\widetilde{b}$ with the initial topology relative to $\{\pi_\mu : \mu \in [0, \infty[\}$ can be obtained by considering an appropriate intersection topology in $\bigcap_{\mu \in [0,\infty[} (S^1_\mu(\mathbb{R}^n))'_b$. See [3, §1, No. 4, Example II] for the notion of intersection topology as a notion strictly related to the notion of projective limit and the notion of initial topology, these two in turn being related to the notion of lattice of topologies on a set (cf. [20, Sect. II.5 and Prerequisites B.3]).

5 The Set $RD(\mathbb{R}^n)$ of Rapidly Decreasing Distributions on $\mathbb{R}^n$ and the Locally Convex Space $(RD(\mathbb{R}^n), \widetilde{b})$

Define the set $RD(\mathbb{R}^n)$ of rapidly decreasing distributions on $\mathbb{R}^n$ by

$$RD(\mathbb{R}^n) = [\mathscr{O}'_C(\mathbb{R}^n)].$$

By adopting the topology $\widetilde{b}$ in $[\mathscr{O}'_C(\mathbb{R}^n)]$, we convert $RD(\mathbb{R}^n)$ into a locally convex space $(RD(\mathbb{R}^n), \widetilde{b})$.

As we have already pointed out, $[\mathscr{O}'_C(\mathbb{R}^n)]$, or what is the same $RD(\mathbb{R}^n)$, is a subset of $\mathscr{S}'(\mathbb{R}^n)$. It turns out that the precise characterization of $RD(\mathbb{R}^n)$ is as follows:

$$RD(\mathbb{R}^n) = \{T \in [\mathscr{S}'(\mathbb{R}^n)] : T \text{ is continuous in the topology of } \mathscr{O}_C(\mathbb{R}^n)\};$$

see [14, Theorem 3]. By [10, Sect. 2.12, Proposition 2] or [20, Sect. II.6, Theorem 6.1], we also have

$$RD(\mathbb{R}^n) = \big\{T \in [\mathscr{S}'(\mathbb{R}^n)] :$$
$$\text{if } \mu \in \mathbb{R}, \text{ then } T \text{ is continuous in the topology of } S^1_\mu(\mathbb{R}^n)\big\}.$$

Since $\mathscr{S}(\mathbb{R}^n)$ is sequentially dense in every $S^1_\mu(\mathbb{R}^n)$, $\mu \in \mathbb{R}$, the latter identity can be rephrased as

$$RD(\mathbb{R}^n) = \big\{T \in [\mathscr{S}'(\mathbb{R}^n)] : \text{ if } \mu \in \mathbb{R}, \text{ then } T \text{ extends uniquely}$$
$$\text{to a continuous functional } T_\mu \text{ on } S^1_\mu(\mathbb{R}^n)\big\}.$$

6 The Subset RD($\mathbb{R}^n$) of $\mathscr{S}'(\mathbb{R}^n)$

If X and Y are topological vector spaces, we denote by $\mathscr{L}(X, Y)$ the space of all continuous linear mappings from X into Y. We abbreviate $\mathscr{L}(X, X)$ to $\mathscr{L}(X)$.

Define

$$\mathbf{RD}(\mathbb{R}^n) := \{T \in \mathscr{S}'(\mathbb{R}^n) : [T*]|_{\mathscr{S}(\mathbb{R}^n)} \in \mathscr{L}(\mathscr{S}(\mathbb{R}^n))\}.$$

The above definition involves the convolution of a distribution T with a test function φ, $T * \varphi$, which is a function belonging to $C^\infty(\mathbb{R}^n)$ whose value at $x \in \mathbb{R}^n$ is $[T * \varphi](x) = T((\varphi^\vee)_{-x})$. Our Theorem 8.4, presented below, shows that requiring a distribution $T \in \mathscr{S}'(\mathbb{R}^n)$ to satisfy the condition $[T*]|_{\mathscr{S}(\mathbb{R}^n)} \in \mathscr{L}(\mathscr{S}(\mathbb{R}^n))$ is a severe restriction. Theorem 6.2, presented later in this section, shows that a similar condition $[T*]|_{\mathscr{S}(\mathbb{R}^n)} \in \mathscr{L}(\mathscr{S}(\mathbb{R}^n), \mathscr{O}_C(\mathbb{R}^n))$ is not a restriction at all.

Theorem 6.1 *We have $RD(\mathbb{R}^n) \subset \mathbf{RD}(\mathbb{R}^n)$.*

Proof The standard topology in the space $\mathscr{S}(\mathbb{R}^n)$ of rapidly decreasing C^∞ functions on $\mathbb{R}^n$ is determined by the system of seminorms $\{\rho_{\mu,\alpha} : \mu \in [0, \infty[, \alpha \in \mathbb{N}_0^n\}$ defined by

$$\rho_{\mu,\alpha}(\varphi) := \sup_{x \in \mathbb{R}^n}(1 + |x|)^\mu |\partial^\alpha \varphi(x)| \quad \text{for every } \varphi \in \mathscr{S}(\mathbb{R}^n).$$

Let $T \in RD(\mathbb{R}^n)$. For transparency, the system of seminorms on $S_\lambda^1(\mathbb{R}^n)$ defined in Sect. 2 is replaced by the equivalent system $\{\pi_{\lambda,\beta}^1 : \beta \in \mathbb{N}_0^n\}$ such that $\pi_{\lambda,\beta}^1(\phi) = \int_{\mathbb{R}^n}(1 + |x|)^{-\lambda}|\partial^\beta \phi(x)|\, dx$ for every $\phi \in S_\lambda^1(\mathbb{R}^n)$. To prove that $T \in \mathbf{RD}(\mathbb{R}^n)$, fix some $\lambda \in]\mu + n, \infty[$. Since T_λ, the extension of T to $S_\lambda^1(\mathbb{R}^n)$, is a continuous linear functional on $S_\lambda^1(\mathbb{R}^n)$, it follows that $S_\lambda^1(\mathbb{R}^n) \ni \phi \mapsto |T_\lambda(\phi)| \in [0, \infty[$ is a continuous seminorm on $S_\lambda^1(\mathbb{R}^n)$, so that there exist $C_\lambda \in]0, \infty[$ and $b_\lambda \in \mathbb{N}$ such that

$$|T_\lambda(\phi)| \leq C_\lambda \max_{|\beta| \leq b_\lambda} \pi_{\lambda,\beta}^1(\phi) \quad \text{for every } \phi \in S_\lambda^1(\mathbb{R}^n).$$

Since $\mathscr{S}(\mathbb{R}^n) \subset S_\lambda^1(\mathbb{R}^n)$ and T_λ is an extension of T, it follows that

$$|T(\psi)| = |T_\lambda(\psi)| \leq C_\lambda \max_{|\beta| \leq b_\lambda} \pi_{\lambda,\beta}^1(\psi) \quad \text{for every } \psi \in \mathscr{S}(\mathbb{R}^n).$$

Consequently, for every $x \in \mathbb{R}^n$, $\alpha \in \mathbb{N}_0^n$ and $\varphi \in \mathscr{S}(\mathbb{R}^n)$,

$$|\partial^\alpha([T * \varphi](x))| = |[T * \partial^\alpha](x)| = |[T_\lambda]_{(y)}((\partial^\alpha \varphi)(x - y))|$$
$$\leq C_\lambda \max_{|\beta| \leq b_\lambda}[\pi_{\lambda,\beta}^1]_{(y)}((\partial^\alpha \varphi)(x - y))$$
$$= C_\lambda \max_{|\beta| \leq b_\lambda} \int_{\mathbb{R}^n}(1 + |y|)^{-\lambda}|(\partial^{\alpha+\beta}\varphi)(x - y)|\, dy.$$

Since $\rho_{\mu,\alpha+\beta}(\varphi) = \sup_{x \in \mathbb{R}^n} (1+|x|)^{\mu} |(\partial^{\alpha+\beta}\varphi)(x)|$, it follows that

$$|(\partial^{\alpha+\beta}\varphi)(x-y)| \leq \rho_{\mu,\alpha+\beta}(\varphi) \cdot (1+|x-y|)^{-\mu},$$

so

$$|\partial^\alpha(T*\varphi)(x)| \leq C_\lambda \max_{|\beta| \leq b_\lambda} \rho_{\mu,\alpha+\beta}(\varphi) \int_{\mathbb{R}^n} (1+|y|)^{-\lambda}(1+|x-y|)^{-\mu} \, dy$$

$$= C_\lambda \max_{|\beta| \leq b_\lambda} \rho_{\mu,\alpha+\beta}(\varphi) \int_{\mathbb{R}^n} (1+|y|)^{-(\lambda-\mu)}(1+|x-y|)^{-\mu}(1+|y|)^{-\mu} \, dy.$$

Since $(1+|x-y|)(1+|y|) \geq 1+|x-y|+|y| \geq 1+|x|$, we have

$$|\partial^\alpha(T*\varphi)(x)| \leq C_\lambda \max_{|\beta| \leq b_\lambda} \rho_{\mu,\alpha+\beta}(\varphi) \cdot \left(\int_{\mathbb{R}^n} (1+|y|)^{-(\lambda-\mu)} \, dy \right) \cdot (1+|x|)^{-\mu},$$

where the integral is finite because $\lambda \in]\mu+n, \infty[$. The last inequality implies that

$$\rho_{\mu,\alpha}(T*\varphi) \leq C_\lambda \max_{|\beta| \leq b_\lambda} \rho_{\mu,\alpha+\beta}(\varphi) \int_{\mathbb{R}^n} (1+|y|)^{-(\lambda-\mu)} \, dy.$$

It follows that $T * \varphi \in \mathscr{S}(\mathbb{R}^n)$ whenever $\varphi \in \mathscr{S}(\mathbb{R}^n)$, and that the mapping $\mathscr{S}(\mathbb{R}^n) \ni \varphi \mapsto T * \varphi \in \mathscr{S}(\mathbb{R}^n)$ is continuous. $\square$

Remark In [15] it is proved that $RD(\mathbb{R}^n) = \mathbf{RD}(\mathbb{R}^n)$. Moreover, if $T \in \mathscr{S}'(\mathbb{R}^n)$, then the equivalent conditions (a) $T \in RD(\mathbb{R}^n)$ and (b) $T \in \mathbf{RD}(\mathbb{R}^n)$ are equivalent to

(c) for every $\mu \in [0, \infty[, \phi \in S^1_\mu(\mathbb{R}^n)$ and $\varphi \in \mathscr{D}(\mathbb{R}^n)$, the function $\mathbb{R}^n \ni z \mapsto T(\phi \cdot \varphi_z) \in \mathbb{C}$ belongs to $\mathscr{S}(\mathbb{R}^n)$; here φ_z denotes the translate of φ by z.

It follows at once that if $T \in \mathscr{D}'(\mathbb{R}^n)$ is compactly supported, then $T \in RD(\mathbb{R}^n)$. These results of [15] will not be used in what follows.

Theorem 6.2 *If $T \in \mathscr{S}'(\mathbb{R}^n)$, then $[T*]|_{\mathscr{S}(\mathbb{R}^n)} \in L(\mathscr{S}(\mathbb{R}^n), \mathscr{O}_C(\mathbb{R}^n))$.*

The above result goes back to Horváth [10, Sect. 4.11, Proposition 7] who defined $\mathscr{O}_C(\mathbb{R}^n)$ as $\lim \mathrm{ind}_{\mu \to \infty} S_\mu(\mathbb{R}^n)$, where the spaces $S_\mu(\mathbb{R}^n)$ and $S^1_\mu(\mathbb{R}^n)$ differ, but are similar.

Corollary 6.3 *If $T \in \mathscr{S}'(\mathbb{R}^n)$ and $T * \varphi \in \mathscr{S}(\mathbb{R}^n)$ for every $\varphi \in \mathscr{S}(\mathbb{R}^n)$, then the linear mapping $\mathscr{S}(\mathbb{R}^n) \ni \varphi \mapsto T * \varphi \in \mathscr{S}(\mathbb{R}^n)$ has closed graph.*

Proof Suppose that $\mathscr{S}(\mathbb{R}^n)\text{-}\lim_{k \to \infty} \varphi_k = \varphi_0$ and $\mathscr{S}(\mathbb{R}^n)\text{-}\lim_{k \to \infty}(T * \varphi_k) = \psi_0$. Since $\mathscr{S}(\mathbb{R}^n) \hookrightarrow \mathscr{O}_C(\mathbb{R}^n)$, we have $\mathscr{O}_C(\mathbb{R}^n)\text{-}\lim_{k \to \infty} \varphi_k = \varphi_0$ and, by Theorem 6.2, $\mathscr{O}_C(\mathbb{R}^n)\text{-}\lim_{k \to \infty}(T * \varphi_k) = T * \psi_0$. Hence $\varphi_0 = \psi_0$, which means that the graph of the mapping $\mathscr{S}(\mathbb{R}^n) \ni \varphi \mapsto T * \varphi \in \mathscr{S}(\mathbb{R}^n)$ is closed. $\square$

Theorem 6.4 (part of [6, Theorem 7.2.2]) *If $T \in \mathscr{S}'(\mathbb{R}^n)$ and $T * \varphi \in \mathscr{S}(\mathbb{R}^n)$ for every $\varphi \in \mathscr{S}(\mathbb{R}^n)$, then the mapping $\mathscr{S}(\mathbb{R}^n) \ni \varphi \mapsto T * \varphi \in \mathscr{S}(\mathbb{R}^n)$ is continuous.*

Proof This is a consequence of Corollary 6.3 and the Closed Graph Theorem. The latter can be applied since the space $\mathscr{S}(\mathbb{R}^n)$ is metrizable and complete. $\square$

7 The Operator Topology in $\mathbf{RD}(\mathbb{R}^n)$

Let $\mathscr{A}$ be the family of all bounded closed, and as such compact, subsets of the Fréchet–Montel space $\mathscr{S}(\mathbb{R}^n)$. According to [24, Sect. IV.7], the locally convex topology of bounded convergence in $\mathscr{L}(\mathscr{S}(\mathbb{R}^n))$ is determined by the system of seminorms $\{p_{\mu,\alpha,A} : \mu \in [0, \infty[, \alpha \in \mathbb{N}_0^n, A \in \mathscr{A}\}$, where $p_{\mu,\alpha,A}(L) = \sup_{\varphi \in A} \rho_{\mu,\alpha}(L(\varphi))$ for every $L \in \mathscr{L}(\mathscr{S}(\mathbb{R}^n))$. The set $\mathscr{L}(\mathscr{S}(\mathbb{R}^n))$ equipped with the topology of bounded convergence constitutes a locally convex space which we denote by $\mathscr{L}(\mathscr{S}(\mathbb{R}^n))_b$.

We introduce in the set $\mathbf{RD}(\mathbb{R}^n)$ the locally convex topology op (abbreviated from "operator topology") as the initial topology induced by the mapping $\mathbf{RD}(\mathbb{R}^n) \ni T \mapsto [T *]|_{\mathscr{S}(\mathbb{R}^n)} \in \mathscr{L}(\mathscr{S}(\mathbb{R}^n))_b$. This means that the topology op in $\mathbf{RD}(\mathbb{R}^n)$ is determined by the system of seminorms $\{r_{\mu,\alpha,A} : \mu \in [0, \infty[, \alpha \in \mathbb{N}_0^n, A \in \mathscr{A}\}$, where $r_{\mu,\alpha,A}(T) := \sup_{\varphi \in A} \rho_{\mu,\alpha}(T * \varphi)$.

8 Locally Convex Space $(\mathbf{RD}(\mathbb{R}^n), op)$ and Fourier Transformation

8.1 *The Fourier Transformation in $\mathscr{S}'(\mathbb{R}^n)$*

The Fourier transformation $\mathscr{F} : \mathscr{S}(\mathbb{R}^n) \to \mathscr{S}(\mathbb{R}^n)$ is a linear topological automorphism of the space $\mathscr{S}(\mathbb{R}^n)$ (see [7, Theorem 5.2.5], [22, Sect. VII.6, Theorem XII] or [24, Sect. VI.1]). Its transpose $\mathscr{F}'$ is a linear topological automorphism of $\mathscr{S}'(\mathbb{R}^n)$ equipped with the *-weak topology. $\mathscr{F}'$ is also a linear topological automorphism of $(\mathscr{S}(\mathbb{R}^n))'_b$ (see [4, Sect. IV.4.2, Proposition 6]). Moreover, since $\mathscr{S}(\mathbb{R}^n)$ is sequentially dense in $\mathscr{S}'(\mathbb{R}^n)$, it follows from the Parseval equality for $\mathscr{F} : \mathscr{S}(\mathbb{R}^n) \to \mathscr{S}(\mathbb{R}^n)$ that $\mathscr{F}'$ is equal to the extension of $\mathscr{F} : \mathscr{S}(\mathbb{R}^n) \to \mathscr{S}(\mathbb{R}^n)$ onto $\mathscr{S}'(\mathbb{R}^n)_b$ by continuity. For this reason in what follows we shall write $\mathscr{F}$ instead of $\mathscr{F}'$. Let us stress that we define $\mathscr{F}$ by the equalities $[\mathscr{F}(\varphi)](\xi) = \int_{\mathbb{R}^n} e^{-i\langle x,\xi \rangle} \varphi(x)\, dx$ for $\varphi \in \mathscr{S}(\mathbb{R}^n)$ and $\langle \mathscr{F}(U), \varphi \rangle = \langle U, \mathscr{F}(\varphi) \rangle$ for $U \in \mathscr{S}'(\mathbb{R}^n)$ and $\varphi \in \mathscr{S}(\mathbb{R}^n)$.

8.2 $\mathscr{O}_M(\mathbb{R}^n)$ as the Algebra of Multipliers of $\mathscr{S}(\mathbb{R}^n)$

The space $\mathscr{O}_M(\mathbb{R}^n)$ of infinitely differentiable slowly increasing functions on $\mathbb{R}^n$ is a locally convex space with the topology determined by Schwartz's system of seminorms defined in [22, Sect. VII.5]. It is almost evident that if $\phi \in \mathscr{O}_M(\mathbb{R}^n)$, then the mapping $m_\phi : \psi \mapsto \phi \cdot \psi$ is in $\mathscr{L}(\mathscr{S}(\mathbb{R}^n))$. What is not obvious is the opposite implication: if $\phi \in C^\infty(\mathbb{R}^n)$ and $m_\phi \in \mathscr{L}(\mathscr{S}(\mathbb{R}^n))$, then $\phi \in \mathscr{O}_M(\mathbb{R}^n)$. An ingenious short proof of this implication can be found in [13, Vol. 2, Chap. CA.III].

Theorem 8.1 ([24, Sect. VI.3, equality (14) of Theorem 6]) *If $T \in \mathscr{S}'(\mathbb{R}^n)$ and $\varphi \in \mathscr{S}(\mathbb{R}^n)$, then*

$$\mathscr{F}(T * \varphi) = \mathscr{F}(\varphi) \cdot \mathscr{F}(T).$$

The proof presented by K. Yosida is short and elementary, yet refined.

Theorem 8.2 *If $T \in \mathbf{RD}(\mathbb{R}^n)$, then $\mathscr{F}(T) = (2\pi)^{n/2} e^{\frac{1}{2}|\cdot|^2} \mathscr{F}(T * e^{-\frac{1}{2}|\cdot|^2}) \in C^\infty(\mathbb{R}^n)$ and $\mathscr{F}|_{\mathbf{RD}(\mathbb{R}^n)}$ is a linear one-to-one mapping of $\mathbf{RD}(\mathbb{R}^n)$ into $\mathscr{O}_M(\mathbb{R}^n)$.*

Proof Since $\mathscr{F}(e^{-\frac{1}{2}|\cdot|^2}) = (2\pi)^{-n/2} e^{-\frac{1}{2}|\cdot|^2}$ (see [19, Sect. 2.2, Example 1] or [23, Sect. I.1, Theorem 1.13]), we infer from Theorem 8.1 that

$$\mathscr{F}(T * e^{-\frac{1}{2}|\cdot|^2}) = \mathscr{F}(T) \cdot \mathscr{F}(e^{-\frac{1}{2}|\cdot|^2}) = \mathscr{F}(T)(2\pi)^{-n/2} e^{-\frac{1}{2}|\cdot|^2},$$

so that

$$\mathscr{F}(T) = \phi_T,$$

where

$$\phi_T = (2\pi)^{n/2} e^{\frac{1}{2}|\cdot|^2} \mathscr{F}(T * e^{-\frac{1}{2}|\cdot|^2}) \in C^\infty(\mathbb{R}^n).$$

In order to prove that if $T \in \mathbf{RD}(\mathbb{R}^n)$, then $\phi_T \in \mathscr{O}_M(\mathbb{R}^n)$, we shall use the implication: if $\phi \in C^\infty(\mathbb{R}^n)$ and $\phi \cdot \mathscr{S}(\mathbb{R}^n) \subset \mathscr{S}(\mathbb{R}^n)$, then $\phi \in \mathscr{O}_M(\mathbb{R}^n)$, that is, all the partial derivatives of ϕ grow slowly at infinity. This implication constitutes a hard part of the characterization of $\mathscr{O}_M(\mathbb{R}^n)$ as the function algebra of multipliers of $\mathscr{S}(\mathbb{R}^n)$. (Let us recall here the proof of V.-K. Khoan, mentioned earlier.) So, in order to prove that $\phi_T \in \mathscr{O}_M(\mathbb{R}^n)$ for every $T \in \mathbf{RD}(\mathbb{R}^n)$, we have only to check that $\phi_T \cdot \mathscr{S}(\mathbb{R}^n) \subset \mathscr{S}(\mathbb{R}^n)$, that is, if $T \in \mathbf{RD}(\mathbb{R}^n)$ and $\varphi \in \mathscr{S}(\mathbb{R}^n)$ then $\phi_T \cdot \varphi \in \mathscr{S}(\mathbb{R}^n)$. But $\phi_T \cdot \varphi = \mathscr{F}(T) \cdot \varphi = \mathscr{F}(T) \cdot \mathscr{F}(\mathscr{F}^{-1}\varphi)$ and, by Theorem 8.1, $\mathscr{F}(T) \cdot \mathscr{F}(\mathscr{F}^{-1}\varphi) = \mathscr{F}(T * \mathscr{F}^{-1}\varphi) \in \mathscr{F}(\mathscr{S}(\mathbb{R}^n)) \subset \mathscr{S}(\mathbb{R}^n)$. □

8.3 The Set of Continuous Linear Operators of $\mathscr{S}(\mathbb{R}^n)$ into Itself as a Locally Convex Space

Let $\mathscr{L}(\mathscr{S}(\mathbb{R}^n))_s$ be the locally convex space obtained by equipping $\mathscr{L}(\mathscr{S}(\mathbb{R}^n))$ with the topology of simple convergence. In accordance with [24, Sect. IV.7], the topology of simple convergence is defined by a system of seminorms $\{s_{\mu,\alpha,\psi} : \mu \in [0, \infty[\, , \alpha \in \mathbb{N}_0^n, \psi \in \mathscr{S}(\mathbb{R}^n)\}$ such that

$$s_{\mu,\alpha,\psi}(L) = \rho_{\mu,\alpha}(L(\psi)) \quad \text{for every operator } L \in \mathscr{L}(\mathscr{S}(\mathbb{R}^n)).$$

The locally convex space $(\mathscr{O}_M(\mathbb{R}^n) \cdot)$ is the subspace of $\mathscr{L}(\mathscr{S}(\mathbb{R}^n))$ comprised of all operators $m_\phi, \phi \in \mathscr{O}_M(\mathbb{R}^n)$. Let us note that, since $\mathscr{S}(\mathbb{R}^n)$ is a Montel space, it follows that in the set $\mathscr{L}(\mathscr{S}(\mathbb{R}^n))$ the topology of simple convergence coincides with the topology of bounded convergence.

Let $\mathscr{F}$ denote the Fourier automorphism of $\mathscr{S}'(\mathbb{R}^n)$ and let $\mathscr{F}_0 := \mathscr{F}_{|\mathscr{S}(\mathbb{R}^n)} \in \mathscr{L}(\mathscr{S}(\mathbb{R}^n))$. Define the action $\widetilde{\mathscr{F}}$ of the Fourier transformation in $\mathscr{L}(\mathscr{S}(\mathbb{R}^n))$ by the equality

$$\widetilde{\mathscr{F}}(L) := \mathscr{F}_0^{-1} \circ L \circ \mathscr{F}_0 \quad \text{for every } L \in \mathscr{L}(\mathscr{S}(\mathbb{R}^n)).$$

Then $\widetilde{\mathscr{F}}$ is a linear topological automorphism of the locally convex space $\mathscr{L}(\mathscr{S}(\mathbb{R}^n))$.

For $T \in \mathscr{S}'(\mathbb{R}^n)$, let the operator $L_T : \mathscr{S}(\mathbb{R}^n) \to C^\infty(\mathbb{R}^n)$ be defined by $L_T(\psi) = T * \psi$ for every $\psi \in \mathscr{S}(\mathbb{R}^n)$. If $T \in \mathbf{RD}(\mathbb{R}^n)$, then $L_T \in \mathscr{L}(\mathscr{S}(\mathbb{R}^n))$ by the third of equivalent condition characterizing the set $[RD(\mathbb{R}^n)]$ in Theorem 1 of [15] (quoted in the Remark in Sect. 6 of the present paper). It follows from Theorems 8.1 and 8.2 that if $T \in [RD(\mathbb{R}^n)]$, then $\mathscr{F}(T) = \phi_T \in \mathscr{O}_M(\mathbb{R}^n)$ and $L_T = \mathscr{F}_0^{-1} \circ L_{\phi_T} \circ \mathscr{F}_0 = \widetilde{\mathscr{F}}(L_{\phi_T})$.

Theorem 8.3 *The linear one-to-one mapping $\mathscr{F}|_\mathbf{RD}$ of $\mathbf{RD}(\mathbb{R}^n)$ into $\mathscr{O}_M(\mathbb{R}^n)$ induces an isomorphism of locally convex spaces of $(\mathbf{RD}, op)$ onto $(\mathscr{O}_M(\mathbb{R}^n) \cdot)$.*

The above theorem resembles a result proved by C. Bargetz and N. Ortner in [1].

Proof *This proof follows suggestions of the Referee.*

Let (μ, α, ψ) range over $[0, \infty[\times \mathbb{N}_0^n \times \mathscr{S}(\mathbb{R}^n)$. The topology op in the set $[\mathbf{RD}(\mathbb{R}^n)]$ is determined by the system of seminorms $\{r_{\mu,\alpha,\psi}\}$, where $r_{\mu,\alpha,\psi}(T) = \rho_{\mu,\alpha}(T * \psi)$ for every $T \in [\mathbf{RD}(\mathbb{R}^n)]$. The topology of the subspace $(\mathscr{O}_M(\mathbb{R}^n) \cdot)$ of $\mathscr{L}(\mathscr{S}(\mathbb{R}^n))$ is determined by the system of seminorms $s_{\mu,\alpha,\psi}$, where $s_{\mu,\alpha,\psi}(m_\phi) = \rho_{\mu,\alpha}(m_\phi(\psi))$ for every $\phi \in \mathscr{O}_M(\mathbb{R}^n)$. By Theorem 8.1, $T * \psi = [\mathscr{F}_0^{-1} \circ m_{\phi_T} \circ \mathscr{F}_0](\psi)$ whenever $T \in [\mathbf{RD}(\mathbb{R}^n)]$ and $\psi \in \mathscr{S}(\mathbb{R}^n)$, whence

$$r_{\mu,\alpha,\psi}(T) = s_{\mu,\alpha,\psi}(\mathscr{F}_0^{-1} \circ m_{\phi_T} \circ \mathscr{F}_0) \quad \text{for every } (\mu, \alpha, \psi) \text{ and } T \in [\mathbf{RD}(\mathbb{R}^n)].$$

This equality shows that $([\mathbf{RD}(\mathbb{R}^n)], op) \ni T \mapsto (m_{\phi_T}) \in (\mathscr{O}_M(\mathbb{R}^n) \cdot)$ is an isomorphism, given that the space $(\mathscr{O}_M(\mathbb{R}^n) \cdot)$ is invariant with respect to $\widetilde{F}$ and the mapping $(\mathscr{O}_M(\mathbb{R}^n) \cdot) \ni \phi \mapsto \mathscr{F}_0^{-1} \circ (m_\phi) \circ \mathscr{F}_0$ is a linear topological automorphism of the space $(\mathscr{O}_M(\mathbb{R}^n) \cdot)$. □

If $T \in \mathbf{RD}(\mathbb{R}^n)$ and $U \in \mathscr{S}'(\mathbb{R}^n)$, then we define the provisional convolution $T \diamond U$ as the distribution belonging to $\mathscr{S}'(\mathbb{R}^n)$ and equal to ${}^t([T^\vee *]|_{\mathscr{S}(\mathbb{R}^n)})$, where the superscript t on the left stands for the transpose operator. This means that

$$\langle T \diamond U, \varphi \rangle = \langle U, T^{\vee} * \varphi \rangle \quad \text{for every } \varphi \in \mathscr{S}(\mathbb{R}^n).$$

The last equality resembles [7, Sect. 4.4, Definition 4.4.1] and [10, Sect. 4.11, Definition 3], and constitutes a provisional definition of convolution in $\mathscr{S}'(\mathbb{R}^n)$, limited to $T \in \mathbf{RD}(\mathbb{R}^n)$ and $U \in \mathscr{S}'(\mathbb{R}^n)$. Since our provisional convolution always leads to $T \diamond U \in \mathscr{S}'(\mathbb{R}^n)$, its cogent ultimate generalization must be an $\mathscr{S}'(\mathbb{R}^n)$-convolution. The author is aware of only one $\mathscr{S}'(\mathbb{R}^n)$-convolution, namely the $\mathscr{S}'(\mathbb{R}^n)$-convolution of Y. Hirata and H. Ogata [8].

Theorem 8.4 *If $T \in \mathbf{RD}(\mathbb{R}^n)$ and $U \in \mathscr{S}'(\mathbb{R}^n)$, then*

$$\mathscr{F}(T \diamond U) = \mathscr{F}(T) \cdot \mathscr{F}(U).$$

Proof Notice first that $\mathscr{F}(T \diamond U)$ makes sense, because $T \diamond U \in \mathscr{S}'(\mathbb{R}^n)$. Also $\mathscr{F}(T) \cdot \mathscr{F}(U)$ makes sense because $\mathscr{F}(T) = \phi_T \in \mathscr{O}_M(\mathbb{R}^n)$. To prove the equality of both expressions, notice first that, by Theorem 8.1, for every $T \in \mathbf{RD}(\mathbb{R}^n)$, $U \in \mathscr{S}'(\mathbb{R}^n)$ and $\varphi \in \mathscr{S}(\mathbb{R}^n)$, we have

$$\begin{aligned}\langle T \diamond U, \varphi \rangle = \langle U, T^{\vee} * \varphi \rangle &= \langle U, \mathscr{F}^{-1}(\mathscr{F}(T^{\vee}) \cdot \mathscr{F}(\varphi)) \rangle \\ &= \langle \mathscr{F}^{-1}(U), \mathscr{F}(T^{\vee}) \cdot \mathscr{F}(\varphi) \rangle \\ &= \langle \mathscr{F}^{-1}(U) \cdot \mathscr{F}(T)^{\vee}, \mathscr{F}(\varphi) \rangle,\end{aligned}$$

whence

$$T \diamond U = \mathscr{F}(\mathscr{F}(T)^{\vee} \cdot \mathscr{F}^{-1}(U)).$$

From the last equality, by the Fourier inversion formula, it follows that

$$\begin{aligned}T \diamond U &= \mathscr{F}\big(\mathscr{F}(T)^{\vee} \cdot (2\pi)^n \mathscr{F}(U)^{\vee}\big) = (2\pi)^n \mathscr{F}\big((\mathscr{F}(T)^{\vee} \cdot \mathscr{F}(U))^{\vee}\big) \\ &= (2\pi)^n \mathscr{F}^{\vee}(\mathscr{F}(T) \cdot \mathscr{F}(U)) = \mathscr{F}^{-1}(\mathscr{F}(T) \cdot \mathscr{F}(U)),\end{aligned}$$

whence

$$\mathscr{F}(T \diamond U) = \mathscr{F}(T) \cdot \mathscr{F}(U).$$

$\square$

Remark Theorem 8.4 means that our provisional convolution $\diamond$ has the property of Fourier exchange between convolution and multiplication. The $\mathscr{S}'(\mathbb{R}^n)$-convolution of Y. Hirata and H. Ogata also has this property, and this permits one to prove that $\diamond$ is the restriction of the H-O-convolution to $\mathbf{RD}(\mathbb{R}^n) \times \mathscr{S}'(\mathbb{R}^n)$. Notice that the H-O-convolution assigns to any convolvable pair $(T, U) \in \mathscr{S}'(\mathbb{R}^n) \times \mathscr{S}'(\mathbb{R}^n)$ a distribution $T * U \in \mathscr{S}'(\mathbb{R}^n)$. Schwartz's convolution [21] assigns to a convolvable pair $(T, U) \in \mathscr{D}'(\mathbb{R}^n) \times \mathscr{D}'(\mathbb{R}^n)$ a distribution $T * U \in \mathscr{D}'(\mathbb{R}^n)$. It is proved in [11, Example 6] that if $(T, U) \in \mathbf{RD}(\mathbb{R}^n) \times \mathscr{S}'(\mathbb{R}^n)$, then the pair (T, U) is Schwartz convolvable. In [5] an example is given of two measures $(T, U) \in \mathscr{S}'(\mathbb{R}^n) \times \mathscr{S}'(\mathbb{R}^n)$

with Schwartz convolution $T * U$ in $\mathscr{D}'(\mathbb{R}^n) \setminus \mathscr{S}'(\mathbb{R}^n)$. This example shows that the $\mathscr{S}'(\mathbb{R}^n)$-convolution, as a distinctive notion, is really needed.

Acknowledgements The author expresses his warmest thanks to the Referee whose remarks and suggestions resulted in fundamental amelioration of the paper. The author is very grateful to Professors Adam Bobrowski and Wojciech Chojnacki for their kind help in correcting the paper and offering valuable linguistic hints.

References

1. Bargetz, C., Ortner, N.: Characterization of L. Schwartz' convolution and multiplier spaces $\mathscr{O}'_C$ and $\mathscr{O}_M$ by the short-time Fourier transform. Rev. R. Acad. Cienc. Exactas Fís. Nat. Ser. A Mat. RACSAM **108**(2), 839–847 (2014)
2. Beattie, R., Butzmann, H.-P.: Convergence and distributions. Atti Sem. Mat. Fis. Univ. Modena **39**(2), 487–494 (1991)
3. Bourbaki, N.: Éléments de Mathématique. Livre III. Topologie Générale. Hermann, Paris (1961). Russian transl.: Nauka, Moscow (1968). English transl.: Springer (2007)
4. Bourbaki, N.: Éléments de Mathématique. Livre V. Espaces Vectoriels Topologiques. Hermann, Paris (1953–1955). Russian transl.: Gos. Izdat. Fiz-Mat. Lit., Moscow (1959). English transl.: Springer (2003)
5. Dierolf, P., Voigt, J.: Convolution and $\mathscr{S}'$-convolution of distributions. Collect. Math. **29**, 185–196 (1978)
6. Gårding, L., Lions, J.-L.: Functional analysis. Nuovo Cimento **14**(Suppl 1), 9–66 (1959)
7. Golse, F.: Distributions, Analyse de Fourier, Équations aux Dérivées Partielles. Les Éditions de l'École Polytechnique, Palaiseau (2012)
8. Hirata, Y., Ogata, H.: On the exchange formula for distributions. J. Sci. Hiroshima Univ. Ser. A **22**, 147–152 (1958)
9. Horváth, J.: Topological Vector Spaces and Distributions. Addison-Wesley (1966)
10. Horváth, J.: Topological Vector Spaces and Distributions. Dover Publications (2012)
11. Horváth, J.: Sur la convolution des distributions. Bull. Sci. Math. **2**(98), 183–192 (1974)
12. Jarchow, H.: Locally Convex Spaces. B. G. Teubner, Stuttgart (1981)
13. Khoan, V.-K.: Distributions, Analyse de Fourier, Opérateurs aux Dérivées Partielles, vols. 1, 2. Vuibert, Paris (1972)
14. Kisyński, J.: On the rapidly decreasing distributions. Preprint, Inst. Math., Polish Acad. Sci., May (2017)
15. Kisyński, J.: Characterization of rapidly decreasing distributions on $\mathbb{R}^n$ by convolutions with test functions. Preprint, Inst. Math., Polish Acad. Sci., March (2019)
16. Larcher, J.: Some remarks concerning the spaces of multipliers and convolutions, $\mathscr{O}_M$ and $\mathscr{O}'_C$, of Laurent Schwartz. Rev. R. Acad. Cienc. Exactas Fís. Nat. Ser. A Mat. RACSAM **108**(2), 407–417 (2012)
17. Larcher, J.: Multiplications and convolutions in L. Schwartz' spaces of test functions and distributions, and their continuity. Analysis (Berlin) **33**(4), 319–332 (2013)
18. Ortner, N., Wagner, P.: On the spaces $\mathscr{O}^m_C$ of John Horváth. J. Math. Anal. Appl. **415**(1), 62–74 (2014)
19. Rauch, J.: Partial Differential Equations. Springer (1991)
20. Schaefer, H.H.: Topological Vector Spaces. Macmillan, New York and Collier-Macmillan, London (1966). Russian transl.: Mir, Moscow (1971)
21. Schwartz, L.: A. Définition intégrale de la convolution de deux distributions. Séminaire Schwartz **1** (1953–1954). Talk no. 22. http://www.numdam.org/item/SLS_1953-1954__1__A23_0

22. Schwartz, L.: Théorie des Distributions, nouvelle éd. Hermann, Paris (1966)
23. Stein, E.M., Weiss, G.: Introduction to Fourier Analysis on Euclidean Spaces. Princeton University Press (1971). Russian transl.: Nauka, Moscow (1974)
24. Yosida, K.: Functional Analysis, 6th edn. Springer (1980)

Theory

The Method of Chernoff Approximation

Yana A. Butko

Abstract This survey describes the method of approximation of operator semigroups, based on the Chernoff theorem. We outline recent results in this domain as well as clarify relations between constructed approximations, stochastic processes, numerical schemes for PDEs and SDEs, path integrals. We discuss Chernoff approximations for operator semigroups and Schrödinger groups. In particular, we consider Feller semigroups in $\mathbb{R}^d$, (semi)groups obtained from some original (semi)groups by different procedures: additive perturbations of generators, multiplicative perturbations of generators (which sometimes corresponds to a random time-change of related stochastic processes), subordination of semigroups/processes, imposing boundary/external conditions (e.g., Dirichlet or Robin conditions), averaging of generators, "rotation" of semigroups. The developed techniques can be combined to approximate (semi)groups obtained via several iterative procedures listed above. Moreover, this method can be implemented to obtain approximations for solutions of some time-fractional evolution equations, although these solutions do not possess the semigroup property.

Keywords Chernoff approximation · Feynman formula · Approximation of operator semigroups · Approximation of transition probabilities · Approximation of solutions of evolution equations · Feynman–Kac formulae · Euler–Maruyama schemes · Feller semigroups · Additive perturbations · Operator splitting · Multiplicative perturbations · Dirichlet boundary/external conditions · Robin boundary conditions · Subordinate semigroups · Time-fractional evolution equations · Schrödinger type equations

---

Y. A. Butko (✉)
Saarland University, Postfach 151150, 66041 Saarbrücken, Germany
e-mail: yanabutko@yandex.ru

© Springer Nature Switzerland AG 2020
J. Banasiak et al. (eds.), *Semigroups of Operators – Theory and Applications*,
Springer Proceedings in Mathematics & Statistics 325,
https://doi.org/10.1007/978-3-030-46079-2_2

1 Introduction

Let $(X, \|\cdot\|_X)$ be a Banach space. A family $(T_t)_{t\geq 0}$ of bounded linear operators on X is called a *strongly continuous semigroup* (denoted as C_0-*semigroup*) if $T_0 = \mathrm{Id}$, $T_t \circ T_s = T_{t+s}$ for all $t, s \geq 0$, and $\lim_{t\to 0}\|T_t\varphi - \varphi\|_X = 0$ for all $\varphi \in X$. The *generator* of the semigroup $(T_t)_{t\geq 0}$ is an operator $(L, \mathrm{Dom}(L))$ in X which is given by $L\varphi := \lim_{t\to 0} t^{-1}(T_t\varphi - \varphi)$, $\mathrm{Dom}(L) := \{\varphi \in X : \lim_{t\to 0} t^{-1}(T_t\varphi - \varphi)$ exists in $X\}$. In the sequel, we denote the semigroup with a given generator L both as $(T_t)_{t\geq 0}$ and as $(e^{tL})_{t\geq 0}$. The following fundamental result of the theory of operator semigroups connects C_0-semigroups and evolution equations: *Let $(L, \mathrm{Dom}(L))$ be a densely defined linear operator in X with a nonempty resolvent set. The Cauchy problem $\frac{\partial f}{\partial t} = Lf$, $f(0) = f_0$ in X for every $f_0 \in \mathrm{Dom}(L)$ has a unique solution $f(t)$ which is continuously differentiable on $[0, +\infty)$ if and only if $(L, \mathrm{Dom}(L))$ is the generator of a C_0-semigroup $(T_t)_{t\geq 0}$ on X. And the solution is given by $f(t) := T_t f_0$.*

Let now Q be a locally compact metric space. Let $(\xi_t)_{t\geq 0}$ be a temporally homogeneous Markov process with the state space Q and with transition probability $P(t, x, dy)$. The family $(T_t)_{t\geq 0}$, given by $T_t\varphi(x) := \int_Q \varphi(y) P(t, x, dy)$, is a semigroup which, for several important classes of Markov processes, happens to be strongly continuous on some suitable Banach spaces of functions on Q. Hence, in this case, we have three equivalent problems:

(1) to construct the C_0-semigroup $(T_t)_{t\geq 0}$ with a given generator $(L, \mathrm{Dom}(L))$ on a given Banach space X;
(2) to solve the Cauchy problem $\frac{\partial f}{\partial t} = Lf$, $f(0) = f_0$ in X;
(3) to determine the transition kernel $P(t, x, dy)$ of an underlying Markov process $(\xi_t)_{t\geq 0}$.

The basic example is given by the operator $(L, \mathrm{Dom}(L))$ which is the closure of $(\frac{1}{2}\Delta, S(\mathbb{R}^d))$ in the Banach space[1] $X = C_\infty(\mathbb{R}^d)$ or in $X = L^p(\mathbb{R}^d)$, $p \in [1, \infty)$. The operator $(L, \mathrm{Dom}(L))$ generates a C_0-semigroup $(T_t)_{t\geq 0}$ on X; this semigroup is given for each $f_0 \in X$ by

$$T_t f_0(x) = (2\pi t)^{-d/2} \int_{\mathbb{R}^d} f_0(y) \exp\left\{-\frac{|x-y|^2}{2t}\right\} dy; \qquad (1)$$

the function $f(t, x) := T_t f_0(x)$ solves the corresponding Cauchy problem for the heat equation $\frac{\partial f}{\partial t} = \frac{1}{2}\Delta f$; and

$$P(t, x, dy) := (2\pi t)^{-d/2} \exp\left\{-\frac{|x-y|^2}{2t}\right\} dy \qquad (2)$$

---

[1] We denote the space of continuous functions on $\mathbb{R}^d$ vanishing at infinity by $C_\infty(\mathbb{R}^d)$ and the Schwartz space by $S(\mathbb{R}^d)$.

is the transition probability of a d-dimensional Brownian motion. However, it is usually not possible to determine a C_0-semigroup in an explicit form, and one has to approximate it. In this note, we demonstrate the method of approximation based on the Chernoff theorem [28, 29]. In the sequel, we use the following (simplified) version of the Chernoff theorem, assuming that the existence of the semigroup under consideration is already established.

Theorem 1.1 *Let $(F(t))_{t \geq 0}$ be a family of bounded linear operators on a Banach space X. Assume that*

(i) $F(0) = \mathrm{Id}$,
(ii) $\|F(t)\| \leq e^{wt}$ *for some* $w \in \mathbb{R}$ *and all* $t \geq 0$,
(iii) *the limit* $L\varphi := \lim\limits_{t \to 0} \frac{F(t)\varphi - \varphi}{t}$ *exists for all* $\varphi \in D$, *where D is a dense subspace in X such that (L, D) is closable and the closure $(L, \mathrm{Dom}(L))$ of (L, D) generates a C_0-semigroup $(T_t)_{t \geq 0}$.*

Then the semigroup $(T_t)_{t \geq 0}$ is given by

$$T_t \varphi = \lim_{n \to \infty} [F(t/n)]^n \varphi \qquad (3)$$

for all $\varphi \in X$, and the convergence is locally uniform with respect to $t \geq 0$.

Any family $(F(t))_{t \geq 0}$ satisfying the assumptions of the Chernoff theorem 1.1 with respect to a given C_0-semigroup $(T_t)_{t \geq 0}$ is called *Chernoff equivalent*, or *Chernoff tangential* to the semigroup $(T_t)_{t \geq 0}$. And the formula (3) is called *Chernoff approximation* of $(T_t)_{t \geq 0}$. Evidently, in the case of a bounded generator L, the family $F(t) := \mathrm{Id} + tL$ is Chernoff equivalent to the semigroup $(e^{tL})_{t \geq 0}$. And we get a classical formula

$$e^{tL} = \lim_{n \to \infty} \left[\mathrm{Id} + \frac{t}{n} L \right]^n. \qquad (4)$$

Moreover, for an arbitrary generator L, one considers $F(t) := (\mathrm{Id} - tL)^{-1} \equiv \frac{1}{t} R_L(1/t)$ (if $(0, \infty)$ is in the resolvent set of L) and obtains the *Post–Widder inversion formula*:

$$T_t \varphi = \lim_{n \to \infty} \left(\mathrm{Id} - \frac{t}{n} L \right)^{-n} \varphi \equiv \lim_{n \to \infty} \left[\frac{n}{t} R_L(n/t) \right]^n \varphi, \quad \forall \varphi \in X.$$

A well-developed functional calculus approach to Chernoff approximation of C_0-semigroups by families $(F(t))_{t \geq 0}$, which are given by (bounded completely monotone) functions of the generators (as, e.g., in the case of the Post–Widder inversion formula above), can be found in [36]. We use another approach. We are looking for arbitrary families $(F(t))_{t \geq 0}$ which are Chernoff equivalent to a given C_0-semigroup (i.e., the only connection of $F(t)$ to the generator L is given via the assertion (iii) of the Chernoff theorem). But we are especially interested in families $(F(t))_{t \geq 0}$ which are given explicitly (e.g., as integral operators with explicit kernels or pseudo-differential operators with explicit symbols). This is useful both for practical calculations and for further interpretations of Chernoff approximations as path integrals (see, e.g., [12, 19, 25] and references therein). Moreover, we consider different operations on generators (what sometimes corresponds to operations on Markov processes) and find out, how to construct Chernoff approximations for C_0-semigroups with modified generators on the base of Chernoff approximations for the original ones.

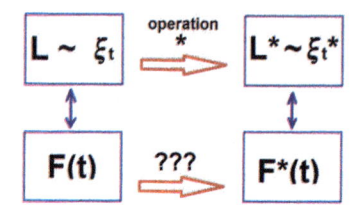

This approach allows to create a kind of a LEGO-constructor: we start with a C_0-semigroup which is already known[2] or Chernoff approximated[3]; then, applying different operations on its generator, we consider more and more complicated C_0-semigroups and construct their Chernoff approximations.

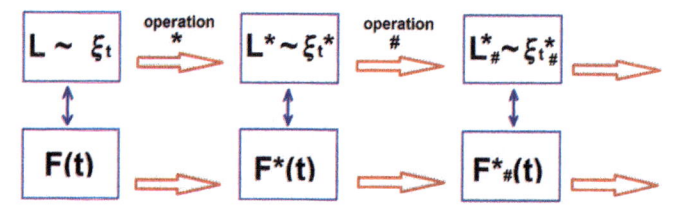

Chernoff approximations are available for the following operations:

- Operator splitting; additive perturbations of a generator (Sect. 2.1, [20, 21, 27]);
- Multiplicative perturbations of a generator/random time change of a process via an additive functional (Sect. 2.3, [20, 21, 27]);
- killing of a process upon leaving a given domain/imposing Dirichlet boundary (or external) conditions (Sect. 2.4, [21, 22, 24]);
- imposing Robin boundary conditions (Sect. 2.4, [52]);
- subordination of a semigroup/process (Sect. 2.5, [21, 23]);
- "rotation" of a semigroup (see Sect. 2.7, [62, 64]);
- averaging of semigroups (see Sect. 2.7, [9, 10, 57]);

Moreover, Chernoff approximations have been obtained for some stochastic Schrödinger type equations in [37, 54–56]; for evolution equations with the Vladimirov operator (this operator is a p-adic analogue of the Laplace operator) in [65–69];

---

[2]E.g., the semigroup generated by a Brownian motion on a star graph with Wentzell boundary conditions at the vertex [42]; see also [11, 18, 30] for further examples.

[3]E.g., the semigroup generated by a Brownian motion on a compact Riemannian manifold (see [72] and references therein) and (Feller) semigroups generated by Feller processes in $\mathbb{R}^d$ (see [27] and Sect. 2.2).

for evolution equations containing Lévy Laplacians in [1, 2]; for some nonlinear equations in [58].

Chernoff approximation can be interpreted as a numerical scheme for solving evolution equations. Namely, for the Cauchy problem $\frac{\partial f}{\partial t} = Lf$, $f(0) = f_0$, we have:

$$u_0 := f_0, \qquad u_k := F(t/n)u_{k-1}, \quad k = 1, \ldots, n, \qquad f(t) \approx u_n.$$

In some particular cases, Chernoff approximations are an abstract analogue of the *operator splitting method* known in the numerics of PDEs (see Remark 2.1). And the Chernoff theorem itself can be understood as a version of the *"Meta-theorem of numerics": consistency and stability imply convergence*. Indeed, conditions (i) and (iii) of Theorem 1.1 are consistency conditions, whereas condition (ii) is a stability condition. Moreover, in some cases, the families $(F(t))_{t \geq 0}$ give rise to Markov chain approximations for $(\xi_t)_{t \geq 0}$ and provide Euler–Maruyama schemes for the corresponding SDEs (see Example 2.1).

If all operators $F(t)$ are integral operators with elementary kernels or pseudo-differential operators with elementary symbols, the identity (3) leads to representation of a given semigroup by n-folds iterated integrals of elementary functions when n tends to infinity. This gives rise to *Feynman formulae*. A *Feynman formula* is a representation of a solution of an initial (or initial-boundary) value problem for an evolution equation (or, equivalently, a representation of the semigroup solving the problem) by a limit of n-fold iterated integrals of some functions as $n \to \infty$. One should not confuse the notions of Chernoff approximation and Feynman formula. On the one hand, not all Chernoff approximations can be directly interpreted as Feynman formulae since, generally, the operators $(F(t))_{t \geq 0}$ do not have to be neither integral operators, nor pseudo-differential operators. On the other hand, representations of solutions of evolution equations in the form of Feynman formulae can be obtained by different methods, not necessarily via the Chernoff Theorem. And such Feynman formulae may have no relations to any Chernoff approximation, or their relations may be quite indirect. Richard Feynman was the first who considered representations of solutions of evolution equations by limits of iterated integrals [33, 34]. He has, namely, introduced a construction of a path integral (known nowadays as *Feynman path integral*) for solving the Schrödinger equation. And this path integral was defined exactly as a limit of iterated finite dimensional integrals. Feynman path integrals can be also understood as integrals with respect to Feynman type pseudomeasures. Analogously, one can sometimes obtain representations of a solution of an initial (or initial-boundary) value problem for an evolution equation (or, equivalently, a representation of an operator semigroup resolving the problem) by functional (or, path) integrals with respect to probability measures. Such representations are usually called *Feynman–Kac formulae*. It is a usual situation that limits in Feynman formulae coincide with (or in some cases define) certain path integrals with respect to probability measures or Feynman type pseudomeasures on a set of paths of a physical system. Hence the iterated integrals in Feynman formulae for some problem give approximations to path integrals representing the solution of the same problem. Therefore, representations of evolution semigroups by Feynman formulae, on the one hand, allow to establish new path-integral-representations and, on the other hand, provide an additional tool to calculate path integrals numerically. Note that different Feynman formulae for the same semigroup allow to establish relations between different path integrals (see, e.g., [21]).

The result of Chernoff has diverse generalizations. Versions, using arbitrary partitions of the time interval $[0, t]$ instead of the equipartition $(t_k)_{k=0}^n$ with $t_k - t_{k-1} = t/n$, are presented, e.g., in [60, 71]. Versions, providing stronger type of convergence, can be found in [78]. The analogue of the Chernoff theorem for multivalued generators can be found, e.g., in [32]. Analogues of Chernoff's result for semigroups, which are continuous in a weaker sense, are obtained, e.g., in [3, 43]. For analogues of the Chernoff theorem in the case of nonlinear semigroups, see, e.g., [5, 16, 17]. The Chernoff Theorem for two-parameter families of operators can be found in [56, 61].

2 Chernoff Approximations for Operator Semigroups and Further Applications

2.1 Chernoff Approximations for the Procedure of Operator Splitting

Theorem 2.1 *Let $(T_t)_{t \geq 0}$ be a strongly continuous semigroup on a Banach space X with generator $(L, \text{Dom}(L))$. Let D be a core for L. Let $L = L_1 + \cdots + L_m$ hold on D for some linear operators L_k, $k = 1, \ldots, m$, in X. Let $(F_k(t))_{t \geq 0}$, $k = 1, \ldots, m$, be families of bounded linear operators on X such that for all $k \in \{1, \ldots, m\}$ holds: $F_k(0) = \text{Id}$, $\|F_k(t)\| \leq e^{a_k t}$ for some $a_k \geq 0$ and all $t \geq 0$, $\lim_{t \to 0} \left\| \frac{F_k(t)\varphi - \varphi}{t} - L_k \varphi \right\|_X = 0$ for all $\varphi \in D$. Then the family $(F(t))_{t \geq 0}$, with $F(t) := F_1(t) \circ \cdots \circ F_m(t)$, is Chernoff equivalent to the semigroup $(T(t))_{t \geq 0}$. And hence the Chernoff approximation*

$$T_t \varphi = \lim_{n \to \infty} \left[F(t/n) \right]^n \varphi \equiv \lim_{n \to \infty} \left[F_1(t/n) \circ \cdots \circ F_m(t/n) \right]^n \varphi \quad (5)$$

holds for each $\varphi \in X$ locally uniformly with respect to $t \geq 0$.

Note that we do not require from summands L_k to be generators of C_0-semigroups. For example, L_1 can be a leading term (which generates a C_0-semigroup) and $L_2, \ldots, L_m$ can be L_1-bounded additive perturbations such that $L := L_1 + L_2 + \cdots + L_m$ again generates a strongly continuous semigroup. Or L may even be a sum of operators L_k, none of which generates a strongly continuous semigroup itself.

Proof Obviously, the family $(F(t))_{t \geq 0}$ satisfies the conditions $F(0) = \text{Id}$ and $\|F(t)\| \leq \|F_1(t)\| \cdot \ldots \cdot \|F_m(t)\| \leq e^{(a_1 + \cdots + a_m)t}$. Further, for each $\varphi \in D$, we have

$$\lim_{t\to 0}\left\|\frac{F(t)\varphi-\varphi}{t}-L\varphi\right\|_X = \lim_{t\to 0}\left\|\frac{F_1(t)\circ\cdots\circ F_m(t)\varphi-\varphi}{t}-L_1\varphi-\cdots-L_m\varphi\right\|_X$$

$$= \lim_{t\to 0}\left\|F_1(t)\circ\cdots\circ F_{m-1}(t)\left(\frac{F_m(t)\varphi-\varphi}{t}-L_m\varphi\right)\right.$$

$$+ (F_1(t)\circ\cdots\circ F_{m-1}(t)-\operatorname{Id})L_m\varphi + \frac{F_1(t)\circ\cdots\circ F_{m-1}(t)\varphi-\varphi}{t}-L_1\varphi-\cdots-L_{m-1}\varphi\bigg\|_X$$

$$\leq \lim_{t\to 0}\left\|\frac{F_1(t)\circ\cdots\circ F_{m-1}(t)\varphi-\varphi}{t}-L_1\varphi-\cdots-L_{m-1}\varphi\right\|_X$$

$$\leq \cdots \leq \lim_{t\to 0}\left\|\frac{F_1(t)\varphi-\varphi}{t}-L_1\varphi\right\|_X = 0.$$

Therefore, all requirements of the Chernoff theorem 1.1 are fulfilled and hence $(F(t))_{t\geq 0}$ is Chernoff equivalent to $(T(t))_{t\geq 0}$. □

Remark 2.1 Let all the assumptions of Theorem 2.1 be fulfilled. Consider for simplicity the case $m=2$. Let $\theta, \tau \in [0,1]$. Similarly to the proof of Theorem 2.1, one shows that the following families $(H^\theta(t))_{t\geq 0}$ and $(G^\tau(t))_{t\geq 0}$ are Chernoff equivalent to the semigroup $(T_t)_{t\geq 0}$ generated by $L = L_1 + L_2$:

$$H^\theta(t) := F_1(\theta t)\circ F_2(t)\circ F_1((1-\theta)t),$$
$$G^\tau(t) := \tau F_1(t)\circ F_2(t) + (1-\tau)F_2(t)\circ F_1(t).$$

Note that we have $H^0(t) = F_2(t)\circ F_1(t)$, and $H^1(t) = F_1(t)\circ F_2(t)$. Hence the parameter θ corresponds to different orderings of non-commuting terms $F_1(t)$ and $F_2(t)$. Further, $G^{1/2}(t) = \frac{1}{2}\left(H^1(t) + H^0(t)\right)$. In the case when both L_1 and L_2 generate C_0-semigroups and $F_k(t) := e^{tL_k}$, Chernoff approximation (5) with families $(H^\theta(t))_{t\geq 0}$, $\theta = 1$ or $\theta = 0$, reduces to the classical Daletsky–Lie–Trotter formula. Moreover, Chernoff approximation (5) can be understood as an abstract analogue of the *operator splitting* known in numerical methods of solving PDEs (see [48] and references therein). If $\theta = 0$ and $\theta = 1$, the families $(H^\theta(t))_{t\geq 0}$ correspond to first order splitting schemes. Whereas the family $(H^{1/2}(t))_{t\geq 0}$ corresponds to the symmetric Strang splitting and, together with $(G^{1/2}(t))_{t\geq 0}$, represents second order splitting schemes.

2.2 Chernoff Approximations for Feller Semigroups

We consider the Banach space $X = C_\infty(\mathbb{R}^d)$ of continuous functions on $\mathbb{R}^d$, vanishing at infinity. A semigroup of bounded linear operators $(T_t)_{t\geq 0}$ on the Banach space X is called *Feller semigroup* if it is a strongly continuous semigroup, it is *positivity preserving* (i.e. $T_t\varphi \geq 0$ for all $\varphi \in X$ with $\varphi \geq 0$) and it is *sub-Markovian* (i.e. $T_t\varphi \leq 1$ for all $\varphi \in X$ with $\varphi \leq 1$). A Markov process, whose semigroup is Feller, is called *Feller process*. Let $(L, \operatorname{Dom}(L))$ be the generator of a Feller semigroup $(T_t)_{t\geq 0}$. Assume that $C_c^\infty(\mathbb{R}^d) \subset \operatorname{Dom}(L)$ (this assumption is quite standard and holds in many cases, see, e.g., [13]). Then we have also[4] $C_\infty^2(\mathbb{R}^d) \subset \mathrm{Dom}(L)$. And $L\varphi(x)$ is given for each $\varphi \in C_\infty^2(\mathbb{R}^d)$ and each $x \in \mathbb{R}^d$ by the following formula:

$$L\varphi(x) = -C(x)\varphi(x) - B(x) \cdot \nabla\varphi(x) + \mathrm{tr}(A(x) \mathrm{Hess}\,\varphi(x)) \\ + \int_{y \neq 0} \left(\varphi(x+y) - \varphi(x) - \frac{y \cdot \nabla\varphi(x)}{1 + |y|^2}\right) N(x, dy), \quad (6)$$

where Hess φ is the Hessian matrix of second order partial derivatives of φ; as well as $C(x) \geq 0$, $B(x) \in \mathbb{R}^d$, $A(x) \in \mathbb{R}^{d \times d}$ is a symmetric positive semidefinite matrix and $N(x, \cdot)$ is a Radon measure on $\mathbb{R}^d \setminus \{0\}$ with $\int_{y \neq 0} |y|^2 (1 + |y|^2)^{-1} N(x, dy) < \infty$ for each $x \in \mathbb{R}^d$. Therefore, L is an integro-differential operator on $C_\infty^2(\mathbb{R}^d)$ which is non-local if $N \neq 0$. This class of generators L includes, in particular, fractional Laplacians $L = -(-\Delta)^{\alpha/2}$ and relativistic Hamiltonians $\sqrt[\alpha]{(-\Delta)^{\alpha/2} + m(x)}$, $\alpha \in (0, 2)$, $m > 0$. Note that the restriction of L onto $C_c^\infty(\mathbb{R}^d)$ is given by a pseudo-differential operator (PDO)

$$L\varphi(x) := -(2\pi)^{-d} \int_{\mathbb{R}^d} \int_{\mathbb{R}^d} e^{ip \cdot (x-q)} H(x, p) \varphi(q) \, dq \, dp, \quad x \in \mathbb{R}^d, \quad (7)$$

with the symbol $-H$ such that

$$H(x, p) = C(x) + iB(x) \cdot p + p \cdot A(x)p + \int_{y \neq 0}\left(1 - e^{iy \cdot p} + \frac{iy \cdot p}{1 + |y|^2}\right) N(x, dy). \quad (8)$$

If the symbol H does not depend on x, i.e. $H = H(p)$, then the semigroup $(T_t)_{t \geq 0}$ generated by $(L, \mathrm{Dom}(L))$ is given by (extensions of) PDOs with symbols $e^{-tH(p)}$:

$$T_t\varphi(x) = (2\pi)^{-d} \int_{\mathbb{R}^d} \int_{\mathbb{R}^d} e^{ip \cdot (x-q)} e^{-tH(p)} \varphi(q) \, dq \, dp, \quad x \in \mathbb{R}^d, \quad \varphi \in C_c^\infty(\mathbb{R}^d).$$

If the symbol H depends on both variables x and p then $(T_t)_{t \geq 0}$ are again PDOs. However their symbols do not coincide with $e^{-tH(x,p)}$ and are not known explicitly. The family $(F(t))_{t \geq 0}$ of PDOs with symbols $e^{-tH(x,p)}$ is not a semigroup any more. However, this family is Chernoff equivalent to $(T_t)_{t \geq 0}$. Namely, the following theorem holds (see [26, 27]):

Theorem 2.2 *Let $H : \mathbb{R}^d \times \mathbb{R}^d \to \mathbb{C}$ be measurable, locally bounded in both variables (x, p), satisfy for each fixed $x \in \mathbb{R}^d$ the representation (8) and the following assumptions:*

---

[4] $C_\infty^m(\mathbb{R}^d) := \{\varphi \in C^m(\mathbb{R}^d) : \partial^\alpha \varphi \in C_\infty(\mathbb{R}^d), |\alpha| \leq m\}.$

(i) $\sup_{q \in \mathbb{R}^d} |H(q, p)| \leq \kappa(1 + |p|^2)$ for all $p \in \mathbb{R}^d$ and some $\kappa > 0$,

(ii) $p \mapsto H(q, p)$ is uniformly (w.r.t. $q \in \mathbb{R}^d$) continuous at $p = 0$,

(iii) $q \mapsto H(q, p)$ is continuous for all $p \in \mathbb{R}^d$.

Assume that the function $H(x, p)$ is such that the PDO with symbol $-H$ defined on $C_c^\infty(\mathbb{R}^d)$ is closable and the closure (denoted by $(L, \mathrm{Dom}(L))$) generates a strongly continuous semigroup $(T_t)_{t \geq 0}$ on $X = C_\infty(\mathbb{R}^d)$. Consider now for each $t \geq 0$ the PDO $F(t)$ with the symbol $e^{-tH(x,p)}$, i.e. for $\varphi \in C_c^\infty(\mathbb{R}^d)$

$$F(t)\varphi(x) = (2\pi)^{-d} \int_{\mathbb{R}^d} \int_{\mathbb{R}^d} e^{ip\cdot(x-q)} e^{-tH(x,p)} \varphi(q) dq dp. \qquad (9)$$

Then the family $(F(t))_{t \geq 0}$ extends to a strongly continuous family on X and is Chernoff equivalent to the semigroup $(T_t)_{t \geq 0}$.

Note that the extensions of $F(t)$ are given (via integration with respect to p in (9)) by integral operators:

$$F(t)\varphi(x) = \int_{\mathbb{R}^d} \varphi(y) v_t^x(dy), \qquad (10)$$

where, for each $x \in \mathbb{R}^d$ and each $t \geq 0$, the sub-probability measure v_t^x is given via its Fourier transform $\mathcal{F}[v_t^x](p) = (2\pi)^{-d/2} e^{-tH(x,-p)-ip\cdot x}$.

Example 2.1 Let in formula (6) additionally $N(x, dy) \equiv 0$, the coefficients A, B, C be bounded and continuous, and

there exist $a_0, A_0 \in \mathbb{R}$ with $0 < a_0 \leq A_0 < \infty$ such that
$$a_0|z|^2 \leq z \cdot A(x)z \leq A_0|z|^2 \quad \text{for all } x, z \in \mathbb{R}^d. \qquad (11)$$

Then L is a second order uniformly elliptic operator and the family $(F(t))_{t \geq 0}$ in (10) has the following view: $F(0) := \mathrm{Id}$ and for all $t > 0$ and all $\varphi \in X$

$$F(t)\varphi(x) := \frac{e^{-tC(x)}}{\sqrt{(4\pi t)^d \det A(x)}} \int_{\mathbb{R}^d} e^{-\frac{A^{-1}(x)(x-tB(x)-y)\cdot(x-tB(x)-y)}{4t}} \varphi(y) dy. \qquad (12)$$

Moreover, it has been shown in [24] that $F'(0) = L$ on a bigger core $C_c^{2,\alpha}(\mathbb{R}^d)$ what is important for further applications (e.g., in Sect. 2.4).

Let now $C \equiv 0$. The evolution equation

$$\frac{\partial f}{\partial t}(t, x) = -B(x) \cdot \nabla f(t, x) + \mathrm{tr}(A(x) \mathrm{Hess}\, f(t, x))$$

is the backward Kolmogorov equation for a d-dimesional Itô diffusion process $(\xi_t)_{t\geq 0}$ satisfying the SDE

$$d\xi_t = -B(\xi_t)dt + \sqrt{2A(\xi_t)}dW_t, \qquad (13)$$

with a d-dimensional Wiener process $(W_t)_{t\geq 0}$. Consider the Euler–Maruyama scheme for the SDE (13) on $[0, t]$ with time step t/n:

$$X_0 := \xi_0, \qquad X_{k+1} := X_k - B(X_k)\frac{t}{n} + \sqrt{\frac{2t}{n}} A(X_k) Z_k, \quad k = 0, \ldots, n-1, \quad (14)$$

where $(Z_k)_{k=0,\ldots,n-1}$ are i.i.d. d-dimensional $N(0, \text{Id})$ Gaussian random variables such that X_k and Z_k are independent for all $k = 0, \ldots, n-1$. Then, for all $k = 0, \ldots, n-1$ holds:

$$\mathbb{E}[f_0(X_{k+1}) \mid X_k] = \mathbb{E}\left[f_0\left(x - B(x)\frac{t}{n} + \sqrt{\frac{2t}{n}} A(x) Z_k \right) \right]\bigg|_{x:=X_k} = F(t/n) f_0(X_k).$$

By the tower property of conditional expectation, one has

$$\mathbb{E}[f_0(X_n) \mid X_0 = x] = \mathbb{E}[\mathbb{E}[f_0(X_n) \mid X_{n-1}] \mid X_0 = x] = \ldots =$$
$$= \mathbb{E}[\ldots \mathbb{E}[\mathbb{E}[f_0(X_n) \mid X_{n-1}] \mid X_{n-2}] \ldots \mid X_0 = x] = F^n(t/n) f_0(x).$$

Hence, by Theorem 2.2, it holds for all $x \in \mathbb{R}^d$

$$\mathbb{E}[f_0(\xi_t) \mid \xi_0 = x] = T_t f_0(x) = \lim_{n\to\infty} F^n(t/n) f_0(x) = \lim_{n\to\infty} \mathbb{E}[f_0(X_n) \mid X_0 = x].$$

And, therefore, the Euler–Maruyama scheme (14) converges weakly.[5] The same holds in the general case of Feller processes satisfying assumptions of Theorem 2.2 (see [15]). And the corresponding Markov chain approximation $(X_k)_{k=0,\ldots,n-1}$ of ξ_t consists of increments of Lévy processes, obtained form the original Feller process by "freezing the coefficients" in the generator in a suitable way (see [14]).

Let us investigate the family $(F(t))_{t\geq 0}$ in (12) more carefully. We have actually

$$F(t)\varphi(x) = e^{-tC(x)} \int_{\mathbb{R}^d} e^{\frac{A^{-1}(x)B(x)\cdot(x-y)}{2}} e^{-t\frac{|A^{-1/2}(x)B(x)|^2}{4}} \varphi(y) p_A(t, x, y) dy, \qquad (15)$$

where $p_A(t, x, y) := ((4\pi t)^d \det A(x))^{-1/2} \exp\left(-\frac{A^{-1}(x)(x-y)\cdot(x-y)}{4t} \right)$. Therefore, Theorem 2.2 yields the following Feynman formula for all $t > 0$, $\varphi \in X$ and $x_0 \in \mathbb{R}^d$:

---

[5] The weak convergence of this Euler–Maruyama scheme is, of course, a classical result, cf. [41].

$$T_t\varphi(x_0) = \lim_{n\to\infty} \int_{\mathbb{R}^{dn}} e^{-\frac{t}{n}\sum_{k=1}^n C(x_{k-1})} e^{\frac{1}{2}\sum_{k=1}^n A^{-1}(x_{k-1})B(x_{k-1})\cdot(x_{k-1}-x_k)} \times \quad (16)$$

$$\times e^{-\frac{t}{4n}\sum_{k=1}^n |A^{-1/2}(x_{k-1})B(x_{k-1})|^2} \varphi(x_n) p_A(t/n, x_0, x_1)\ldots p_A(t/n, x_{n-1}, x_n)\, dx_1\cdots dx_n.$$

And the convergence is uniform with respect to $x_0 \in \mathbb{R}^d$ and $t \in (0, t^*]$ for all $t^* > 0$. The limit in the right hand side of formula (16) coincides with the following path integral (compare with the formula (34) in [47] and formula (3) in [44]):

$$T_t\varphi(x_0) = \mathbb{E}^{x_0}\left[\exp\left(-\int_0^t C(X_s)ds\right) \exp\left(-\frac{1}{2}\int_0^t A^{-1}(X_s)B(X_s)\cdot dX_s\right) \times \right.$$

$$\left. \times \exp\left(-\frac{1}{4}\int_0^t A^{-1}(X_s)B(X_s)\cdot B(X_s)ds\right)\varphi(X_t)\right].$$

Here the stochastic integral $\int_0^t A^{-1}(X_s)B(X_s)\cdot dX_s$ is an Itô integral. And $\mathbb{E}^{x_0}$ is the expectation of a (starting at x_0) diffusion process $(X_t)_{t\geq 0}$ with the variable diffusion matrix A and without any drift, i.e $(X_t)_{t\geq 0}$ solves the stochastic differential equation

$$dX_t = \sqrt{2A(X_t)}dW_t.$$

Remark 2.2 Let now $N(x, dy) := N(dy)$ in formula (6), i.e. N does not depend on x. Let the coefficients A, B, C be bounded and continuous, and the property (11) hold. Then $L = L_1 + L_2$, where L_1 is the local part of L, given in the first line of (6) and L_2 is the non-local part of L, given in the second line of (6). And, respectively, $H(x, p) = H_1(x, p) + H_2(p)$ in (8), where $H_1(x, p)$ is a quadratic polynomial with respect to p with variable coefficients and H_2 does not depend on x. Then the closure of $(L_2, C_c^\infty(\mathbb{R}^d))$ in X generates a C_0–semigroup $(e^{tL_2})_{t\geq 0}$ and operators e^{tL_2} are PDOs with symbols e^{-tH_2} on $C_c^\infty(\mathbb{R}^d)$. Let the family of probability measures $(\eta_t)_{t\geq 0}$ be such that $\mathcal{F}[\eta_t] = (2\pi)^{-d/2}e^{-tH_2}$. Then we have $e^{tL_2}\varphi = \varphi * \eta_t$ on X. Assume that $H_2 \in C^\infty(\mathbb{R}^d)$. Then the family $(F(t))_{t\geq 0}$ in (9) can be represented (for $\varphi \in C_c^\infty(\mathbb{R}^d)$) in the following way (cf. with formula (10)):

$$F(t)\varphi(x) = \left[\mathcal{F}^{-1} \circ e^{-tH(x,\cdot)} \circ \mathcal{F}\varphi\right](x) = \left[\mathcal{F}^{-1}\circ e^{-tH_1(x,\cdot)} \circ \mathcal{F} \circ \mathcal{F}^{-1} \circ e^{-tH_2}\circ \mathcal{F}\varphi\right](x)$$

$$= \left(\varphi * \eta_t * \rho_t^x\right)(x),$$

where $\rho_t^x(z) := e^{-tC(x)}\left((4\pi t)^d \det A(x)\right)^{-1/2} \exp\left\{-\frac{A^{-1}(x)(z-tB(x))\cdot(z-tB(x))}{4t}\right\}$, i.e. the family $(F_1(t))_{t\geq 0}$, $F_1(t)\varphi(x) := (\varphi * \rho_t^x)(x)$, is actually given by formula (12). The representation

$$F(t)\varphi(x) = \left(\varphi * \eta_t * \rho_t^x\right)(x) \quad (17)$$

holds even for all $\varphi \in X$, $x \in \mathbb{R}^d$ and without the assumption that $H_2 \in C^\infty(\mathbb{R}^d)$. Denoting e^{tL_2} as $F_2(t)$, we obtain that $F(t) = F_1(t) \circ F_2(t)$. Due to Theorem 2.2, $F'(0) = L$ on a core $D := C_c^\infty(\mathbb{R}^d)$. Using Theorem 2.1 and Example 2.1, one shows that $F'(0) = L$ even on $D = C_c^{2,\alpha}(\mathbb{R}^d)$ as soon as $C_c^{2,\alpha}(\mathbb{R}^d) \subset \mathrm{Dom}(L_2)$ (without the assumption $H_2 \in C^\infty(\mathbb{R}^d)$). The bigger core D is more suitable for further applications of the family $(F(t))_{t \geq 0}$ in the form of (17) in Sect. 2.4.

Example 2.2 Consider the symbol $H(x, p) := a(x)|p|$, where $a \in C^\infty(\mathbb{R}^d)$ is a strictly positive bounded function. The closure of the PDO $(L, C_c^\infty(\mathbb{R}^d))$ with symbol $-H$ acts as $L\varphi(x) := a(x)\left(-(-\Delta)^{1/2}\right)\varphi(x)$, generates a Feller semigroup $(T_t)_{t \geq 0}$ and, by Theorem 2.2, the following family $(F(t))_{t \geq 0}$ is Chernoff equivalent to $(T_t)_{t \geq 0}$:

$$F(t)\varphi(x) := (2\pi)^{-d} \int_{\mathbb{R}^d} \int_{\mathbb{R}^d} e^{ip\cdot(x-q)} e^{-ta(x)|p|} \varphi(q) dq dp$$

$$= \Gamma\left(\frac{d+1}{2}\right) \int_{\mathbb{R}^d} \varphi(q) \frac{a(x)t}{\left(\pi|x-q|^2 + a^2(x)t^2\right)^{\frac{d+1}{2}}} dq,$$

where Γ is the Euler gamma-function. We see that the multiplicative perturbation $a(x)$ of the fractional Laplacian contributes actually to the time parameter in the definition of the family $(F(t))_{t \geq 0}$. This motivates the result of the following subsection.

2.3 Chernoff Approximations for Multiplicative Perturbations of a Generator

Let Q be a metric space. Consider the Banach space $X = C_b(Q)$ of bounded continuous functions on Q with supremum-norm $\|\cdot\|_\infty$. Let $(T_t)_{t \geq 0}$ be a strongly continuous semigroup on X with generator $(L, \mathrm{Dom}(L))$. Consider a function $a \in C_b(Q)$ such that $a(q) > 0$ for all $q \in Q$. Then the space X is invariant under the multiplication operator a, i.e. $a(X) \subset X$. Consider the operator $\widehat{L}$, defined for all $\varphi \in \mathrm{Dom}(\widehat{L})$ and all $q \in Q$ by

$$\widehat{L}\varphi(q) := a(q)(L\varphi)(q), \quad \text{where} \quad \mathrm{Dom}(\widehat{L}) := \mathrm{Dom}(L). \tag{18}$$

Assumption 2.1 We assume that $(\widehat{L}, \mathrm{Dom}(\widehat{L}))$ generates a strongly continuous semigroup (which is denoted by $(\widehat{T}_t)_{t \geq 0}$) on the Banach space X.

Some conditions assuring the existence and strong continuity of the semigroup $(\widehat{T}_t)_{t \geq 0}$ can be found, e.g., in [31, 45]. The operator $\widehat{L}$ is called a *multiplicative perturbation* of the generator L and the semigroup $(\widehat{T}_t)_{t \geq 0}$, generated by $\widehat{L}$, is called a *semigroup with the multiplicatively perturbed* with the function *a generator*. The following result has been shown in [21] (cf. [20, 27]).

Theorem 2.3 *Let Assumption 2.1 hold. Let* $(F(t))_{t\geq 0}$ *be a strongly continuous family*[6] *of bounded linear operators on the Banach space X, which is Chernoff equivalent to the semigroup* $(T_t)_{t\geq 0}$. *Consider the family of operators* $(\widehat{F}(t))_{t\geq 0}$ *defined on X by*

$$\widehat{F}(t)\varphi(q) := (F(a(q)t)\varphi)(q) \quad \text{for all } \varphi \in X, \ q \in Q. \tag{19}$$

The operators $\widehat{F}(t)$ *act on the space X, the family* $(\widehat{F}(t))_{t\geq 0}$ *is again strongly continuous and is Chernoff equivalent to the semigroup* $(\widehat{T}_t)_{t\geq 0}$ *with multiplicatively perturbed with the function a generator, i.e. the Chernoff approximation*

$$\widehat{T}_t\varphi = \lim_{n\to\infty} \left[\widehat{F}(t/n)\right]^n \varphi$$

is valid for all $\varphi \in X$ *locally uniformly with respect to* $t \geq 0$.

Remark 2.3 (i) The statement of Theorem 2.3 remains true for the following Banach spaces (cf. [21]):
(a) $X = C_\infty(Q) := \{\varphi \in C_b(Q) : \lim_{\rho(q,q_0)\to\infty} \varphi(q) = 0\}$, where q_0 is an arbitrary fixed point of Q and the metric space Q is unbounded with respect to its metric ρ;
(b) $X = C_0(Q) := \{\varphi \in C_b(Q) : \forall \varepsilon > 0 \ \exists \text{ a compact } K_\varphi^\varepsilon \subset Q \text{ such that } |\varphi(q)| < \varepsilon \text{ for all } q \notin K_\varphi^\varepsilon\}$, where the metric space Q is assumed to be locally compact.
(ii) As it follows from the proof of Theorem 2.3, if $\lim_{t\to 0} \left\| \frac{F(t)\varphi-\varphi}{t} - L\varphi \right\|_X = 0$ for all $\varphi \in D$ then also $\lim_{t\to 0} \left\| \frac{\widehat{F}(t)\varphi-\varphi}{t} - \widehat{L}\varphi \right\|_X = 0$ for all $\varphi \in D$.

Corollary 2.1 *Let* $(X_t)_{t\geq 0}$ *be a Markov process with the state space Q and transition probability* $P(t,q,dy)$. *Let the corresponding semigroup* $(T_t)_{t\geq 0}$,

$$T_t\varphi(q) = \mathbb{E}^q[\varphi(X_t)] \equiv \int_Q \varphi(y)P(t,q,dy),$$

be strongly continuous on the Banach space X, where $X = C_b(Q)$, $X = C_\infty(Q)$ *or* $X = C_0(Q)$, *and Assumption 2.1 hold. Then by Theorem 2.3 and Remark 2.3 the family* $(\widehat{F}(t))_{t\geq 0}$ *defined by*

$$\widehat{F}_t\varphi(q) := \int_Q \varphi(y)P(a(q)t,q,dy),$$

---

[6] The family $(F(t))_{t\geq 0}$ of bounded linear operators on a Banach space X is called strongly continuous if $\lim_{t\to t_0} \|F(t)\varphi - F(t_0)\varphi\|_X = 0$ for all $t, t_0 \geq 0$ and all $\varphi \in X$.

is strongly continuous and is Chernoff equivalent to the semigroup $(\widehat{T}_t)_{t\geq 0}$ with multiplicatively perturbed (with the function a) generator. Therefore, the following Chernoff approximation is true for all $t > 0$ and all $q_0 \in Q$:

$$\widehat{T}_t\varphi(q_0) = \lim_{n\to\infty} \int_Q \cdots \int_Q \varphi(q_n) P(a(q_0)t/n, q_0, dq_1) P(a(q_1)t/n, q_1, dq_2) \times \cdots$$
$$\times P(a(q_{n-1})t/n, q_{n-1}, dq_n), \quad (20)$$

where the order of integration is from q_n to q_1 and the convergence is uniform with respect to $q_0 \in Q$ and locally uniform with respect to $t \geq 0$.

Remark 2.4 A multiplicative perturbation of the generator of a Markov process is equivalent to some random time change of the process (see [32, 76, 77]). Note that $\widehat{P}(t, q, dy) := P(a(q)t, q, dy)$ is not a transition probability any more. Nevertheless, if the transition probability $P(t, q, dy)$ of the original process is known, formula (20) allows to approximate the unknown transition probability of the modified process.

2.4 Chernoff Approximations for Semigroups Generated by Processes in a Domain with Prescribed Behaviour at the Boundary of/Outside the Domain

Let $(\xi_t)_{t\geq 0}$ be a (sub-) Markov process in $\mathbb{R}^d$. Assume that the corresponding semigroup $(T_t)_{t\geq 0}$ is strongly continuous on some Banach space X of functions on $\mathbb{R}^d$, e.g. $X = C_\infty(\mathbb{R}^d)$ or $X = L^p(\mathbb{R}^d)$, $p \in [1, \infty)$. Let $(L, \mathrm{Dom}(L))$ be the generator of $(T_t)_{t\geq 0}$ in X. Assume that a Chernoff approximation of $(T_t)_{t\geq 0}$ via a family $(F(t))_{t\geq 0}$ is already known (and hence we have a core D for L such that $\lim_{t\to 0} \left\|\frac{F(t)\varphi - \varphi}{t} - L\varphi\right\|_X = 0$ for all $\varphi \in D$). Consider now a domain $\Omega \subset \mathbb{R}^d$. Let $(\xi_t)_{t\geq 0}$ start in Ω and impose some reasonable "Boundary Conditions" (**BC**), i.e. conditions on the behaviour of $(\xi_t)_{t\geq 0}$ at the boundary $\partial\Omega$, or (if the generator L is non-local) outside Ω. This procedure gives rise to a (sub-) Markov process in Ω which we denote by $(\xi_t^*)_{t\geq 0}$. In some cases, the corresponding semigroup $(T_t^*)_{t\geq 0}$ is strongly continuous on some Banach space Y of functions on Ω (e.g. $Y = C(\overline{\Omega})$, $Y = C_0(\Omega)$ or $Y = L^p(\Omega)$, $p \in [1, \infty)$). The question arises: *how to construct a Chernoff approximation of $(T_t^*)_{t\geq 0}$ on the base of the family $(F(t))_{t\geq 0}$, i.e. how to incorporate **BC** into a Chernoff approximation?* A possible strategy to answer this question is to construct a proper extension E^* of functions from Ω to $\mathbb{R}^d$ such that, first, $E^* : Y \to X$ is a linear contraction and, second, there exists a core D^* for the generator $(L^*, \mathrm{Dom}(L^*))$ of $(T_t^*)_{t\geq 0}$ with $E^*(D^*) \subset D$. Then it is easy to see that the family $(F^*(t))_{t\geq 0}$ with

$$F^*(t) := R_t \circ F(t) \circ E^* \quad (21)$$

is Chernoff equivalent to the semigroup $(T_t^*)_{t\geq 0}$. Here R_t is, in most cases, just the restriction of functions from $\mathbb{R}^d$ to Ω, and, for the case of Dirichlet **BC**, it is a multiplication with a proper cut-off function ψ_t having support in Ω such that $\psi_t \to 1_\Omega$ as $t \to 0$ (see [22, 24]). This strategy has been successfully realized in the following cases (note that extensions E^* are obtained in a constructive way and can be implemented in numerical schemes):

Case 1: $X = C_\infty(\mathbb{R}^d)$, $(\xi_t)_{t\geq 0}$ is a Feller process whose generator L is given by (6) with A, B, C of the class $C^{2,\alpha}$, A satisfies (11), and either $N \equiv 0$ or $N \neq 0$ and the non-local term of L is a relatively bounded perturbation of the local part of L with some extra assumption on jumps of the process (see details in [22, 24]). The family $(F(t))_{t\geq 0}$ is given by (10) (see also (17), or (12) in the corresponding particular cases) and $D = C_c^{2,\alpha}(\mathbb{R}^d)$. Further, Ω is a bounded $C^{4,\alpha}$–smooth domain, $Y = C_0(\Omega)$, **BC** are the homogeneous Dirichlet boundary/external conditions corresponding to killing of the process upon leaving the domain Ω. A proper extension E^* has been constructed in [6], and it maps $\mathrm{Dom}(L^*) \cap C^{2,\alpha}(\overline{\Omega})$ into D.

One can further simplify the Chernoff approximation constructed via the family $(F^*(t))_{t\geq 0}$ of (21) and show that the following Feynman formula solves the considered Cauchy–Dirichlet problem (see [22]):

$$T_t^*\varphi(x_0) = \lim_{n\to\infty} \int_\Omega \cdots \int_\Omega \int_\Omega \varphi(x_n) v_{t/n}^{x_{n-1}}(dx_n) v_{t/n}^{x_{n-2}}(dx_{n-1}) \cdots v_{t/n}^{x_0}(dx_1). \quad (22)$$

The convergence in this formula is however only locally uniform with respect to $x_0 \in \Omega$ (and locally uniform with respect to $t \geq 0$). Similar results hold also for non-degenerate diffusions in domains of a compact Riemannian manifold M with homogeneous Dirichlet **BC** (see, e.g. [19]), what can be shown by combining approaches described in Sects. 2.1, 2.3, 2.4 and using families $(F(t))_{t\geq 0}$ of [72] which are Chernoff equivalent to the heat semigroup on $C(M)$.

Case 2: $X = C_\infty(\mathbb{R}^d)$, $(\xi_t)_{t\geq 0}$ is a Brownian motion, the family $(F(t))_{t\geq 0}$ is the heat semigroup (1) (hence $D = \mathrm{Dom}(L)$), Ω is a bounded C^∞-smooth domain, $Y = C(\overline{\Omega})$, **BC** are the Robin boundary conditions

$$\frac{\partial \varphi}{\partial \nu} + \beta \varphi = 0 \quad \text{on } \partial\Omega, \quad (23)$$

where ν is the outer unit normal, β is a smooth bounded nonnegative function on $\partial\Omega$. A proper extension E^* (and the corresponding Chernoff approximation itself) has been constructed in [52], and it maps $D^* := \mathrm{Dom}(L^*) \cap C^\infty(\overline{\Omega})$ into the $\mathrm{Dom}(L)$.

This result can be further generalized for the case of diffusions, using the techniques of Sects. 2.1 and 2.3. This will be demonstrated in Example 2.3. Note, however, that the extension E^* of [52] maps D^* into the set of functions which do not belong to $C^2(\mathbb{R}^d)$. Hence it is not possible to use the family (12) (and $D = C_c^{2,\alpha}(\mathbb{R}^d)$) in a straightforward manner for approximation of diffusions with Robin **BC**.

Example 2.3 Let $X = C_\infty(\mathbb{R}^d)$. Consider $(L_1, \text{Dom}(L_1))$ being the closure of $\left(\frac12\Delta, S(\mathbb{R}^d)\right)$ in X. Then $\text{Dom}(L_1)$ is continuously embedded in $C^{1,\alpha}(\mathbb{R}^d)$ for every $\alpha \in (0,1)$ by Theorem 3.1.7 and Corollary 3.1.9 (iii) of [46], and $(L_1, \text{Dom}(L_1))$ generates a C_0-semigroup $(T_1(t))_{t\geq}$ on X, this is the heat semigroup given by (1). Let $a \in C_b(\mathbb{R}^d)$ be such that $a(x) \geq a_0$ for some $a_0 > 0$ and all $x \in \mathbb{R}^d$. Then $(\widehat{L_1}, \text{Dom}(L_1))$, $\widehat{L_1}\varphi(x) := a(x)L_1\varphi(x)$, generates a C_0-semigroup $(\widehat{T_1}(t))_{t\geq 0}$ on X by [31]. Therefore, the family $(\widehat{F}(t))_{t\geq 0}$ with

$$\widehat{F}(t)\varphi(x) := \int_{\mathbb{R}^d} \varphi(y) P(a(x)t, x, dy),$$

where $P(t, x, dy)$ is given by (2), is Chernoff equivalent to $(\widehat{T_1}(t))_{t\geq 0}$ by Corollary 2.1. And $\left\|\frac{\widehat{F}(t)\varphi - \varphi}{t} - \widehat{L_1}\varphi\right\|_X \to 0$ as $t \to 0$ for each $\varphi \in \text{Dom}(L_1)$. Let now $C \in C_b(\mathbb{R}^d)$ and $B \in C_b(\mathbb{R}^d; \mathbb{R}^d)$. Then the operator $(L, \text{Dom}(L))$,

$$L\varphi(x) := \frac{a(x)}{2}\Delta\varphi(x) - B(x) \cdot \nabla\varphi(x) - C(x)\varphi(x), \quad \text{Dom}(L) := \text{Dom}(L_1),$$

(obtained by a relatively bounded additive perturbation of $(\widehat{L_1}, \text{Dom}(L_1))$) generates a C_0-semigroup $(T_t)_{t\geq 0}$ on X (e.g., by Theorem 4.4.3 of [40]). Motivated by Sect. 2.3 and the view of the translation semigroup, consider the family $(F_2(t))_{t\geq 0}$ of contractions on X given by $F_2(t)\varphi(x) := \varphi(x - tB(x))$. Then, for all $\varphi \in \text{Dom}(L_1) \subset C^{1,\alpha}(\mathbb{R}^d)$, holds $\left\|\frac{F_2(t)\varphi - \varphi}{t} + B \cdot \nabla\varphi\right\|_X \leq \text{const} \cdot t^\alpha |B|^{\alpha+1} \to 0$, $t \to 0$, and $\|F_2(t)\| \leq 1$ for all $t \geq 0$. Therefore, by Theorem 2.1, the family $(F(t))_{t\geq 0}$ with

$$F(t)\varphi(x) := \left[e^{-tC} \circ F_2(t) \circ \widehat{F}(t)\right]\varphi(x)$$
$$\equiv e^{-tC(x)} \int_{\mathbb{R}^d} \varphi(y) P(a(x - tB(x))t, x - tB(x), dy)$$

is Chernoff equivalent to the semigroup $(T_t)_{t\geq 0}$. Let now Ω be a bounded C^∞-smooth domain, $Y = C(\overline{\Omega})$. Consider $(L^*, \text{Dom}(L^*))$ in Y with[7]

$$\text{Dom}(L^*) := \Big\{\varphi \in Y \cap H^1(\Omega) \,:\, L\varphi \in Y,$$
$$\int_\Omega \Delta\varphi u\, dx + \int_\Omega \nabla\varphi \nabla u\, dx + \int_{\partial\Omega} \beta\varphi u\, d\sigma = 0 \,\forall u \in H^1(\Omega)\Big\},$$
$$L^*\varphi := L\varphi, \quad \forall \varphi \in \text{Dom}(L^*).$$

Then $(L^*, \text{Dom}(L^*))$ generates a C_0-semigroup $(T_t^*)_{t\geq 0}$ on Y (cf. [53]). Consider $R : X \to Y$ being the restriction of a function from $\mathbb{R}^d$ to $\overline{\Omega}$. Consider the extension

---

[7] Here we consider the Robin **BC** (23) given in a weaker form via the first Green's formula; $d\sigma$ is the surface measure on $\partial\Omega$.

$E^* : Y \to X$ constructed in [52]. This extension is a linear contraction, obtained via an orthogonal reflection at the boundary and multiplication with a suitable cut-off function, whose behaviour at $\partial\Omega$ is prescribed (depending on β) in such a way that the weak Laplacian of the extension $E^*(\varphi)$ is continuous for each $\varphi \in D^* :=$ Dom$(L^*) \cap C^\infty(\Omega)$ and $E^*(D^*) \subset$ Dom(L_1). We omit the explicit description of E^*, in order to avoid corresponding technicalities. We consider the family $(F^*(t))_{t \geq 0}$ on Y given by $F^*(t) := R \circ F(t) \circ E^*$, i.e.

$$F^*(t)\varphi(x) := e^{-tC(x)} \int_{\mathbb{R}^d} E^*[\varphi](y) P(a(x - tB(x))t, x - tB(x), dy), \qquad x \in \overline{\Omega}.$$

Then $F^*(0) = $ Id, $\|F^*(t)\| \leq e^{t\|C\|_\infty}$, and we have for all $\varphi \in D^*$

$$\lim_{t \to 0} \left\| \frac{F^*(t)\varphi - \varphi}{t} - L^*\varphi \right\|_Y = \lim_{t \to 0} \left\| R \circ \left(\frac{F(t)E^*[\varphi] - E^*[\varphi]}{t} - LE^*[\varphi] \right) \right\|_Y$$
$$\leq \lim_{t \to 0} \left\| \frac{F(t)E^*[\varphi] - E^*[\varphi]}{t} - LE^*[\varphi] \right\|_X = 0.$$

Therefore, the family $(F^*(t))_{t \geq 0}$ is Chernoff equivalent to the semigroup $(T_t^*)_{t \geq 0}$ by Theorem 1.1, i.e. $T_t^*\varphi = \lim_{n \to \infty} [F^*(t/n)]^n \varphi$ for each $\varphi \in Y$ locally uniformly with respect to $t \geq 0$.

2.5 Chernoff Approximations for Subordinate Semigroups

One of the ways to construct strongly continuous semigroups is given by the procedure of subordination. From two ingredients: an original C_0 contraction semigroup $(T_t)_{t \geq 0}$ on a Banach space X and a convolution semigroup[8] $(\eta_t)_{t \geq 0}$ supported by $[0, \infty)$, this procedure produces the C_0 contraction semigroup $(T_t^f)_{t \geq 0}$ on X with

$$T_t^f \varphi := \int_0^\infty T_s \varphi \, \eta_t(ds), \qquad \forall \varphi \in X.$$

If the semigroup $(T_t)_{t \geq 0}$ corresponds to a stochastic process $(X_t)_{t \geq 0}$, then subordination is a random time-change of $(X_t)_{t \geq 0}$ by an independent increasing Lévy process (subordinator) with distributions $(\eta_t)_{t \geq 0}$. If $(T_t)_{t \geq 0}$ and $(\eta_t)_{t \geq 0}$ both are known explicitly, so is $(T_t^f)_{t \geq 0}$. But if, e.g., $(T_t)_{t \geq 0}$ is not known, neither $(T_t^f)_{t \geq 0}$ itself, nor even the generator of $(T_t^f)_{t \geq 0}$ are known explicitly any more. This impedes the construction of a family $(F(t))_{t \geq 0}$ with a prescribed (but unknown explicitly) derivative at $t = 0$. This

---

[8] A family $(\eta_t)_{t \geq 0}$ of bounded Borel measures on $\mathbb{R}$ is called a convolution semigroup if $\eta_t(\mathbb{R}) \leq 1$ for all $t \geq 0$, $\eta_t * \eta_s = \eta_{t+s}$ for all $t, s \geq 0$, $\eta_0 = \delta_0$, and $\eta_t \to \delta_0$ vaguely as $t \to 0$, i.e. $\lim_{t \to 0} \int_\mathbb{R} \varphi(x) \eta_t(dx) = \int_\mathbb{R} \varphi(x) \delta_0(dx) \equiv \varphi(0)$ for all $\varphi \in C_c(\mathbb{R})$. A convolution semigroup $(\eta_t)_{t \geq 0}$ is supported by $[0, \infty)$ if supp $\eta_t \subset [0, \infty)$ for all $t \geq 0$.

difficulty is overwhelmed below by construction of families $(\mathcal{F}(t))_{t\geq 0}$ and $(\mathcal{F}_\mu(t))_{t\geq 0}$ which incorporate approximations of the generator of $(T_t^f)_{t\geq 0}$ itself. Recall that each convolution semigroup $(\eta_t)_{t\geq 0}$ supported by $[0,\infty)$ corresponds to a Bernstein function f via the Laplace transform $\mathcal{L}$: $\mathcal{L}[\eta_t] = e^{-tf}$ for all $t > 0$. Each Bernstein function f is uniquely defined by a triplet (σ, λ, μ) with constants $\sigma, \lambda \geq 0$ and a Radon measure μ on $(0,\infty)$, such that $\int_{0+}^\infty \frac{s}{1+s}\mu(ds) < \infty$, through the representation $f(z) = \sigma + \lambda z + \int_{0+}^\infty (1-e^{-sz})\mu(ds)$, $\forall z : \operatorname{Re} z \geq 0$. Let $(L, \operatorname{Dom}(L))$ be the generator of $(T_t)_{t\geq 0}$ and $(L^f, \operatorname{Dom}(L^f))$ be the generator of $(T_t^f)_{t\geq 0}$. Then each core for L is also a core for L^f and, for $\varphi \in \operatorname{Dom}(L)$, the operator L^f has the representation

$$L^f\varphi = -\sigma\varphi + \lambda L\varphi + \int_{0+}^\infty (T_s\varphi - \varphi)\mu(ds).$$

Let $(F(t))_{t\geq 0}$ be a family of contractions on $(X, \|\cdot\|_X)$ which is Chernoff equivalent to $(T_t)_{t\geq 0}$, i.e. $F(0) = \operatorname{Id}$, $\|F(t)\| \leq 1$ for all $t \geq 0$ and there is a set $D \subset \operatorname{Dom}(L)$, which is a core for L, such that $\lim_{t\to 0}\left\|\frac{F(t)\varphi - \varphi}{t} - L\varphi\right\|_X = 0$ for each $\varphi \in D$. The first candidate for being Chernoff equivalent to $(T_t^f)_{t\geq 0}$ could be the family of operators $(F^*(t))_{t\geq 0}$ given by $F^*(t)\varphi := \int_0^\infty F(s)\varphi\, \eta_t(ds)$ for all $\varphi \in X$. However, its derivative at zero does not coincide with L^f on D. Nevertheless, with suitable modification of $(F^*(t))_{t\geq 0}$, Theorem 2.1 and the discussion below Theorem 1.1, the following has been proved in [23].

Theorem 2.4 *Let $m : (0,\infty) \to \mathbb{N}_0$ be a monotone function[9] such that $m(t) \to +\infty$ as $t \to 0$. Let the mapping $[F(\cdot/m(t))]^{m(t)}\varphi : [0,\infty) \to X$ be Bochner measurable as the mapping from $([0,\infty), \mathcal{B}([0,\infty)), \eta_t^0)$ to $(X, \mathcal{B}(X))$ for each $t > 0$ and each $\varphi \in X$.*

Case 1: *Let $(\eta_t^0)_{t\geq 0}$ be the convolution semigroup (supported by $[0,\infty)$) associated to the Bernstein function f_0 defined by the triplet $(0, 0, \mu)$. Assume that the corresponding operator semigroup $(S_t)_{t\geq 0}$, $S_t\varphi := \varphi * \eta_t^0$, is strong Feller.[10] Consider the family $(\mathcal{F}(t))_{t\geq 0}$ of operators on $(X, \|\cdot\|_X)$ defined by $\mathcal{F}(0) := \operatorname{Id}$ and*

$$\mathcal{F}(t)\varphi := e^{-\sigma t} \circ F(\lambda t) \circ \mathcal{F}_0(t)\varphi, \quad t > 0,\ \varphi \in X, \tag{24}$$

with $\mathcal{F}_0(0) = \operatorname{Id}$ and[11]

$$\mathcal{F}_0(t)\varphi := \int_{0+}^\infty [F(s/m(t))]^{m(t)}\varphi\, \eta_t^0(ds), \quad t > 0,\ \varphi \in X. \tag{25}$$

---

[9] One can take, e.g., $m(t) := \lfloor 1/t \rfloor$ = the largest integer $n \leq 1/t$. Recall that $\mathbb{N}_0 := \mathbb{N} \cup \{0\}$.

[10] The semigroup $(S_t)_{t\geq 0}$ is strong Feller iff all the measures η_t^0 admit densities of the class $L^1([0,\infty))$ with respect to the Lebesgue measure (cf. Example 4.8.21 of [40]).

[11] For any bounded operator B, its zero degree B^0 is considered to be the identity operator. For each $t > 0$, a non-negative integer $m(t)$ and a bounded Bochner measurable mapping $[F(\cdot/m(t))]^{m(t)}\varphi : [0,\infty) \to X$, the integral in the right hand side of formula (25) is well defined.

The family $(\mathcal{F}(t))_{t\geq 0}$ is Chernoff equivalent to the semigroup $(T_t^f)_{t\geq 0}$, and hence

$$T_t^f \varphi = \lim_{n\to\infty} \left[\mathcal{F}(t/n)\right]^n \varphi$$

for all $\varphi \in X$ locally uniformly with respect to $t \geq 0$.

Case 2: Assume that the measure μ is bounded. Consider a family $(\mathcal{F}_\mu(t))_{t\geq 0}$ of operators on $(X, \|\cdot\|_X)$ defined for all $\varphi \in X$ and all $t \geq 0$ by

$$\mathcal{F}_\mu(t)\varphi := e^{-\sigma t} F(\lambda t)\left(\varphi + t \int_{0+}^{\infty} (F^{m(t)}(s/m(t))\varphi - \varphi)\mu(ds)\right).$$

The family $(\mathcal{F}_\mu(t))_{t\geq 0}$ is Chernoff equivalent to the semigroup $(T_t^f)_{t\geq 0}$, and hence

$$T_t^f \varphi = \lim_{n\to\infty} \left[\mathcal{F}_\mu(t/n)\right]^n \varphi$$

for all $\varphi \in X$ locally uniformly with respect to $t \geq 0$.

The constructed families $(\mathcal{F}(t))_{t\geq 0}$ and $(\mathcal{F}_\mu(t))_{t\geq 0}$ can be used (in combination with the techniques of Sects. 2.1, 2.3, 2.4 and results of [42, 72]), e.g., to approximate semigroups generated by subordinate Feller diffusions on star graphs and Riemannian manifolds. Note that the family (24) can be used when the convolution semigroup $(\eta_t^0)_{t\geq 0}$ is known explicitly. This is the case of inverse Gaussian (including 1/2-stable) subordinator, Gamma subordinator and some others (see, e.g., [11, 18, 30] for examples).

2.6 Approximation of Solutions of Time-Fractional Evolution Equations

We are interested now in distributed order time-fractional evolution equations of the form

$$\mathcal{D}^\mu f(t) = Lf(t), \tag{26}$$

where $(L, \mathrm{Dom}(L))$ is the generator of a C_0- contraction semigroup $(T_t)_{t\geq 0}$ on some Banach space $(X, \|\cdot\|_X)$ and $\mathcal{D}^\mu$ is the distributed order fractional derivative with respect to the time variable t:

$$\mathcal{D}^\mu u(t) := \int_0^1 \frac{\partial^\beta}{\partial t^\beta} u(t)\mu(d\beta), \quad \text{where} \quad \frac{\partial^\beta}{\partial t^\beta} u(t) := \frac{1}{\Gamma(1-\beta)} \int_0^t \frac{u'(r)}{(t-r)^\beta} dr,$$

μ is a finite Borel measure with supp $\mu \in (0, 1)$. Equations of such type are called *time-fractional Fokker–Planck–Kolmogorov equations* (tfFPK-equations) and arise in the framework of continuous time random walks and fractional kinetic theory [35, 49, 51, 73, 79]. As it is shown in papers [38, 39, 50], such tfFPK-equations are governing equations for stochastic processes which are time-changed Markov processes, where the time-change $(E_t^\mu)_{t\geq 0}$ arises as the first hitting time of level $t > 0$ (or, equivalently, as the inverse process) for a mixture $(D_t^\mu)_{t\geq 0}$ of independent stable subordinators with the mixing measure μ.[12] Existence and uniqueness of solutions of initial and initial-boundary value problems for such tfFPK-equations are considered, e.g., in [74, 75]. The process $(E_t^\mu)_{t\geq 0}$ is sometimes called *inverse subordinator*. However, note that it is not a Markov process. Nevertheless, $(E_t^\mu)_{t\geq 0}$ possesses a nice marginal density function $p^\mu(t, x)$ (with respect to the Lebesgue measure dx). It has been shown in [38, 50] that the family of linear operators $(\mathcal{T}_t)_{t\geq 0}$ from X into X given by

$$\mathcal{T}_t \varphi := \int_0^\infty T_\tau \varphi \, p^\mu(t, \tau) \, d\tau, \quad \forall \varphi \in X, \tag{27}$$

is uniformly bounded, strongly continuous, and the function $f(t) := \mathcal{T}_t f_0$ is a solution of the Cauchy problem

$$\mathcal{D}^\mu f(t) = Lf(t), \quad t > 0,$$
$$f(0) = f_0. \tag{28}$$

This result shows that solutions of tfFPK-equations are a kind of subordination of solutions of the corresponding time-non-fractional evolution equations with respect to "subordinators" $(E_t^\mu)_{t\geq 0}$. The non-Markovity of $(E_t^\mu)_{t\geq 0}$ implies that the family $(\mathcal{T}_t)_{t\geq 0}$ is not a semigroup any more. Hence we have no chances to construct Chernoff approximations for $(\mathcal{T}_t)_{t\geq 0}$. Nevertheless, the following is true (see [22]).

Theorem 2.5 *Let the family $(F(t))_{t\geq 0}$ of contractions on X be Chernoff equivalent to $(T_t)_{t\geq 0}$. Let $f_0 \in \text{Dom}(L)$. Let the mapping $F(\cdot) f_0 : [0, \infty) \to X$ be Bochner measurable as a mapping from $([0, \infty), \mathcal{B}([0, \infty)), dx)$ to $(X, \mathcal{B}(X))$. Let μ be a finite Borel measure with supp $\mu \in (0, 1)$ and the family $(\mathcal{T}_t)_{t\geq 0}$ be given by formula (27). Let $f : [0, \infty) \to X$ be defined via $f(t) := \mathcal{T}_t f_0$. For each $n \in \mathbb{N}$ define the mappings $f_n : [0, \infty) \to X$ by*

$$f_n(t) := \int_0^\infty F^n(\tau/n) f_0 \, p^\mu(t, \tau) \, d\tau. \tag{29}$$

Then it holds locally uniformly with respect to $t \geq 0$ that

$$\|f_n(t) - f(t)\|_X \to 0, \quad n \to \infty.$$

---

[12] Hence $(D_t^\mu)_{t\geq 0}$ is a subordinator corresponding to the Bernstein function $f^\mu(s) := \int_0^1 s^\beta \mu(d\beta)$, $s > 0$, and $E_t^\mu := \inf\{\tau \geq 0 : D_\tau^\mu > t\}$.

Of course, similar approximations are valid also in the case of "ordinary subordination" (by a Lévy subordinator) considered in Sect. 2.5. Note also that there exist different Feynman-Kac formulae for the Cauchy problem (28). In particular, the function

$$f(t, x) := \mathbb{E}\left[f_0\left(\xi\left(E_t^\mu\right)\right) \mid \xi(E_0^\mu) = x\right], \tag{30}$$

where $(\xi_t)_{t\geq 0}$ is a Markov process with generator L, solves the Cauchy problem (28) (cf. Theorem 3.6 in [38], see also [75]). Furthermore, the considered equations (with $\mu = \delta_{\beta_0}$, $\beta_0 \in (0, 1)$) are related to some time-non-fractional evolution equations of higher order (see, e.g., [4, 59]). Therefore, the approximations f_n constructed in Theorem 2.5 can be used simultaneously to approximate path integrals appearing in different stochastic representations of the same function $f(t, x)$ and to approximate solutions of corresponding time-non-fractional evolution equations of higher order.

Example 2.4 Let $\mu = \delta_{1/2}$, i.e. $\mathcal{D}^\mu$ is the Caputo derivative of 1/2-th order and $(E_t^{1/2})_{t\geq 0}$ is a 1/2-stable inverse subordinator whose marginal probability density is known explicitly: $p^{1/2}(t, \tau) = \frac{1}{\sqrt{\pi t}} e^{-\frac{\tau^2}{4t}}$. Let $X = C_\infty(\mathbb{R}^d)$ and $(L, \text{Dom}(L))$ be the Feller generator given by (6). Let all the assumptions of Theorem 2.2 be fulfilled. Hence we can use the family $(F(t))_{t\geq 0}$ given by (10) (or by (12) if $N \equiv 0$). Therefore, by Theorem 2.2 and Theorem 2.5, the following Feynman formula solves the Cauchy problem (28):

$$f(t, x_0) = \lim_{n\to\infty} \int_0^\infty \int_{\mathbb{R}^d} \cdots \int_{\mathbb{R}^d} \frac{1}{\sqrt{\pi t}} e^{-\frac{\tau^2}{4t}} \varphi(x_n) v_{\tau/n}^{x_{n-1}}(dx_n) \cdots v_{\tau/n}^{x_0}(dx_1) d\tau.$$

2.7 Chernoff Approximations for Schrödinger Groups

Case 1: PDOs. In Sect. 2.2, we have used the technique of pseudo-differential operators (PDOs). Namely, (with a slight modification of notations) we have considered operator semigroups $(e^{-t\widehat{H}})_{t\geq 0}$ generated by PDOs $-\widehat{H}$ with symbols $-H$ (see formula (7)). We have approximated semigroups via families of PDOs $(F(t))_{t\geq 0}$ with symbols e^{-tH}, i.e. $F(t) = \widehat{e^{-tH}}$. Note again that $e^{-t\widehat{H}} \neq \widehat{e^{-tH}}$ in general. It was established in Theorem 2.2 that

$$e^{-t\widehat{H}} = \lim_{n\to\infty} \left[\widehat{e^{-tH/n}}\right]^n \tag{31}$$

for a class of symbols H given by (8). The same approach can be used to construct Chernoff approximations for Schrödinger groups $(e^{-it\widehat{H}})_{t\in\mathbb{R}}$ describing quantum evolution of systems obtained by a quantization of classical systems with Hamilton functions H. Namely, it holds under certain conditions

$$e^{-it\widehat{H}} = \lim_{n\to\infty} \left[\widehat{e^{-itH/n}}\right]^n. \tag{32}$$

On a heuristic level, such approximations have been considered already in works [7, 8]. A rigorous mathematical treatment and some conditions, when (32) holds, can be found in [70]. Note that right hand sides of both (31) and (32) can be interpreted as phase space Feynman path integrals [7, 8, 12, 25, 70].

Case 2: "rotation". Another approach to construct Chernoff approximations for Schrödinger groups $(e^{itL})_{t\in\mathbb{R}}$ is based on a kind of "rotation" of families $(F(t))_{t\geq 0}$ which are Chernoff equivalent to semigroups $(e^{tL})_{t\geq 0}$ (see [62]). Namely, let $(L, \text{Dom}(L))$ be a self-adjoint operator in a Hilbert space X which generates a C_0–semigroup $(e^{tL})_{t\geq 0}$ on X. Let a family $(F(t))_{t\geq 0}$ be Chernoff equivalent[13] to $(e^{tL})_{t\geq 0}$. Let the operators $F(t)$ be self-adjoint for all $t \geq 0$. Then the family $(F^*(t))_{t\geq 0}$,

$$F^*(t) := e^{i(F(t)-\text{Id})},$$

is Chernoff equivalent to the Schrödinger (semi)group $(e^{itL})_{t\geq 0}$. Indeed, $F^*(0) = \text{Id}$, $\|F^*(t)\| \leq 1$ since all $F^*(t)$ are unitary operators, and $(F^*)'(0) = iF'(0)$. Hence the following Chernoff approximation holds

$$e^{itL}\varphi = \lim_{n\to\infty} e^{in(F(t/n)-\text{Id})}\varphi, \quad \forall\, \varphi \in X. \tag{33}$$

Since all $F(t)$ are bounded operators, one can calculate $e^{in(F(t/n)-\text{Id})}$ via Taylor expansion or via formula (4). Let us illustrate this approach with the following example.

Example 2.5 Consider the function H given by (8), Assume that H does not depend on x, i.e. $H = H(p)$, and H is real-valued (hence $B \equiv 0$ and $N(dy)$ is symmetric). Such symbols H correspond to symmetric Lévy processes. It is well-known[14] that the closure $(L, \text{Dom}(L))$ of $(-\widehat{H}, C_c^\infty(\mathbb{R}^d))$ generates a C_0– semigroup $(T_t)_{t\geq 0}$ on $L^2(\mathbb{R}^d)$; operators T_t are self-adjoint and coincide with operators $F(t)$ given in (9), i.e. $T_t = \widehat{e^{-tH}}$ on $C_c^\infty(\mathbb{R}^d)$. Therefore, the Chernoff approximation (33) holds for the Schrödinger (semi)group $(e^{itL})_{t\geq 0}$ resolving the Cauchy problem

$$-i\frac{\partial f}{\partial t}(t,x) = Lf(t,x), \qquad f(0,x) = f_0(x)$$

---

[13] Actually, $(F(t))_{t\geq 0}$ does not need to fulfill the condition (ii) of the Chernoff Theorem 1.1 in this construction.

[14] See, e.g., Example 4.7.28 in [40].

in $L^2(\mathbb{R}^d)$ with L being the generator of a symmetric Lévy process. Note that this class of generators contains symmetric differential operators with constant coefficients (with $H(p) = C + iB \cdot p + p \cdot Ap$), fractional Laplacians (with $H(p) := |p|^\alpha$, $\alpha \in (0, 2)$) and relativistic Hamiltonians (with $H(p) := \sqrt[\alpha]{|p|^\alpha + m}$, $m > 0$, $\alpha \in (0, 2]$). Assume additionally that $H \in C^\infty(\mathbb{R}^d)$. Then $F(t) : S(\mathbb{R}^d) \to S(\mathbb{R}^d)$ and it holds on $S(\mathbb{R}^d)$ (with $\mathcal{F}$ and $\mathcal{F}^{-1}$ being Fourier and inverse Fourier transforms respectively):

$$F(t) = \mathcal{F}^{-1} \circ e^{-tH} \circ \mathcal{F}; \qquad in(F(t/n) - \mathrm{Id}) = \mathcal{F}^{-1} \circ \left(in\left(e^{-tH/n} - 1\right)\right) \circ \mathcal{F};$$

$$[F^*(t/n)]^n = e^{in(F(t/n) - \mathrm{Id})} = \sum_{k=0}^\infty \frac{1}{k!} \left(\mathcal{F}^{-1} \circ \left(in\left(e^{-tH/n} - 1\right)\right) \circ \mathcal{F}\right)^k$$

$$= \mathcal{F}^{-1} \circ \left[\sum_{k=0}^\infty \frac{1}{k!} \left(in\left(e^{-tH/n} - 1\right)^k\right)\right] \circ \mathcal{F} = \mathcal{F}^{-1} \circ \exp\left\{in\left(e^{-tH/n} - 1\right)\right\} \circ \mathcal{F}.$$

Therefore, $[F^*(t/n)]^n$ is a PDO with symbol $\exp\left\{in\left(e^{-tH/n} - 1\right)\right\}$ on $S(\mathbb{R}^d)$. Hence we have obtained the following representation for the Schrödinger (semi)group $(e^{itL})_{t \geq 0}$:

$$e^{itL}\varphi(x) = \lim_{n \to \infty} (2\pi)^{-d} \int_{\mathbb{R}^d} \int_{\mathbb{R}^d} e^{ip \cdot (x-q)} \exp\left\{in\left(e^{-tH(p)/n} - 1\right)\right\} \varphi(q) \, dq dp, \quad (34)$$

for all $\varphi \in S(\mathbb{R}^d)$ and all $x \in \mathbb{R}^d$. The convergence in (34) is in $L^2(\mathbb{R}^d)$ and is locally uniform with respect to $t \geq 0$.

Case 3: shifts and averaging. One more approach to construct Chernoff approximations for semigroups and Schrödinger groups generated by differential and pseudo-differential operators is based on shift operators (see [63, 64]), averaging (see [10, 57]) and their combination (see [9, 10]). Let us demonstrate this method by means of simplest examples. So, consider $X = C_\infty(\mathbb{R})$ or $X = L^p(\mathbb{R})$, $p \in [1, \infty)$. Consider $(L, \mathrm{Dom}(L))$ in X being the closure of $(\Delta, S(\mathbb{R}))$. Let $(T_t)_{t \geq 0}$ be the corresponding C_0-semigroup on X. Consider the family of shift operators $(S_t)_{t \geq 0}$,

$$S_t \varphi(x) := \frac{1}{2}\left(\varphi(x + \sqrt{t}) + \varphi(x - \sqrt{t})\right), \qquad \forall \varphi \in X, \quad x \in \mathbb{R}. \quad (35)$$

Then all S_t are bounded linear operators on X, $\|S_t\| \leq 1$ and for all $\varphi \in S(\mathbb{R})$ holds (via Taylor expansion):

$$S_t \varphi(x) - \varphi(x) = \frac{1}{2}\left(\varphi(x + \sqrt{t}) - \varphi(x)\right) + \frac{1}{2}\left(\varphi(x - \sqrt{t}) - \varphi(x)\right)$$

$$= \frac{1}{2}\left(\sqrt{t}\varphi'(x) + \frac{1}{2}t\varphi''(x) + o(t)\right) + \frac{1}{2}\left(-\sqrt{t}\varphi'(x) + \frac{1}{2}t\varphi''(x) + o(t)\right)$$

$$= t\varphi''(x) + o(t) = L\varphi(x) + o(t).$$

Moreover, it holds that $\lim_{t \to 0} \|t^{-1}(S_t\varphi - \varphi) - L\varphi\|_X = 0$ for all $\varphi \in S(\mathbb{R}^d)$. Hence the family $(S_t)_{t \geq 0}$ is Chernoff equivalent to the heat semigroup (1) on X. Extending $(S_t)_{t \geq 0}$ to the d-dimensional case and applying the "rotation" techniques in $X = L^2(\mathbb{R}^d)$, one obtains Chernoff approximation for the Schrödinger group $(e^{it\Delta})_{t \geq 0}$ [64]. Further, one can apply the techniques of Sects. 2.1–2.5, to construct Chernoff approximations for Schrödinger groups generated by more complicated differential and pseudo-differential operators.

Let us now combine this techniques with averaging. Averaging is an extension of the classical Daletsky-Lie-Trotter formula (see Sect. 2.1) for the case when the generator $(L, \text{Dom}(L))$ in a Banach space X is not just a finite sum of linear operators L_k, but an integral:

$$L := \int_{\mathcal{E}} L_\varepsilon d\mu(\varepsilon), \tag{36}$$

where $\mathcal{E}$ is a set and μ is a suitable probability measure on (a σ-algebra of subsets of) $\mathcal{E}$, and L_ε are linear operators in X for all $\varepsilon \in \mathcal{E}$. It turns out that (under suitable assumptions, see [10, 57]) the family $(F(t))_{t \geq 0}$,

$$F(t)\varphi := \int_{\mathcal{E}} e^{tL_\varepsilon} \varphi \, d\mu(\varepsilon), \qquad \varphi \in X,$$

is Chernoff equivalent to the semigroup $(e^{tL})_{t \geq 0}$ on X. Moreover, it is easy to see that the following generalization of Theorem 2.1 holds.

Theorem 2.6 *Let $(T_t)_{t \geq 0}$ be a strongly continuous semigroup on a Banach space X with generator $(L, \text{Dom}(L))$. Let D be a core for L. Let μ be a probability measure on a (measurable) space $\mathcal{E}$. Let the representation (36) holds on D with some linear operators $(L_\varepsilon)_{\varepsilon \in \mathcal{E}}$ in X. Let $(F_\varepsilon(t))_{t \geq 0}$, $\varepsilon \in \mathcal{E}$, be families of bounded linear operators on X such that $F_\varepsilon(0) = \text{Id}$ for each $\varepsilon \in \mathcal{E}$; $\|F_\varepsilon(t)\| \leq e^{at}$ for some $a \geq 0$, all $t \geq 0$ and all $\varepsilon \in \mathcal{E}$; and for each $\varphi \in D$ holds*

$$\limsup_{t \to 0 \ \varepsilon \in \mathcal{E}} \left\| \frac{F_\varepsilon(t)\varphi - \varphi}{t} - L_\varepsilon \varphi \right\|_X = 0.$$

Then the family $(F(t))_{t \geq 0}$ of bounded linear operators $F(t)$ on X, with

$$F(t)\varphi := \int_{\mathcal{E}} F_\varepsilon(t)\varphi \, d\mu(\varepsilon), \quad \varphi \in X,$$

is Chernoff equivalent to the semigroup $(T(t))_{t \geq 0}$.

Let us now combine the techniques of shifts and averaging in the following way. Consider $X = C_\infty(\mathbb{R}^d)$ or $X = L^p(\mathbb{R}^d)$, $p \in [1, \infty)$. We generalize the family $(S_t)_{t \geq 0}$ of (35) to the following family $(U_\mu(t))_{t \geq 0}$: consider the family $(S_\varepsilon(t))_{t \geq 0}$, $S_\varepsilon(t)\varphi(x) := \varphi(x + \sqrt{t}\varepsilon)$ for all $\varphi \in X$ and for a fixed $\varepsilon \in \mathbb{R}^d$; define the family $(U_\mu(t))_{t \geq 0}$ by

$$U_\mu(t)\varphi(x) := \int_{\mathbb{R}^d} S_\varepsilon(t)\varphi(x)\, d\mu(\varepsilon) \equiv \int_{\mathbb{R}^d} \varphi(x + \varepsilon\sqrt{t})\, d\mu(\varepsilon).$$

Assume that μ is a symmetric measure with finite (mixed) moments up to the third order and positive second moments $a_j := \int_{\mathbb{R}^d} \varepsilon_j^2 \mu(d\varepsilon) > 0$, $j = 1, \ldots, d$. Then one can show that the family $(U_\mu(t))_{t \geq 0}$ is Chernoff equivalent to the heat semigroup $(e^{t \Delta_A})_{t \geq 0}$, where $\Delta_A := \frac{1}{2} \sum_{j=1}^d a_j \frac{\partial^2}{\partial x_j^2}$. Substituting $(S_\varepsilon(t))_{t \geq 0}$ by the family $(S_\varepsilon^\sigma(t))_{t \geq 0}$, $S_\varepsilon^\sigma(t)\varphi(x) := \varphi(x + \varepsilon t^\sigma)$, for some suitable $\sigma > 0$, and choosing proper measures μ, one can construct analogous Chernoff approximations for semigroups generated by fractional Laplacians and relativistic Hamiltonians. This approach can be further generalized by considering pseudomeasures μ, what leads to Chernoff approximations for Schrödinger groups.

Acknowledgements I would like to thank Christian Bender whose kind support made this work possible. I would like to thank Markus Kunze for useful discussions and communication of references [52, 53].

References

1. Accardi, L., Smolyanov, O.G.: Feynman formulas for evolution equations with the Lévy Laplacian on infinite-dimensional manifolds. Dokl. Akad. Nauk **407**(5), 583–588 (2006)
2. Accardi, L., Smolyanov, O.G.: Feynman formulas for evolution equations with Levy Laplacians on manifolds. In: Quantum Probability and Infinite Dimensional Analysis, QP–PQ: Quantum Probability White Noise Analysis, vol. 20, pp. 13–25. World Scientific Publishing, Hackensack, NJ (2007)
3. Albanese, A.A., Mangino, E.: Trotter-Kato theorems for bi-continuous semigroups and applications to Feller semigroups. J. Math. Anal. Appl. **289**(2), 477–492 (2004)
4. Baeumer, B., Meerschaert, M.M., Nane, E.: Brownian subordinators and fractional Cauchy problems. Trans. Am. Math. Soc. **361**(7), 3915–3930 (2009)
5. Barbu, V.: Nonlinear semigroups and differential equations in Banach spaces. Editura Academiei Republicii Socialiste România, Bucharest; Noordhoff International Publishing, Leiden (1976). Translated from the Romanian
6. Baur, B., Conrad, F., Grothaus, M.: Smooth contractive embeddings and application to Feynman formula for parabolic equations on smooth bounded domains. Comm. Statist. Theory Methods **40**(19–20), 3452–3464 (2011)
7. Berezin, F.A.: Non-Wiener path integrals. Theoret. Math. Phys. **6**(2), 141–155 (1971)
8. Berezin, F.A.: Feynman path integrals in a phase space. Sov. Phys. Usp. **23**, 763–788 (1980)
9. Borisov, L.A., Orlov, Y.N., Sakbaev, V.J.: Chernoff equivalence for shift operators, generating coherent states in quantum optics. Lobachevskii J. Math. **39**(6), 742–746 (2018)
10. Borisov, L.A., Orlov, Y.N., Sakbaev, V.Z.: Feynman averaging of semigroups generated by Schrödinger operators. Infin. Dimens. Anal. Quantum Probab. Relat. Top. **21**(2), 1850010, 13 (2018)
11. Borodin, A.N., Salminen, P.: Handbook of Brownian Motion–facts and Formulae. Probability and Its Applications, 2nd edn. Birkhäuser Verlag, Basel (2002)
12. Böttcher, B., Butko, Y.A., Schilling, R.L., Smolyanov, O.G.: Feynman formulas and path integrals for some evolution semigroups related to τ-quantization. Russ. J. Math. Phys. **18**(4), 387–399 (2011)

13. Böttcher, B., Schilling, R., Wang, J.: Lévy matters. III, Lecture Notes in Mathematics, vol. 2099. Springer, Cham (2013). Lévy-type processes: construction, approximation and sample path properties. With a short biography of Paul Lévy by Jean Jacod, Lévy Matters
14. Böttcher, B., Schilling, R.L.: Approximation of Feller processes by Markov chains with Lévy increments. Stoch. Dyn. **9**(1), 71–80 (2009)
15. Böttcher, B., Schnurr, A.: The Euler scheme for Feller processes. Stoch. Anal. Appl. **29**(6), 1045–1056 (2011)
16. Brézis, H., Pazy, A.: Semigroups of nonlinear contractions on convex sets. J. Funct. Anal. **6**, 237–281 (1970)
17. Brézis, H., Pazy, A.: Convergence and approximation of semigroups of nonlinear operators in Banach spaces. J. Funct. Anal. **9**, 63–74 (1972)
18. Burridge, J., Kuznetsov, A., Kwaśnicki, M., Kyprianou, A.E.: New families of subordinators with explicit transition probability semigroup. Stoch. Process. Appl. **124**(10), 3480–3495 (2014)
19. Butko, Y.A.: Feynman formulas and functional integrals for diffusion with drift in a domain on a manifold. Mat. Zametki **83**(3), 333–349 (2008)
20. Butko, Y.A.: Feynman formulae for evolution semigroups. Scientific Periodical Bauman MSTU "Science and Education" **3**, 95–132 (2014)
21. Butko, Y.A.: Chernoff approximation of evolution semigroups generated by Markov processes. Feynman formulae and path integrals. Habilitationsschrift. Fakultaet fuer Mathematik und Informatik, Universitaet des Saarlandes (2017). https://www.math.uni-sb.de/ag/fuchs/Menupkte/Arbeitsgruppe/yana.html
22. Butko, Y.A.: Chernoff approximation for semigroups generated by killed Feller processes and Feynman formulae for time-fractional Fokker-Planck-Kolmogorov equations. Fract. Calc. Appl. Anal. **21**(5), 1203–1237 (2018)
23. Butko, Y.A.: Chernoff approximation of subordinate semigroups. Stoch. Dyn. **18**(3), 1850021, 19 (2018)
24. Butko, Y.A., Grothaus, M., Smolyanov, O.G.: Lagrangian Feynman formulas for second-order parabolic equations in bounded and unbounded domains. Infin. Dimens. Anal. Quantum Probab. Relat. Top. **13**(3), 377–392 (2010)
25. Butko, Y.A., Grothaus, M., Smolyanov, O.G.: Feynman formulae and phase space Feynman path integrals for tau-quantization of some Lévy-Khintchine type Hamilton functions. J. Math. Phys. **57**(2), 023508, 22 (2016)
26. Butko, Y.A., Schilling, R.L., Smolyanov, O.G.: Feynman formulas for Feller semigroups. Dokl. Akad. Nauk **434**(1), 7–11 (2010)
27. Butko, Y.A., Schilling, R.L., Smolyanov, O.G.: Lagrangian and Hamiltonian Feynman formulae for some Feller semigroups and their perturbations. Infin. Dimens. Anal. Quantum Probab. Relat. Top. **15**(3), 26 (2012)
28. Chernoff, P.R.: Note on product formulas for operator semigroups. J. Funct. Anal. **2**, 238–242 (1968)
29. Chernoff, P.R.: Product formulas, nonlinear semigroups, and addition of unbounded operators. American Mathematical Society, Providence, R.I. (1974). Memoirs of the American Mathematical Society, No. 140
30. Cont, R., Tankov, P.: Financial Modelling with Jump Processes. Chapman & Hall/CRC Financial Mathematics Series. Chapman & Hall/CRC, Boca Raton, FL
31. Dorroh, J.R.: Contraction semi-groups in a function space. Pac. J. Math. **19**, 35–38 (1966)
32. Ethier, S.N., Kurtz, T.G.: Markov processes. Wiley Series in Probability and Mathematical Statistics: Probability and Mathematical Statistics. Wiley, New York (1986). Characterization and Convergence
33. Feynman, R.P.: Space-time approach to non-relativistic quantum mechanics. Rev. Mod. Phys. **20**, 367–387 (1948)
34. Feynman, R.P.: An operator calculus having applications in quantum electrodynamics. Phys. Rev. **2**(84), 108–128 (1951)

35. Gillis, J.E., Weiss, G.H.: Expected number of distinct sites visited by a random walk with an infinite variance. J. Math. Phys. **11**, 1307–1312 (1970)
36. Gomilko, A., Kosowicz, S., Tomilov, Y.: A general approach to approximation theory of operator semigroups. Journal de Mathematiques Pures et Appliquees
37. Gough, J., Obrezkov, O.O., Smolyanov, O.G.: Randomized Hamiltonian Feynman integrals and stochastic Schrödinger-Itô equations. Izv. Ross. Akad. Nauk Ser. Mat. **69**(6), 3–20 (2005)
38. Hahn, M., Kobayashi, K., Umarov, S.: SDEs driven by a time-changed Lévy process and their associated time-fractional order pseudo-differential equations. J. Theor. Probab. **25**(1), 262–279 (2012)
39. Hahn, M., Umarov, S.: Fractional Fokker-Planck-Kolmogorov type equations and their associated stochastic differential equations. Fract. Calc. Appl. Anal. **14**(1), 56–79 (2011)
40. Jacob, N.: Pseudo Differential Operators and Markov Processes, vol. I. Imperial College Press, London (2001). Fourier Analysis and Semigroups
41. Kloeden, P.E., Platen, E.: Numerical Solution of Stochastic Differential Equations. Applications of Mathematics, vol. 23 (New York). Springer, Berlin (1992)
42. Kostrykin, V., Potthoff, J., Schrader, R.: Construction of the paths of Brownian motions on star graphs II. Commun. Stoch. Anal. **6**(2), 247–261 (2012)
43. Kúhnemund, F.: Bicontinuous semigroups on spaces with two topologies: theory and applications (2001). Dissertation der Mathematischen Fakultát der Eberhard Karls Universität Túbingen zur Erlangung des Grades eines Doktors der Naturwissenschaften
44. Lejay, A.: A probabilistic representation of the solution of some quasi-linear PDE with a divergence form operator. Application to existence of weak solutions of FBSDE. Stoch. Process. Appl. **110**(1), 145–176 (2004)
45. Lumer, G.: Perturbation de générateurs infinitésimaux, du type "changement de temps". Ann. Inst. Fourier (Grenoble) **23**(4), 271–279 (1973)
46. Lunardi, A.: Analytic Semigroups and Optimal Regularity in Parabolic Problems. Modern Birkhäuser Classics. Birkhäuser/Springer Basel AG, Basel (1995)
47. Lunt, J., Lyons, T.J., Zhang, T.S.: Integrability of functionals of Dirichlet processes, probabilistic representations of semigroups, and estimates of heat kernels. J. Funct. Anal. **153**(2), 320–342 (1998)
48. MacNamara, S., Strang, G.: Operator splitting. In: Splitting Methods in Communication, Imaging, Science, and Engineering, Scientific Computation, pp. 95–114. Springer, Cham (2016)
49. Metzler, R., Klafter, J.: The random walk's guide to anomalous diffusion: a fractional dynamics approach. Phys. Rep. **339**(1), 77 (2000)
50. Mijena, J.B., Nane, E.: Strong analytic solutions of fractional Cauchy problems. Proc. Am. Math. Soc. **142**(5), 1717–1731 (2014)
51. Montroll, E.W., Shlesinger, M.F.: On the wonderful world of random walks. In: Nonequilibrium Phenomena, II, Stud. Statist. Mech., XI, pp. 1–121. North-Holland, Amsterdam (1984)
52. Nittka, R.: Approximation of the semigroup generated by the Robin Laplacian in terms of the Gaussian semigroup. J. Funct. Anal. **257**(5), 1429–1444 (2009)
53. Nittka, R.: Regularity of solutions of linear second order elliptic and parabolic boundary value problems on Lipschitz domains. J. Differ. Equ. **251**(4–5), 860–880 (2011)
54. Obrezkov, O.O.: Representation of a solution of a stochastic Schrödinger equation in the form of a Feynman integral. Fundam. Prikl. Mat. **12**(5), 135–152 (2006)
55. Obrezkov, O.O., Smolyanov, O.G.: Representations of the solutions of Lindblad equations with the help of randomized Feynman formulas. Dokl. Akad. Nauk **466**(5), 518–521 (2016)
56. Obrezkov, O.O., Smolyanov, O.G., Trumen, A.: A generalized Chernoff theorem and a randomized Feynman formula. Dokl. Akad. Nauk **400**(5), 596–601 (2005)
57. Orlov, Y.N., Sakbaev, V.Z., Smolyanov, O.G., Feynman formulas as a method of averaging random Hamiltonians. Proc. Steklov Inst. Math. **285**(1), 222–232: Translation of Tr. Mat. Inst. Steklova **285**(2014), 232–243 (2014)
58. Orlov, Y.N., Sakbaev, V.Z., Smolyanov, O.G.: Feynman formulas for nonlinear evolution equations. Dokl. Akad. Nauk **477**(3), 271–275 (2017)

59. Orsingher, E., D'Ovidio, M.: Probabilistic representation of fundamental solutions to $\frac{\partial u}{\partial t} = \kappa_m \frac{\partial^m u}{\partial x^m}$. Electron. Commun. Probab. **17**(1885), 12 (2012)
60. Pazy, A.: Semigroups of linear operators and applications to partial differential equations. Applied Mathematical Sciences, vol. 44. Springer, New York (1983)
61. Plyashechnik, A.S.: Feynman formula for Schrödinger-type equations with time- and space-dependent coefficients. Russ. J. Math. Phys. **19**(3), 340–359 (2012)
62. Remizov, I.D.: Quasi-Feynman formulas—a method of obtaining the evolution operator for the Schrödinger equation. J. Funct. Anal. **270**(12), 4540–4557 (2016)
63. Remizov, I.D.: Solution of the Schrödinger equation by means of the translation operator. Mat. Zametki **100**(3), 477–480 (2016)
64. Remizov, I.D., Starodubtseva, M.F.: Quasi-Feynman formulas providing solutions of the multidimensional Schrödinger equation with unbounded potential. Mat. Zametki **104**(5), 790–795 (2018)
65. Smolyanov, O.G., Shamarov, N.N.: Feynman and Feynman-Kac formulas for evolution equations with the Vladimirov operator. Dokl. Akad. Nauk **420**(1), 27–32 (2008)
66. Smolyanov, O.G., Shamarov, N.N.: Feynman formulas and path integrals for evolution equations with the Vladimirov operator. Tr. Mat. Inst. Steklova, 265(Izbrannye Voprosy Matematicheskoy Fiziki i p-adicheskogo Analiza), 229–240 (2009)
67. Smolyanov, O.G., Shamarov, N.N.: Hamiltonian Feynman integrals for equations with the Vladimirov operator. Dokl. Akad. Nauk **431**(2), 170–174 (2010)
68. Smolyanov, O.G., Shamarov, N.N.: Hamiltonian Feynman formulas for equations containing the Vladimirov operator with variable coefficients. Dokl. Akad. Nauk **440**(5), 597–602 (2011)
69. Smolyanov, O.G., Shamarov, N.N., Kpekpassi, M.: Feynman-Kac and Feynman formulas for infinite-dimensional equations with the Vladimirov operator. Dokl. Akad. Nauk **438**(5), 609–614 (2011)
70. Smolyanov, O.G., Tokarev, A.G., Truman, A.: Hamiltonian Feynman path integrals via the Chernoff formula. J. Math. Phys. **43**(10), 5161–5171 (2002)
71. Smolyanov, O.G., Weizsäcker, H.v., Wittich, O.: Chernoff's theorem and the construction of semigroups. In: Evolution Equations: Applications to Physics, Industry, Life Sciences and Economics (Levico Terme, 2000), Progress in Nonlinear Differential Equations Applications, vol. 55, pp. 349–358. Birkhäuser, Basel (2003)
72. Smolyanov, O.G., Weizsäcker, H.v., Wittich, O.: Chernoff's theorem and discrete time approximations of Brownian motion on manifolds. Potential Anal. **26**(1), 1–29 (2007)
73. Umarov, S.: Continuous time random walk models associated with distributed order diffusion equations. Fract. Calc. Appl. Anal. **18**(3), 821–837 (2015)
74. Umarov, S.: Fractional Fokker-Planck-Kolmogorov equations associated with SDEs on a bounded domain. Fract. Calc. Appl. Anal. **20**(5), 1281–1304 (2017)
75. Umarov, S., Hahn, M., Kobayashi, K.: Beyond the Triangle: Brownian Motion, Ito Calculus, and Fokker-Planck Equation—Fractional Generalizations. World Scientific Publishing Co. Pte. Ltd., Hackensack, NJ (2018)
76. Volkonskiĭ, V.A.: Random substitution of time in strong Markov processes. Teor. Veroyatnost. i Primenen **3**, 332–350 (1958)
77. Volkonskiĭ, V.A.: Additive functionals of Markov processes. Trudy Moskov. Mat. Obšč. **9**, 143–189 (1960)
78. Zagrebnov, V.A.: Note on the Chernoff product formula. Preprint, 20 pp. (2019). https://hal.archives-ouvertes.fr/hal-02373231
79. Zaslavsky, G.M.: Chaos, fractional kinetics, and anomalous transport. Phys. Rep. **371**(6), 461–580 (2002)

Laplacians with Point Interactions—Expected and Unexpected Spectral Properties

Amru Hussein and Delio Mugnolo

Abstract We study the one-dimensional Laplace operator with point interactions on the real line identified with two copies of the half-line $[0, \infty)$. All possible boundary conditions that define generators of C_0-semigroups on $L^2([0, \infty)) \oplus L^2([0, \infty))$ are characterized. Here, the Cayley transform of the matrices that describe the boundary conditions plays an important role and using an explicit representation of the Green's functions, it allows us to study invariance properties of semigroups.

Keywords Operator semigroups · Point interactions · Spectra of non-selfadjoint operators · Asymptotically positive semigroups

1 Introduction

Here, point interactions for the Laplacian on the real line are considered. The real line is realized here as two half-lines $[0, \infty) \sqcup [0, \infty)$ coupled at the two boundary points. More concretely, we consider realizations $-\Delta(A, B)$ of

$$-\frac{d^2}{dx^2} \quad \text{in} \quad L^2([0, \infty); \mathbb{C}) \oplus L^2([0, \infty); \mathbb{C})$$

A. Hussein
Department of Mathematics, TU Kaiserslautern, Paul-Ehrlich-Straße, 67663 Kaiserslautern, Germany
e-mail: hussein@mathematik.uni-kl.de

D. Mugnolo (✉)
Chair of Analysis, Faculty of Mathematics and Computer Science, FernUniversität Hagen, Hagen, Germany
e-mail: delio.mugnolo@fernuni-hagen.de

with boundary conditions of the form

$$A\begin{pmatrix}\psi_1(0)\\ \psi_2(0)\end{pmatrix} + B\begin{pmatrix}\psi_1'(0)\\ \psi_2'(0)\end{pmatrix} = 0 \quad \text{for} \quad A, B \in \mathbb{C}^{2\times 2}, \tag{1}$$

where one has $\psi = (\psi_1, \psi_2)^T \in L^2([0, \infty); \mathbb{C}) \oplus L^2([0, \infty); \mathbb{C})$. Like in [19, Sect. 4], we regard this setting as a toy model of more complicated quantum graphs.

There are many studies on self-adjoint boundary conditions, cf. [4] and references therein, boundary conditions leading to so-called spectral operators, cf. [7], or boundary conditions related to quadratic forms, cf. [20]. However, a study of all possible boundary conditions of this form seems to be lacking so far. In this note, we turn to classical Hille–Yosida theory and address the issue of semigroup generation by realizations of the Laplacian with point interactions of the above type. It turns out that resolvent estimates for $\Delta(A, B)$ are closely related to the behavior of the Cayley transform.

One could naively expect that imposing two linearly independent boundary conditions is both necessary and sufficient to induce a realization that generates a semigroup, because there are two boundary points and this leads to the rank condition

$$\text{Rank}(A\ B) = 2;$$

and in fact if $\text{Rank}(A\ B) \neq 2$, then $\sigma(-\Delta(A, B)) = \mathbb{C}$, see [12, Prop. 4.2]. However, this rank condition is not yet sufficient to establish basic spectral properties and it turns out that the question of determining when A, B induce a semigroup generator is not trivial. In a previous work Krejčiřík, Siegl and the first author, see [12], pointed out the importance of the Cayley transform

$$\mathfrak{S}(k; A, B) := -(A + ikB)^{-1}(A - ikB), \quad k \in \mathbb{C}, \tag{2}$$

for basic spectral properties. The condition that $A + ikB$ is invertible for some $k \in \mathbb{C}$ has been used in [12] as definition for the notion of *regular boundary conditions*: on general metric graphs irregular boundary conditions can produce very wild spectral features, ranging from empty spectrum—as in the situation considered here—to empty resolvent set. For the case of one boundary point this cannot occur: the easiest non-trivial case features two boundary points and will be investigated in detail in the following.

In the present setting we find out that realizations $\Delta(A, B)$ with irregular boundary conditions have empty resolvent set and thus fail to be generators of C_0-semigroups; surprisingly, it turns out that there are even some regular boundary conditions that do not define generators of C_0-semigroups. We will see that not only the mere existence of the Cayley transform is relevant, but also its asymptotic behavior. The crucial point is that the Cayley transform $\mathfrak{S}(k; A, B)$ appears in a natural way in an explicit formula for the resolvent of $\Delta(A, B)$, which in turn easily allows us to check the conditions of the Hille–Yosida Theorem in its version for analytic semigroups.

Once generation is assessed, we turn to the issue of qualitative properties of the semigroup generated by $\Delta(A, B)$, again in dependence of A, B. It is well-known that relevant features of a semigroup—in particular, whether it is positive and/or L^∞-contractive—is tightly related to analogous invariance properties of its generator's resolvent. Using again our machinery, we are then able to formulate sufficient conditions for invariance in terms of properties of $\mathfrak{S}(k; A, B)$. In the context of general metric graphs, positivity and Markovian features of semigroups in dependence of the boundary conditions have been studied already in [16, Sect. 5]—however only for self-adjoint boundary conditions (1) and giving only sufficient conditions—and in [5, Sects. 5–6] for the case of only m-sectorial boundary conditions for which a complete characterization is obtained, see also [14, 18, 19] for related results. The notion of m-sectorial boundary conditions is explained in Sect. 4 below: roughly speaking, these are boundary conditions that induce realizations of $\Delta(A, B)$ associated with sesquilinear forms. One step beyond the hitherto discussed invariance properties, we are finally also able to characterize asymptotic positivity of the semigroup – a rather weak property recently introduced in [6].

Our note is organized as follows: In Sect. 2 we are going to present our setting, including relevant function spaces and the parametrization of our boundary conditions. Section 3 contains our main result, Theorem 1 as well as a few examples that show its applicability. The proof of Theorem 1 is given in Sect. 6, it is based on a number of technical lemmata, which will be proved in Sects. 4 and 5. Finally, we are going to discuss positivity, asymptotic positivity, and further invariance issues in Sect. 7.

2 Function Spaces, Operators and Boundary Conditions

Whenever $I \subset \mathbb{R}$ is an interval, denote by $L^2(I)$ the usual space of complex-valued square integrable function with scalar product $\langle \cdot, \cdot \rangle_{L^2}$. Moreover, let $H^1(I)$ and $H^2(I)$ be the Sobolev spaces of order one and two, and set $H_0^2(I) := \{\psi \in H^2(I) : \psi, \psi'|_{\partial I} = 0\}$. Then one defines minimal and maximal operators in

$$L^2([0, \infty)) \oplus L^2([0, \infty))$$

by

$$\Delta_{\max}\psi = \psi'', \quad D(\Delta_{\max}) = H^2([0, \infty)) \oplus H^2([0, \infty)),$$
$$\Delta_{\min}\psi = \psi'', \quad D(\Delta_{\min}) = H_0^2([0, \infty)) \oplus H_0^2([0, \infty)).$$

Since $D(\Delta_{\max})/D(\Delta_{\min}) \cong \mathbb{C}^4$, any realization

$$\Delta_{\min} \subset \Delta \subset \Delta_{\max}$$

is determined by a subspace $\mathcal{M} \subset \mathbb{C}^4$ and $\Delta = \Delta_{\mathcal{M}}$ with

$$D(\Delta_{\mathcal{M}}) = \left\{ \psi \in D(\Delta_{\max}) : (\underline{\psi}, \underline{\psi}')^T \in \mathcal{M} \right\},$$

where

$$\underline{\psi} := \begin{pmatrix} \psi_1(0) \\ \psi_2(0) \end{pmatrix}, \quad \underline{\psi}' := \begin{pmatrix} \psi_1'(0) \\ \psi_2'(0) \end{pmatrix},$$

and one sets

$$[\underline{\psi}] := \begin{pmatrix} \underline{\psi} \\ \underline{\psi}' \end{pmatrix}.$$

For $\dim \mathcal{M} = 2$, $\mathcal{M}$ can be represented as kernel of a surjective linear map from $\mathbb{C}^4 \to \mathbb{C}^2$, and hence the condition $\dim \mathcal{M} = 2$ is equivalent to existence of matrices $A, B \in \mathbb{C}^{2\times 2}$ with $\mathcal{M} = \mathcal{M}(A, B) = \operatorname{Ker}(A\ B)$ and $\operatorname{Rank}(A\ B) = 2$. With respect to our goal of studying the generator property of different realizations of Laplacians on $L^2([0, \infty)) \oplus L^2([0, \infty))$, this is the only case which provides enough boundary conditions, and we will restrict to it throughout this note. For simplicity, we refer to boundary conditions defined by $[\underline{\psi}] \in \mathcal{M}(A, B) = \operatorname{Ker}(A\ B)$ in short as *boundary conditions A, B*.

Boundary conditions A, B and A', B' are equivalent if $\mathcal{M}(A, B) = \mathcal{M}(A', B')$, and one sets

$$\Delta(A, B) := \Delta_{\mathcal{M}(A,B)}.$$

Note that $A' = CA$ and $B' = CB$ define equivalent boundary conditions whenever $C \in \mathbb{C}^{2\times 2}$ is invertible, since $\operatorname{Ker}(A'\ B') = \operatorname{Ker}(A\ B)$.

The following notion of regularity of boundary conditions has been introduced in [12, Sect. 3.2]. Note that there are also further notions of regularity, in particular the one introduced by Birkhoff, cf. [2, 3], see also [7]. This regularity assumption does not agree with the one used here, see [12, Sect. 3.3].

Definition 1 (*Regular and irregular boundary conditions*) Let A, B be boundary conditions with $\operatorname{Rank}(A\ B) = 2$. These are called *regular* if $A + ikB$ is invertible for some $k \in \mathbb{C}$, and *irregular* otherwise.

Remark 1 It can be shown that A, B are irregular if and only if $\operatorname{Rank}(A\ B) = 2$ and $\operatorname{Ker} A \cap \operatorname{Ker} B \neq \{0\}$, cf. [12, Prop. 3.3].

3 Generator Properties and Examples

The following is the main result of our paper. Here σ_r and σ_p denote as usual the residual, and point spectrum, respectively. Note that for non-self-adjoint operators there are various notions of the essential spectrum. Five types, defined in terms of Fredholm properties and denoted by σ_{e_j} for $j = 1, \ldots, 5$, are discussed in detail in [8, Chap. IX]. All these essential spectra coincide for self-adjoint T, but for closed non-self-adjoint T in general one has only the inclusions $\sigma_{e_j}(T) \subset \sigma_{e_i}(T)$ for $j < i$. However, in the case of our operators $\Delta(A, B)$ these five concepts coincide and we thus set for the essential spectrum $\sigma_{ess}(-\Delta(A, B)) := \sigma_{e_j}(-\Delta(A, B))$ for $i = 1, \ldots, 5$, see the proof of [12, Prop. 4.11].

Theorem 1 *Let the boundary conditions A, B be regular. Then the following assertions hold.*

(a) $\sigma_{ess}(-\Delta(A, B)) = [0, \infty)$ *and* $\sigma_r(-\Delta(A, B)) = \emptyset$. *Furthermore,* $\lambda = k^2 \in \sigma_p(-\Delta(A, B))$ *if and only if k with $\operatorname{Im} k > 0$ solves $\det(A + ikB) = 0$, and in this case the geometric multiplicity of λ is given by $\dim \operatorname{Ker}(A + ikB)$.*

(b) $\Delta(A, B)$ *is not the infinitesimal generator of a C_0-semigroup on $L^2\big([0, \infty)\big) \oplus L^2\big([0, \infty)\big)$ if and only if $\dim \operatorname{Ker} A = 0$, $\dim \operatorname{Ker} B = 1$, and $P^\perp A^{-1} B P^\perp = 0$. (Here we denote by P the orthogonal projection onto $\operatorname{Ker} B$ and $P^\perp := \mathbb{1} - P$, where $\mathbb{1} = \operatorname{diag}(1, 1)$).*

(c) *If $\Delta(A, B)$ is a generator, then the C_0-semigroup extends to an analytic semigroup.*

(d) *If $\Delta(A, B)$ is a generator and furthermore if any pole s of $k \mapsto \mathfrak{S}(k; A, B)$ or $k \mapsto \mathfrak{S}(-\bar{k}; A, B)^*$ with $\operatorname{Im} s > 0$ satisfies $\operatorname{Re} s > 0$, and $s = 0$ is not a pole of any of these functions, then the semigroup generated by $\Delta(A, B)$ is uniformly bounded.*

(e) *If $A = L + P$ and $B = P^\perp$ for an orthogonal projection P in $\mathbb{C}^2$, $P^\perp = \mathbb{1} - P$, and $L \in \mathbb{C}^{2 \times 2}$ with $P^\perp L P^\perp = L$, then $\Delta(A, B)$ is even the generator of a cosine operator function and hence of an analytic semigroup of angle $\frac{\pi}{2}$ on $L^2\big([0, \infty)\big) \oplus L^2\big([0, \infty)\big)$. This semigroup is always quasi-contractive, and in fact contractive if the numerical range of L is contained in $\{z : \operatorname{Re} z \leq 0\}$.*

If $\Delta(A, B)$ is a generator, then the semigroup consists of operators that are bounded on $L^2\big([0, \infty)\big) \oplus L^2\big([0, \infty)\big)$ and map $L^2\big([0, \infty)\big) \oplus L^2\big([0, \infty)\big)$ into $H^2\big([0, \infty)\big) \oplus H^2\big([0, \infty)\big) \hookrightarrow L^\infty\big([0, \infty)\big) \oplus L^\infty\big([0, \infty)\big)$, hence these are integral operators by the Kantorovich–Vulikh Theorem.

Remark 2 (*Irregular boundary conditions do not define generators*) If A, B are irregular, $\Delta(A, B)$ cannot be a generator of a C_0-semigroup since $\sigma(\Delta(A, B)) = \mathbb{C}$. For the case of general finite metric graphs, determining spectra and resolvent estimates for irregular boundary conditions is more involved.

Remark 3 (*Multiplicity of eigenvalues*) For regular boundary conditions A, B, the geometric multiplicity of an eigenvalue $-k^2$ of $\Delta(A, B)$ is at most two, and equal to two if and only if $A + ikB = 0$. This implies that $\operatorname{Ker} B = \operatorname{Ker} A = \{0\}$, and that there are equivalent boundary conditions $A' = l \cdot \mathbb{1}$ and $B' = \mathbb{1}$ with $\operatorname{Re} l = \operatorname{Im} k > 0$.

Unfortunately, we are not able to determine the semigroup's analyticity angle in the general case; as a matter of fact, we cannot exclude that $\Delta(A, B)$ is always the generator of a cosine operator function. Indeed, the proof of (c) shows that the spectrum of $\Delta(A, B)$ is always contained in a parabola centered around the real axis; this is a necessary condition for generation of a cosine operator function, cf. [1, Thm. 3.14.18]. Generation of a cosine operator function under the assumptions of Theorem 1(e) is known at least since [19], see also [9, Sect. 3.6]; a possibly more general sufficient condition for generation of a cosine operator function (also in $L^p([0, \infty)) \oplus L^p([0, \infty))$, for $p \neq 2$) has been proposed in [9, Thm. 2.3], but it looks rather difficult to check in practice.

The proof of Theorem 1 relies upon certain fundamental asymptotic properties of the Cayley transform $\mathfrak{S}(\cdot, A, B)$ depending on the different kinds of boundary conditions. Here, we will distinguish three different cases which are discussed in detail in Sects. 4 and 5 below. We defer the proof of Theorem 1 to Sect. 6. The generator property is traced back to the uniform boundedness of the Cayley transform $k \mapsto \mathfrak{S}(k; A, B)$ outside a compact set containing its poles, where for irregular boundary conditions one might set $\mathfrak{S}(k; A, B) = \infty$. Some cases for the possible behavior of the Cayley transform are illustrated in the following examples.

Example 1 (*Boundary conditions defining operators associated with sectorial forms*) Let

$$A = \begin{pmatrix} A_{11} & 0 \\ A_{21} & 0 \end{pmatrix} \quad \text{and} \quad B = \begin{pmatrix} 1 & 0 \\ 0 & 1 \end{pmatrix}.$$

for any $A_{11}, A_{22} \in \mathbb{C}$: the boundary conditions A, B correspond to $A\underline{\psi} + B\underline{\psi}' = 0$, i.e.,

$$A_{11}\psi_1(0) + \psi_1'(0) = 0, \quad A_{21}\psi_1(0) + \psi_2'(0) = 0.$$

Then A, B are regular, since $\det(A + ikB) = ik(A_{11} + ik) \neq 0$ for $k \notin \{0, iA_{11}\}$. The Cayley transform

$$\mathfrak{S}(k; A, B) = \begin{pmatrix} -(A_{11} + ik)^{-1}(A_{11} - ik) & 0 \\ (ik)^{-1} A_{12}[(A_{11} + ik)^{-1}(A_{11} - ik) - 1] & 1 \end{pmatrix} \quad (3)$$

is uniformly bounded away from its singularities $\{0, iA_{11}\}$, where 0 is in fact a removable singularity. Since $\dim \operatorname{Ker} B = 0$, by Theorem 1 $\Delta(A, B)$ generates an analytic semigroup; if $\operatorname{Im} i A_{11} > 0$, then $\lambda = A_{11}^2$ is a (simple, by Remark 3) eigenvalue of $\Delta(A, B)$, and $\sigma_{ess}(-\Delta(A, B)) = [0, \infty)$. Note that $-\Delta(A, B)$ is associated with the sesquilinear form defined by

$$\delta_{A,B}[\psi] = \|\psi'\|_{L^2}^2 - \langle A\underline{\psi}, \underline{\psi} \rangle_{\mathbb{C}^2}, \quad \psi \in H^1([0, \infty)) \oplus H^1([0, \infty)) \quad (4)$$

and hence it is sectorial. In particular, the semigroup generated by $\Delta(A, B)$ is contractive if the numerical range of A is contained in the left halfplane: this is the case if and only if $A_{21} = 0$. We will refer to boundary conditions of this type as *m-sectorial*, see Sect. 4.1 below.

In a more general setting the question if $-\Delta(A, B)$ is associated with a form of the type given in (4) is discussed in [13]. The following is a prominent example from the theory of $\mathcal{PT}$-symmetric operators, and it is discussed for instance in [12, Example 3.5] and also in the references given there.

Example 2 (*Boundary conditions defining operators not associated with sectorial forms*) Consider

$$A_\tau = \begin{pmatrix} 1 & -e^{i\tau} \\ 0 & 0 \end{pmatrix} \quad \text{and} \quad B_\tau = \begin{pmatrix} 0 & 0 \\ 1 & e^{-i\tau} \end{pmatrix}, \quad \tau \in [0, \pi/2),$$

leading to the boundary conditions

$$\psi_1(0) = e^{i\tau}\psi_2(0), \quad \psi_1'(0) = -e^{-i\tau}\psi_2'(0).$$

Here, $\det(A_\tau + ikB_\tau) = 2ki\cos\tau \neq 0$ and hence by Theorem 1 the operator $\Delta(A_\tau, B_\tau)$ has no eigenvalues. Integration by parts gives

$$\langle -\Delta(A_\tau, B_\tau)\psi, \psi \rangle_{L^2} = \|\psi'\|_{L^2}^2 + (1 - e^{2i\tau})\psi_2(0)\overline{\psi_2'(0)}, \quad \psi \in D(\Delta(A_\tau, B_\tau)).$$

The trace of the derivative cannot be balanced by $\|\psi'\|_{L^2}^2$, hence in particular $\psi \mapsto \langle -\Delta(A_\tau, B_\tau)\psi, \psi \rangle_{L^2}$ does not define a closed sesquilinear form: indeed, the numerical range of this form is the entire complex plane. Nevertheless, $\Delta(A_\tau, B_\tau)$ does generate an analytic semigroup, as in fact $\Delta(A_\tau, B_\tau)$ is similar to the one-dimensional Laplacian on $\mathbb{R}$. Observe that because $\Delta(A_\tau, B_\tau)$ is not dissipative, the semigroup it generates cannot be contractive; it is bounded, though, due to its similarity with the Gaussian semigroup on $\mathbb{R}$. Observe that A_τ, B_τ define irregular boundary conditions for $\tau = \frac{\pi}{2}$.

The following two examples are slight modifications of cases discussed in [7, Sect. XIX.6, p. 2373], where these boundary conditions are considered for the interval $[0, 1]$.

Example 3 (*Intermediate boundary conditions*) Consider

$$A = \begin{pmatrix} 1 & 0 \\ 0 & 1 \end{pmatrix} \quad \text{and} \quad B = \begin{pmatrix} 0 & 0 \\ -1 & 0 \end{pmatrix},$$

and the boundary conditions $A\underline{\psi} + B\underline{\psi'} = 0$, i.e.,

$$\psi_1(0) = 0, \quad \psi_1'(0) = \psi_2(0).$$

Then $\det(A + ikB) = 1$ for all $k \in \mathbb{C}$, i.e., A, B are regular. Furthermore, dim Ker $A = 0$ and dim Ker $B = 1$, but $P^\perp B P^\perp = 0$ and $PBP^\perp = -1 \neq 0$ and

$$\mathfrak{S}(k; A, B) = -\begin{pmatrix} 1 & 0 \\ 2ik & 1 \end{pmatrix}, \quad k \in \mathbb{C}.$$

We conclude from Theorem 1 that $\Delta(A, B)$ does not generate a C_0-semigroup on $L^2([0, \infty)) \oplus L^2([0, \infty))$, although its (purely essential) spectrum is $[0, \infty)$. This example highlights that $\mathfrak{S}(\cdot; A, B)$ is not uniformly bounded in general.

Example 4 (*Totally degenerate boundary conditions*) Consider

$$A = \begin{pmatrix} 1 & 0 \\ 0 & 0 \end{pmatrix} \quad \text{and} \quad B = \begin{pmatrix} 0 & 0 \\ 1 & 0 \end{pmatrix}.$$

Then Rank$(A\ B) = 2$, but $\det(A + ikB) = 0$ for any $k \in \mathbb{C}$, and hence A, B are irregular.

4 Cayley Transforms

In this section we are going to derive properties of the Cayley transform that are essential in the proof of Theorem 1.

4.1 M-Sectorial Boundary Conditions

For regular boundary conditions the Cayley transform (2) is well-defined except for at most two singularities $k \in \mathbb{C}$. One important class of boundary conditions are related to quadratic forms.

Definition 2 Boundary conditions A, B are said to be *m-sectorial* if there exist $L, P \in \mathbb{C}^{2\times 2}$ such that P is an orthogonal projection, $P^\perp = \mathbb{1} - P$, and $L = P^\perp L P^\perp$, and such that $A = L + P$ and $B = P^\perp$.

The reason for this name is that whenever A, B are m-sectorial boundary conditions, $-\Delta(A, B)$ is associated with the sectorial sesquilinear form defined by

$$\delta_{P,L}[\psi] = \|\psi'\|_{L^2}^2 - \langle LP^\perp \underline{\psi}, P^\perp \underline{\psi}\rangle_{\mathbb{C}^2}, \ \psi \in \{\psi \in H^1([0,\infty)) \oplus H^1([0,\infty)): P\underline{\psi} = 0\}, \tag{5}$$

cf. e.g. [22, Def. 1.7] for this notion. Hence, $-\Delta(A, B)$ is an m-sectorial operator and $\Delta(A, B)$ generates an analytic semigroup. M-sectorial boundary conditions are in particular regular since

$$A + ikB = \begin{pmatrix} L + ikP^\perp & 0 \\ 0 & P \end{pmatrix}$$

is invertible for $k > \|L\|$. The Cayley transform can be estimated as follows.

We first consider the case dim Ker $B = 1$: then dim Ran $L \leq 1$, and with respect to Ran P and Ran $P^\perp$ one obtains a block decomposition

$$A \pm ikB = \begin{pmatrix} L \pm ik & 0 \\ 0 & \pm ik \end{pmatrix} \quad \text{and} \quad \mathfrak{S}(k; A, B) = \begin{pmatrix} -(L+ik)^{-1}(L-ik) & 0 \\ 0 & 1 \end{pmatrix}.$$

For A invertible and $B = \mathbb{1} = \operatorname{diag}(1, 1)$ one has

$$\mathfrak{S}(k; A, B) = -(A + ik\mathbb{1})^{-1}(A - ik\mathbb{1}) = -(A/ik + \mathbb{1})^{-1}(A/ik - \mathbb{1}),$$

and hence

$$\|\mathfrak{S}(k; A, B)\| \leq \frac{2}{1 - \frac{\|A\|}{|k|}} \quad \text{for } |k| > \|A\|.$$

Therefore, $\mathfrak{S}(k; A, B)$ is uniformly bounded away from its poles, i.e., outside a compact set.

If $B = \mathbb{1}$ and dim Ran $L = 1$, one obtains a canonical block decomposition with respect to Ker L and $(\operatorname{Ker} L)^\perp$: we denote by P_L the orthogonal projection onto $(\operatorname{Ker} L)^\perp$. With respect to it we denote $L_{11} := P_L L P_L$ and likewise $L_{12} := P_L^\perp L P_L$ and we find

$$A \pm ikB = \begin{pmatrix} L_{11} \pm ik & 0 \\ L_{12} & \pm ik \end{pmatrix},$$

and hence

$$\begin{aligned}
\mathfrak{S}(k; A, B) &= -\begin{pmatrix} (L_{11} + ik)^{-1} & 0 \\ -(ik)^{-1}L_{12}(L_{11}+ik)^{-1} & +(ik)^{-1} \end{pmatrix} \begin{pmatrix} L_{11} - ik & 0 \\ L_{12} & -ik \end{pmatrix} \\
&= \begin{pmatrix} -(L_{11}+ik)^{-1}(L_{11}-ik) & 0 \\ (ik)^{-1}L_{12}[(L_{11}+ik)^{-1}(L_{11}-ik) - 1] & 1 \end{pmatrix}.
\end{aligned} \tag{6}$$

Similarly to the case of A invertible and $B = \mathbb{1}$, using (6) one can show that $\mathfrak{S}(k; A, B)$ is uniformly bounded away from its poles for general m-sectorial boundary conditions. This is summarized in the following.

Lemma 1 *Let A, B define m-sectorial boundary conditions. Then $\mathfrak{S}(k; A, B)$ is uniformly bounded outside a compact set.*

Depending now on the dimension of Ker A and Ker B one can distinguish the following cases listed in Table 1, where the cases dim Ker $A = 1$, dim Ker $B = 2$, and dim Ker $A = 2$, dim Ker $B = 1$ collide with the rank condition, and hence are

Table 1 Different cases of boundary conditions

dim Ker A	dim Ker B	equiv. b.c.	$-\Delta(A, B)$
0	0	$A' = B^{-1}A$ and $B' = \mathbb{1}$	m-sectorial
1	0	$A' = B^{-1}A$ and $B' = \mathbb{1}$	m-sectorial
2	0	$A' = B^{-1}A$ and $B' = \mathbb{1}$	m-sectorial
0	2	$A = \mathbb{1}$ and $B = 0$	m-sectorial
0	1	Some block representation	Regular
1	1	Some block representation	Regular or irregular

excluded. We have already remarked that for $\text{Rank}(A\ B) \neq 2$ one has $\sigma(\Delta(A, B)) = \mathbb{C}$.

4.2 The Case dim Ker $a = 0$ and dim Ker $B = 1$

Lemma 2 *Let* $\dim \text{Ker } A = 0$ *and* $\dim \text{Ker } B = 1$. *Then* $\text{Rank}(A\ B) = 2$, *equivalent boundary conditions are given by*

$$A' = \mathbb{1} \quad \text{and} \quad B' = A^{-1}B,$$

and the boundary conditions A, B *are regular.*

Proof First, since A is invertible, its columns are linearly independent and therefore $\text{Rank}(A\ B) \geq 2$, and A', B' define equivalent boundary conditions. Furthermore, since $\text{Ker } A \cap \text{Ker } B = \{0\}$ these boundary conditions are regular. □

Let P be the orthogonal projection onto Ker B and $P^\perp = \mathbb{1} - P$, then without loss of generality, consider

$$A = \mathbb{1} \quad \text{and} \quad B = \begin{pmatrix} P^\perp B P^\perp & 0 \\ P B P^\perp & 0 \end{pmatrix}.$$

Lemma 3 *Let* $\dim \text{Ker } A = 0$ *and* $\dim \text{Ker } B = 1$. *Then the Cayley transform* $\mathfrak{S}(\cdot; A, B)$ *is uniformly bounded outside a compact set containing its only possible pole if and only if* $P^\perp B P^\perp \neq 0$. *If* $P^\perp B P^\perp = 0$, *and hence* $PBP^\perp \neq 0$, *then* $\|\mathfrak{S}(k; A, B)\| = \mathcal{O}(|k|)$ *for* $|k| \to \infty$

Proof Here, since dim Ker $B = 1$, Ran $P^\perp = \text{span}\{p_1\}$ and Ran $P = \text{span}\{p_2\}$, where $\{p_1, p_2\}$ is an orthonormal basis of $\mathbb{C}^2$. In this basis

$$(A \pm ikB) = \begin{pmatrix} 1 \pm ikB_{11} & 0 \\ \pm ikB_{21} & 1 \end{pmatrix}, \quad (A \pm ikB)^{-1} = \frac{1}{1 \pm ikB_{11}} \begin{pmatrix} 1 & 0 \\ \mp ikB_{21} & 1 \pm ik \end{pmatrix}.$$

The Cayley transform is then

$$\mathfrak{S}(k; A, B) = -(A + ikB)^{-1}(A - ikB) = -\frac{1}{1 + ikB_{11}} \begin{pmatrix} 1 - ikB_{11} & 0 \\ -2ikB_{21} & 1 \end{pmatrix}.$$

For $B_{11} \neq 0$ this is uniformly bounded away from the pole $k = i/B_{11}$. For $B_{11} = 0$ there are no poles, and dim Ker $B = 1$ implies that $B_{21} \neq 0$. In this case $\|\mathfrak{S}(k; A, B)\| = \mathcal{O}(|k|)$ for $|k| \to \infty$. □

4.3 The Case dim Ker $a = 1$ and dim Ker $B = 1$

In this subsection we focus on the case of dim Ker $A = $ dim Ker $B = 1$. Denote by $Q^\perp$ the orthogonal projection on Ran A, $Q = \mathbb{1} - Q^\perp$, and as before P the orthogonal projection on Ker B, where each Q, $Q^\perp$ and P, $P^\perp$ has one-dimensional range. Then

$$A = \begin{pmatrix} Q^\perp A P^\perp & Q^\perp A P \\ 0 & 0 \end{pmatrix} \quad \text{and} \quad B = \begin{pmatrix} Q^\perp B P^\perp & 0 \\ Q B P^\perp & 0 \end{pmatrix}. \tag{7}$$

Lemma 4 *Let* dim Ker $A = $ dim Ker $B = 1$. *Then* Rank$(A\ B) = 2$ *if and only if* $QBP^\perp \neq 0$. *The boundary conditions* A, B *are irregular if and only if* Ker $A = $ Ker B, *i.e. if* $Q^\perp AP = 0$.

Proof From (7) one deduces $(A\ B)$ is surjective if and only if $QBP^\perp \neq 0$. Recall that A, B are irregular if and only if Ker $A = $ Ker B, see [12, Prop. 3.3], and here (7) implies that Ker $A = $ Ker B if and only if $Q^\perp AP = 0$. □

Lemma 5 *Let* A, B *define regular boundary conditions with* dim Ker $A = $ dim Ker $B = 1$. *Then the Cayley transform* $\mathfrak{S}(\cdot; A, B)$ *has one possible pole, and away from this* $\mathfrak{S}(\cdot; A, B)$ *is uniformly bounded.*

Proof Note that Ran $P^\perp = \text{span}\{p_1\}$ and Ran $P = \text{span}\{p_2\}$, where $\{p_1, p_2\}$ is an orthonormal basis of $\mathbb{C}^2$. For A, B, regular, in this basis, equivalent boundary conditions are

$$A = \begin{pmatrix} A_{11} & 1 \\ 0 & 0 \end{pmatrix} \quad \text{and} \quad B = \begin{pmatrix} B_{11} & 0 \\ 1 & 0 \end{pmatrix}.$$

Let Ran $P = \mathrm{span}\{p_1\}$, Ran $P^\perp = \mathrm{span}\{p_2\}$, and Ran $Q = \mathrm{span}\{q_1\}$, Ran $Q^\perp = \mathrm{span}\{q_2\}$ where $\{p_1, p_2\}$ and $\{q_1, q_2\}$ are orthonormal basis of $\mathbb{C}^2$. Now, a coordinate change from $\{q_1, q_2\}$ to $\{p_1, p_2\}$ is given by a unitary U, and hence equivalent boundary conditions UA and UB can be written in the basis $\{p_1, p_2\}$ as

$$A = \begin{pmatrix} A_{11} & A_{12} \\ 0 & 0 \end{pmatrix} \quad \text{and} \quad B = \begin{pmatrix} B_{11} & 0 \\ B_{21} & 0 \end{pmatrix}.$$

By Lemma 4, one has $QBP^\perp \neq 0$ and $Q^\perp AP \neq 0$ and hence $B_{21} \neq 0$ and $A_{12} \neq 0$. Therefore equivalent boundary conditions are

$$A = \begin{pmatrix} A_{11} & 1 \\ 0 & 0 \end{pmatrix} \quad \text{and} \quad B = \begin{pmatrix} B_{11} & 0 \\ 1 & 0 \end{pmatrix}.$$

Hence

$$(A \pm ikB) = \begin{pmatrix} A \pm ikB_{11} & 1 \\ 0 & \pm ik \end{pmatrix}, \quad (A \pm ikB)^{-1} = \frac{1}{\pm ik(A_{11} \pm ikB_{11})} \begin{pmatrix} \pm ik & -1 \\ 0 & A_{11} \pm ikB_{11} \end{pmatrix},$$

and

$$\mathfrak{S}(k; A, B) = \begin{pmatrix} -\frac{A_{11} - ikB_{11}}{A_{11} + ikB_{11}} & \frac{-2}{A_{11} + ikB_{11}} \\ 0 & -1 \end{pmatrix}.$$

This is uniformly bounded away from the only possible pole at $k = ik/B_{11}$. $\square$

Our findings are summarized in Table 2, where as before P is the orthogonal projection onto Ker B and $P^\perp = \mathbb{1} - P$, and uniformly bounded refers to the Cayley transform away from its poles.

5 Resolvent Estimates

The keystone of our analysis is that for regular boundary conditions the resolvent $(-\Delta(A, B) - k^2)^{-1}$ is an integral operator, i.e.,

$$(-\Delta(A, B) - k^2)^{-1} f(x) = \int_{[0,\infty) \sqcup [0,\infty)} r_{A,B}(x, y; k) f(y) \, dy,$$

where

$$f = \begin{pmatrix} f_1 \\ f_2 \end{pmatrix} \in L^2([0, \infty)) \oplus L^2([0, \infty)), \quad x = \begin{pmatrix} x_1 \\ x_2 \end{pmatrix}, \quad y = \begin{pmatrix} y_1 \\ y_2 \end{pmatrix} \in [0, \infty) \sqcup [0, \infty),$$

with kernel

Table 2 Cayley transforms

dim Ker A	dim Ker B	Condition	Cayley transform	Ref.
0	0	None	Uniformly bounded	Lemma 1
1	0	None	Uniformly bounded	Lemma 1
2	0	None	Uniformly bounded	Lemma 1
0	2	None	Uniformly bounded	Lemma 1
0	1	$P^\perp B P^\perp \neq 0$	Uniformly bounded	Lemma 3
0	1	$P^\perp B P^\perp = 0$	$\\|\mathfrak{S}(k; A, B)\\| = \mathcal{O}(\|k\|)$	Lemma 3
1	1	Ker $A \neq$ Ker B	Uniformly bounded	Lemma 5
1	1	Ker $A =$ Ker B	$\mathfrak{S}(k; A, B) = \infty$ A, B irregular	Lemma 5

$$r_{A,B}(x, y; k) = \frac{i}{2k}\left\{\begin{pmatrix} e^{ik|x_1-y_1|} & 0 \\ 0 & e^{ik|x_2-y_2|} \end{pmatrix} + \begin{pmatrix} e^{ikx_1} & 0 \\ 0 & e^{ikx_2} \end{pmatrix} \mathfrak{S}(k; A, B) \begin{pmatrix} e^{iky_1} & 0 \\ 0 & e^{iky_2} \end{pmatrix}\right\} \quad (8)$$

whenever $k \in \mathbb{C}$ such that Im $k > 0$ and $A + ikB$ is invertible, cf. [12, Prop. 4.7]. We stress that the first addend on the right hand side corresponds to the kernel of the Laplacian on $\mathbb{R}$ without any point interactions; the second addend can be thus interpreted as a correcting term that mirrors the influence of the point interactions. It is also remarkable that the kernel is bounded and jointly uniformly continuous on $\mathbb{R} \times \mathbb{R}$, regardless of A, B; in particular, it extends to a bounded linear operator from L^1 to L^∞.

Lemma 6 (Resolvent estimate in case of uniformly bounded Cayley transform) *Let the boundary conditions A, B be regular. If there is a neighborhood U of the set of poles of $k \mapsto \mathfrak{S}(k; A, B)$ such that $\mathfrak{S}(\cdot; A, B)$ is uniformly bounded on $\mathbb{C} \setminus U$, then there exists $C > 0$ such that*

$$\\|(-\Delta(A, B) - k^2)^{-1}\\| \leq \frac{1}{\mathrm{dist}(k^2, [0, \infty))} + \frac{C}{|k|\,|\mathrm{Im}\,k|\,\mathrm{dist}(S, k)}, \quad (9)$$

where Im $k > 0$ *with* $\det(A + ikB) \neq 0$ *and*

$S = \{s \in \mathbb{C}\colon \mathrm{Im}\, s > 0 \text{ or } s \in [0, \infty),\ \text{and } s \text{ non-removable singularity of } \mathfrak{S}(\cdot; A, B)\}$.

Proof By using (8) we obtain in $\mathbb{C} \setminus U$ the estimate

$$\|(-\Delta(A, B) - k^2)^{-1} f\| \leq \frac{1}{|k|^2} \|f\| + \frac{1}{|k|} \|\mathfrak{S}(k; A, B)\| \cdot \|f\| \cdot \|e^{ik \cdot}\|^2.$$

The first term follows from the standard resolvent estimate for the Laplacian on $\mathbb{R}$ with no point interactions, while for the second one we have used the product form of the kernel, and moreover $\|e^{ik \cdot}\|^2 = 1/(2\text{Im } k)$. Note that non-removable singularities of $\mathfrak{S}(k; A, B)$ are poles of order one and hence, $\|\mathfrak{S}(k; A, B)\| \leq C \text{dist}(S, k)$. □

In the case of m-sectorial boundary conditions, a stronger estimate holds.

Lemma 7 (Resolvent estimate for m-sectorial boundary conditions) *Let the boundary conditions A, B be m-sectorial. Then there exist $C > 0$ and $\omega \geq 0$ such that*

$$\|\lambda(\Delta(A, B) - \lambda^2)^{-1}\| \leq \frac{C}{\text{Re } \lambda - \omega} \tag{10}$$

all $\lambda \in \mathbb{C}$ with $\text{Re } \lambda > \omega$; in particular, the spectrum of $\Delta(A, B)$ is contained in a parabola centered on the real axis and contained in a left half-plane.

Proof The proof is based on a result due to Lions: If a bounded, H-elliptic sesquilinear form a with form domain V satisfies the additional condition

$$|\text{Im } a(u, u)| \leq M \|u\|_V \|u\|_H \quad \text{for all } u \in V, \tag{11}$$

then the associated operator A generates a cosine operator function on H with associated Kisyński space V and its resolvent satisfies an estimate corresponding to (10), see e.g. [1, Sect. 3.14] and [20, Sect. 6.2]. Hence, it suffices to observe that (11) is satisfied by the form $\delta_{A,B}$ defined in (5): the proof of this fact is analogous to that of [20, Lemma 6.63]. □

The only cases where the Cayley transform is not uniformly bounded have been discussed in Lemma 3.

Lemma 8 (Resolvent estimate for the other cases) *Let $\dim \text{Ker } A = 0$ and $\dim \text{Ker } B = 1$ and let $P^\perp B P^\perp = 0$. Then $\sigma_p(-\Delta(A, B)) = \emptyset$ and for some $c > 0$*

$$\|(-\Delta(A, B) + \kappa^2)^{-1}\| \geq \frac{c}{\kappa^{3/2}} \quad \text{as } \kappa \to \infty.$$

In particular $\Delta(A, B)$ is not a generator of a C_0-semigroup.

Proof As in [12, Sect. 6] one can show that $\Delta(A, B)$ is unitarily equivalent to $\Delta(A', B')$ with

$$\mathfrak{S}(k; A', B') = \begin{pmatrix} -1 & 0 \\ 2ikB_{21} & 1 \end{pmatrix}, \quad B_{21} \neq 0,$$

and there are no eigenvalues nor poles of $\mathfrak{S}(k; A', B')$. Consider the function $u = (u_1, u_2)^T = (\chi_{[0,1]}, 0)^T$, where $\chi_{[0,1]}$ denotes the characteristic function of the unit interval. Then

$$(-\Delta(A, B) - k^2)^{-1} u = \frac{i}{2k} \left\{ \left(\int_0^1 e^{ik|x_1 - y_1|} dy_1 \right) + \left(\frac{e^{ikx_1}}{2ikB_{12}e^{ikx_2}} \right) \int_0^1 e^{iky_1} dy_1 \right\}, \quad (12)$$

and estimating the second component only

$$\|(-\Delta(A, B) - k^2)^{-1} u\| \geq |B_{12}| \cdot \left| \int_0^1 e^{iky_1} dy_1 \right| \cdot \|e^{ikx_2}\| = |B_{12}| \frac{|e^{ik} - 1|}{|k|} \frac{1}{(2\mathrm{Im}\, k)^{1/2}}.$$

In particular for $k = i\kappa$, $\kappa > 0$,

$$\frac{|e^{-\kappa} - 1|}{\sqrt{2}|\kappa|^{3/2}} \to \frac{1}{\sqrt{2}|\kappa|^{3/2}} \quad \text{as } \kappa \to \infty,$$

and therefore

$$\|(\Delta(A, B) - \kappa^2)^{-1}\| \geq \mathcal{O}(|\kappa|^{-3/2}).$$

In particular, assume that $\Delta(A, B)$ is the generator of a C_0-semigroup, then

$$\mathcal{O}(|\kappa|^{-3/2}) \leq \|\Delta(A, B) - \kappa^2)^{-1}\| \leq M/(\kappa^2 - \omega), \quad \text{for } \omega > 0 \text{ and } M > 0,$$

which multiplying by κ^2 and passing to the limit $\kappa \to \infty$ leads to a contradiction. Recall that $\Delta(A, B)$ is closed and densely defined. □

6 Proof of Theorem 1

We are finally in the position to prove our main result.

Proof The proof of (a) can be deduced from [12, Sect. 4]: The statement on the residual spectrum follows from [12, Prop. 4.6] for the case of only external edges. Essential spectra are discussed in [12, Prop. 4.11]. The statement on the eigenvalues follows from the *Ansatz* for the eigenfunctions

$$\psi(x; k) = \begin{pmatrix} \alpha_1(k) e^{ikx_1} \\ \alpha_2(k) e^{ikx_2} \end{pmatrix},$$

which is square integrable only if $\mathrm{Im}\, k > 0$, and there are non-trivial $\alpha_j(k)$, $j = 1, 2$, such that $\psi(\cdot; k) \in D(\Delta(A, B))$ if and only if $\det(A + ikB) = 0$, and the geometric multiplicity is given by $\dim \mathrm{Ker}(A + ikB)$. For part (b), uniform boundedness of the Cayley transform is characterized in Lemmas 1, 3 and 5. The corresponding resolvent estimates are given in Lemmas 6 and 8, where Lemma 8 discusses the case of non-generators. Lemma 6 implies that for $\omega > |S|$, where S is a set of singularities of $\mathfrak{S}(k; A, B)$ defined there, $\Delta(A, B)$ is a closed densely defined operator with $[\omega^2, \infty) \in \rho(\Delta(A, B))$. For any sector

$$\Sigma_\theta := \{k \in \mathbb{C} \colon \operatorname{Im} k > 0, |\operatorname{Re} k| \leq \tan(\theta) \operatorname{Im} k\}, \quad \theta \in (0, \pi/2), \tag{13}$$

one has $|k| \cong |\operatorname{Re} k| + \operatorname{Im} k \leq (1 + \tan(\theta)) \operatorname{Im} k$, and therefore

$$\|(-\Delta(A, B) - k^2)^{-1}\| \leq \frac{1}{|k|^2} + \frac{C_\omega}{|k|^2} \leq \frac{1 + C_\omega}{|k|^2}, \quad k \in \Sigma_c, \ |k| > \omega.$$

In particular, $\Delta(A, B)$ is sectorial on the sector $\Sigma_{2\theta} - \omega^2 \sin(\pi - 2\theta)$. Shifting the sector allows avoiding the two poles of $\mathfrak{S}(k; A, B)$: this finishes the proof of (b) and (c), whereas (e) is proved in Lemma 7.

A necessary condition for boundedness of a semigroup is that the spectrum of its generator $\mathcal{A}$ is contained in $\{z \in \mathbb{C} : \operatorname{Re} z \leq 0\}$. To prove (d), recall that by a celebrated result due to Gomilko [11] (see also [23]), for semigroups acting on a Hilbert space H boundedness is equivalent to said spectral inclusion *and* the additional condition

$$\sup_{\delta > 0} \delta \int_{\delta - i\infty}^{\delta + i\infty} \left(\|(\mathcal{A} - \lambda)^{-1} f\|^2 + \|(\mathcal{A}^* - \lambda)^{-1} f\|^2 \right) |d\lambda| < \infty \quad \text{for all } f \in H.$$

Here, the kernel of $(-\Delta(A, B) - k^2)^{-1}$ is given by (8) and the kernel of $(-\Delta(A, B)^* - \overline{k^2})^{-1}$ is given by the adjoint kernel $r_{A,B}(y, x; -\overline{k})^*$. Analogously to Lemma 6 one can estimate the resolvent norm away from the singularities of $\mathfrak{S}(k; A, B)$ and $\mathfrak{S}(-\overline{k}; A, B)^*$. There are finitely many such singularities: by assumption, they have a finite, strictly positive distance from the imaginary axis. In particular, the estimate in Lemma 6 implies sectoriality in sectors with vertex zero

$$\|(\Delta(A, B) - \lambda)^{-1}\| \leq \frac{C}{|\lambda|} \quad \text{for } \lambda \in \Sigma_\theta := \{z \in \mathbb{C} \setminus \{0\} \colon |\arg z| > \theta\}, \ \theta \in (0, \pi/2),$$

and an analogous estimate holds for $\|(\Delta(A, B)^* - \lambda)^{-1}\|$. Therefore,

$$\sup_{\delta > 0} \delta \int_{\delta - i\infty}^{\delta + i\infty} \left(\|(\Delta(A, B) - \lambda)^{-1} f\|^2 + \|(\Delta(A, B)^* - \lambda)^{-1} f\|^2 \right) |d\lambda|$$

$$\leq C \sup_{\delta > 0} \delta \int_{\delta - i\infty}^{\delta + i\infty} \frac{1}{|\delta + i\lambda|^2} |d\lambda| \leq C \sup_{\delta > 0} \delta \int_{|\lambda| > \delta} \frac{1}{|\lambda|^2} d|\lambda| = 2C < \infty.$$

This completes the proof. $\square$

7 Invariance Properties

Several issues in the qualitative analysis of semigroups associated with sesquilinear forms are made particularly easy by variational methods. In particular, the classical Beurling–Deny criteria have been generalized in [21]; based upon this general criterion, invariance properties for heat equations on metric graphs have been obtained in [5]. We can paraphrase [5, Prop. 4.3] (see also [20, Thms. 6.71 and 6.72]) and obtain the following: given a closed convex subset C of $\mathbb{C}^2$, we denote by $\mathcal{C}$ the induced closed convex subset of $L^2([0, \infty)) \oplus L^2([0, \infty))$ defined by

$$\mathcal{C} := \{ f \in L^2([0, \infty)) \oplus L^2([0, \infty)) : f(x) \in C \text{ for a.e.} x \in [0, \infty) \sqcup [0, \infty) \}$$

Proposition 1 *Let the boundary conditions A, B be m-sectorial. Let C be a closed convex subset of $\mathbb{C}^2$ with $(0, 0) \in C$.*

Then the semigroup generated by $\Delta(A, B)$ leaves $\mathcal{C}$ invariant if and only if both the projection onto $\mathrm{Ran}\, B$ and the semigroup generated by $A + B - \mathbb{1}$ leave C invariant.

The power of our approach lies in the possibility of the explicit representation (8) of the resolvent kernel. From this, some semigroup properties can be derived even for boundary conditions that are not m-sectorial, when Ouhabaz' variational methods are not available.

Lemma 9 *Let the boundary conditions be regular and $\Delta(A, B)$ generate a contractive C_0-semigroup. Let C be a closed convex subset of $\mathbb{C}^2$. Then the semigroup generated by $\Delta(A, B)$ leaves $\mathcal{C}$ invariant provided*

$$\frac{\kappa}{2}\begin{pmatrix} e^{-\kappa|x_1-y_1|} + e^{-\kappa(x_1+y_1)}\sigma_{11}(i\kappa) & e^{-\kappa(x_1+y_2)}\sigma_{12}(i\kappa) \\ e^{-\kappa(x_2+y_1)}\sigma_{21}(i\kappa) & e^{-\kappa|x_2-y_2|} + e^{-\kappa(x_1+y_2)}\sigma_{22}(i\kappa) \end{pmatrix}$$

leaves C invariant for all $\kappa > 0$ and all $x_i, y_j \in [0, \infty)$, where $\sigma_{ij}(i\kappa) = (\mathfrak{S}(i\kappa; A, B))_{ij}$, $1 \leq i, j \leq 2$. In particular, the semigroup generated by $\Delta(A, B)$ is L^∞-contractive provided $\mathbb{1} + \mathfrak{S}(i\kappa; A, B)$ leaves $\{\xi \in \mathbb{C}^2 : |\xi_1| + |\xi_2| \leq 1\}$ invariant for all $\kappa > 0$.

Proof It is well-known that the semigroup leaves $\mathcal{C}$ invariant if and only if so does $\lambda(\lambda - \Delta(A, B))^{-1}$ for all $\lambda > 0$, see [21, Prop. 2.3]. By (8), $\kappa^2 r_{A,B}(x, y; i\kappa)$ is the kernel of $\lambda(\lambda - \Delta(A, B))^{-1}$ for $\lambda = -(i\kappa)^2$; a direct computation shows that for all $\mu \in \mathbb{R}$

$$\mu^2 r_{A,B}(x, y; i\kappa) = \frac{\mu^2}{2\kappa}\begin{pmatrix} e^{-\kappa|x_1-y_1|} + e^{-\kappa(x_1+y_1)}\sigma_{11}(i\kappa) & e^{-\kappa(x_1+y_2)}\sigma_{12}(i\kappa) \\ e^{-\kappa(x_2+y_1)}\sigma_{21}(i\kappa) & e^{-\kappa|x_2-y_2|} + e^{-\kappa(x_1+y_2)}\sigma_{22}(i\kappa) \end{pmatrix} \quad (14)$$

and the main claim now follows from closedness of $\mathcal{C}$, taking $\mu = \kappa$.

Finally, in order to prove the assertion about L^∞-contractivity, it suffices to observe that the matrix in (14) is in absolute value no larger than $\mathbb{1} + \mathfrak{S}(i\kappa; A, B)$. □

If $\mathcal{C}$ is the positive cone, then the assertion can be sharpened as follows.

Corollary 1 *Let the boundary conditions be regular and $\Delta(A, B)$ generate a quasi-contractive C_0-semigroup. Then the semigroup generated by $\Delta(A, B)$ is real if and only if $\mathfrak{S}(i\kappa; A, B)$ has real entries; in this case, the semigroup is additionally positive whenever $\mathbb{1} + \mathfrak{S}(i\kappa; A, B)$ leaves $\{\xi \in \mathbb{R}^2 : \xi_1 \geq 0, \xi_2 \geq 0\}$ invariant for some κ_0 and all $\kappa \geq \kappa_0$.*

Proof Positivity and reality of a semigroup is unaffected by scalar (real) perturbations of its generator. Furthermore, reality (resp., positivity) of a positive operator is equivalent to reality (resp., positivity) of its kernel, cf. [17, Thm. 5.2]. Finally, (14) shows that the entries of the resolvent's kernel at $i\kappa$ are real if and only if so are the entries of $\mathfrak{S}(i\kappa; A, B)$. The proof of the positivity follows essentially the proof of [16, Thm. 4.6] although there only self-adjoint boundary conditions are considered: we only need to observe that if $\mathbb{1} + \mathfrak{S}(i\kappa; A, B)$ has real and positive entries, then for $\mu = \kappa$ the matrix in (14) is entry-wise no smaller than

$$\frac{\kappa}{2} \begin{pmatrix} e^{-\kappa(x_1+y_1)}\left(1 + \sigma_{11}(i\kappa)\right) & e^{-\kappa(x_1+y_2)}\sigma_{12}(i\kappa) \\ e^{-\kappa(x_2+y_1)}\sigma_{21}(i\kappa) & e^{-\kappa(x_2+y_2)}\left(1 + \sigma_{22}(i\kappa)\right) \end{pmatrix}, \tag{15}$$

whence the claim follows. □

Example 5 By Proposition 1, the boundary conditions in Example 1 define a semigroup that leaves invariant $\mathcal{C}$ if and only if the semigroup generated by A leaves C invariant: e.g., the semigroup generated by $\Delta(A, B)$ is positive if and only if A_{11} is real and $A_{21} \geq 0$. That this is a sufficient condition can be deduced from Corollary 1, too.

Furthermore, the semigroup is L^∞-contractive (and in this case automatically L^p-contractive for all $p \in [1, \infty]$) if and only if $\operatorname{Re} A_{11} \leq 0$ and $A_{12} = 0$, cf. [18, Lemma 6.1]. (Observe that the latter condition induces a decoupling of our system, as we are left with two Laplacians on $[0, \infty)$ with Neumann and Robin boundary conditions, respectively.)

Remark 4 One criterion for contractivity of the semigroup is that A, B are m-sectorial with $\operatorname{Re} L \leq 0$, see [15, Thm. 2.4], or equivalently $\operatorname{Re} AB^* \leq 0$, but as we know the case of m-sectorial boundary conditions can be treated more directly by Proposition 1, without invoking Lemma 9. Example 2 shows that non-(quasi-)contractive semigroups can actually arise; we note in passing that in the setting of Example 2

$$\mathfrak{S}(i\kappa; A_\tau, B_\tau) = \frac{1}{\cos \tau} \begin{pmatrix} i \sin \tau & 1 \\ 1 & -i \sin \tau \end{pmatrix}, \quad \kappa > 0, \tag{16}$$

(see [12, Example 3.5]), i.e., $\mathbb{1} + \mathfrak{S}(i\kappa; A, B)$ does not leave either the positive cone of $\mathbb{R}^2$ or the unit ball of $\ell^\infty \times \ell^\infty$ invariant.

An interesting consequence of Theorem 1(a) is that if there is a simple pole k_0 of $k \mapsto (A + ikB)^{-1}$ with $\operatorname{Im} k_0 > |\operatorname{Re} k_0|$, then the peripheral spectrum of $\Delta(A, B)$

is finite and consists of simple poles of the resolvent. This paves the way to study semigroups that are merely *asymptotically* positive; i.e., those semigroups whose orbits starting at positive initial data tend to the lattice's positive cone, see [6, Def. 8.1].

Proposition 2 *Let $\Delta(A, B)$ generate a C_0-semigroup. Assume the zero k of $\{k : \mathrm{Im}\, k > 0\} \ni k \mapsto \det(A + ikB) \in \mathbb{C}$ of larger magnitude lies on $i(0, \infty)$, i.e., $k = i\kappa_0$ for some $\kappa_0 > 0$, and let $A \neq \kappa_0 B$. Consider the following assertions:*

(i) *the semigroup generated by $\Delta(A, B)$ is asymptotically positive;*
(ii) *the spectral projection of $\Delta(A, B)$ associated with κ_0^2 is positive;*
(iii) *the distance from the set $[0, \infty)$ of each entry of $(\kappa - \kappa_0)^2 \mathfrak{S}(i\kappa; A, B)$ tends to 0 as $\kappa \searrow \kappa_0$;*
(iv) $\lim_{\kappa \searrow \kappa_0} \frac{(\kappa - \kappa_0)^2}{\det(A - \kappa B)} = 0.$

Then $(i) \Leftrightarrow (ii) \Leftarrow (iii) \Leftarrow (iv)$.

We refer to [10, Sect. IV.1] for the definition of spectral projections of possibly non-selfadjoint operators. Observe that, e.g., the boundary conditions in Example 2 do not satisfy the assumptions of Proposition 2.

Proof By Theorem 1 and Remark 3, the main assumptions imply that the peripheral spectrum contains precisely one eigenvalue, which is simple: hence the spectral bound is a dominant spectral value. We are thus in the position to apply [6, Thm. 1.2]: in view of (14) we conclude that asymptotic positivity (in the sense of [6, Def. 7.2]) of $\lambda \mapsto (\lambda - \Delta(A, B))^{-1}$ at $\lambda_0 := -(i\kappa_0)^2 = \kappa_0^2$ (which is in turn equivalent to (ii), by [6, Thm. 7.6]) is equivalent to the condition that the distance from $[0, \infty)$ of each entry of

$$\frac{(\kappa - \kappa_0)^2}{2\kappa} \begin{pmatrix} e^{-\kappa|x_1 - y_1|} + e^{-\kappa(x_1 + y_1)}\sigma_{11}(i\kappa) & e^{-\kappa(x_1 + y_2)}\sigma_{12}(i\kappa) \\ e^{-\kappa(x_2 + y_1)}\sigma_{21}(i\kappa) & e^{-\kappa|x_2 - y_2|} + e^{-\kappa(x_1 + y_2)}\sigma_{22}(i\kappa) \end{pmatrix} \quad (17)$$

tends to 0 as $\kappa \searrow \kappa_0$ for each $x_1, x_2, y_1, y_2 \in [0, \infty)$. Now, observe that if K is a cone in a lattice X, $\delta \in (0, 1]$, and $a \in K$, then for any $b \in X$ $\mathrm{dist}(b, K) \geq \mathrm{dist}(a + \delta b, K)$. We conclude that the distance of the matrix in (17) from the positive cone of $\mathbb{C}^2$ is no larger than the distance of the matrix $(\kappa - \kappa_0)^2 \mathfrak{S}(i\kappa; A, B)$ from the same cone: this proves that (i) is implied by (iii). To conclude the proof, it suffices to observe that the poles of $k \mapsto \mathfrak{S}(i\kappa; A, B)$ are the zeros of $k \mapsto \det(A + ikB)$, with equal multiplicity. □

Example 6 Let us come back to the setting in Example 1. We have already seen that if $A_{11} > 0$, then A_{11}^2 is a dominant eigenvalue of $\Delta(A, B)$ and we can hence apply Proposition 2 to (3) and conclude that the semigroup generated by $\Delta(A, B)$ is asymptotically positive for any A_{11}, A_{12}.

Example 7 Consider

$$A = \begin{pmatrix} 0 & -1 \\ -1 & 0 \end{pmatrix} \quad \text{and} \quad B = \begin{pmatrix} 1 & 0 \\ 0 & 1 \end{pmatrix}.$$

Because $\det(A + ikB) = -k^2 - 1$, we obtain that $1 = -(i\kappa_0)^2$ with $\kappa_0 = 1$ is the only eigenvalue of $\Delta(A, B)$; and it is simple, since $A \neq B$. Accordingly, by Proposition 1 the semigroup generated by $\Delta(A, B)$ is not positive (nor it is ℓ^∞-contractive). Let us sharpen this assertion:

$$\frac{(\kappa - \kappa_0)^2}{\det(A - \kappa B)} = \frac{(\kappa - 1)^2}{\kappa^2 - 1} = \frac{\kappa - 1}{\kappa + 1} \xrightarrow{\kappa \searrow 1} 0,$$

and we conclude from Proposition 2 that the semigroup generated by $\Delta(A, B)$ is asymptotically positive.

Acknowledgements The second author would like to thank Jochen Glück (Passau) for helpful discussions.
D.M. was partially supported by the Deutsche Forschungsgemeinschaft (Grant 397230547).

References

1. Arendt, W., Batty, C.J.K., Hieber, M., Neubrander, F.: Vector-Valued Laplace Transforms and Cauchy Problems, Monographs in Mathematics, vol. 96. Birkhäuser, Basel (2001)
2. Birkhoff, G.D.: Boundary value and expansion problems of ordinary linear differential equations. Trans. Am. Math. Soc. **9**(4), 373–395 (1908)
3. Birkhoff, G.D.: On the asymptotic character of the solutions of certain linear differential equations containing a parameter. Trans. Am. Math. Soc. **9**(2), 219–231 (1908)
4. Berkolaiko, G., Kuchment, P.: Introduction to Quantum Graphs. Mathematical Surveys and Monographs, vol. 186. American Mathematical Society, Providence, RI (2013)
5. Cardanobile, S., Mugnolo, D.: Parabolic systems with coupled boundary conditions. J. Differ. Equ. **247**, 1229–1248 (2009)
6. Daners, D., Glück, J., Kennedy, J.B.: Eventually and asymptotically positive semigroups on Banach lattices. J. Differ. Equ. **261**(5), 2607–2649 (2016)
7. Dunford, N., Schwartz, J.T.: Linear Operators. Part III: Spectral Operators. Interscience Publishers [Wiley], New York-London-Sydney (1971)
8. Edmunds, D.E., Evans, W.D.: Spectral Theory and Differential Operators. Oxford Mathematical Monographs. The Clarendon Press, Oxford University Press, New York (1987). Oxford Science Publications
9. Engel, K.-J., Kramar Fijavž, M.: Waves and diffusion on metric graphs with general vertex conditions. Evol. Equ. Control Theory **8**, 633–661 (2019)
10. Engel, K.-J., Nagel, R.: One-Parameter Semigroups for Linear Evolution Equations. Graduate Texts in Mathematics, vol. 194. Springer, New York (2000)
11. Gomilko, A.M.: Conditions on the generator of a uniformly bounded C_0-semigroup. Funct. Anal. Appl. **33**, 294–296 (1999)
12. Hussein, A., Krejčiřík, D., Siegl, P.: Non-self-adjoint graphs. Trans. Am. Math. Soc. **367**(4), 2921–2957 (2015)
13. Hussein, A.: Maximal quasi-accretive Laplacians on finite metric graphs. J. Evol. Equ. **14**(2), 477–497 (2014)
14. Kant, U., Klauß, T., Voigt, J., Weber, M.: Dirichlet forms for singular one-dimensional operators and on graphs. J. Evol. Equ. **9**, 637–659 (2009)
15. Kostrykin, V., Potthoff, J., Schrader, R.: Contraction semigroups on metric graphs. In: Analysis on Graphs and Its Applications, Proceedings of Symposia in Pure Mathematics, vol. 77, pp. 423–458. American Mathematical Society, Providence, RI (2008)

16. Kostrykin, V., Schrader, R.: Laplacians on metric graphs: eigenvalues, resolvents and semigroups. In: Quantum Graphs and Their Applications, Contemporary Mathematics, vol. 415, pp. 201–225. American Mathematical Society, Providence, RI (2006)
17. Mugnolo, D., Nittka, R.: Properties of representations of operators acting between spaces of vector-valued functions. Positivity **15**, 135–154 (2011)
18. Mugnolo, D.: Gaussian estimates for a heat equation on a network. Netw. Het. Media **2**, 55–79 (2007)
19. Mugnolo, D.: Vector-valued heat equations and networks with coupled dynamic boundary conditions. Adv. Diff. Equ. **15**, 1125–1160 (2010)
20. Mugnolo, D.: Semigroup Methods for Evolution Equations on Networks. Understanding Complex Systems. Springer, Cham (2014)
21. Ouhabaz, E.M.: Invariance of closed convex sets and domination criteria for semigroups. Potential Anal. **5**, 611–625 (1996)
22. Ouhabaz, E.M.: Analysis of Heat Equations on Domains. London Mathematical Society Monograph Series, vol. 30. Princeton University Press, Princeton, NJ (2005)
23. Shi, D.H., Feng, D.X.: Characteristic conditions of the generation of C_0 semigroups in a Hilbert space. J. Math. Anal. Appl. **247**, 356–376 (2000)

Remarks on a Characterization of Generators of Bounded C_0-Semigroups

Sylwia Kosowicz

Abstract We provide and discuss an integral resolvent criterion for generation of bounded C_0-semigroups on Banach spaces.

Keywords Generator · C_0-semigroup · Resolvent · UMD-space

2010 Mathematics Subject Classification 47D03 · 47A60

1 Introduction

In [9, 11] integral resolvent conditions for generation of bounded C_0-semigroups on Banach spaces were discussed: it was proved that if A is a densely defined, closed operator on a complex Banach space X such that (a) the resolvent set $\rho(A)$ of A contains the open right half plane $\mathbb{C}_+$, and (b) for all $x \in X$ and $y \in X^*$,

$$\sup_{\sigma>0} \sigma \int_{\sigma-i\infty}^{\sigma+i\infty} |\langle R^2(\lambda, A)x, y\rangle| \, |d\lambda| < \infty, \tag{1}$$

where, as customary, $R(\lambda, A) = (\lambda - A)^{-1}$ is the resolvent of A and $\langle x, y \rangle$ denotes the value of a continuous linear functional $y \in X^*$ on a vector $x \in X$, then A generates a bounded C_0-semigroup on X. Moreover, if X is a Hilbert space, then (1) is a necessary condition for A to be a bounded semigroup generator. In general Banach spaces this is not the case: Gomilko in [11] remarked that the characterization of scalar-type spectral operators given in [15] suggests that (1) may not be satisfied even if A is a bounded operator in a reflexive Banach space, and the semigroup $(e^{tA})_{t\geq 0}$ is bounded. A complete proof of this conjecture was provided in [4]. Related issues

---

S. Kosowicz (✉)
Faculty of Mathematics and Computer Science, Nicolaus Copernicus University, 12/18 Chopin Str., 87-100 Toruń, Poland
e-mail: sylwiak@mat.umk.pl

concerning perturbation theory for generators of not necessarily strongly continuous operator semigroups on Hilbert spaces were discussed in [1, 14].

As it transpires, integral resolvent characterizations of generators of bounded semigroups are very much related to a complex inversion formula for the abstract Laplace transforms, and its version for semigroups. It was proved in [6] (see also [12]) that if X is a UMD space and if A is the generator of a bounded C_0-semigroup $(e^{tA})_{t\geq 0}$ on X, then for all $\sigma > 0$ and $x \in X$,

$$e^{tA}x = \lim_{\alpha \to \infty} \frac{1}{2\pi i} \int_{\sigma-i\alpha}^{\sigma+i\alpha} e^{\lambda t} R(\lambda, A) x \, d\lambda, \qquad (2)$$

where the integral converges uniformly in $t \in [a, b]$, $0 < a < b < \infty$. The UMD-property of X is essential here, since it was shown in [6] that formula (2) ceases to be true for a shift semigroup on $L_1(\mathbb{R})$. A counterpart of this result for cosine operator functions was obtained by Krol in [16]; Krol also proposed new integral characterizations of generators of cosine functions and C_0-groups on UMD spaces, based on the inversion of the Laplace transform.

We also note that in [2] a holomorphic functional calculus for operators of half-plane type was constructed. All key facts of semigroup theory, including the complex inversion formula, can be elegantly obtained in this framework. The functional calculus approach to operator semigroups will be crucial for the sequel.

From [13, Sect. 11.6] it follows that if A is the generator of a bounded C_0-semigroup $(e^{tA})_{t\geq 0}$ on X, then for any $\sigma > 0$,

$$e^{tA}x = \lim_{\alpha \to \infty} \frac{1}{2\pi i} \int_{\sigma-i\alpha}^{\sigma+i\alpha} \left(1 - \frac{|\operatorname{Im} \lambda|}{\alpha}\right) e^{\lambda t} R(\lambda, A) x \, d\lambda, \quad x \in X, \qquad (3)$$

uniformly in $0 < a \leq t \leq b < \infty$. Hence

$$\limsup_{\alpha \to \infty} \left\| \int_{\sigma-i\alpha}^{\sigma+i\alpha} \left(1 - \frac{|\operatorname{Im} \lambda|}{\alpha}\right) e^{\lambda t} R(\lambda, A) x \, d\lambda \right\| \leq C \|x\|, \quad t > 0, \quad x \in X, \qquad (4)$$

with $C = 2\pi \sup_{t \geq 0} \|e^{tA}\|$.

In this paper we provide a related sufficient condition for A to generate a bounded semigroup: we prove that if the resolvent $R(\lambda, A)$ is bounded in any half-plane $\{\lambda : \operatorname{Re} \lambda > \sigma\}$, $\sigma > 0$, and (4) holds for any $t > 0$ with $\sigma = \sigma_t$ depending possibly on t, then A generates a bounded C_0-semigroup on X. If X is a UMD space, then the condition becomes necessary. We also analyze relations between a couple of conditions involved in the resolvent characterization of the generation property, and show their relevance.

2 Main Result

Let $\mathbb{C}_+$ stand for the set $\{\lambda \in \mathbb{C} : \operatorname{Re} \lambda > 0\}$. Our goal in this paper is to prove the following result.

Theorem 1 *Let X be a Banach space, and let A be a densely defined, closed operator on X with $\mathbb{C}_+ \subset \rho(A)$. Suppose that the following conditions hold.*

(i) *For every $\sigma > 0$ there exists an $M_\sigma > 0$ that*

$$\|R(\lambda, A)\| \leqslant M_\sigma, \quad \lambda \in \mathbb{C}_+, \ \operatorname{Re} \lambda \geq \sigma, \tag{5}$$

(ii) *There exists a $C > 0$ with the following property: for every $t > 0$ there is a $\sigma > 0$ such that*

$$\limsup_{\alpha \to \infty} \left\| \int_{\sigma - i\alpha}^{\sigma + i\alpha} \left(1 - \frac{|\operatorname{Im} \lambda|}{\alpha}\right) e^{\lambda t} R(\lambda, A) x \, d\lambda \right\| \leqslant C \|x\|, \quad x \in X. \tag{6}$$

Then A is the generator of a bounded C_0-semigroup $(e^{tA})_{t \geq 0}$ on X, satisfying $\|e^{tA}\| \leq C/(2\pi)$ and (3) holds.

For the proof we recall from [8, p. 234] that for all $\sigma > 0$ and $t > 0$ one has

$$\lim_{\alpha \to \infty} \frac{1}{2\pi i} \int_{\sigma - i\alpha}^{\sigma + i\alpha} \frac{e^{\lambda t}}{\lambda} \, d\lambda = 1 \quad \text{and} \quad \lim_{\alpha \to \infty} \frac{1}{2\pi i} \int_{\sigma - i\alpha}^{\sigma + i\alpha} \frac{e^{\lambda t}}{\lambda^2} \, d\lambda = t \tag{7}$$

uniformly on $t \in [a, b]$. Moreover, for $\sigma > 0$ and $t < 0$,

$$\lim_{\alpha \to \infty} \frac{1}{2\pi i} \int_{\sigma - i\alpha}^{\sigma + i\alpha} \frac{e^{\lambda t}}{\lambda} \, d\lambda = \lim_{\alpha \to \infty} \frac{1}{2\pi i} \int_{\sigma - i\alpha}^{\sigma + i\alpha} \frac{e^{\lambda t}}{\lambda^2} \, d\lambda = 0. \tag{8}$$

Relations (7) will be used in the proof of the following proposition which is a stepping stone for our main theorem.

Proposition 1 *Let A be a closed, densely defined operator on X. Suppose there is a $\sigma > 0$ such that*

$$\sup_{\operatorname{Re} \lambda = \sigma} \|R(\lambda, A)\| < \infty, \tag{9}$$

and let

$$S_{\alpha,\sigma}(t)x := \frac{1}{2\pi i} \int_{\sigma-i\alpha}^{\sigma+i\alpha} \left(1 - \frac{|\operatorname{Im}\lambda|}{\alpha}\right) e^{\lambda t} R(\lambda, A)x\, d\lambda,$$

$$S_{\alpha,\sigma}^{(0)}(t)x := \frac{1}{2\pi i} \int_{\sigma-i\alpha}^{\sigma+i\alpha} e^{\lambda t} R(\lambda, A)x\, d\lambda, \quad t > 0,\, x \in X.$$

Then, for $x \in D(A^2)$,

$$\lim_{\alpha \to \infty} S_{\alpha,\sigma}(t)x = \lim_{\alpha \to \infty} S_{\alpha,\sigma}^{(0)}(t)x = x + tAx + \frac{1}{2\pi i} \int_{\sigma-i\infty}^{\sigma+i\infty} \frac{e^{\lambda t} R(\lambda, A) A^2 x}{\lambda^2}\, d\lambda, \tag{10}$$

uniformly in $0 < a \le t \le b < \infty$.

Proof Using the representation

$$R(\lambda, A)x = \frac{1}{\lambda} x + \frac{1}{\lambda^2} Ax + \frac{1}{\lambda^2} R(\lambda, A) A^2 x, \quad x \in D(A^2), \tag{11}$$

we obtain that

$$S_{\alpha,\sigma}^{(0)}(t)x = \frac{1}{2\pi i} \int_{\sigma-i\alpha}^{\sigma+i\alpha} \frac{e^{\lambda t}}{\lambda} \left(x + \frac{Ax}{\lambda}\right) d\lambda + \frac{1}{2\pi i} \int_{\sigma-i\alpha}^{\sigma+i\alpha} \frac{e^{\lambda t} R(\lambda, A) A^2 x}{\lambda^2}\, d\lambda.$$

Thus, the second part of formula (10) is a consequence of (7). On the other hand, a simple calculation yields

$$\int_{\sigma-i\alpha}^{\sigma+i\alpha} \left(1 - \frac{|\operatorname{Im}\lambda|}{\alpha}\right) e^{\lambda t} R(\lambda, A)x\, d\lambda = \int_{\sigma-i\alpha}^{\sigma+i\alpha} \left(\frac{1}{\alpha} \int_{|\operatorname{Im}\lambda|}^{\alpha} d\beta\right) e^{\lambda t} R(\lambda, A)x\, d\lambda$$

$$= \frac{1}{\alpha} \int_0^\alpha \int_{\sigma-i\beta}^{\sigma+i\beta} e^{\lambda t} R(\lambda, A)x\, d\lambda\, d\beta, \tag{12}$$

i.e., $S_{\alpha,\sigma}(t) = \frac{1}{\alpha} \int_0^\alpha S_{\beta,\sigma}^{(0)}(t)\, d\beta$. It follows that $S_{\alpha,\sigma}(t)$, being the average of $S_{\beta,\sigma}^{(0)}(t)$ over β, converges, as $\alpha \to \infty$, whenever $S_{\beta,\sigma}^{(0)}(t)$ converges as $\beta \to \infty$, and the two limits coincide. This completes the proof. Since the latter convergence has already been established, we are done. □

Remark 1 Let $f_t(z) := e^{-tz}$, $z \in \mathbb{C}$. If (5) holds, then a closed operator $f_t(-A)$ can be defined by means of the half-plane holomorphic functional calculus developed in [2]. Namely, by [2, Lemma 2.4],

$$f_t(-A)x := (A-\delta)^2 \frac{1}{2\pi i} \int_{\sigma-i\infty}^{\sigma+i\infty} \frac{e^{t\lambda}}{(\lambda-\delta)^2} R(\lambda, A)x\, d\lambda, \quad x \in D(A^2),\, \delta > \sigma > 0,$$

and the definition does not depend on $\sigma > 0$ and $\delta > \sigma$. Moreover, by the same [2, Lemma 2.4], we have

$$f_t(-A)x = \lim_{\alpha \to \infty} S^{(0)}_{\alpha,\sigma}(t)x, \qquad x \in D(A^2).$$

A crucial fact from [2, Proposition 2.5], linking the semigroup theory with the theory of functional calculi, is that A is the generator of a C_0-semigroup $(e^{tA})_{t\geq 0}$ if and only if $f_t(-A)$ is a bounded operator on X for all $t \in [0,1]$ and $\sup_{t\in[0,1]} \|f_t(-A)\| < \infty$. In this case, $e^{tA} = f_t(-A)$, $t \geq 0$.

Proof (Proof of Theorem 1) Let $t > 0$ be fixed and let $\sigma = \sigma_t > 0$ be such that condition (6) holds. The equiboundedness assumption (b) coupled with Proposition 1 allows us to conclude that the limit

$$S_\sigma(t)x := \lim_{\alpha \to \infty} \int_{\sigma-i\alpha}^{\sigma+i\alpha} \left(1 - \frac{|\operatorname{Im}\lambda|}{\alpha}\right) e^{\lambda t} R(\lambda, A)x\, d\lambda, \qquad x \in X, \tag{13}$$

exists for all $x \in X$, because the set $D(A^2)$ is dense in X. Moreover, this formula defines a bounded linear operator $S_\sigma(t)$ on X with norm not exceeding $C/(2\pi)$.

Proposition 1 also says that

$$S_\sigma(t)x = S^{(0)}_\sigma(t)x \qquad x \in D(A^2). \tag{14}$$

Then the fact that A is the generator of a bounded C_0-semigroup $(e^{tA})_{t\geq 0}$ on X with $\|e^{tA}\| \leq C/(2\pi)$ follows from Remark 1. $\square$

The next result follows immediately from Theorem 1.

Corollary 1 *Let A be a densely defined, closed operator on X with $\mathbb{C}_+ \subset \rho(A)$. Then A is the generator of a bounded C_0-semigroup on X if and only if (5) holds and there exists a $C > 0$ such that for any $t > 0$ the estimate (6) is satisfied for some $\sigma = \sigma_t > 0$ (and then for any $\sigma > 0$).*

Indeed, Theorem 1 implies the "if" part. For the proof of the "only if" part it suffices to note that $R(\cdot, A)$ is the Laplace transform of a bounded semigroup, and thus (5) holds. Since, as noted in the introduction, (4) is true as well, the "only if" part follows.

Note that the property

$$\limsup_{\alpha \to \infty} \left\| \int_{\sigma-i\alpha}^{\sigma+i\alpha} e^{\lambda t} R(\lambda, A)x\, d\lambda \right\| \leq C \|x\|, \qquad x \in X, \tag{15}$$

for some $C > 0$, implies (6). This is an easy consequence of the relation (12). So, (5) and (15) together provide a sufficient condition for A to be the generator of a bounded C_0-semigroup.

On the other hand, as noted in the introduction, if X has the UMD property and $(e^{tA})_{t\geq 0}$ is a bounded C_0-semigroup on X, then (2) is true. Taking into account (1) we thus obtain the following statement.

Corollary 2 *Let X be a UMD Banach space, and let A be a closed, densely defined operator in X with $\mathbb{C}_+ \subset \rho(A)$ satisfying (5). Then A is the generator of a bounded C_0-semigroup on X if and only if there is a $C > 0$ such that for any $t > 0$ and for some $\sigma = \sigma_t > 0$ (and then for any $\sigma > 0$) the estimate (15) holds.*

3 Comments on Conditions (1) and (5)

In this section we would like to discuss the role of conditions (1) and (5) in the light of the results presented in the previous section.

3.1 Condition (1)

Starting from the first of these conditions, studied in [11] and [9], we will show directly that if an operator A satisfies (1), then A satisfies (5) and (15) as well. This provides a new proof of the generation result from [11] and [9], mentioned in the introduction. We will also show that the converse is not true: there are operators which satisfy (5) and (15) but do not satisfy (1).

Proposition 2 *Condition (1) implies (5) and (15).*

Proof $(1) \Rightarrow (5)$

Suppose that a densely defined, closed operator A satisfies (1). Then, by the closed graph theorem, there exists a $c > 0$ such that for all $\sigma > 0$ we have

$$\int_{\sigma-i\infty}^{\sigma+i\infty} |\langle R^2(\lambda, A)x, y\rangle| |d\lambda| \leq \frac{c}{\sigma} \|x\| \|y\|, \quad x \in X, \quad y \in X^*. \tag{16}$$

We will show that this relation implies (5). To this end, for $\delta > 0$, we define the function

$$f_\delta(\lambda) := \langle R(\lambda + \delta, A)x, y\rangle,$$

and remark that f_δ is analytic on $\mathbb{C}_+$, and continuous on $\overline{\mathbb{C}_+}$. Inequality (16) says that the derivative f'_δ belongs to the Hardy space $H^1(\mathbb{C}_+)$ and

$$\|f'_\delta\|_{H^1} \leq \frac{c}{\delta} \|x\| \|y\|.$$

Then, by the Carleson embedding theorem [10, Section II.3], we have that for any $\varphi \in [-\pi/2, \pi/2]$,

$$\int_0^\infty |f'_\delta(\rho e^{i\varphi})| d\rho \leq \sqrt{2} \|f_\delta\|_{H^1(\mathbb{C})}, \tag{17}$$

and in particular,

$$|f_\delta(z)| \leq |f_\delta(0)| + \sqrt{2}\|f'_\delta\|_{H^1} \leq \left(\|R(\delta, A)\| + \frac{\sqrt{2}c}{\delta}\right)\|x\|\|y\|, \qquad z \in \mathbb{C}_+.$$

Hence we obtain (5).

((1) & (5)) $\Rightarrow$ (15)

Next, by (16), for all $x \in X$, $\alpha > 0$, $\sigma > 0$ and $t > 0$,

$$\left\|\int_{\sigma-i\alpha}^{\sigma+i\alpha} e^{\lambda t} R^2(\lambda, A) x \, d\lambda\right\| \leq \sup_{\|y\|\leq 1} e^{\sigma t} \int_{\sigma-i\alpha}^{\sigma+i\alpha} |\langle R^2(\lambda, A)x, y\rangle| |d\lambda| \leq \frac{c}{\sigma} e^{\sigma t} \|x\|. \quad (18)$$

Moreover, integrating by parts we infer that for every $x \in X$ and $t > 0$:

$$t \int_{\sigma-i\alpha}^{\sigma+i\alpha} e^{\lambda t} R(\lambda, A) x \, d\lambda = e^{\sigma t}(e^{i\alpha t} R(\sigma+i\alpha, A) - e^{-i\alpha t} R(\sigma-i\alpha, A))x \quad (19)$$

$$+ \int_{\sigma-i\alpha}^{\sigma+i\alpha} e^{\lambda t} R^2(\lambda, A) x \, d\lambda.$$

Since $D(A)$ is dense X, the resolvent identity together with (5) implies that $\lim_{\alpha \to +\infty} \|R(\sigma \pm i\alpha)x\| = 0$ for all $x \in X$. Therefore, by (18) and (19) with $\sigma = 1/t$, it follows that for all $x \in X$ and $t > 0$,

$$\limsup_{\alpha \to \infty} \left\|\int_{1/t-i\alpha}^{1/t+i\alpha} e^{\lambda t} R(\lambda, A) x \, d\lambda\right\| \leq ce\|x\|, \qquad x \in X.$$

Hence (15) holds also. $\square$

On the other hand, (1) needs not hold if one assumes (5) and (15). As noted in [6] (see also our Introduction and formula (2) there) to prove this it suffices to provide an example of a generator A of a bounded C_0-semigroup on a UMD space X, which does not satisfy (1). To this end, we use an idea from [11] and recall the following statement on scalar spectral type operators proved in [4, Theorem 3.3]. For the definition and basic properties of scalar-type spectral operators we refer e.g. to [7].

Proposition 3 *Let A be a bounded operator on a reflexive Banach space X such that $\sigma(A) \subset i\mathbb{R}$. Then A is a scalar type spectral operator if and only if for all $x \in X$ and $y \in X^*$,*

$$\sup_{\sigma>0} \sigma \int_{\sigma-i\infty}^{\sigma+i\infty} |\langle R^2(\lambda, A)x, y\rangle| \, |d\lambda| < \infty,$$

and for all $x \in X$,

$$\sup_{t>0} \frac{1}{t} \int_0^t \|e^{-sA}x\| \, ds < \infty.$$

We are now ready to present our example.

Example 1 It was observed in [5] that for any infinite dimensional Hilbert space H and any $1 \leq p < \infty$, $p \neq 2$, the von Neumann-Schatten ideal $C_p(H)$ is an UMD space, and moreover there exist commuting bounded scalar-type spectral operators A_1 and A_2 on $C_p(H)$ with spectrums contained in the real axis such that $A_1 + A_2$ is not scalar-type spectral. Clearly, the operator $A := i(A_1 + A_2)$ generates a C_0-group $(e^{tA})_{t\in\mathbb{R}}$ on $C_p(H)$, and the group is uniformly bounded on $\mathbb{R}$, since A_1 and A_2 are commuting. Thus, the operator A satisfies (5) and (15). On the other hand, by Proposition 3, A does not satisfy (1). Clearly, the same holds for the operator $-A$.

3.2 Condition (5)

Finally, we also comment on condition (5): we will show that condition (15) alone does not imply that A is the generator of a bounded semigroup; in other words, there are operators satisfying (15) which, however, do not satisfy (5) and, consequently, are not generators of bounded semigroups. In our analysis we will use the following simple assertion; its proof is similar to that presented in [3, Theorem 5.1].

Proposition 4 *Let A be the generator of a bounded C_0-semigroup $(e^{tA})_{t\geq 0}$ on a Hilbert space X. Then for all $\sigma > 0$, $x \in X$ and $t \in (-\infty, 0)$,*

$$\lim_{\alpha \to \infty} \int_{\sigma-i\alpha}^{\sigma+i\alpha} e^{\lambda t} R(\lambda, A)x \, d\lambda = 0. \tag{20}$$

Proof Let $M := \sup_{t \geq 0} \|e^{tA}\|$, so that $\|R(\lambda, A)\| \leq M/\sigma$, $\text{Re}\,\lambda \geq \sigma > 0$. Using the representations of $R(\lambda, A)$ and $R(\lambda, A^*)$ as Laplace transforms and Plancherel's theorem, we infer that (16) holds for all $\sigma > 0$, $x, y \in X$, with $c_\sigma = \pi M^2/\sigma$. So, setting $f_{x,y}(s) := \langle R^2(\sigma + is, A)x, y\rangle$, $s \in \mathbb{R}$ and noting that $f_{x,y} \in L_1(\mathbb{R})$, we have

$$f_{x,y}(s) = \int_{-\infty}^{\infty} e^{-ist} L_{x,y}(t) \, dt, \tag{21}$$

for all $x, y \in X$ and $\sigma > 0$, where

$$L_{x,y}(t) = \begin{cases} te^{-\sigma t}\langle e^{tA}x, y\rangle, & t \geq 0, \\ 0, & t < 0, \end{cases}$$

satisfies $L_{x,y} \in C_0(\mathbb{R}) \cap L_1(\mathbb{R})$. Then, inverting the Fourier transform of $L_{x,y}$, we have
$$\lim_{\alpha\to\infty}\int_{-\alpha}^{\alpha} e^{its} f_{x,y}(s)\,ds = \int_{-\infty}^{\infty} e^{its} f_{x,y}(s)\,ds = 0, \quad t \in (-\infty, 0). \tag{22}$$

So, by (19), (21), and (22), for all $x \in X$,
$$\text{weak} - \lim_{\alpha\to\infty}\int_{\sigma-i\alpha}^{\sigma+i\alpha} e^{\lambda t} R(\lambda, A) x\, d\lambda = 0, \quad t \in (-\infty, 0). \tag{23}$$

On the other hand, from (8) and (11) it follows that for any $t < 0$ there exists
$$\lim_{\alpha\to 0}\int_{\sigma-i\alpha}^{\sigma+i\alpha} e^{\lambda t} R(\lambda, A) x\, d\lambda = \int_{\sigma-i\infty}^{\sigma+i\infty} \frac{e^{t\lambda} R(\lambda, A) A^2 x}{\lambda^2}\, d\lambda, \quad x \in D(A^2).$$

Then, using density of $D(A^2)$ in X and (23), we obtain the assertion (20). $\square$

The next example shows that (5) can be omitted neither in Theorem 1 nor Corollary 2.

Example 2 Let X be a Hilbert space, and let D be the generator of a nilpotent C_0-semigroup $(e^{tD})_{t\geq 0}$ on X. (For instance, one may consider the left shift semigroup on $X = L_2[0, 1]$.) Then $\sigma(D) = \emptyset$, for every $\delta > 0$ the operator $A + \delta$ generates a bounded C_0-semigroup $(e^{\delta t} e^{tD})_{t\geq 0}$, and for every $\sigma > 0$ there exists $M_\sigma > 0$ such that
$$\|R(\lambda, D)\| \leq M_\sigma, \quad \operatorname{Re}\lambda \geq \sigma. \tag{24}$$

Setting $\lambda = 2\sigma - \mu$, $\delta = 2\sigma$ and $\tau = -t < 0$, we have
$$\int_{\sigma-i\alpha}^{\sigma+i\alpha} e^{\lambda t} R(\lambda, -D) x\, d\lambda = -e^{\delta t}\int_{\sigma-i\alpha}^{\sigma+i\alpha} e^{\mu\tau} R(\mu, D+\delta) x\, d\mu$$

for all $\sigma > 0$, $t > 0$ and $x \in X$. Since the choice of $\delta > 0$ is arbitrary, applying Proposition 4 to $D + \delta$, we infer that
$$\lim_{\alpha\to\infty}\int_{\sigma-i\alpha}^{\sigma+i\alpha} e^{\lambda t} R(\lambda, -D) x\, d\lambda = 0, \quad t > 0,$$

so that (15) holds for $A := -D$ with $C = 0$. Note that $R(\cdot, A)$ is an operator-valued entire function, thus it can not be bounded in $\mathbb{C}$. Hence, in view of (24), A does not satisfy (5), and A does not generate a C_0-semigroup on X for the same reason.

Acknowledgements The author is very much grateful to the referee for careful reading of the manuscript and suggestions helping to significantly improve the presentation.

References

1. Batty, C.J.K.: On a perturbation theorem of Kaiser and Weis. Semigroup Forum **70**, 471–474 (2005)
2. Batty, C.J.K., Haase, M., Mubeen, J.: The holomorphic functional calculus approach to operator semigroups. Acta Sci. Math. (Szeged) **79**, 289–323 (2013)
3. Chill, R., Tomilov, Y.: Stability of C_0-semigroups and geometry of Banach spaces. Math. Proc. Camb. Philos. Soc. **135**, 493–511 (2003)
4. Cojuhari, P.A., Gomilko, A.M.: On the characterization of scalar type spectral operators. Studia Math. **184**, 121–131 (2008)
5. Doust, I., Gillespie, T.A.: Well-boundedness of sums and products of operators. J. Lond. Math. Soc. **68**, 183–192 (2013)
6. Driouich, A., El-Mennaoui, O.: On the inverse Laplace transform of C_0-semigroups in UMD-spaces. Archiv Math. (Basel) **72**, 56–63 (1999)
7. Dunford, N., Schwartz, J.: Linear Operators, Part III. Interscience (1971)
8. Engel, K.-J., Nagel, R.: One-Parameter Semigroups for Linear Evolution Equations, Graduate Texts in Mathematics, vol. 194. Springer, New York (2000)
9. Feng, D.X., Shi, D.H.: Characteristic conditions of the generation of C_0-semigroups in a Hilbert space. J. Math. Anal. Appl. **247**, 356–376 (2000)
10. Garnett, J.B.: Bounded Analytic Functions. Academic Press (1981)
11. Gomilko, A.M.: Conditions on the generator of a uniformly bounded C_0-semigroup, Funktsional. Anal. i Prilozhen. **33**, 66–69 (in Russian); transl. Funct. Anal. Appl. **33**(1999), 294–296 (1999)
12. Haase, M.: The complex inversion formula revisited. J. Aust. Math. Soc. **84**, 73–83 (2008)
13. Hille, E., Phillips, R.S.: Functional analysis and semi-groups, Third printing of the revised ed. of 1957, AMS Colloq. Publ. vol. 31. AMS, Providence, RI (1974)
14. Kaiser, C., Weis, L.: A perturbation theorem for operator semigroups in Hilbert spaces. Semigroup Forum **67**, 63–75 (2003)
15. Kantorovitz, S.: On the characterization of spectral operators. Trans. Am. Math. Soc. **111**, 152–181 (1964)
16. Król, S.: Resolvent characterisation of generators of cosine functions and C_0-groups. J. Evol. Equ. **13**, 281–309 (2013)

Semigroups Associated with Differential-Algebraic Equations

Sascha Trostorff

Abstract We consider differential-algebraic equations in infinite dimensional state spaces, and study under which conditions we can associate a C_0-semigroup with such equations. We determine the right space of initial values and characterise the existence of a C_0-semigroup in the case of operator pencils with polynomially bounded resolvents.

Keywords Differential-algebraic equations · C_0-semigroups · Wong sequence · Operator pencil · Polynomially bounded resolvent

2010 MSC 46N20 · 34A09 · 35M31

1 Introduction

In the case of matrices the study of differential-algebraic equations (DAEs), i.e. equations of the form

$$(Eu)'(t) + Au(t) = f(t),$$
$$u(0) = u_0$$

for matrices $E, A \in \mathbb{R}^{n \times n}$, is a very active field in mathematics (see e.g. [7–10] and the references therein). The main difference to classical differential equations is that the matrix E is allowed to have a nontrivial kernel. Thus, one cannot expect to solve the equation for each right hand side f and each initial value u_0. In case of matrices one can use normal forms (see e.g. [2, 12, Theorem 2.7]) to determine the 'right' space of initial values, so-called *consistent initial values*. However, this approach cannot be used in case of operators on infinite dimensional spaces. Another approach uses so-called Wong sequences associated with matrices E and A (see e.g.

S. Trostorff (✉)
Mathematisches Seminar, CAU Kiel, Kiel, Germany
e-mail: trostorff@math.uni-kiel.de

[3]) and this turns out to be applicable also in the operator case.

In contrast to the finite dimensional case, the case of infinite dimensions is not that well studied. It is the aim of this article, to generalise some of the results in the finite dimensional case to infinite dimensions. For simplicity, we restrict ourselves to homogeneous problems. More precisely, we consider equations of the form

$$(Eu)'(t) + Au(t) = 0 \quad (t > 0), \tag{1}$$
$$u(0) = u_0,$$

where $E \in L(X; Y)$ for some Banach spaces X, Y and $A : \text{dom}(A) \subseteq X \to Y$ is a densely defined closed linear operator. We will define the notion of mild and classical solutions for such equations and determine the 'right' space of initial conditions for which a mild solution could be obtained.

For doing so, we start with the definition of Wong sequences associated with (E, A) in Sect. 3 which turns out to yield the right spaces for initial conditions. In Sect. 4 we consider the space of consistent initial values and provide some necessary conditions for the existence of a C_0-semigroup associated with the above problem under the assumption that the space of consistent initial values is closed and the mild solutions are unique (Hypotheses A). In Sect. 5 we consider operators (E, A) such that $(zE + A)$ is boundedly invertible on a right half plane and the inverse is polynomially bounded on that half plane. In this case it is possible to determine the space of consistent initial values in terms of the Wong sequence and we can characterise the conditions for the existence of a C_0-semigroup yielding the mild solutions of (1) at least in the case of Hilbert spaces. One tool needed in the proof are the Fourier-Laplace transform and the Theorem of Paley-Wiener, which will be recalled in Sect. 2.

As indicated above, the study of DAEs in infinite dimensions is not such an active field of study as for the finite dimensional case. We mention [22] where in Hilbert spaces the case of selfadjoint operators E is treated using positive definiteness of the operator pencil. Similar approaches in Hilbert spaces were used in [16] for more general equations. However, in both references the initial condition was formulated as $(Eu)(0) = u_0$. We also mention the book [6], where such equations are studied with the focus on maximal regularity. Another approach for dealing with such degenerated equations uses the framework of set-valued (or multi-valued) operators, see [5, 11]. Furthermore we refer to [18, 19], where sequences of projectors are used to decouple the system. Moreover, there exist several references in the Russian literature, where the equations are called Sobolev type equations (see e.g. [23] an the references therein). Finally, we mention the articles [24, 25], which are closely related to the present work, but did not consider the case of operator pencils with polynomially bounded resolvents. In case of bounded operators E and A, equations of the form (1) were studied by the author in [27, 28], where the concept of Wong sequences associated with (E, A) was already used.

We assume that the reader is familiar with functional analysis and in particular with the theory of C_0-semigroups and refer to the monographs [4, 14, 29]. Throughout, if not announced differently, X and Y are Banach spaces.

2 Preliminaries

We collect some basic knowledge on the so-called Fourier-Laplace transformation and weak derivatives in exponentially weighted L_2-spaces, which is needed in Sect. 5. We remark that these concepts were successfully used to study a broad class of partial differential equations (see e.g. [15–17] and the references therein).

Definition Let $\rho \in \mathbb{R}$ and H a Hilbert space. Define

$$L_{2,\rho}(\mathbb{R}; H) := \{f : \mathbb{R} \to H\,;\, f \text{ measurable},\, \int_{\mathbb{R}} \|f(t)\| e^{-2\rho t} \, dt < \infty\}$$

with the usual identification of functions which are equal almost everywhere. Moreover, we define the Sobolev space

$$H_{\rho}^1(\mathbb{R}; H) := \{f \in L_{2,\rho}(\mathbb{R}; H)\,;\, f' \in L_{2,\rho}(\mathbb{R}; H)\},$$

where the derivative is meant in the distributional sense. Finally, we define $\mathcal{L}_\rho$ as the unitary extension of the mapping

$$C_c(\mathbb{R}; H) \subseteq L_{2,\rho}(\mathbb{R}; H) \to L_2(\mathbb{R}; H), \quad f \mapsto \left(t \mapsto \frac{1}{\sqrt{2\pi}} \int_{\mathbb{R}} e^{-(it+\rho)s} f(s) \, ds\right).$$

We call $\mathcal{L}_\rho$ the *Fourier-Laplace transform*. Here, $C_c(\mathbb{R}; H)$ denotes the space of H-valued continuous functions with compact support.

Remark 2.1 It is a direct consequence of Plancherels theorem, that $\mathcal{L}_\rho$ becomes unitary.

The connection of $\mathcal{L}_\rho$ and the space $H_\rho^1(\mathbb{R}; H)$ is explained in the next proposition.

Proposition 2.2 (see e.g. [26, Proposition 1.1.4]) *Let $u \in L_{2,\rho}(\mathbb{R}; H)$ for some $\rho \in \mathbb{R}$. Then $u \in H_\rho^1(\mathbb{R}; H)$ if and only if $\big(t \mapsto (it + \rho)\big(\mathcal{L}_\rho u\big)(t)\big) \in L_2(\mathbb{R}; H)$. In this case we have*

$$\big(\mathcal{L}_\rho u'\big)(t) = (it + \rho)\big(\mathcal{L}_\rho u\big)(t) \quad (t \in \mathbb{R} \text{ a.e.}).$$

Moreover, we have the following variant of the classical Sobolev embedding theorem.

Proposition 2.3 (Sobolev-embedding theorem, see [16, Lemma 3.1.59] or [26, Proposition 1.1.8]) *Let $u \in H_\rho^1(\mathbb{R}; H)$ for some $\rho \in \mathbb{R}$. Then, u has a continuous representative with $\sup_{t \in \mathbb{R}} \|u(t)\| e^{-\rho t} < \infty$.*

Finally, we need the Theorem of Paley-Wiener allowing to characterise those L_2-functions supported on the positive real axis in terms of their Fourier-Laplace transform.

Theorem 2.4 (Paley-Wiener, [13] or [20, 19.2 Theorem]) *Let $\rho \in \mathbb{R}$. We define the Hardy space*

$$\mathcal{H}_2(\mathbb{C}_{\mathrm{Re} > \rho}; H) := \left\{ f : \mathbb{C}_{\mathrm{Re} > \rho} \to H ; f \text{ holomorphic}, \sup_{\mu > \rho} \int_\mathbb{R} \|f(\mathrm{i}t + \mu)\|^2 \mathrm{d}t < \infty \right\}.$$

Let $u \in L_{2,\rho}(\mathbb{R}; H)$. Then $\mathrm{spt}\, u \subseteq \mathbb{R}_{\geq 0}$ if and only if

$$\big((\mathrm{i}t + \mu) \mapsto (\mathcal{L}_\mu u)(t)\big) \in \mathcal{H}_2(\mathbb{C}_{\mathrm{Re} > \rho}; H).$$

3 Wong Sequences

Throughout, let $E \in L(X; Y)$ and $A : \mathrm{dom}(A) \subseteq X \to Y$ densely defined closed linear.

Definition For $k \in \mathbb{N}$ we define the spaces $\mathrm{IV}_k \subseteq X$ recursively by

$$\mathrm{IV}_0 := \mathrm{dom}(A),$$
$$\mathrm{IV}_{k+1} := A^{-1}[E[\mathrm{IV}_k]].$$

This sequence of subspaces is called the *Wong sequence associated with (E, A)*.

Remark 3.1 We have

$$\mathrm{IV}_{k+1} \subseteq \mathrm{IV}_k \quad (k \in \mathbb{N}).$$

Indeed, for $k = 0$ this follows from $\mathrm{IV}_1 = A^{-1}[E[\mathrm{IV}_0]] \subseteq \mathrm{dom}(A) = \mathrm{IV}_0$ and hence, the assertion follows by induction.

Definition We define

$$\rho(E, A) := \{z \in \mathbb{C}; (zE + A)^{-1} \in L(Y; X)\}$$

the *resolvent set associated with (E, A)*.

We start with some useful facts on the Wong sequence. The following result was already given in [27] in case of a bounded operator A.

Lemma 3.2 *Let $k \in \mathbb{N}$. Then*

$$E(zE + A)^{-1}A \subseteq A(zE + A)^{-1}E$$

and

$$(zE + A)^{-1}E[\mathrm{IV}_k] \subseteq \mathrm{IV}_{k+1}$$

for each $z \in \rho(E, A)$. Moreover, for $x \in \mathrm{IV}_k$ we find elements $x_1, \ldots, x_k \in X$, $x_{k+1} \in \mathrm{dom}(A)$ such that

$$(zE + A)^{-1}Ex = \frac{1}{z}x + \sum_{\ell=1}^{k} \frac{1}{z^{\ell+1}} x_\ell + \frac{1}{z^{k+1}}(zE + A)^{-1}Ax_{k+1} \quad (z \in \rho(E, A) \setminus \{0\}).$$

Proof For $x \in \mathrm{dom}(A)$ we compute

$$\begin{aligned} E(zE + A)^{-1}Ax &= E\left(x - (zE + A)^{-1}zEx\right) \\ &= Ex - zE(zE + A)^{-1}Ex \\ &= Ex - Ex + A(zE + A)^{-1}Ex \\ &= A(zE + A)^{-1}Ex. \end{aligned}$$

We prove the second and third claim by induction. Let $k = 0$ and $x \in \mathrm{IV}_0 = \mathrm{dom}(A)$. Then $(zE + A)^{-1}Ex \in \mathrm{dom}(A)$ with

$$A(zE + A)^{-1}Ex = E(zE + A)^{-1}Ax \in E[\mathrm{dom}(A)] = E[\mathrm{IV}_0]$$

and thus, $(zE + A)^{-1}Ex \in \mathrm{IV}_1$. Moreover

$$(zE + A)^{-1}Ex = \frac{1}{z}(x - (zE + A)^{-1}Ax)$$

showing the equality with $x_1 = -x \in \mathrm{dom}(A)$. Assume now that both assertions hold for $k \in \mathbb{N}$ and let $x \in \mathrm{IV}_{k+1}$. Then $Ax = Ey$ for some $y \in \mathrm{IV}_k$ and we infer

$$A(zE + A)^{-1}Ex = E(zE + A)^{-1}Ey \in E[\mathrm{IV}_{k+1}]$$

by induction hypothesis. Hence, $(zE + A)^{-1}Ex \in \mathrm{IV}_{k+2}$. Moreover, by assumption we find $y_1, \ldots, y_k \in X$ and $y_{k+1} \in \mathrm{dom}(A)$ such that

$$\begin{aligned} \frac{1}{z}y + \sum_{\ell=1}^{k} \frac{1}{z^{\ell+1}} y_\ell + \frac{1}{z^{k+1}}(zE + A)^{-1}Ay_{k+1} &= (zE + A)^{-1}Ey \\ &= (zE + A)^{-1}Ax \\ &= x - z(zE + A)^{-1}Ex. \end{aligned}$$

Thus, we obtain the desired formula with $x_1 := -y$, $x_j = -y_{j-1}$ for $j \in \{2, \ldots, k+2\}$. $\square$

Lemma 3.3 *Assume $\rho(E, A) \neq \emptyset$. Then for each $k \in \mathbb{N}$ we have that*

$$A^{-1}\left[E\left[\overline{\mathrm{IV}_k}\right]\right] \subseteq \overline{\mathrm{IV}_{k+1}}.$$

Proof We prove the claim by induction. For $k = 0$, let $x \in \mathrm{dom}(A)$ such that $Ax = Ey$ for some $y \in \overline{\mathrm{IV}_0}$. Hence, we find a sequence $(y_n)_{n \in \mathbb{N}}$ in IV_0 with $y_n \to y$ and since E is bounded, we derive $Ey_n \to Ey = Ax$. For $z \in \rho(E, A)$ set

$$x_n := (zE + A)^{-1} zEx + (zE + A)^{-1} Ey_n.$$

By Lemma 3.2 we have that $x_n \in \mathrm{IV}_1$ and

$$\begin{aligned} \lim_{n \to \infty} x_n &= (zE + A)^{-1} zEx + (zE + A)^{-1} Ey \\ &= (zE + A)^{-1} zEx + (zE + A)^{-1} Ax \\ &= x, \end{aligned}$$

hence $x \in \overline{\mathrm{IV}_1}$.

Assume now that the assertion holds for some $k \in \mathbb{N}$ and let $x \in A^{-1}\left[E\left[\overline{\mathrm{IV}_{k+1}}\right]\right]$. Then clearly $x \in A^{-1}\left[E\left[\overline{\mathrm{IV}_k}\right]\right] \subseteq \overline{\mathrm{IV}_{k+1}}$ and hence, we find a sequence $(w_n)_{n \in \mathbb{N}}$ in IV_{k+1} with $w_n \to x$. For $z \in \rho(A, E)$ we infer

$$(zE + A)^{-1} zEw_n \to (zE + A)^{-1} zEx$$

and by Lemma 3.2 we have $(zE + A)^{-1} zEx \in \overline{\mathrm{IV}_{k+2}}$. Moreover, we find a sequence $(y_n)_{n \in \mathbb{N}}$ in IV_{k+1} with $Ax = \lim_{n \to \infty} Ey_n$. As above, we set

$$x_n := (zE + A)^{-1} zEx + (zE + A)^{-1} Ey_n$$

and obtain a sequence in $\overline{\mathrm{IV}_{k+2}}$ converging to x. Hence $x \in \overline{\mathrm{IV}_{k+2}}$. $\square$

4 Necessary Conditions for C_0-Semigroups

In this section we focus on the differential-algebraic problem

$$\begin{aligned} Eu'(t) + Au(t) &= 0 \quad (t > 0) \\ u(0) &= u_0, \end{aligned} \tag{2}$$

where again $E \in L(X; Y)$ and $A : \text{dom}(A) \subseteq X \to Y$ is a linear closed densely defined operator and $u_0 \in X$. We begin with the notion of classical solutions and mild solutions of the above problem.

Definition Let $u : \mathbb{R}_{\geq 0} \to X$ be continuous.

(a) u is called a *classical solution* of (2), if u is continuously differentiable on $\mathbb{R}_{\geq 0}$, $u(t) \in \text{dom}(A)$ for each $t \geq 0$ and (2) holds.
(b) u is called a *mild solution* of (2), if $u(0) = u_0$ and for all $t > 0$ we have $\int_0^t u(s)ds \in \text{dom}(A)$ and

$$Eu(t) + A \int_0^t u(s)ds = Eu_0.$$

Obviously, a classical solution of (2) is also a mild solution of (2). The main question is now to determine a natural space, where one should seek for (mild) solutions. In particular, we have to find the initial values. We define the space of such values by

$$U := \{u_0 \in X \,;\, \exists u : \mathbb{R}_{\geq 0} \to X \text{ mild solution of (2)}\}.$$

Clearly, U is a subspace of X.

Proposition 4.1 *Assume $\rho(E, A) \neq \emptyset$. Let $x \in U$ and u_x be a mild solution of (2) with initial value x. Then $u_x(t) \in \bigcap_{k \in \mathbb{N}} \overline{IV_k}$ for each $t \geq 0$. In particular, $U \subseteq \bigcap_{k \in \mathbb{N}} \overline{IV_k}$.*

Proof Let $t \geq 0$. Obviously, we have that $u_x(t) \in \overline{IV_0} = \overline{\text{dom}(A)} = X$. Assume now that we know $u_x(t) \in \overline{IV_k}$ for all $t \geq 0$. We then have

$$A \int_t^{t+h} u_x(s)ds = Eu_x(t+h) - Eu_x(t) \in E[\overline{IV_k}] \quad (h > 0)$$

and thus,

$$\int_t^{t+h} u_x(s)ds \in A^{-1}\left[E\left[\overline{IV_k}\right]\right] \subseteq \overline{IV_{k+1}}$$

by Lemma 3.3. Hence,

$$u_x(t) = \lim_{h \to 0} \int_t^{t+h} u_x(s)ds \in \overline{IV_{k+1}}.$$
□

We state the following hypothesis, which we assume to be valid throughout the whole section.

Hypotheses A The space U is closed and for each $u_0 \in U$ the mild solution of (2) is unique.

As in the case of Cauchy problems, we can show that we can associate a C_0-semigroup with (2). The proof follows the lines of [1, Theorem 3.1.12].

Proposition 4.2 *Denote for $x \in U$ the unique mild solution of (2) by u_x. Then the mappings*
$$T(t) : U \to X, \quad x \mapsto u_x(t)$$
for $t \geq 0$ define a C_0-semigroup on U. In particular, $\operatorname{ran} T(t) \subseteq U$ for each $t \geq 0$.

Proof Consider the mapping
$$\Phi : U \to C(\mathbb{R}_{\geq 0}; X), \quad x \mapsto u_x.$$
We equip $C(\mathbb{R}_{\geq 0}; X)$ with the topology induced by the seminorms
$$p_n(f) := \sup_{t \in [0,n]} \|f(t)\| \quad (n \in \mathbb{N})$$
for which $C(\mathbb{R}_{\geq 0}; X)$ becomes a Fréchet space. Then Φ is linear and closed. Indeed, if $(x_n)_{n \in \mathbb{N}}$ is a sequence in U such that $x_n \to x$ and $u_{x_n} \to u$ as $n \to \infty$ for some $x \in U$ and $u \in C(\mathbb{R}_{\geq 0}, X)$ we derive $\int_0^t u_{x_n}(s)\,\mathrm{d}s \to \int_0^t u(s)\,\mathrm{d}s$ for each $t \geq 0$ since $u_{x_n} \to u$ uniformly on $[0, t]$. Moreover,
$$A \int_0^t u_{x_n}(s)\,\mathrm{d}s = Ex_n - Eu_{x_n}(t) \to Ex - Eu(t) \quad (n \to \infty)$$
for each $t \geq 0$ and hence, $\int_0^t u(s)\,\mathrm{d}s \in \operatorname{dom}(A)$ with
$$A \int_0^t u(s)\,\mathrm{d}s = Ex - Eu(t) \quad (t \geq 0).$$
Finally, since $u(0) = \lim_{n \to \infty} u_{x_n}(0) = x$, we infer that $u = u_x$ and hence, Φ is closed. By the closed graph theorem (see e.g. [21, III, Theorem 2.3]), we derive that Φ is continuous. In particular, for each $t \geq 0$ the operator
$$T(t)x = u_x(t) = \Phi(x)(t)$$
is bounded and linear. Moreover, $T(t)x = u_x(t) \to x$ as $t \to 0$ for each $x \in U$. We are left to show that $\operatorname{ran} T(t) \subseteq U$ and that T satisfies the semigroup law. For doing so, let $x \in U$ and $t \geq 0$. We define the function $u : \mathbb{R}_{\geq 0} \to X$ by $u(s) := u_x(t+s) = T(t+s)x$. Then clearly, u is continuous with $u(0) = u_x(t) = T(t)x$ and
$$\int_0^s u(r)\,\mathrm{d}r = \int_0^s u_x(t+r)\,\mathrm{d}r = \int_t^{s+t} u_x(r)\,\mathrm{d}r = \int_0^{s+t} u_x(r)\,\mathrm{d}r - \int_0^t u_x(r)\,\mathrm{d}r \in \operatorname{dom}(A)$$
for each $s \geq 0$ with

$$A \int_0^s u(r)dr = A \int_0^{s+t} u_x(r)dr - A \int_0^t u_x(r)dr$$
$$= Eu_x(s+t) - Eu_x(t)$$
$$= Eu(s) - Eu_x(t) \quad (s \geq 0).$$

Hence, u is a mild solution of (2) with initial value $u_x(t)$ and thus, $u_x(t) \in U$. This proves ran $T(t) \subseteq U$ and

$$T(t+s)x = u(s) = T(s)u_x(t) = T(s)T(t)x \quad (s, t \geq 0, x \in U). \qquad \square$$

We want to inspect the generator of T a bit closer.

Proposition 4.3 *Let B denote the generator of the C_0-semigroup T. Then we have $-EB \subseteq A$.*

Proof Let $x \in \mathrm{dom}(B)$. Consequently, $u_x \in C^1(\mathbb{R}_{\geq 0}; X)$ and thus,

$$A\frac{1}{h}\int_t^{t+h} u_x(s)ds = -\frac{1}{h}E\left(u_x(t+h) - u_x(t)\right) \to -Eu_x'(t) \quad (h \to 0)$$

for each $t \geq 0$. Since $\frac{1}{h}\int_t^{t+h} u_x(s)ds \to u_x(t)$ as $h \to 0$ we infer that $u_x(t) \in \mathrm{dom}(A)$ for each $t \geq 0$ and

$$Eu_x'(t) + Au_x(t) = 0 \quad (t \geq 0),$$

i.e. u is a classical solution of (2). Choosing $t = 0$, we infer $x \in \mathrm{dom}(A)$ and $EBx = -Ax$. $\qquad \square$

5 Pencils with Polynomially Bounded Resolvent

Let $E \in L(X; Y)$ and $A : \mathrm{dom}(A) \subseteq X \to Y$ densely defined closed and linear. Throughout this section we assume the following.

Hypotheses B *There exist $\rho_0 \in \mathbb{R}$, $C \geq 0$ and $k \in \mathbb{N}$ such that:*

(a) $\mathbb{C}_{\mathrm{Re} \geq \rho_0} \subseteq \rho(E, A)$,
(b) $\forall z \in \mathbb{C}_{\mathrm{Re} \geq \rho_0} : \|(zE + A)^{-1}\| \leq C|z|^k$.

Definition We call the minimal $k \in \mathbb{N}$ such that there exists $C \geq 0$ with

$$\|(zE + A)^{-1}\| \leq C|z|^k \quad (z \in \mathbb{C}_{\mathrm{Re} \geq \rho_0})$$

the *index of* (E, A), denoted by $\mathrm{ind}(E, A)$.

Proposition 5.1 *Consider the Wong sequence* $(\mathrm{IV}_k)_{k\in\mathbb{N}}$ *associated with* (E,A). *Then*
$$\overline{\mathrm{IV}_k} = \overline{\mathrm{IV}_{k+1}}$$
for all $k > \mathrm{ind}(E,A)$.

Proof Since we clearly have $\overline{\mathrm{IV}_{k+1}} \subseteq \overline{\mathrm{IV}_k}$ it suffices to prove $\mathrm{IV}_k \subseteq \overline{\mathrm{IV}_{k+1}}$ for $k > \mathrm{ind}(E,A)$. So, let $x \in \mathrm{IV}_k$ for some $k > \mathrm{ind}(E,A)$. By Lemma 3.2 there exist $x_1, \ldots, x_k \in X$, $x_{k+1} \in \mathrm{dom}(A)$ such that

$$(zE+A)^{-1}Ex = \frac{1}{z}x + \sum_{\ell=1}^{k} \frac{1}{z^{\ell+1}} x_\ell + \frac{1}{z^{k+1}}(zE+A)^{-1}Ax_{k+1}. \quad (z \in \rho(E,A)).$$

We define $x_n := (nE+A)^{-1}nEx$ for $n \in \mathbb{N}_{\geq \rho_0}$. Then $x_n \in \mathrm{IV}_{k+1}$ by Lemma 3.2 and by what we have above

$$x_n = x + \sum_{\ell=1}^{k} \frac{1}{n^\ell} x_\ell + \frac{1}{n^k}(nE+A)^{-1}Ax_{k+1} \quad (n \in \mathbb{N}_{\geq \rho_0}).$$

Since $k > \mathrm{ind}(E,A)$, we have that $\frac{1}{n^k}(nE+A)^{-1} \to 0$ as $n \to \infty$ and hence, $x_n \to x$ as $n \to \infty$, which shows the claim. $\square$

Our next goal is to determine the space U. For doing so, we restrict ourselves to Hilbert spaces X.

Proposition 5.2 *Assume Hypotheses A and let X be a Hilbert space. Then* $U = \overline{\mathrm{IV}_{\mathrm{ind}(E,A)+1}}$.

Proof By Propositions 4.1 and 5.1 we have that $U \subseteq \overline{\mathrm{IV}_{\mathrm{ind}(E,A)+1}}$. We now prove that $\mathrm{IV}_{\mathrm{ind}(E,A)+1} \subseteq U$, which would yield the assertion. Let $x \in \mathrm{IV}_{\mathrm{ind}(E,A)+1}$ and $\rho > \max\{0, \rho_0\}$. We define

$$v(z) := 1/\sqrt{2\pi}(zE+A)^{-1}Ex \quad (z \in \mathbb{C}_{\mathrm{Re}\geq\rho})$$

and show that $v \in \mathcal{H}_2(\mathbb{C}_{\mathrm{Re}\geq\rho}; X)$. For doing so, we use Lemma 3.2 to find $x_1, \ldots, x_k \in X$, $x_{k+1} \in \mathrm{dom}(A)$, $k := \mathrm{ind}(E,A)+1$, such that

$$\sqrt{2\pi}v(z) = (zE+A)^{-1}Ex = \frac{1}{z}x + \sum_{\ell=1}^{k} \frac{1}{z^{\ell+1}} x_\ell + \frac{1}{z^{k+1}}(zE+A)^{-1}Ax_{k+1} \quad (z \in \mathbb{C}_{\mathrm{Re}\geq\rho}).$$

Then we have
$$\|v(z)\| \leq \frac{K}{|z|} \quad (z \in \mathbb{C}_{\mathrm{Re}\geq\rho})$$

for some constant $K \geq 0$ and hence, $v \in \mathcal{H}_2(\mathbb{C}_{\mathrm{Re}\geq\rho}; X)$, since obviously v is holomorphic. Setting

$$u := \mathcal{L}_\rho^* v(\mathrm{i} \cdot + \rho)$$

we thus have $u \in L_{2,\rho}(\mathbb{R}_{\geq 0}; X)$ by the Theorem of Paley-Wiener, Theorem 2.4. Moreover,

$$\sqrt{2\pi}\,zv(z) - x = \sum_{\ell=1}^{k} \frac{1}{z^\ell} x_\ell + \frac{1}{z^k}(zE + A)^{-1} A x_{k+1} \quad (z \in \mathbb{C}_{\mathrm{Re}\geq\rho})$$

and thus, $z \mapsto zv(z) - \frac{1}{\sqrt{2\pi}} x \in \mathcal{H}_2(\mathbb{C}_{\mathrm{Re}\geq\rho}; X)$ which yields

$$(u - \chi_{\mathbb{R}_{\geq 0}} x)' = \mathcal{L}_\rho^* \left((\mathrm{i} \cdot + \rho)\, v - \frac{1}{\sqrt{2\pi}} x \right) \in L_{2,\rho}(\mathbb{R}_{\geq 0}; X),$$

i.e. $u - \chi_{\mathbb{R}_{\geq 0}} x \in H_\rho^1(\mathbb{R}; X)$, which shows that u is continuous on $\mathbb{R}_{\geq 0}$ by the Sobolev embedding theorem, Proposition 2.3. We now prove that u is indeed a mild solution. Since $u - \chi_{\mathbb{R}_{\geq 0}} x$ is continuous on $\mathbb{R}$, we infer that

$$u(0+) - x = 0,$$

and thus u attains the initial value x. Moreover,

$$\begin{aligned}
\left(\mathcal{L}_\rho E(u - \chi_{\mathbb{R}_{\geq 0}} x)'\right)(t) &= (\mathrm{i}t + \rho) E \left(\mathcal{L}_\rho (u - \chi_{\mathbb{R}_{\geq 0}} x)\right)(t) \\
&= (\mathrm{i}t + \rho) E v(\mathrm{i}t + \rho) - \frac{1}{\sqrt{2\pi}} Ex \\
&= \frac{1}{\sqrt{2\pi}}((\mathrm{i}t + \rho)E((\mathrm{i}t + \rho)E + A)^{-1} Ex - Ex) \\
&= -\frac{1}{\sqrt{2\pi}} A((\mathrm{i}t + \rho)E + A)^{-1} Ex \\
&= -A \left(\mathcal{L}_\rho u\right)(t)
\end{aligned}$$

for almost every $t \in \mathbb{R}$. Hence, $u(t) \in \mathrm{dom}(A)$ almost everywhere and

$$-Au(t) = \left(E(u - \chi_{\mathbb{R}_{\geq 0}} x)'\right)(t)$$

for almost every $t \in \mathbb{R}$. By integrating over an interval $[0, t]$, we derive

$$-\int_0^t Au(s)\,ds = Eu(t) - Ex \quad (t \geq 0)$$

and hence, u is a mild solution of (2). Thus, $x \in U$ and so, $U = \overline{\mathrm{IV}_{\mathrm{ind}(E,A)+1}}$. □

For sake of readability, we introduce the following notion.

Definition We define the space

$$V := A^{-1}\left[E\left[\overline{IV_{ind(E,A)+1}}\right]\right].$$

Remark 5.3 Note that $IV_{ind(E,A)+2} \subseteq V \subseteq \overline{IV_{ind(E,A)+2}} = \overline{IV_{ind(E,A)+1}}$ by Lemma 3.3 and Proposition 5.1.

Lemma 5.4 *Assume that $E : \overline{IV_{ind(E,A)+1}} \to Y$ is injective. Then*

$$C := E^{-1}A : V \subseteq \overline{IV_{ind(E,A)+1}} \to \overline{IV_{ind(E,A)+1}}$$

is well-defined and closed.

Proof Note that $A[V] \subseteq E[\overline{IV_{ind(E,A)+1}}]$ and thus, C is well-defined. Let $(x_n)_{n\in\mathbb{N}}$ be a sequence in V such that $x_n \to x$ and $Cx_n \to y$ in $\overline{IV_{ind(E,A)+1}}$ for some $x, y \in \overline{IV_{ind(E,A)+1}}$. We then have

$$Ax_n = ECx_n \to Ey$$

and hence, $x \in \text{dom}(A)$ with $Ax = Ey \in E\left[\overline{IV_{ind(E,A)+1}}\right]$. This shows, $x \in V$ and $Cx = E^{-1}Ax = y$, thus C is closed. □

Proposition 5.5 *Assume Hypotheses A and let X be a Hilbert space. Denote by B the generator of T. Then $E : \overline{IV_{ind(E,A)+1}} \to Y$ is injective and $B = -C$, where C is the operator defined in Lemma 5.4.*

Proof By Proposition 4.3 we have $-EB \subseteq A$. Hence, for $x \in U = \overline{IV_{ind(E,A)+1}}$ (see Proposition 5.2) and $z \in \rho(B) \cap \rho(E, A)$ we obtain

$$(zE + A)(z - B)^{-1}x = (zE - EB)(z - B)^{-1}x = Ex$$

and hence,

$$(z - B)^{-1}x = (zE + A)^{-1}Ex.$$

Thus, if $Ex = 0$ for some $x \in \overline{IV_{ind(E,A)+1}}$, we infer that $(z - B)^{-1}x = 0$ and thus, $x = 0$. Hence, E is injective and thus, C is well defined. Moreover, we observe that for $x, y \in \overline{IV_{ind(E,A)+1}}$ and $z \in \rho(E, A) \cap \rho(B)$ we have

$$x \in \text{dom}(C) \wedge (z + C)x = y \Leftrightarrow x \in \text{dom}(A) : (zE + A)x = Ey$$
$$\Leftrightarrow x = (zE + A)^{-1}Ey = (z - B)^{-1}y$$

and thus, $z \in \rho(-C)$ with $(z + C)^{-1} = (z - B)^{-1}$, which in turn implies $B = -C$. □

The converse statement also holds true, even in the case of a Banach space X.

Proposition 5.6 *Let $E : \overline{IV_{ind(E,A)+1}} \to Y$ be injective and $-C$ generate a C_0-semigroup on $\overline{IV_{ind(E,A)+1}}$, where C is the operator defined in Lemma 5.4. Then Hypotheses A holds.*

Proof Denote by T the semigroup generated by $-C$. By Proposition 4.1 and Proposition 5.1 we know that $U \subseteq \overline{\text{IV}_{\text{ind}(E,A)+1}}$. We first prove equality here. For doing so, we need to show that $T(\cdot)x$ is a mild solution of (2) for $x \in \overline{\text{IV}_{\text{ind}(E,A)+1}}$. We have

$$T(t)x + C\int_0^t T(s)x\,ds = x \quad (t \geq 0).$$

Since $EC \subseteq A$, we know that

$$\int_0^t T(s)x\,ds \in \text{dom}(A) \quad (t \geq 0)$$

and that

$$A\int_0^t T(s)x\,ds = EC\int_0^t T(s)x\,ds = Ex - ET(t)x$$

and thus, $T(\cdot)x$ is a mild solution of (2), which in turn implies $x \in U$. So, we indeed have $U = \overline{\text{IV}_{\text{ind}(E,A)+1}}$ and hence, U is closed. It remains to prove the uniqueness of mild solutions for initial values in U. So, let u_x be a mild solution for some $x \in U$. By Proposition 4.1 we know that $u_x(t) \in \overline{\text{IV}_{\text{ind}(E,A)+1}}$ for each $t \geq 0$. Hence,

$$A\int_0^t u_x(s)\,ds = Ex - Eu_x(t) \in E\left[\overline{\text{IV}_{\text{ind}(E,A)+1}}\right] \quad (t \geq 0),$$

which shows $\int_0^t u_x(s)\,ds \in V = \text{dom}(C)$. Hence,

$$C\int_0^t u_x(s)\,ds = E^{-1}A\int_0^t u_x(s)\,ds = x - u_x(t) \quad (t \geq 0),$$

i.e. u_x is a mild solution of the Cauchy problem associated with $-C$. Hence, $u_x = T(\cdot)x$, which shows the claim. □

We summarise our findings of this section in the following theorem.

Theorem 5.7 *We consider the following two statements.*

(a) *Hypotheses A holds,*
(b) $E : \overline{\text{IV}_{\text{ind}(E,A)+1}} \to Y$ *is injective and* $-C$ *generates a* C_0-*semigroup on* $\overline{\text{IV}_{\text{ind}(E,A)+1}}$, *where C is the operator defined in Lemma 5.4.*

Then (b) $\Rightarrow$ (a) and if X is a Hilbert space, then (b) $\Leftrightarrow$ (a).

The crucial condition for Hypotheses A to hold is the injectivity of $E : \overline{\text{IV}_{\text{ind}(E,A)+1}} \to Y$. It is noteworthy that $E|_{\text{IV}_{\text{ind}(E,A)+1}}$ is always injective. Indeed, if $Ex = 0$ for some $x \in \text{IV}_{\text{ind}(E,A)+1}$, we can use Lemma 3.2 to find x_1, $\ldots$, $x_{\text{ind}(E,A)+1} \in X$, $x_{\text{ind}(E,A)+2} \in \text{dom}(A)$ such that

$$(zE + A)^{-1}Ex = \frac{1}{z}x + \sum_{\ell=1}^{\text{ind}(E,A)+1} \frac{1}{z^{\ell+1}}x_\ell + \frac{1}{z^{\text{ind}(E,A)+2}}(zE + A)^{-1}Ax_{\text{ind}(E,A)+2} \quad (z \in \rho(E, A)).$$

Thus, we have $0 = z(zE + A)^{-1}Ex \to x$ as $z \to \infty$ and hence, $x = 0$. However, it is not true in general, that the injectivity carries over to the closure $\overline{\mathrm{IV}_{\mathrm{ind}(E,A)+1}}$ as the following example shows.

Example 5.8 Consider the Hilbert space $L_2(-2, 2)$ and define the operator
$$\partial^\# : \mathrm{dom}(\partial^\#) \subseteq L_2(-2, 2) \to L_2(-2, 2), \quad u \mapsto u',$$
where
$$\mathrm{dom}(\partial^\#) := \{u \in H^1(-2, 2)\,;\, u(-2) = u(2)\}.$$
It is well-known that this operator is skew-selfadjoint. We set
$$E := \chi_{[-1,1]}(\mathrm{m}), \quad A := \chi_{[-2,2]\setminus[-1,1]}(\mathrm{m}) + \partial^\#,$$
where $\chi_I(\mathrm{m})$ denotes the multiplication operator with the function χ_I on $L_2(-2, 2)$. Clearly, E is linear and bounded and A is closed linear and densely defined. Moreover, for $z \in \mathbb{C}_{\mathrm{Re}>0}$ and $u \in \mathrm{dom}(\partial^\#)$ we obtain

$$\begin{aligned}
\mathrm{Re}\langle(zE + A)u, u\rangle &= \mathrm{Re}\langle z\chi_{[-1,1]}(\mathrm{m})u + \chi_{[-2,2]\setminus[-1,1]}(\mathrm{m})u, u\rangle \\
&= \mathrm{Re}\, z\|\chi_{[-1,1]}(\mathrm{m})u\|_{L_2(-2,2)}^2 + \|\chi_{[-2,2]\setminus[-1,1]}(\mathrm{m})u\|_{L_2(-2,2)}^2 \\
&\geq \min\{\mathrm{Re}\, z, 1\}\|u\|_{L_2(-2,2)}^2,
\end{aligned}$$

where we have used the skew-selfadjointness of $\partial^\#$ in the first equality. Hence, we have
$$\|u\|_{L_2(-2,2)} \leq \frac{1}{\min\{\mathrm{Re}\, z, 1\}} \|(zE + A)u\|_{L_2(-2,2)},$$
which proves the injectivity of $(zE + A)$ and the continuity of its inverse. Since the same argumentation works for the adjoint $(zE + A)^*$, it follows that $(zE + A)^{-1} \in L(L_2(-2, 2))$ with
$$\|(zE + A)^{-1}\| \leq \frac{1}{\min\{\mathrm{Re}\, z, 1\}} \quad (z \in \mathbb{C}_{\mathrm{Re}>0}).$$
Hence, (E, A) satisfies Hypotheses B on $\mathbb{C}_{\mathrm{Re}\geq\rho_0}$ for each $\rho_0 > 0$ with $\mathrm{ind}(E, A) = 0$. Moreover, we have
$$u \in \mathrm{IV}_1 = A^{-1}[E[\mathrm{dom}(A)]]$$
if and only if $u \in \mathrm{dom}(\partial^\#)$ and
$$\chi_{[-2,2]\setminus[-1,1]}(\mathrm{m})u + u' = \chi_{[-1,1]}(\mathrm{m})v$$
for some $v \in \mathrm{dom}(\partial^\#)$. The latter is equivalent to $u \in \mathrm{dom}(\partial^\#) \cap H^2(-1, 1)$ and

$$u(t) + u'(t) = 0 \quad (t \notin [-1, 1] \text{ a.e.}).$$

Thus, we have

$$\text{IV}_1 = \Big\{ u \in \text{dom}(\partial^\#) \cap H^2(-1,1)\,;\ \exists c \in \mathbb{R}\,: $$
$$u(t) = c\left(\chi_{[-2,-1]}(t)e^{-t} + \chi_{[1,2]}(t)e^{4-t}\right) \quad (t \notin [-1,1]\text{ a.e.})\Big\}.$$

In particular, we obtain that

$$v(t) := \chi_{[-2,-1]}(t)e^{-t} + \chi_{[1,2]}(t)e^{4-t} \quad (t \in (-2, 2))$$

belongs to $\overline{\text{IV}_1}$. But this function satisfies $Ev = 0$ and hence, E is not injective on $\overline{\text{IV}_1}$.

Remark 5.9 In the case $E, A \in L(X; Y)$ and $\text{ind}(E, A) = 0$, the injectivity of E carries over to $\overline{\text{IV}_1}$. Indeed, we observe that the operators

$$(nE + A)^{-1} nE = 1 - (nE + A)^{-1}$$

for $n \in \mathbb{N}$ large enough are uniformly bounded. Moreover, for $x \in \text{IV}_1$ we have

$$(nE + A)^{-1} nEx \to x \quad (n \to \infty)$$

and hence, the latter convergence carries over to $x \in \overline{\text{IV}_1}$. In particular, if $Ex = 0$ for some $x \in \overline{\text{IV}_1}$, we infer $x = 0$ and thus, E is indeed injective on $\overline{\text{IV}_1}$. So far, the author is not able to prove or disprove that the injectivity also holds for $\text{ind}(E, A) > 0$ if E and A are bounded.

Acknowledgements We thank Felix Schwenninger for pointing our attention to the concept of Wong sequences in matrix calculus. Moreover, we thank Florian Pannasch for the observation in Remark 5.9 and the anonymous referee for drawing our attention to the subject of Sobolev type equations in the Russian literature.

References

1. Arendt, W., Batty, C.J., Hieber, M., Neubrander, F.: Vector-Valued Laplace Transforms and Cauchy Problems, 2nd edn. Birkhäuser, Basel (2011)
2. Berger, T., Ilchmann, A., Trenn, S.: The quasi-Weierstrass form for regular matrix pencils. Linear Algebra Appl. **436**(10), 4052–4069 (2012)
3. Berger, T., Trenn, S.: The quasi-Kronecker form for matrix pencils. SIAM J. Matrix Anal. Appl. **33**(2), 336–368 (2012)
4. Engel, K.J., Nagel, R.: One-Parameter Semigroups for Linear Evolution Equations. Graduate Texts in Mathematics. Springer (2000)
5. Favini, A., Yagi, A.: Multivalued linear operators and degenerate evolution equations. Ann. Mat. Pura Appl. **4**(163), 353–384 (1993)

6. Favini, A., Yagi, A.: Degenerate Differential Equations in Banach Spaces. Marcel Dekker, New York, NY (1999)
7. Ilchmann, A., Reis, T. (eds.): Surveys in Differential-Algebraic Equations I. Springer, Berlin (2013)
8. Ilchmann, A., Reis, T. (eds.): Surveys in Differential-Algebraic Equations II. Springer, Cham (2015)
9. Ilchmann, A., Reis, T. (eds.): Surveys in Differential-Algebraic Equations III. Springer, Cham (2015)
10. Ilchmann, A., Reis, T. (eds.): Surveys in Differential-Algebraic Equations IV. Springer, Cham (2017)
11. Knuckles, C., Neubrander, F.: Remarks on the Cauchy problem for multi-valued linear operators. In: Partial Differential Equations. Models in Physics and Biology. Contributions to the Conference, Held in Han-sur-Lesse, Belgium, in December 1993, pp. 174–187. Akademie Verlag, Berlin (1994)
12. Kunkel, P., Mehrmann, V.: Differential-Algebraic Equations. Analysis and Numerical Solution. European Mathematical Society Publishing House, Zürich (2006)
13. Paley, R.E., Wiener, N.: Fourier transforms in the complex domain. (American Mathematical Society Colloquium Publications, vol. 19). American Mathematical Society, New York, VIII (1934)
14. Pazy, A.: Semigroups of linear operators and applications to partial differential equations. Applied Mathematical Sciences, vol. 44. Springer, New York etc., VIII (1983)
15. Picard, R.: A structural observation for linear material laws in classical mathematical physics. Math. Methods Appl. Sci. **32**(14), 1768–1803 (2009)
16. Picard, R., McGhee, D.: Partial differential equations. A unified Hilbert space approach. de Gruyter Expositions in Mathematics, vol. 55. de Gruyter, Berlin, xviii (2011)
17. Picard, R., Trostorff, S., Waurick, M.: Well-posedness via Monotonicity. An Overview. In: Arendt, W., Chill, R., Tomilov, Y. (eds.) Operator Semigroups Meet Complex Analysis, Harmonic Analysis and Mathematical Physics, Operator Theory: Advances and Applications, vol. 250, pp. 397–452. Springer International Publishing (2015)
18. Reis, T.: Consistent initialization and perturbation analysis for abstract differential-algebraic equations. Math. Control Signals Syst. **19**(3), 255–281 (2007)
19. Reis, T., Tischendorf, C.: Frequency domain methods and decoupling of linear infinite dimensional differential algebraic systems. J. Evol. Equ. **5**(3), 357–385 (2005)
20. Rudin, W.: Real and Complex Analysis. Mathematics Series. McGraw-Hill (1987)
21. Schaefer, H.H.: Topological Vector Spaces, vol. 3. Springer, New York, NY (1971)
22. Showalter, R.E.: Degenerate parabolic initial-boundary value problems. J. Differ. Equ. **31**, 296–312 (1979)
23. Sviridyuk, G.A., Fedorov, V.E.: Linear Sobolev Type Equations and Degenerate Semigroups of Operators. Inverse and Ill-posed Problems Series. VSP, Utrecht (2003)
24. Thaller, B., Thaller, S.: Factorization of degenerate Cauchy problems: the linear case. J. Oper. Theory **36**(1), 121–146 (1996)
25. Thaller, B., Thaller, S.: Semigroup theory of degenerate linear Cauchy problems. Semigroup Forum **62**(3), 375–398 (2001)
26. Trostorff, S.: Exponential Stability and Initial Value Problems for Evolutionary Equations. Habilitation thesis, TU Dresden (2018). arXiv: 1710.08750
27. Trostorff, S., Waurick, M.: On higher index differential-algebraic equations in infinite dimensions. In: The Diversity and Beauty of Applied Operator Theory, Operator Theory: Advances and Applications, vol. 268, pp. 477–486. Birkhäuser/Springer, Cham (2018)
28. Trostorff, S., Waurick, M.: On differential-algebraic equations in infinite dimensions. J. Differ. Equ. **266**(1), 526–561 (2019)
29. Yosida, K.: Functional Analysis, 6th edn. Springer (1995)

Positive Degenerate Holomorphic Groups of the Operators and Their Applications

Sophiya A. Zagrebina and Natalya N. Solovyova

Abstract In this paper, we study degenerate holomorphic groups of the operators generated by the linear and continuous operators L and M, where the operator M is (L, p)-bounded. Necessary and sufficient conditions for the positivity of such groups are found. Using these groups, we obtain positive solutions to linear homogeneous and inhomogeneous Sobolev type equations. As an example, positive degenerate holomorphic groups are considered in Sobolev spaces of sequences.

Keywords Positive degenerate holomorphic groups of operators · Linear Sobolev type equations · Positive solutions · Sobolev spaces of sequences

1 Introduction

The paper [1] was the first to consider one-parameter degenerate holomorphic groups of operators $\{U^t : t \in \mathbf{R}\}$ as the subject of investigation. The developed theory of such groups is presented in [2, Ch. 4]. These groups are generated by a pair of the operators L and M, where the operator $L \in \mathscr{L}(\mathscr{U}; \mathscr{F})$ (i.e., L is linear and bounded), and the operator $M \in \mathscr{C}l(\mathscr{U}; \mathscr{F})$ (i.e., M is linear, closed, and densely defined), $\mathscr{U}$ and $\mathscr{F}$ are Banach spaces. Moreover, the operator M must be (L, p)-bounded, i.e. the L-spectrum $\sigma^L(M) = \mathbf{C} \setminus \rho^L(M)$ is bounded, and the point ∞ is a pole of the order p of the L-resolvent $(\mu L - M)^{-1}$ of the operator M. (Here $\rho^L(M) = \{\mu \in \mathbf{C} : (\mu L - M)^{-1} \in \mathscr{L}(\mathscr{F}; \mathscr{U})\}$ is the L-resolvent set of the operator M).

S. A. Zagrebina (✉) · N. N. Solovyova
South Ural State University, Lenina, 76, Chelyabinsk, Russia
e-mail: zagrebinasa@susu.ru

N. N. Solovyova
e-mail: nsolowjowa@mail.ru

The main properties of degenerate holomorphic groups are as follows:

(i) the group $\{U^t : t \in \mathbf{R}\}$ is holomorphically extendable to the entire complex plane while the group property $U^t U^s = U^{t+s}$ holds true;

(ii) the unit of the group is the projector $P \in \mathscr{L}(\mathscr{U})$ (i.e. $U^0 = P$), which is equal to the identity operator $\mathbf{I}$, if the operator $L^{-1} \in \mathscr{L}(\mathscr{F}; \mathscr{U})$ exists;

(iii) the group $\{U^t : t \in \mathbf{R}\}$ is a resolving group of the linear homogeneous Sobolev type equation

$$L\dot{u} = Mu, \qquad (1)$$

i.e. the vector-function $u(t) = U^t u_0$ is a solution to equation (1) for any $u_0 \in \mathscr{U}$.

The scheme for constructing a positive degenerate group and its connection with the operators L and M will be formulated in the next section.

The article describes all the properties of degenerate holomorphic groups of operators and gives conditions under which the Cauchy problem

$$u(0) = u_0 \qquad (2)$$

and the Showalter–Sidorov problem [3]

$$P(u(0) - u_0) = 0, \qquad (3)$$

are uniquely solvable not only for Eq. (1), but also for the linear inhomogeneous Sobolev type equation

$$L\dot{u} = Mu + f. \qquad (4)$$

The main result of the article is the description of the necessary and sufficient conditions under which degenerate holomorphic groups of operators are positive. Here we use the approach proposed in the fundamental monograph [4]. We also note that the work [5] was the first to study in this direction.

As applications of the obtained abstract results, we construct positive solutions to problems (2), (3) for Eq. (4), where the operators $L = \mathrm{diag}\{L_k(\lambda)\}$, $M = \mathrm{diag}\{M_k(\lambda)\}$, while $L_k(\lambda)$ and $M_k(\lambda)$ are polynomials with real coefficients such that

$$\deg L_k(\lambda) \geq \deg M_k(\lambda) \quad \forall k \in \mathbf{N}. \qquad (5)$$

Moreover, we take the Sobolev spaces of sequences [6, 7]

$$l_q^m = \left\{ u = \{u_k\} : \sum_{k=0}^{\infty} \lambda_k^{\frac{mq}{2}} |u_k|^q < \infty \right\}, \quad m \in \mathbf{R}, \ q \in [1, +\infty) \qquad (6)$$

as the Banach spaces $\mathscr{U}$ and $\mathscr{F}$, $\{\lambda_k\} \subset \mathbf{R}_+$ is a monotone sequence.

Note that many mathematical models (for instance, see [9–12]) satisfy relation (5) after applying the Fourier transform. In the future, we intend to extend the approach applied here to Sobolev type stochastic equations [13] and Sobolev type equations defined on sets of different geometric structures [14, 15], as well as to consider more general initial-final value conditions [16] instead of conditions (2), (3).

2 Positive Degenerate Holomorphic Groups of Operators

Let $\mathscr{U}$ and $\mathscr{F}$ be Banach spaces, the operators $L \in \mathscr{L}(\mathscr{U}; \mathscr{F})$ (i.e., L is linear and bounded), $M \in \mathscr{C}l(\mathscr{U}; \mathscr{F})$ (i.e., M is linear, closed, and densely defined). The sets $\rho^L(M) = \{\mu \in \mathbf{C} : (\mu L - M)^{-1} \in \mathscr{L}(\mathscr{F}; \mathscr{U})\}$ and $\sigma^L(M) = \mathbf{C} \setminus \rho^L(M)$ are called the *L-resolvent set* and the *L-spectrum of the operator M*, respectively. The operator M is called (L, σ)-bounded, if

$$\exists a \in \mathbf{R}_+ \; \forall \mu \in \mathbf{C} \; (|\mu| > a) \Rightarrow (\mu \in \rho^L(M)). \tag{7}$$

If the operator M is (L, σ)-bounded, then there exist the projectors

$$P = \frac{1}{2\pi i} \int_\gamma R^L_\mu(M) d\mu \in \mathscr{L}(\mathscr{U}), \tag{8}$$

$$Q = \frac{1}{2\pi i} \int_\gamma L^L_\mu(M) d\mu \in \mathscr{L}(\mathscr{F}). \tag{9}$$

Here $R^L_\mu(M) = (\mu L - M)^{-1} L$ and $L^L_\mu(M) = L(\mu L - M)^{-1}$ are the *right* and *left L-resolvents of the operator M*, respectively; the contour $\gamma = \{\mu \in \mathbf{C} : |\mu| = r > a\}$. Hereinafter, the contour integrals are understood in the sense of Riemann. Consider the subspaces $\mathscr{U}^0 = \ker P$, $\mathscr{U}^1 = \operatorname{im} P$, $\mathscr{F}^0 = \ker Q$, $\mathscr{F}^1 = \operatorname{im} Q$; and denote by L_k (M_k) the restriction of the operator L (M) on $\mathscr{U}^k$ ($\mathscr{U}^k \cap \operatorname{dom} M$), $k = 0, 1$. The following theorem is true.

Theorem 1 [2, Ch. 4.1] *Let the operator M be (L, σ)-bounded. Then*
(i) the operator $L_k \in \mathscr{L}(\mathscr{U}^k; \mathscr{F}^k)$, $k = 0, 1$; moreover, there exists the operator $L_1^{-1} \in \mathscr{L}(\mathscr{F}^1; \mathscr{U}^1)$;
(ii) the operator $M_k \in \mathscr{C}l(\mathscr{U}^k; \mathscr{F}^k)$, $k = 0, 1$; moreover, there exists the operator $M_0^{-1} \in \mathscr{L}(\mathscr{F}^0; \mathscr{U}^0)$.

Let us construct the operators $H = M_0^{-1} L_0 \in \mathscr{L}(\mathscr{U}^0)$, $S = L_1^{-1} M_1 \in \mathscr{L}(\mathscr{U}^1)$. The operator M is called (L, p)-bounded, $p \in \mathbf{N}$, if $H^p \neq \mathbf{O}$, while $H^{p+1} = \mathbf{O}$, and the operator M is called $(L, 0)$-bounded, if $H = \mathbf{O}$.

Corollary 1 [2, Ch. 4.1] *Let the operator M be (L, p)-bounded, $p \in \{0\} \cup \mathbf{N}$. Then the L-resolvent $(\mu L - M)^{-1}$ of the operator M can be expanded in the Laurent series in the ring $|\mu| > a$:*

$$(\mu L - M)^{-1} = \sum_{k=1}^{\infty} \mu^{-k} S^{k-1} L_1^{-1} Q - \sum_{k=0}^{p} \mu^k H^k M_0^{-1} (\mathbf{I} - Q). \tag{10}$$

Let $\mathscr{B}$ be a Banach space. Recall (see [1], [2, Ch. 4]) that the map $V^\bullet \in C^\infty(\mathbf{R}; \mathscr{B})$ is called a *group*, if

$$V^s V^t = V^{s+t} \quad for \ any \ s, t \in \mathbf{R}. \tag{11}$$

The group $V^\bullet$ is usually identified with its graph $\{V^t : t \in \mathbf{R}\}$. The group $\{V^t : t \in \mathbf{R}\}$ is called holomorphic if the group is holomorphically extendable to the entire complex plane $\mathbf{C}$ such that property (11) holds true. The group $\{V^t : t \in \mathbf{R}\}$ is called degenerate if its unit V^0 is a projector.

Theorem 2 *Let the operator M be (L, p)-bounded, $p \in \{0\} \cup \mathbf{N}$, and $\ker L \neq \{0\}$. Then there exist degenerate holomorphic groups of the operators*

$$U^t = \frac{1}{2\pi i} \int_\gamma R_\mu^L(M) e^{\mu t} d\mu, \quad t \in \mathbf{R}, \tag{12}$$

$$F^t = \frac{1}{2\pi i} \int_\gamma L_\mu^L(M) e^{\mu t} d\mu, \quad t \in \mathbf{R}. \tag{13}$$

U^t and F^t are defined on the spaces $\mathscr{U}$ and $\mathscr{F}$, respectively. Here contour $\gamma \in \mathbf{C}$ is the same as in (8),(9).

Note that the units of these groups are projectors (8), (9), i.e. $U^0 = P$ and $F^0 = Q$.

Corollary 2 *Let the conditions of Theorem 2 be satisfied. Then*

$$U^t = \lim_{n \to \infty} (n(nL - tM)^{-1} L)^n, \quad F^t = \lim_{n \to \infty} (nL(nL - tM)^{-1})^n$$

for all $t \in \mathbf{R} \setminus \{0\}$.

Proof It follows from (10) and (12)–(13) that

$$U^t = \frac{1}{2\pi i} \int_\gamma (\mathbf{I} - S)^{-1} P e^{\mu t} d\mu = \sum_{k=0}^\infty \frac{(St)^k}{k!} P = e^{St} P.$$

Further,

$$(R_\mu^L(M))^{p+1} = (\mu \mathbf{I} - S)^{-p-1} P.$$

Hence, using $\mu = \dfrac{n}{t}$, we obtain

$$\lim_{n \to \infty} (n(nL - tM)^{-1} L)^n = \lim_{n \to \infty} \left(\mathbf{I} - \frac{t}{n} S\right)^{-n} P = e^{St} P.$$

The statement for the group $\{F^t : t \in \mathbf{R}\}$ is proved similarly. ●

Note that the proof of Corollary 2 implies

$$((p+1)((p+1)L - tM)^{-1}L)^{p+1} = \left(\mathbf{I} - \frac{t}{p+1}S\right)^{-p-1} P.$$

Therefore,

$$\lim_{t \to 0}(n(nL - tM)^{-1}L)^n = \lim_{t \to 0}\left(\mathbf{I} - \frac{t}{n}S\right)^{-n} P = P$$

for all $n \geq p + 1$. Similarly,

$$\lim_{t \to 0}(nL(nL - tM)^{-1})^n = \lim_{t \to 0}\left(\mathbf{I} - \frac{t}{n}T\right)^{-n} Q = Q$$

for all $n \geq p + 1$. Here the operator $T = M_1 L_1^{-1} \in \mathscr{L}(\mathscr{F}^1)$. We also note that the proofs of all unproven statements can be found in [1] and [2, Ch. 4].

Further, the vector space $\mathscr{B}$ is called a *Riesz space* [6, Ch. 2], if $\mathscr{B}$ is endowed with the order relation $\geq$ (satisfying the axioms of reflexivity, transitivity, and antisymmetry), which is consistent with the vector structure, i.e.

$$(x \geq y) \Rightarrow (x + z \geq y + z) \quad \text{for all} \quad x, y, z \in \mathscr{B}, \quad \text{and}$$

$$(x \geq y) \Rightarrow (\alpha x \geq \alpha y) \quad \text{for all} \quad x, y \in \mathscr{B} \quad \text{and} \quad \alpha \in \overline{\mathbf{R}_+}, \quad \text{where} \quad \overline{\mathbf{R}_+} = \{0\} \cup \mathbf{R}_+.$$

The Riesz space $\mathscr{B}$ is called a *functional Riesz space*, if the elements $x_+ = max\{x, 0\}$ and $x_- = min\{-x, 0\}$ can be defined for any $x \in \mathscr{B}$, where $x_+, x_- \in \mathscr{B}$. The functional Riesz space $\mathscr{B}$ is called a *normalized functional Riesz space*, if $\mathscr{B}$ is endowed with the norm $||\cdot||_\mathscr{B}$ such that

$$(|x| \geq |y|) \Rightarrow (||x||_\mathscr{B} \geq ||y||_\mathscr{B}) \quad \text{for all} \quad x, y \in \mathscr{B}. \tag{14}$$

Here $|x| = x_+ + x_-$. A complete normalized functional Riesz space is called a *Banach lattice*.

Let $\mathscr{B}$ be a Banach space. The convex set $\mathscr{C} \subset \mathscr{B}$ is called a *cone*, if $\mathscr{C} + \mathscr{C} = \mathscr{C}$ and $\alpha \mathscr{C} \subset \mathscr{C}$ for all $\alpha \in \overline{\mathbf{R}_+}$. The cone $\mathscr{C}$ is called *proper*, if $\mathscr{C} \cap (-\mathscr{C}) = \{0\}$, and *generative*, if $\mathscr{B} = \mathscr{C} - \mathscr{C}$, i.e. for each vector $x \in \mathscr{B}$ there exist the vectors $y, z \in \mathscr{C}$ such that $x = y - z$. If there exists the proper generative cone $\mathscr{C}$ in the Banach space $\mathscr{B}$, then the order relation $\geq$ such that $(x \geq y) \Leftrightarrow (x - y \in \mathscr{C})$ can be defined on $\mathscr{B}$. Due to the properties of the cone $\mathscr{C}$, the order relation $\geq$ is consistent with the vector structure of the space $\mathscr{B}$, and if the vector $|x| \in \mathscr{B}$ satisfying (14) can be defined for each vector $x \in \mathscr{B}$, then the space $\mathscr{B}$ is a Banach lattice. On the other hand, if $\mathscr{B}$ is a Banach lattice, then the set $\mathscr{B}_+ = \{x \in \mathscr{B} : x \geq 0\}$ is a proper generative cone.

Let $\mathscr{B} = (\mathscr{B}, \mathscr{C})$ be a Banach lattice. Here $\mathscr{C}$ is a proper generative cone. Note that, generally speaking, $\mathscr{C}$ may not coincide with the *canonical cone* $\mathscr{B}_+$. The operator $A \in \mathscr{L}(\mathscr{B})$ is called positive, if $Ax \geq 0$ for any $x \in \mathscr{C}$. The group of operators $V^\bullet = \{V^t : t \in \mathbf{R}\}$ acting on the space $\mathscr{B}$ is called *positive*, if $V^t x \geq 0$ for any $x \in \mathscr{C}$ and $t \in \mathbf{R}$. If $V^\bullet$ is a degenerate group, then its unit V^0 is a projector that splits the space $\mathscr{B}$ into the direct sum $\mathscr{B} = \mathscr{B}^0 \oplus \mathscr{B}^1$, where $\mathscr{B}^0 = \ker V^0$ and $\mathscr{B}^1 = \operatorname{im} V^0$. Since $V^t = V^0 V^t V^0$, then $\mathscr{B}^0 = \ker V^t$, and $\mathscr{B}^1 = \operatorname{im} V^t$. Therefore, we can define $\ker V^\bullet = \mathscr{B}^0$ and $\operatorname{im} V^\bullet = \mathscr{B}^1$. If, in addition, the degenerate group $V^\bullet$ is positive, then $\mathscr{B}^1$ is a Banach lattice with the proper generative cone $\mathscr{C}^1 = \{x \in \mathscr{C} : V^0 x = x\} = \mathscr{B}^1 \cap \mathscr{C}$. If the space $\mathscr{B}^0$ is also a Banach lattice with the proper generative cone $\mathscr{C}^0$ and the order relation $\succeq$, then the cone $\mathscr{C}^* = \mathscr{C}^0 \oplus \mathscr{C}^1$ can generate a new Banach lattice of the space $\mathscr{B}$ with the order relation $\ominus$, i.e.

$$(x \ominus y) \leftrightarrow (x_0 \succeq y_0) \wedge (x_1 \geq y_1).$$

Finally, we arrived at the main goal of this section. Let us describe the conditions under which the degenerate holomorphic groups $U^\bullet = \{U^t : t \in \mathbf{R}\}$ and $F^\bullet = \{F^t : t \in \mathbf{R}\}$ are positive. Further, we consider the Banach space $\mathscr{U}(\mathscr{F})$ to be the Banach lattice $\mathscr{U} = (\mathscr{U}, \mathscr{C}_\mathscr{U})$ ($\mathscr{F} = (\mathscr{F}, \mathscr{C}_\mathscr{F})$), where $\mathscr{C}_\mathscr{U}$ ($\mathscr{C}_\mathscr{F}$) is a proper generative cone.

Theorem 3 *Let the operator M be (L, p)-bounded, $p \in \{0\} \cup \mathbf{N}$. Then the following statements are equivalent.*
  (i) $(\mu R_\mu^L(M))^{p+1}$ $((\mu L_\mu^L(M))^{p+1})$ *is positive for all sufficiently large* $\mu \in \mathbf{R}_+$.
  (ii) *The degenerate holomorphic group $U^\bullet$ ($F^\bullet$) is positive.*

Proof It follows from Corollary 2 that

$$U^t = \begin{cases} \lim_{n\to\infty} (n(nL - tM)^{-1} L)^n & \text{for } t \in \mathbf{R} \setminus \{0\}; \\ P & \text{for } t = 0. \end{cases} \quad (15)$$

Use $\mu = \dfrac{n}{t}$ to obtain the desired statement. For the group $F^\bullet$, the statement is proved similarly. ∎

3 Positive Solutions to Linear Sobolev Type Equation

Let $\mathscr{U} = (\mathscr{U}, \mathscr{C}_\mathscr{U})$ and $\mathscr{F} = (\mathscr{F}, \mathscr{C}_\mathscr{F})$ be Banach lattices, where $\mathscr{C}_\mathscr{U}$ and $\mathscr{C}_\mathscr{F}$ are proper generative cones. Let the operators $L \in \mathscr{L}(\mathscr{U}, \mathscr{F})$ and $M \in \mathscr{Cl}(\mathscr{U}, \mathscr{F})$. Consider the linear homogeneous Sobolev type equation

$$L\dot{u} = Mu. \quad (16)$$

The vector-function $u \in C^1(\mathbf{R}, \mathscr{U})$ satisfying this equation is called a *solution to equation* (16). The solution $u = u(t)$ is called a *solution to the Cauchy problem*, if $u = u(t)$ satisfies the condition

$$u(0) = u_0 \qquad (17)$$

for some $u_0 \in \mathscr{U}$. The solution $u = u(t)$ is called a *solution to the Showalter–Sidorov problem*, if $u = u(t)$ satisfies the condition

$$P(u(0) - u_0) = 0. \qquad (18)$$

It is easy to see that the vector-function $u(t) = U^t u_0$ is a solution to equation (16) and to problem (18) for any $u_0 \in \mathscr{U}$, where $\{U^t : t \in \mathbf{R}\}$ is a degenerate holomorphic group of operators of the form (12), (13). The question on the existence and uniqueness of solution to problem (16), (17) and the question of the uniqueness of solution to problem (16), (18) arise. In order to answer these questions, recall that the set $\mathscr{B} \subset \mathscr{U}$ is called the phase space of equation (16) if any its solution $u(t) \in \mathscr{B}$ for any $t \in \mathbf{R}$; and there exists the unique solution $u \in C^1(\mathbf{R}, \mathscr{U})$ to problem (17) for Eq. (16) for any $u_0 \in \mathscr{B}$. The following theorem is true.

Theorem 4 [2, Ch. 4] *Let the operator M be (L, p)-bounded, $p \in \{0\} \cup \mathbf{N}$. Then*
(i) *the phase space of equation (16) is the subspace $\mathscr{U}^1$;*
(ii) *for any $u_0 \in \mathscr{U}$, there exists the unique solution $u = u(t)$ to problem (16), (18), and the solution has the form $u(t) = U^t u_0$.*

Theorem 4 gives a complete answer to both questions posed. However, we give a remark about the uniqueness of solution to problem (16), (18). By virtue of Statement (i) of Theorem 4, any solution to equation (16) belongs to the space $\mathscr{U}^1$ pointwise, i.e. $u(t) \in \mathscr{U}^1$ for all $t \in \mathbf{R}$. Therefore, if we represent an arbitrary initial vector u_0 as $u_0 = u_0^0 + u_0^1$, where $u_0^k \in \mathscr{U}^k$, $k = 0, 1$ (according to Theorem 1), then the solution $u(t) = U^t u_0$ to problem (16), (18) is also the unique solution to this problem under the initial condition $v_0 = v_0^0 + u_0^1$, where the vector $v_0^0 \in \mathscr{U}^0$ is arbitrary. This remark is important in order to correctly understand the reasoning given below.

Let $\mathscr{U} = (\mathscr{U}, \mathscr{C}_\mathscr{U})$ be a Banach lattice and the group $U^\bullet = \{U^t : t \in \mathbf{R}\}$ be positive. It follows from Statement (i) of Theorem 4 that the phase space of equation (16) is also a Banach lattice, i.e. $\mathscr{U}^1 = (\mathscr{U}^1, \mathscr{C}_\mathscr{U}^1)$, where the proper generative cone $\mathscr{C}_\mathscr{U}^1 = \mathscr{U}^1 \cap \mathscr{C}_\mathscr{U} \neq \{0\}$. Therefore, the following corollary is true.

Corollary 3 *Let the condition of Theorem 4 be satisfied, $\mathscr{U}$ be a Banach lattice, and the degenerate holomorphic group $U^\bullet$ be positive. Then, for any $u_0 \in \mathscr{C}_\mathscr{U}^1$, there exists the unique positive solution $u = u(t)$ to problem (16), (17), and the solution has the form $u(t) = U^t u_0$.*

Consider positive solutions to problem (16), (18). According to the conditions of Corollary 3, the solution $u(t) = U^t u_0$ to problem (16), (18) is positive for any initial vector $u_0 \in \mathscr{U}$ such that $P u_0 \in \mathscr{C}_\mathscr{U}^1$. Therefore, the following corollary is true.

Corollary 4 *Let the conditions of Corollary 3 be satisfied. Then, for any $u_0 \in \mathscr{U}$ such that $P u_0 \in \mathscr{C}_\mathscr{U}^1$, there exists the unique positive solution $u = u(t)$ to problem (16), (18), and the solution has the form $u(t) = U^t u_0$.*

In order to illustrate this result, we give an example.

Example 1 [17] Let the spaces $\mathscr{U} = \mathscr{F} = \mathbf{R}^2$, and the operators $L, M \in \mathscr{L}(\mathscr{U}, \mathscr{F})$ be given by the matrices

$$L = \begin{pmatrix} 1 & 1 \\ 1 & 1 \end{pmatrix}, \quad M = \begin{pmatrix} 0 & 1 \\ 1 & 0 \end{pmatrix}.$$

Construct the L-resolvent $(\mu L - M)^{-1}$ of the operator M:

$$(\mu L - M)^{-1} = (2\mu - 1)^{-1} \begin{pmatrix} \mu & 1 - \mu \\ 1 - \mu & \mu \end{pmatrix}.$$

Therefore, the operator M is $(L, 0)$-bounded, and the L-spectrum $\sigma^L(M) = \left\{\frac{1}{2}\right\}$. Moreover, since the right L-resolvent $R_\mu^L(M) = (\mu L - M)^{-1} L$ has the form

$$R_\mu^L(M) = (2\mu - 1)^{-1} \begin{pmatrix} 1 & 1 \\ 1 & 1 \end{pmatrix},$$

then the degenerate holomorphic group of operators $U^\bullet = \{U^t : t \in \mathbf{R}\}$ has the form

$$U^t = \frac{1}{2\pi i} \int_{|\mu - \frac{1}{2}|=1} R_\mu^L(M) e^{\mu t} d\mu = \frac{e^{\frac{t}{2}}}{2} \begin{pmatrix} 1 & 1 \\ 1 & 1 \end{pmatrix},$$

and its unit is the projector

$$P = \frac{1}{2} \begin{pmatrix} 1 & 1 \\ 1 & 1 \end{pmatrix}.$$

(By the way, the same is true for the degenerate holomorphic group $F^\bullet = \{F^t : t \in \mathbf{R}\}$).

Next, we use the relation $Pu = u$ in order to find the phase space $\mathscr{U}^1$ of equation (16) in this case, and use the relation $Pu = 0$ in order to find the kernel $\mathscr{U}^0$ of the group $U^\bullet$:

$$\mathscr{U}^1 = \{u \in \mathbf{R}^2 : u_2 - u_1 = 0\}, \quad \mathscr{U}^0 = \{u \in \mathbf{R}^2 : u_2 + u_1 = 0\}.$$

Introduce the order relation $\geq$ on the space $\mathscr{U}$, which defines the canonical cone $\mathscr{C}_{\mathscr{U}} = \overline{\mathbf{R}_2} \times \overline{\mathbf{R}_2}$, by the formula $(u = col(u_1, u_2) \geq v = col(v_1, v_2)) \Leftrightarrow ((u_1 \geq v_1) \wedge (u_2 \geq v_2))$. It is easy to see that the operator $R_\mu^L(M)$ is positive for $\mu \in \left(\frac{1}{2}, +\infty\right)$, and the group $U^\bullet$ is also positive (this fact is not surprising due to Theorem 3!). Therefore, on $\mathscr{U}^1$, there exists the order relation $\geq$ generated by the cone $\mathscr{C}_{\mathscr{U}}^1 = \mathscr{U}^1 \cap \mathscr{C}_{\mathscr{U}} = \{u \in \mathbf{R}^2 : u_2 = u_1 \geq 0\}$. However, on the subspace $\mathscr{U}^0$, we can also define the cone $\mathscr{C}_{\mathscr{U}}^0 = \{u \in \mathbf{R}^2 : u_2 = -u_1 \geq 0\}$. The direct sum $\mathscr{C}_{\mathscr{U}}^0 \oplus \mathscr{C}_{\mathscr{U}}^1$ is the proper generative cone $\mathscr{C}_\mathscr{U}^*$, which generates the order relation $\succeq$ on $\mathbf{R}^2$ such that $(u \succeq 0) \Leftrightarrow ((u_2 - u_1 \geq 0) \wedge (u_2 + u_1 \geq 0))$. Note that both the right L-resolvent $R_\mu^L(M)$ and the group $U^\bullet$ are positive in the sense of the order relation $\succeq$.

Example 2 [17] Let the operators

$$L = \begin{pmatrix} 1 & 1 \\ 1 & 1 \end{pmatrix}, \quad M = \begin{pmatrix} -1 & 0 \\ 0 & -1 \end{pmatrix}$$

be given in space $\mathbf{R}^2$ ordered by the canonical cone $\overline{\mathbf{R}_+} \times \overline{\mathbf{R}_+}$. In this case, it is easy to see that neither the right L-resolvent of the operator M

$$R_\mu^L(M) = (2\mu - 1)^{-1} \begin{pmatrix} 1 & -1 \\ -1 & 1 \end{pmatrix},$$

nor the resolving degenerate holomorphic group

$$U^t = \frac{e^{\frac{t}{2}}}{2} \begin{pmatrix} 1 & -1 \\ -1 & 1 \end{pmatrix}$$

are positive in the sense of the canonical order relation $(u \geq 0) \Leftrightarrow (u_1 \geq 0) \wedge (u_2 \geq 0)$. Moreover, the phase space $\mathscr{U}^1 = \{u \in \mathscr{U} : u_2 + u_1 = 0\}$ intersects the canonical cone $\overline{\mathbf{R}_+} \times \overline{\mathbf{R}_+} = \mathscr{C}_\mathscr{U}$ at only one point $\mathscr{U}^1 \cap \mathscr{C}_\mathscr{U} = \{0\}$. However, as shown in Example 1, the canonical cone $\overline{\mathbf{R}_+} \times \overline{\mathbf{R}_+}$ can be replaced by another. Moreover, both the right L-resolvent $R_\mu^L(M)$ of the operator M and the degenerate holomorphic group are positive with respect to the new order relation $\succeq$. Indeed, define the cone $\mathscr{C}_\mathscr{U}^0 = \{u \in \mathbf{R}^2 : -u_2 = -u_1 \geq 0\}$ on the kernel $\ker U^\bullet = \{u \in \mathbf{R}^2 : u_2 - u_1 = 0\} = \mathscr{U}_0$. The direct sum $\mathscr{C}_\mathscr{U}^0 \oplus \mathscr{C}_\mathscr{U}^1 = \mathscr{C}_\mathscr{U}^* = \{u \in \mathbf{R}^2 : (u \succeq 0) \Leftrightarrow ((-u_2 - u_1 \geq 0) \wedge (u_2 - u_1 \geq 0))\}$. It is easy to see that both the right L-resolvent $R_\mu^L(M)$ of the operator M and the group of the operators $U^\bullet$ are positive with respect to the order relation $\succeq$.

We considered the positive solutions to problems (17), (18) for linear homogeneous equation (16) in sufficient detail. Let us find positive solutions to the linear inhomogeneous Sobolev type equation

$$L\dot{u} = Mu + f. \tag{19}$$

Let $f \in C(\mathscr{I}; \mathscr{F})$ be a vector-function defined on the interval $\mathscr{I} \subset \mathbf{R}$ containing the zero point, and taking values in the Banach space $\mathscr{F}$. The vector-function $u \in C^1(\mathscr{I}; \mathscr{U})$ is called a *solution* to equation (19), if u satisfies this equation. The solution $u = u(t)$ to Eq. (19) is called a solution to problem (19), (17) (to problem (19), (18)), if $u = u(t)$ satisfies condition (17) (condition (18)). The following theorem is true.

Theorem 5 [1], [2, Ch. 5] *Let the operator M be (L, p)-bounded, $p \in \{0\} \cup \mathbf{N}$. Then for any initial vector $u_0 \in U$ and any vector-function $f = f(t)$ such that $(\mathbf{I} - Q)f \in$*

$C^{p+1}(\mathcal{I}; \mathcal{U}^0)$ and $Qf \in C(\overline{\mathcal{I}}; \mathcal{U}^1)$, there exists the unique solution $u = u(t)$ to problem (19), (18), and the solution has the form

$$u(t) = U^t u_0 + \int_0^t U^{t-\tau} L_1^{-1} f^1(\tau) d\tau - \sum_{k=0}^{p} H^k M_0^{-1} f^{0(k)}(t), \quad t \in \mathcal{I}. \quad (20)$$

If, in addition, the initial vector $u_0 \in U$ is such that

$$(\mathbf{I} - P)u_0 = -\sum_{k=0}^{p} H^k M_0^{-1} f^{0(k)}(0), \quad (21)$$

then there exists the unique solution $u = u(t)$ to problem (19),(17), and the solution has the form (20).

Here

$$f^{0(k)}(t) = \frac{d^k}{dt^k}(\mathbf{I} - Q)f(t), \quad k = \overline{0, p}, \quad f^1(t) = Qf(t).$$

Let us describe conditions under which solution (20) is positive. Let $\mathcal{U} = (\mathcal{U}, \mathcal{C}_\mathcal{U})$ and $\mathcal{F} = (\mathcal{F}, \mathcal{C}_\mathcal{F})$ be Banach lattices, where the proper generative cones $\mathcal{C}_\mathcal{U}$ and $\mathcal{C}_\mathcal{F}$ generate the order relations $\succeq_\mathcal{U}$ and $\succeq_\mathcal{F}$, respectively. Let the operators $L, M \in \mathcal{L}(\mathcal{U}, \mathcal{F})$, and the operator M be (L, p)-bounded, $p \in \{0\} \cup \mathbf{N}$. The order relations $\succeq_\mathcal{U}$ and $\succeq_\mathcal{F}$ are said to be *consistent with respect to the pair of the operators* (L, M) (in short, (L, M)-consistent), if

(i) the cones $\mathcal{C}_\mathcal{U} = \mathcal{C}_\mathcal{U}^0 \oplus \mathcal{C}_\mathcal{U}^1$ and $\mathcal{C}_\mathcal{F} = \mathcal{C}_\mathcal{F}^0 \oplus \mathcal{C}_\mathcal{F}^1$, moreover, $\mathcal{C}_\mathcal{U}^k$ and $\mathcal{C}_\mathcal{F}^k$ are proper generative cones in the spaces $\mathcal{U}^k$ and $\mathcal{F}^k$, $k = 0, 1$, respectively;

(ii) the operators $L_k, M_k \in \mathcal{L}(\mathcal{C}_\mathcal{U}^k, \mathcal{C}_\mathcal{F}^k)$, moreover, the operators $M_0^{-1} \in \mathcal{L}(\mathcal{C}_\mathcal{F}^0, \mathcal{C}_\mathcal{U}^0)$ and $L_1^{-1} \in \mathcal{L}(\mathcal{C}_\mathcal{F}^1, \mathcal{C}_\mathcal{U}^1)$.

Theorem 6 *Let $\mathcal{U} = (\mathcal{U}, \mathcal{C}_\mathcal{U})$ and $\mathcal{F} = (\mathcal{F}, \mathcal{C}_\mathcal{F})$ be Banach lattices, the operators $L, M \in \mathcal{L}(\mathcal{U}, \mathcal{F})$, moreover, the operator M be (L, p)-bounded, $p \in \{0\} \cup \mathbf{N}$. Let the order relation $\succeq_\mathcal{U}$ and $\succeq_\mathcal{F}$ be (L, M)-consistent, and $U^\bullet$ be a positive degenerate holomorphic group of operators. Then for any vector-function $f = f^0 + f^1$ such that $-f^{0(k)} \in C^1(\mathcal{I}; \mathcal{C}_\mathcal{U}^0)$ for all $k = \overline{0, p+1}$ and $f^1 \in C(\overline{\mathcal{I}}; \mathcal{C}_\mathcal{U}^1)$, and for any initial vector $u_0 \in \mathcal{U}$ such that $Pu_0 \in \mathcal{C}_\mathcal{U}^1$, there exists the unique positive solution $u = u(t)$ to problem (19), (17), and the solution has the form (20). If, in addition, the initial vector u_0 satisfies condition (21), then there exists the unique positive solution $u = u(t)$ to problem (19), (17), and the solution has the form (20).*

By virtue of Theorem 5, the proof of Theorem 6 is elementary. Let us give some remarks. First, of course, the positivity of the solution $u = u(t)$ to both problems is understood in the sense of the order relation $\succeq_\mathcal{U}$, i.e. $u(t) \succeq_\mathcal{U} 0$ for $t \in \mathcal{I}$. Second, by construction, the operator $H = M_0^{-1} L_0 \in \mathcal{L}(\mathcal{C}_\mathcal{U}^0)$. Finally, when solving problem (19), (18), the vector $(\mathbf{I} - P)u_0 \in \mathcal{U}^0$ can be any, i.e. such that $(\mathbf{I} - P)u_0 \notin \mathcal{C}_\mathcal{U}^0$.

In order to illustrate Theorem 6, we consider the following example.

Example 3 Let L^* and M^* be square matrices of the order n, and $det(\alpha L^* - M^*) \neq 0$ for some $\alpha \in \mathbf{C}$. Then there exist non-degenerate square matrices A and B with real elements such that [18, Ch. 12]

$$AL^*B = \mathrm{diag}\{\mathscr{I}_{p_1}^0, \mathscr{I}_{p_2}^0, \ldots, \mathscr{I}_{p_k}^0, \mathbf{I_{n-m}}\},$$

$$AM^*B = \mathrm{diag}\{\mathbf{I_m}, \mathbf{S}\}.$$

Here $\mathscr{I}_{p_l}^0$ is a Jordan cell of the order p_l, $\sum_{l=1}^{k} p_l = m$, with zeros on the main diagonal, $\mathbf{I_k}$ is the identity matrix of the order k, and S is a square matrix of the order $n - m$ with real elements. Consider the spaces $\mathscr{U} = \mathscr{F} = \mathbf{R^n}$ and define the operators L and M by the matrices AL^*B and AM^*B, respectively. It is easy to see that the operator M is (L, m)-bounded. If, in addition, the degenerate holomorphic group of operators $U^t = e^{tS}P$ is positive, where $P = \mathrm{diag}\{\mathbf{O_m}, \mathbf{I_{n-m}}\}$ is a projector, then we obtain positive solutions to problems (19), (17) and (19), (18) under the conditions of Theorem 6 on u_0 and f. Note that the operator $H = \mathrm{diag}\{\mathscr{I}_{p_1}^0, \mathscr{I}_{p_2}^0, \ldots, \mathscr{I}_{p_k}^0, \mathbf{O_{n-m}}\}$ in this case.

4 Positive Solutions in Sobolev Spaces of Sequences

Example 4 [17] In the strip $(0, \pi) \times \mathbf{R}$, consider the solutions $u = u(x, t)$ to the equation

$$\lambda u_t - u_{txx} = \alpha u_{xx}, \tag{22}$$

which simulates many natural processes (for instance, see [1–3]). As usual, we endow (22) with the boundary

$$u(0, t) = u(\pi, t) = 0, \quad t \in \mathbf{R}, \tag{23}$$

and initial

$$u(x, 0) = u_0(x), \quad x \in (0, \pi), \tag{24}$$

conditions. In order to consider problem (22)–(24) from the point of view proposed above, we define the spaces $\mathscr{U} = \{u \in W_2^2(0, \pi) : u(0) = u(\pi) = 0\}$ and $\mathscr{F} = L_2(0, \pi)$, where $W_2^2(0, \pi)$ is a Sobolev space, and $L_2(0, \pi)$ is a Lebesgue space. Denote by $A = \frac{\partial^2}{\partial x^2}$ the operator $A : \mathscr{U} \to \mathscr{F}$. It is easy to show that $A \in \mathscr{L}(\mathscr{U}; \mathscr{F})$. There exists the operator $A^{-1} \in \mathscr{L}(\mathscr{F}; \mathscr{U})$. (Detailed coverage of all these questions can be found in [19, 20]). Consider the operator A as a closed and densely defined operator in the space $\mathscr{F}$, i.e. $A \in \mathscr{Cl}(\mathscr{F})$, $\mathrm{dom}\, A = \mathscr{U}$ (this is possible due to the continuity and density of the embedding $\mathscr{U} \hookrightarrow \mathscr{F}$). The spectrum $\sigma(A) = \{-k^2\}$ and the eigenfunctions $\phi_k(x) = \sin kx$ can be found. By construction, $\mathscr{U}$ and $\mathscr{F}$ are Hilbert spaces with scalar products $[\cdot, \cdot]$ and $(\cdot, \cdot)$, respectively. Transform the set $\{\phi_k\}$ to be orthonormal in $\mathscr{U}$ and $\mathscr{F}$, and denote the results by $\{\xi_k\}$ and $\{\eta_k\}$, respectively. Note that $\{\xi_k\}$ ($\{\eta_k\}$) is the orthonormal basis of the space $\mathscr{U}(\mathscr{F})$.

Construct the operators $L = \lambda \mathbf{I} - A$ and $M = \alpha A$, where $\mathbf{I} : \mathscr{U} \to \mathscr{F}$ is the embedding operator. Find the L-spectrum $\sigma^L(M)$ of the operator M:

$$\sigma^L(M) = \begin{cases} \left\{ \frac{-\alpha k^2}{\lambda + k^2} : \alpha \in \mathbf{R} \setminus \{0\}, \ \lambda \notin \sigma(A), \ k \in \mathbf{N} \right\}; \\ \left\{ \frac{-\alpha k^2}{\lambda + k^2} : \alpha \in \mathbf{R} \setminus \{0\}, \ k \in \mathbf{N} \setminus \{l\}, \ \lambda = -l^2 \right\}; \\ \{0\}, \quad \text{if} \ \alpha = 0. \end{cases}$$

Further, we consider only the case of $\alpha \in \mathbf{R} \setminus \{0\}$, since this case is the most interesting. The operator M is (L, σ)-bounded. Find the projectors

$$P = \begin{cases} \displaystyle\sum_{k=1}^{\infty} [\cdot, \xi_k]\xi_k, & \text{if} \ \lambda \notin \sigma(A); \\ \displaystyle\sum_{k \neq l} [\cdot, \xi_k]\xi_k, & \text{if} \ \lambda = -l^2; \end{cases}$$

$$Q = \begin{cases} \displaystyle\sum_{k=1}^{\infty} (\cdot, \eta_k)\eta_k, & \text{if} \ \lambda \notin \sigma(A); \\ \displaystyle\sum_{k \neq l} (\cdot, \eta_k)\eta_k, & \text{if} \ \lambda = -l^2. \end{cases}$$

Further, we consider only the degenerate case of $\lambda \in \sigma(A)$, since this case is the most interesting.

Let $\alpha \in \mathbf{R} \setminus \{0\}$ and $\lambda \in \sigma(A)$, then the operator M is $(L, 0)$-bounded, and the degenerate holomorphic groups of the operators $U^\bullet$ and $F^\bullet$ are given by the formulas

$$U^t = \sum_{k \neq l} e^{\mu_k t}[\cdot, \xi_k]\xi_k, \quad F^t = \sum_{k \neq l} e^{\mu_k t}(\cdot, \eta_k)\eta_k,$$

where $\mu_k = \dfrac{-\alpha k^2}{(\lambda + k^2)}$, $k \neq l$. It is clear that, in order to study the positivity of the groups $U^\bullet$ and $F^\bullet$, we need rather difficult technique (for instance, see [21]), which is hardly possible to develop here. Therefore, in order to illustrate the results given in Sects. 2 and 3, we consider the Sobolev spaces of sequences, which can be interpreted as the spaces of Fourier coefficients of the functional spaces $\mathscr{U}$ and $\mathscr{F}$ in this example. The use of the Sobolev spaces of sequences are also justified by the fact that it was possible to transfer the theory [2] to the quasi-Banach spaces of sequences l_q, $q \in (0, 1)$, [6–8] exactly in these spaces. For the functional spaces L_q, $q \in (0, 1)$, such a transfer is impossible.

Let us describe Sobolev spaces of sequences. Let $\{\lambda_k\} \subset \mathbf{R}_+$ be a monotone sequence such that $\lim_{k\to\infty} \lambda_k = +\infty$. Construct the space

$$l_q^m = \left\{ (u_k) : \sum_{k=1}^{\infty} \lambda_k^{\frac{mq}{2}} |u_k|^q < +\infty \right\}, \quad q \in [1, +\infty).$$

The space l_q^m is a Banach space with the norm

$$\|u\|_{m,q} = \left(\sum_{k=1}^{\infty} \lambda_k^{\frac{mq}{2}} |u_k|^q \right)^{1/q}, \quad \text{where} \quad q \in [1, +\infty), \quad m \in \overline{\mathbf{R}}_+.$$

Note the continuous and dense embedding $l_q^m \hookrightarrow l_q^n$ for $m \geq n$. Let $L_k = L_k(\lambda)$ and $M_k = M_k(\lambda)$ be polynomials with real coefficients of degrees r_k and s_k, respectively, i.e. $L_k(\lambda) = \sum_{j=0}^{r_k} a_j^k \lambda^j$ and $M_k(\lambda) = \sum_{j=0}^{s_k} b_j^k \lambda^j$, $r_k \geq s_k$, $k \in \mathbf{N}$. Construct the operators $L = \text{diag}\{L_k(\lambda_k)\}$ and $M = \text{diag}\{M_k(\lambda_k)\}$. The operators $L \in \mathscr{L}(l_q^{m+2r}; l_q^m)$, $r = \max_k r_k$, $M \in \mathscr{L}(l_q^{m+2s}; l_q^m)$, $s = \max_k s_k$, $m \in \mathbf{R}_+$, $q \in [1, +\infty)$. Indeed,

$$\|Lu\|_{m,q} \leq \left(\sum_{k=1}^{\infty} \lambda_k^{\frac{mq}{2}} |L_k(\lambda_k)|^q |u_k|^q \right)^{1/q} \leq \left(\sum_{k=1}^{\infty} \lambda_k^{\frac{mq}{2}+rq} |u_k|^q \right)^{1/q} = \|u\|_{m+2r,q}.$$

Due to the limited volume of the paper, we consider only the simplest case $L_k(\lambda) = L^*(\lambda)$ and $M_k(\lambda) = M^*(\lambda)$, $k \in \mathbf{N}$. Let the operators

$$L = \text{diag}\{L^*(\lambda_k)\}, \quad M = \text{diag}\{M^*(\lambda_k)\}, \tag{25}$$

and

$$\deg L^*(\lambda) = r \geq s = \deg M^*(\lambda). \tag{26}$$

Find the L-spectrum $\sigma^L(M)$ of the operator M

$$\sigma^L(M) = \left\{ \mu_k = \frac{M^*(\lambda_k)}{L^*(\lambda_k)} : L^*(\lambda_k) \neq 0 \right\}.$$

Since the set of roots of the equation $L^*(\lambda) = 0$ is finite, it follows from (26) that the set $\sigma^L(M)$ is bounded. Therefore, the operator M is (L, σ)-bounded. Construct the right L-resolvent of the operator M

$$R_\mu^L(M) = \sum_{k=1}^{\infty} \frac{L^*(\lambda_k) <\cdot, e_k> e_k}{\mu L^*(\lambda_k) - M^*(\lambda_k)} = {\sum_k}' \frac{<\cdot, e_k> e_k}{\mu - \mu_k}.$$

The sign $\ll '\gg$ at $\ll \sum \gg$ means that the sum is taken over k such that $L^*(\lambda_k) \neq 0$; $e_k = (e_{kl})$, where $e_{kl} = 0$ for $l \neq k$, and $e_{kl} = 1$ for $l = k$; the functional $< \cdot, e_k >$ has the form $< u, e_k > = u_k$. It is easy to see that $\ker L = \ker R_\mu^L(M)$ for $\mu \neq \mu_k$. Therefore, the operator M is $(L, 0)$-bounded. Construct the degenerate holomorphic group of operators

$$U^t = \sum_k{}' e^{\mu_k t} < \cdot, e_k > e_k. \qquad (27)$$

By construction of (25), the operators $L, M \in \mathscr{L}(l_q^{m+2r}; l_q^m), m \in \overline{\mathbf{R}}_+, q \in [1, +\infty)$. In the space l_q^{m+2r}, we define the canonical cone $\mathscr{C} = \{u \in l_q^{m+2r} : u_k \geq 0, \quad k \in \mathbf{N}\}$. It is easy to see that the operator $R_\mu^L(M)$ is positive for $\mu > \max_k \mu_k$, if degenerate holomorphic group (27) is positive.

Example 5 Consider the formula $\Lambda u = (\lambda_k u_k)$ to define the operator $\Lambda \in \mathscr{L}(l_q^{m+2}, l_q^m)$, the prototype of which is the operator A considered in Example 4. Construct the operators $L = (\lambda - \lambda_k) < \cdot, e_k > e_k$ and $M = \alpha < \cdot, e_k > e_k$. The operator M is $(L, 0)$-bounded, and $\dim \ker L \in \mathbf{N}$. The positive degenerate holomorphic group of operators has the form

$$U^t = \sum_k{}' e^{\mu_k t} < \cdot, e_k > e_k,$$

where $\mu_k = \dfrac{\alpha \lambda_k}{\lambda - \lambda_k}$ for all $k \in \mathbf{N}$ except as $\lambda = \lambda_k$.

Acknowledgements The work was supported by Act 211 Government of the Russian Federation, contract no 02.A03.21.0011.

References

1. Sviridyuk, G.A.: On the general theory of operator semigroups. Russ. Math. Surv. **49**(4), 47–74 (1994)
2. Sviridyuk, G.A., Fedorov, V.E.: Linear Sobolev Type Equations and Degenerate Semigroups of Operators. Utrecht, Boston, Köln, Tokyo, VSP (2003)
3. Sviridyuk, G.A., Zagrebina, S.A.: The Showalter–Sidorov problem as a phenomena of the Sobolev-type equations. Izvestiya Irkutskogo Gosudarstvennogo Universiteta. Seriya "Matematika" **3**(1), 104–125 (2010) (in Russ.)
4. Banasiak, J., Arlotti, L.: Perturbations of Positive Semigroups with Applications. Springer, London Ltd. (2006)
5. Solovyova, N.N., Zagrebina, S.A., Sviridyuk, G.A.: Sobolev type mathematical models with relatively positive operators in the sequence spaces. Bulletin of the South Ural State University. Series of "Mathematics. Mechanics. Physics", vol. 9, no. 4, pp. 27–35. https://doi.org/10.14529/mmph170404 (2017)
6. Keller, A.V., Al-Delfi, J.K.: Holomorphic degenerate groups of operators in quasi-Banach spaces. Bulletin of the South Ural State University. Series of "Mathematics. Mechanics. Physics", vol. 7, no. 1, pp. 20–27 (2015) (in Russ.)

7. Zamyshlyaeva, A.A., Al-Isawi J.K.T.: On some properties of solutions to one class of evolution Sobolev type mathematical models in quasi-Sobolev spaces. Bulletin of the South Ural State University, Series "Mathematical Modelling, Programming & Computer Software", vol. 8, no. 4, pp. 113–119. https://doi.org/10.14529/mmp150410/ (2015)
8. Sagadeeva, M.A., Hasan, F.L.: Bounded solutions of Barenblatt–Zheltov–Kochina model in quasi-Sobolev spaces. Bulletin of the South Ural State University. Series "Mathematical Modelling, Programming & Computer Software", vol. 8, Issue 4, pp. 138–144 (2015)
9. Barenblatt, G.I., Zheltov, Iu.P., Kochina, I.N.: Basic concepts in the theory of seepage of homogeneous liquids in fissured rocks [strata]. J. Appl. Math. Mech. **24**(5), 1286–1303 (1960)
10. Hallaire M.: Soil water movement in the film and vapor phase under the influence of evapotranspiration. Water and its conduction in soils. In: Proceedings of XXXVII Annual Meeting of the Highway Research Board, Highway Research Board Special Report, vol. 40, pp. 88–105 (1958)
11. Chen, P.J., Gurtin, M.E.: On a theory of heat conduction involving two temperatures. J. Appl. Math. Phys. (ZAMP) **19**(4), 614–627 (1968)
12. Hoff, N.J.: Creep buckling. Aeronaut. Q. **7**(1), 1–20 (1956)
13. Sviridyuk, G.A., Manakova, N.A.: Dynamic models of Sobolev type with the Showalter–Sidorov condition and additive "noises". Bulletin of the South Ural State University. Series "Mathematical Modelling, Programming & Computer Software", vol. 7, no. 1, pp. 90–103 (2014) (in Russ.)
14. Kitaeva, O.G., Shafranov, D.E., Sviridyuk, G.A.: Exponential dichotomies in Barenblatt–Zheltov–Kochina model in spaces of differential forms with "noise". Bulletin of the South Ural State University. Series "Mathematical Modelling, Programming & Computer Software", vol. 12, no. 2, pp. 47–57 (2019)
15. Sviridyuk, G.A., Zagrebina, S.A., Konkina, A.S.: The Oskolkov equations on the geometric graphs as a mathematical model of the traffic flow. Bulletin of the South Ural State University. Series "Mathematical Modelling, Programming & Computer Software", vol. 8, issue 3, pp. 148–154 (2015)
16. Favini, A., Zagrebina, S.A., Sviridyuk, G.A.: Multipoint initial-final value problems for dynamical Sobolev-type equations in the space of noises. Electron. J. Differ. Equ. **2018**, 128 (2018)
17. Banasiak J.: Private Communication. References
18. Gantmacher, F.R.: The Theory of Matrices, p. 660. AMS Chelsea Publishing, Reprinted by American Mathematical Society (2000)
19. Sobolev, S.L.: Applications of Functional Analysis in Mathematical Physics, p. 239. American Mathematical Society (1963)
20. Ladyzhenskaya, O.A., Ural'tseva, N.N.: Quasi-linear elliptic equations and variational problems with many independent variables. Russ. Math. Surv. **16**(1), 17–91 (1961)
21. Chekorun, M.D., Park, E., Temam, R.: The Stampacchia maximum principle for stochastic partial equations and applications. J. Differ. Equ. **260**(3), 2926–2972 (2016)

Applications

Microscopic Selection of Solutions to Scalar Conservation Laws with Discontinuous Flux in the Context of Vehicular Traffic

Boris Andreianov and Massimiliano D. Rosini

Abstract In the context of road traffic modeling we consider a scalar hyperbolic conservation law with the flux (fundamental diagram) which is discontinuous at $x = 0$, featuring variable velocity limitation. The flow maximization criterion for selection of a unique admissible weak solution is generally admitted in the literature, however justification for its use can be traced back to the irrelevant vanishing viscosity approximation. We seek to assess the use of this criterion on the basis of modeling proper to the traffic context. We start from a first order microscopic follow-the-leader (FTL) model deduced from basic interaction rules between cars. We run numerical simulations of FTL model with large number of agents on truncated Riemann data, and observe convergence to the flow-maximizing Riemann solver. As an obstacle towards rigorous convergence analysis, we point out the lack of order-preservation of the FTL semigroup.

Keywords Conservation laws · Traffic flow · Discontinuous flux · Riemann solvers · First order follow-the-leader model · Order-preservation · Point constraint on the flux · Follow-the-leader semigroup

---

MDR is member of GNAMPA. MDR acknowledges the support of the National Science Centre, Poland, Project "Mathematics of multi-scale approaches in life and social sciences" No. 2017/25/B/ST1/00051, by the INdAM-GNAMPA Project 2019 "Equazioni alle derivate parziali di tipo iperbolico o non locale ed applicazioni" and by University of Ferrara, FIR Project 2019 "Leggi di conservazione di tipo iperbolico: teoria ed applicazioni".

B. Andreianov
Institut Denis Poisson (CNRS UMR7013), Université de Tours, Université d'Orléans, 37200 Tours, Parc Grandmont, France
e-mail: boris.andreianov@univ-tours.fr

Peoples Friendship University of Russia (RUDN University), 6 Miklukho-Maklaya St, Moscow 117198, Russian Federation

M. D. Rosini (✉)
Department of Mathematics and Computer Science, University of Ferrara, I-44121 Ferrara, Italy
e-mail: rsnmsm@unife.it

Instytut Matematyki, Uniwersytet Marii Curie-Skłodowskiej, pl. Marii Curie-Skłodowskiej 1, 20-031 Lublin, Poland

1 Introduction

1.1 Discontinuous-Flux Scalar Conservation Law

The main object of study in this paper is the first order macroscopic model for vehicular traffic

$$\partial_t \rho + \partial_x f(\rho, x) = 0, \qquad f(\rho, x) \doteq \begin{cases} f_-(\rho) & \text{if } x < 0, \\ f_+(\rho) & \text{if } x \geqslant 0, \end{cases} \tag{1}$$

where $\rho = \rho(t, x) \in [0, 1]$ is the (normalized) density at time $t \geqslant 0$ and position $x \in \mathbb{R}$, while $f_- \geqslant 0$ and $f_+ \geqslant 0$ are the fundamental diagrams (fluxes) corresponding to the two sections $x < 0$ and $x > 0$ of the road separated by $x = 0$, called "interface" in the sequel. We assume that f_- and f_+ are convex and satisfy the *matching conditions*

$$f_-(0) = f_+(0), \qquad f_-(1) = f_+(1).$$

Our concern is with admissibility of solution discontinuities at the interface location $x = 0$, which is a question asked also for other models of vehicular traffic with point singularities, such as [14, 34, 54]. Our paper shares this primary motivation with the recent work [51]. The author of [51] studies admissible jumps at the flux discontinuity location by analyzing (both analytically and numerically) *stationary wave profiles*. We pursue (numerically only, at the present stage) the complementary line of investigation, by looking at solutions of the Riemann problems and interpreting our results in terms of global properties of the flow (we put forward the flux maximization principle).

Equation (1) is a scalar hyperbolic conservation law in one space dimension. It arises in road traffic but also in many other application problems, such as two-phase flows through porous media [5, 36], continuous sedimentation in a clarifier-thickener unit [16, 32] or ion etching in the semiconductor industry [49].

The derivatives in (1) are interpreted in the weak sense. Indeed, even for smooth data classical solutions may not exist globally in time since discontinuities can arise in finite time. Yet weak solutions are in general not unique. This motivates the introduction of additional admissibility criteria able to single out a unique solution. In principle, these conditions depend on the physical phenomena under consideration. However it turns out that in the setting of continuous flux (the case $f_- \equiv f_+$ in (1)) the same notion of entropy admissible solution ([41], see also [42, 46, 55]) is relevant for all traditional applications. In different fluid mechanics applications the widely accepted justification for the choice of this solution notion is based upon the vanishing viscosity method that can be traced back as far as [37, 47]. The use of the same solution notion in the road traffic (the Lighthill–Whitham–Richards model [44, 48], LWR model in the sequel) is rather based upon the coincidence of admissible discontinuities: those observed in the traffic context turn out to be the admissible jumps in the entropy sense [41, 46].

The situation is drastically different in the discontinuous-flux setting of (1). Coexistence of infinitely many admissibility criteria—equally consistent from the pure mathematical standpoint—has been discovered in [1] and further developed e.g. in [5, 6, 10, 17, 35]. To be precise, there exist infinitely many $\mathbf{L}^1$-contractive semigroups of solutions to the formal problem (1), moreover, different semigroups correspond to different modeling contexts or different model parameters. It is highlighted in the recent survey paper [4] that choosing admissibility condition for (1) at the interface $x = 0$ means prescribing an interface coupling condition; this is very much similar to prescribing different boundary conditions (Dirichlet, Robin, Neumann, obstacle, etc.) for classical elliptic or parabolic PDEs. As a matter of fact, interface coupling conditions are part of the model and have to be prescribed independently from the differential relations applied in open space-time domains. E.g. in the context of flows in "two-rocks" porous media, the physically relevant coupling depends on additional nonlinearities ϕ_-, ϕ_+ accounting for capillary pressures that are not even present in the formal PDE formulation (1), see [5]. Thus, considering (1) without precise indication of the desired interface coupling means working with an incomplete model, similar to speaking of "Laplacian operator on a bounded domain" without specific boundary conditions!

In the context of hyperbolic models of road traffic, many nontrivial models are postulated by prescribing Riemann solvers. The choice of admissibility conditions at $x = 0$, i.e. the prescription of precise interface coupling conditions, is equivalent to the choice of a Riemann solver at the interface location [10, 35]. Making this choice is not always straightforward. In the traffic context, the lack of an established method to single out physically reasonable discontinuities leads to the introduction of different notions of solutions for the same problem (sometimes in the same paper!), see for instance [14, 34, 54]. In [35], it is even stated that "there is no a priori preferable physical solution". In this respect, let us stress that the vanishing viscosity (or, more generally, vanishing capillarity, cf. [5]) admissibility criterion is successful in fluid dynamics applications, but it is not recognized as instrumental in the context of traffic modeling.

1.2 The Flow Maximization Principle, Models with a Point Constraint on the Flux and Their Justification

There is a number of works on trafic modeling where the jump admissibility criterion is taken according to the principle of flow maximization: we refer, for instance, to [18, 20, 26, 40, 45]. It also appeared in modeling of flows in porous media [1, 5, 19, 38], the resulting solution being referred to as the "optimal entropy solution" and justified via the vanishing capillarity approach in specific flux configurations.

The flow maximization principle is appealing but not always justified; as a matter of fact, quite often it is implicitly assumed. This naturally occurs when one takes a successful—e.g., having a well-developed well-posedness theory—solution notion from the related literature, even if it may actually address a quite different application context. A further motivation stems from the fact that one and the same notion of solution was recognized as the correct one for very different applications, at least in the classical continuous-flux case of (1). However, we have seen that somewhat unexpectedly, the discontinuous-flux setting reveals much more diversity. From the accurate modeling perspective, in traffic applications the flow maximization should result as a collective behavior emerging from individual-driver behavior. However, flow maximization does not always occur in practice, as it is evident in the capacity drop phenomenon at crossroads [52] and in the Braess paradox phenomenon [15]. In particular, it was demonstrated in [7] that the capacity drop and the Braess paradox can be reproduced starting from the concept of flux limitation (as opposed to the flow maximization) introduced in [21, 22].

Let us briefly focus on the flux limitation model of [21]. It formally takes the form (1) with continuous flux (i.e., $f_- = f_+ \doteq f$) and with explicit (though formal) interface condition $f(\rho(t, 0^\pm)) \leq F(t)$ for some map F given beforehand. This model is introduced essentially in terms of the intuitively appealing Riemann solver at the interface $x = 0$; but the authors of [21] also provide a global entropy formulation and deduce it from the asymptotic analysis involving a discontinuous-flux conservation law similar to (1), but with two flux discontinuities situated at $x = \pm\varepsilon$, with $\varepsilon \to 0^+$. Doing so, they rely upon the global entropy formulation for (1) appearing in the influential works for the case where the graphs f_-, f_+ can be ordered, such as [39, 50, 53]. However, the admissibility formulation in use in [39, 50, 53] can be traced back to—or connected to, see e.g. [11, 33]—the vanishing viscosity approximation, not suitable for road traffic applications. Another source of justification is the "minimal jump" condition of [36], which relies on a mathematical rather than on a modeling selection principle. As a matter of fact, a byproduct of our analysis is a more realistic justification of the model introduced in [21].

We point out that at the present stage our justification is not rigorous because our conclusions are based only upon accurate modeling and numerical experiments lacking, at the present stage, the full convergence proof.

1.3 Microscopic-to-Macroscopic Limit in Traffic Modeling. the Purpose and the Contents of the Paper

Our main goal is to propose a method to select an appropriate solution. We rely on the paradigm of passage to the limit from a well-assessed microscopic model to a macroscopic model, similar to taking the hydrodynamical limit of Boltzmann equations. For the LWR traffic model, such passage to the limit was fully realized for the first time in [29]. The main advantage of this approach is that it requires to set assumptions on the interacting behavior of the cars only at the microscopic level. Since traffic dynamics are essentially microscopic, it is much easier to physically motivate the microscopic rather than some macroscopic assumptions. Thus, we first encode the interacting rules in a (first order) microscopic follow-the-leader (FTL) approximation of (1). We then obtain a macroscopic Riemann solver as many particle limit by applying, at the level of numerical simulations and for carefully identified sets of data, the approximation procedure adapted from [29]. It turns out that from this procedure, we indeed observe the emergence of the Riemann solver which maximizes the flow at $x = 0$. Let us stress that we do not require this property at the microscopic level, as we only prescribe elementary car interaction rules. Therefore our results further assess the relevance of the flow maximization admissibility criterion, at least in the present setting of discontinuous-flux LWR model (1), in the related LWR model with point constraint on the flow and its variants [8, 9, 21]. A closely related investigation of the FTL approximation of the discontinuous-flux LWR model, with a focus on emerging standing-wave profiles (which is a way of understanding admissibility of discontinuities), is conducted in [51]; our results are complementary of those of [51].

Further, we identify specific difficulties on the way of rigorously proving the observed numerical convergence results. First, the techniques of [29] and subsequent works [27, 28, 30, 31] on the same subject do not extend readily to discontinuous-flux LWR model (1). Second, note that while dealing with discontinuous-flux scalar conservation laws, common techniques of passage to the limit arose that are equally applicable to vanishing viscosity (or vanishing capillarity), Finite Volume and Wave-Front Tracking approximations. These techniques are essentially based upon identification of a small number of explicit limits that take the form of stationary solutions, possibly discontinuous at $x = 0$ (such discontinuities are therefore claimed admissible at the limit), and on the $\mathbf{L}^1$-contraction property valid already at the level of approximations. We refer to [10] for a formalization of this technique. Note that the latter contraction principle is equivalent to the order-preservation of the solution semigroup, by virtue of the conservativity. We raise the question whether the FTL approximation of car densities fulfills the order-preservation property, and answer this question by a counterexample. This assesses the necessity of developing a non-standard approach to the limit analysis of the FTL approximations. Moreover, while one expects on heuristic grounds that the admissibility at $x = 0$ can be disconnected from admissibility (in the standard entropy sense) in the regions $\{x > 0\}$, $\{x < 0\}$, the localization argument based on the finite speed of propagation in LWR models does not apply neither since the FTL model possess an infinite speed of propagation. We illustrate and discuss these two properties of the FTL semigroup, which make it qualitatively different from its expected limit which is the LWR model (1) with interface coupling prescribed by the flow maximization principle.

1.4 Outline of the Paper

The paper is organized as follows. In the next section we introduce the model and the definition of admissible solutions corresponding to a Riemann solver in Definition 4. In Sect. 3 we recall the definition of the Riemann solver RS_M which maximizes the flow at $x = 0$ and then show how it can be deduced from a FTL model. In Sect. 4 we show that, differently from the macroscopic model, the microscopic model does not satisfy the order-preservation property and the finite speed of propagation. In the last section we explain how the Riemann solver introduced in [21] for a scalar conservation law with point constraint on the flow can be deduced from a FTL model.

2 The Model and General Definition of Solutions

Consider a road network parametrized by $x \in \mathbb{R}$ and composed of two single lane roads $\mathfrak{R}_- \doteq (-\infty, 0)$ and $\mathfrak{R}_+ \doteq [0, +\infty)$ connected together at the junction $x = 0$. Assume that each driver has the same behavior, all cars are identical and have maximal speed $V_{\max} > 0$. Let $V_-, V_+ > 0$ be respectively the speed limits in $\mathfrak{R}_-$ and $\mathfrak{R}_+$, with $V_-, V_+ \leqslant V_{\max}$. The natural adaptation of the Lighthill, Whitham [44] and Richards [48] (LWR) model to the present case is expressed by the Cauchy problem for a one-dimensional scalar conservation law

$$\begin{cases} \partial_t \rho + \partial_x f(\rho, x) = 0, & x \in \mathbb{R}, \ t > 0, \\ \rho(0, x) = \bar{\rho}(x), & x \in \mathbb{R}, \end{cases} \quad (2)$$

where $\rho = \rho(t, x) \in [0, 1]$ is the (normalized) density of cars at time $t > 0$ and position $x \in \mathbb{R}, \bar{\rho} \in \mathbf{L}^1_{\text{loc}}(\mathbb{R}; [0, 1])$ is the initial density and the flux $f : [0, 1] \times \mathbb{R} \to [0, +\infty)$ is defined by

$$f(r, x) \doteq \begin{cases} f_-(r) & \text{if } x \in \mathfrak{R}_-, \\ f_+(r) & \text{if } x \in \mathfrak{R}_+, \end{cases}$$

where

$$f_-(r) \doteq r \, v_-(r), \qquad f_+(r) \doteq r \, v_+(r).$$

Here $v_-, v_+ : [0, 1] \to [0, V_{\max}]$ are the velocity functions. We assume that v_- and v_+ are continuous non-increasing functions with

$$v_-(0) = V_-, \quad v_+(0) = V_+, \quad v_-(1) = 0 = v_+(1),$$

and such that f_- and f_+ are bell-shaped. Notice that $f(r, x) = r \, v(r, x)$, where $v : [0, 1] \times \mathbb{R} \to [0, V_{\max}]$ is defined by

$$v(r, x) \doteq \begin{cases} v_-(r) & \text{if } x \in \mathfrak{R}_-, \\ v_+(r) & \text{if } x \in \mathfrak{R}_+. \end{cases}$$

Typical choice for v_- and v_+ is

$$v_-(r) \doteq \min\{V_-, V_{\max}(1-r)\}, \qquad v_+(r) \doteq \min\{V_+, V_{\max}(1-r)\}. \qquad (3)$$

Since (2) is a conservation law, it is necessary to consider solutions in the sense of distributions, the so-called weak solutions. Define $\mathbb{R}_+ \doteq [0, +\infty)$.

Definition 1 We say that $\rho \in C^0(\mathbb{R}_+; \mathbf{BV}(\mathbb{R}; [0, 1]))$ is a **weak solution** to Cauchy problem (2) if the following conditions are satisfied:

(**w.1**) For any test function $\varphi \in C_c^\infty(\mathbb{R}_+ \times \mathbb{R}; \mathbb{R})$ with compact support in $x < 0$, we have

$$\iint_{\mathbb{R}_+ \times \mathfrak{R}_-} \left(\rho\, \partial_t \varphi + f_-(\rho)\, \partial_x \varphi\right) dx\, dt + \int_{\mathfrak{R}_-} \bar{\rho}(x)\, \varphi(0, x)\, dx = 0.$$

(**w.2**) For any test function $\varphi \in C_c^\infty(\mathbb{R}_+ \times \mathbb{R}; \mathbb{R})$ with compact support in $x > 0$, we have

$$\iint_{\mathbb{R}_+ \times \mathfrak{R}_+} \left(\rho\, \partial_t \varphi + f_+(\rho)\, \partial_x \varphi\right) dx\, dt + \int_{\mathfrak{R}_+} \bar{\rho}(x)\, \varphi(0, x)\, dx = 0.$$

(**w.3**) The Rankine-Hugoniot condition is satisfied along $x = 0$, namely

$$f_-\bigl(\rho(t, 0^-)\bigr) = f_+\bigl(\rho(t, 0^+)\bigr), \qquad \text{for a.e. } t > 0. \qquad (4)$$

We stress that coupling condition (4) guarantees the conservation of the total number of cars.

Weak solutions are in general not unique. A well known criterion to select a unique weak solution was proposed by Kruzhkov [41]. However Kruzhkov solutions are not always physically reasonable [43]. This is evident, for instance, when dealing with traffic flows through crossroads [20, 25, 26, 35] or point-wise bottlenecks [21]. However, we stick to the notion of Kruzhkov admissibility away from the flux discontinuity location $x = 0$. Namely, following [35] we postulate

Definition 2 We say that $\rho \in C^0(\mathbb{R}_+; \mathbf{BV}(\mathbb{R}; [0, 1]))$ is an **admissible solution** to Cauchy problem (2) if it satisfies the following conditions:

(**a.1**) ρ is a weak solution to Cauchy problem (2) in the sense of Definition 1.
(**a.2**) For any test function $\varphi \in C_c^\infty\bigl((0, +\infty) \times (-\infty, 0); \mathbb{R}\bigr)$ with $\varphi \geq 0$ and $\kappa \in [0, 1]$, we have

$$\iint_{\mathbb{R}_+ \times \mathfrak{R}_-} \Big(|\rho - \kappa| \, \partial_t \varphi + \text{sign}(\rho - \kappa) \left(f_-(\rho) - f_-(\kappa) \right) \partial_x \varphi \Big) \, dx \, dt \geq 0.$$

(a.3) For any test function $\varphi \in \mathbf{C}_c^\infty((0, +\infty) \times (0, +\infty); \mathbb{R})$ with $\varphi \geq 0$ and $\kappa \in [0, 1]$, we have

$$\iint_{\mathbb{R}_+ \times \mathfrak{R}_+} \Big(|\rho - \kappa| \, \partial_t \varphi + \text{sign}(\rho - \kappa) \left(f_+(\rho) - f_+(\kappa) \right) \partial_x \varphi \Big) \, dx \, dt \geq 0.$$

Conditions listed in Definition 2 do not select a unique weak solution. However, to do so it is sufficient to uniquely characterize the physically reasonable discontinuities along $x = 0$. This can be achieved by choosing a Riemann solver. Let $\mathbb{1}_A$ be the indicator function of the set A.

Definition 3 We say that $\mathsf{RS} \colon [0, 1]^2 \to \mathbf{BV}(\mathbb{R}; [0, 1])$ is a **Riemann solver** if for any $(\rho_L, \rho_R) \in [0, 1]^2$ the self-similar function

$$\rho(t, x) \doteq \mathsf{RS}[\rho_L, \rho_R](x/t) \tag{5}$$

together with its traces along $x = 0$

$$\rho^- \doteq \rho(t, 0^-) = \mathsf{RS}[\rho_L, \rho_R](0^-), \qquad \rho^+ \doteq \rho(t, 0^+) = \mathsf{RS}[\rho_L, \rho_R](0^+),$$

satisfy the following conditions:

(r.1) ρ is an admissible solution to Cauchy problem (2) with initial density

$$\bar{\rho}(x) \doteq \rho_L \, \mathbb{1}_{\mathfrak{R}_-}(x) + \rho_R \, \mathbb{1}_{\mathfrak{R}_+}(x). \tag{6}$$

(r.2) The self-similar function

$$r^-(t, x) \doteq \begin{cases} \rho(t, x) & \text{if } x < 0, \ t \geq 0, \\ \rho^- & \text{if } x \geq 0, \ t \geq 0, \end{cases}$$

is the Kruzhkov solution to the Riemann problem

$$\begin{cases} \partial_t r^- + \partial_x f_-(r^-) = 0, & x \in \mathbb{R}, \ t > 0, \\ r^-(0, x) = \rho_L \, \mathbb{1}_{\mathfrak{R}_-}(x) + \rho^- \, \mathbb{1}_{\mathfrak{R}_+}(x), & x \in \mathbb{R}, \end{cases}$$

namely for any test function $\varphi \in \mathbf{C}_c^\infty(\mathbb{R}_+ \times \mathbb{R}; \mathbb{R})$ with $\varphi \geq 0$ and $\kappa \in [0, 1]$

$$\iint_{\mathbb{R}_+\times\mathbb{R}}\Big(|r^- - \kappa|\,\partial_t\varphi + \mathrm{sign}(r^- - \kappa)\big(f_-(r^-) - f_-(\kappa)\big)\partial_x\varphi\Big)\,\mathrm{d}x\,\mathrm{d}t$$
$$+\rho_L\int_{\mathfrak{R}_-}\varphi(0,x)\,\mathrm{d}x + \rho^-\int_{\mathfrak{R}_+}\varphi(0,x)\,\mathrm{d}x \geqslant 0.$$

(r.3) The self-similar function

$$r^+(t,x) \doteq \begin{cases} \rho^+ & \text{if } x<0,\ t\geqslant 0, \\ \rho(t,x) & \text{if } x\geqslant 0,\ t\geqslant 0, \end{cases}$$

is the Kruzhkov solution to the Riemann problem

$$\begin{cases} \partial_t r^+ + \partial_x f_+(r^+) = 0, & x\in\mathbb{R},\ t>0, \\ r^+(0,x) = \rho^+\,\mathbb{1}_{\mathfrak{R}_-}(x) + \rho_R\,\mathbb{1}_{\mathfrak{R}_+}(x), & x\in\mathbb{R}, \end{cases}$$

namely for any test function $\varphi \in C_c^\infty(\mathbb{R}_+ \times \mathbb{R};\mathbb{R})$ with $\varphi\geqslant 0$ and $\kappa\in[0,1]$

$$\iint_{\mathbb{R}_+\times\mathbb{R}}\Big(|r^+ - \kappa|\,\partial_t\varphi + \mathrm{sign}(r^+ - \kappa)\big(f_+(r^+) - f_+(\kappa)\big)\partial_x\varphi\Big)\,\mathrm{d}x\,\mathrm{d}t$$
$$+\rho^+\int_{\mathfrak{R}_-}\varphi(0,x)\,\mathrm{d}x + \rho_R\int_{\mathfrak{R}_+}\varphi(0,x)\,\mathrm{d}x \geqslant 0.$$

(r.4) We have

$$\mathsf{RS}[\rho^-,\rho^+](\xi) = \begin{cases} \rho^- & \text{if } x<0, \\ \rho^+ & \text{if } x\geqslant 0. \end{cases} \qquad (7)$$

(r.5) The function $(\rho_L,\rho_R) \mapsto \big(f_-(\rho^-), f_+(\rho^+)\big)$ is continuous.

Some comments on the conditions listed in Definition 3 are in order. Condition (r.1) ensures that a Riemann solver associates to any Riemann datum (6) a self-similar function ρ through (5), which is an admissible solution to Riemann problem (2), (6). In particular it satisfies coupling condition (4)

$$f_-(\rho^-) = f_+(\rho^+). \qquad (8)$$

Conditions (r.2), (r.3) can be rephrased as follows: r^- and r^+ are the solutions to the initial-boundary value problems [2, 13, 23]

$$\begin{cases} \partial_t r^- + \partial_x f_-(r^-) = 0, & x<0,\ t>0, \\ r^-(0,x) = \rho_L, & x<0, \\ r^-(t,0) = \rho^-, & t>0, \end{cases} \qquad \begin{cases} \partial_t r^+ + \partial_x f_+(r^+) = 0, & x>0,\ t>0, \\ r^+(0,x) = \rho_R, & x>0, \\ r^+(t,0) = \rho^+, & t>0, \end{cases}$$

coupled through condition (8). Notice that r^- has only waves generated at $x = 0$ and with negative speed, whereas r^+ has only waves generated at $x = 0$ and with positive speed. By (r.4) the ordered pair of the traces of the admissible solution selected by RS is in a sense a fixed point of RS by (7). It is a minimal requirement to develop a numerical scheme with a time discretization based on RS, see [24, 25]. Condition (r.4) is therefore a stability condition. Condition (r.5) is a regularity property for RS.

We are now in the position to give the definition of admissible solutions to Cauchy problem (2) associated to a Riemann solver RS.

Definition 4 Fix a Riemann solver RS in the sense of Definition 3. We say that $\rho \in \mathbf{C}^0(\mathbb{R}_+; \mathbf{BV}(\mathbb{R}; [0, 1]))$ is an RS-**admissible solution** to Cauchy problem (2) if it is an admissible solution in the sense of Definition 2 and

$$\mathsf{RS}[\rho(t, 0^-), \rho(t, 0^+)](\xi) = \begin{cases} \rho(t, 0^-) & \text{if } \xi < 0, \\ \rho(t, 0^+) & \text{if } \xi \geq 0, \end{cases} \quad \text{for a.e. } t > 0.$$

In [35] it is proved that the above definition selects a unique admissible solution, given a fixed Riemann solver. We are therefore left to define a physically reasonable Riemann solver. This is achieved in the next section.

3 Microscopic Deduction of the Riemann Solver RS_M

In this section we select a Riemann solver. The choice is motivated by basic well established physically reasonable rules on the microscopic interactions between the cars. We first deduce from the microscopic rules a FTL model. We then obtain an empirical Riemann solver as a many particle limit in analogy to what is done in [29]. It turns out that the obtained Riemann solver coincides with the Riemann solver RS_M that maximizes the flow at $x = 0$, even if we do not require any flow maximization property for the FTL model.

For completeness we recall in the next subsection the definition of RS_M. We then introduce the FTL model underlying the simulations which are used to show that the corresponding Riemann solver is indeed RS_M. In Figs. 1 and 2 we present the results of some selected simulations.

3.1 Riemann Solver RS_M

In this subsection we recall the Riemann solver RS_M which maximizes the flow at $x = 0$. This means that if RS is a Riemann solver, then by (8)

$$f_-\bigl(\mathsf{RS}[\rho_L,\rho_R](0^-)\bigr) = f_+\bigl(\mathsf{RS}[\rho_L,\rho_R](0^+)\bigr)$$
$$\leqslant f_-\bigl(\mathsf{RS}_M[\rho_L,\rho_R](0^-)\bigr) = f_+\bigl(\mathsf{RS}_M[\rho_L,\rho_R](0^+)\bigr).$$

We introduce some notations. Let $\mathsf{L}_-, \mathsf{L}_+ \colon [0,1]^2 \to \mathbf{BV}(\mathbb{R};[0,1])$ be the Lax Riemann solvers associated to the fluxes f_- and f_+. Notice that $\mathsf{L}_-[\rho_L,\rho_R] \equiv \mathsf{L}_+[\rho_L,\rho_R]$ if and only if $\rho_L, \rho_R \geqslant \max\{R_-, R_+\}$, where $R_-, R_+ \in (0,1)$ are implicitly defined by
$$v(R_-) = V_-, \quad v(R_+) = V_+.$$

Let M_- and M_+ be the densities at which f_- and f_+ reach their maxima. Let
$$Q_-(\rho_L) \doteq \begin{cases} f_-(\rho_L) & \text{if } \rho_L < M_-, \\ f_-(M_-) & \text{if } \rho_L \geqslant M_-, \end{cases} \qquad Q_+(\rho_R) \doteq \begin{cases} f_+(M_+) & \text{if } \rho_R < M_+, \\ f_+(\rho_R) & \text{if } \rho_R \geqslant M_+, \end{cases}$$

be the maximal flows at $x = 0$ attained by $\mathsf{L}_-[\rho_L,\rho]$ and $\mathsf{L}_+[\rho,\rho_R]$ for ρ varying in $[0,1]$. Let then $\hat{\rho}(\rho_L,\rho_R) \in [M_-,1]$ and $\check{\rho}(\rho_L,\rho_R) \in [0,M_+]$ be implicitly defined by
$$f_-\bigl(\hat{\rho}(\rho_L,\rho_R)\bigr) = \min\bigl\{Q_-(\rho_L), Q_+(\rho_R)\bigr\} = f_+\bigl(\check{\rho}(\rho_L,\rho_R)\bigr).$$

Then $\mathsf{RS}_M \colon [0,1]^2 \to \mathbf{BV}(\mathbb{R};[0,1])$ is defined by
$$\mathsf{RS}_M[\rho_L,\rho_R](v) \doteq \begin{cases} \mathsf{L}_-[\rho_L,\hat{\rho}(\rho_L,\rho_R)](v) & \text{if } v < 0, \\ \mathsf{L}_+[\check{\rho}(\rho_L,\rho_R),\rho_R](v) & \text{if } v \geqslant 0. \end{cases}$$

3.2 FTL Model

In this subsection we introduce the FTL model corresponding to Riemann problem (2), (6). Fix $(\rho_L,\rho_R) \in [0,1]^2$ and consider the approximate Riemann problem

$$\begin{cases} \partial_t \rho + \partial_x f(\rho, x) = 0, & x \in \mathbb{R},\ t > 0, \\ \rho(0,x) = \rho_L\, \mathbb{1}_{[-\delta,0)}(x) + \rho_R\, \mathbb{1}_{[0,\delta]}(x), & x \in \mathbb{R}. \end{cases} \qquad (9)$$

Here $\delta > 0$ is introduced so that the traffic has *finite* total length of cars, which is $\delta(\rho_L + \rho_R)$. The case $\rho_L = 0 = \rho_R$ is trivial; assume therefore that $\rho_L + \rho_R \neq 0$. Let $[x_{\min}, x_{\max}]$ be the support of the initial datum, namely

$$x_{\min} \doteq \begin{cases} -\delta & \text{if } \rho_L \neq 0, \\ 0 & \text{if } \rho_L = 0, \end{cases} \qquad x_{\max} \doteq \begin{cases} \delta & \text{if } \rho_R \neq 0, \\ 0 & \text{if } \rho_R = 0. \end{cases}$$

In analogy with [29], for any fixed $n \in \mathbb{N}$, we let $\ell \doteq \delta(\rho_L + \rho_R)/n$ and consider the approximate discrete density

$$r(t,x) \doteq \sum_{i=1}^{n} \frac{\ell}{x_i(t) - x_{i+1}(t)} \mathbb{1}_{[x_{i+1}(t), x_i(t))}(x), \qquad (10)$$

where x_i are the solutions to the FTL model

$$\begin{cases} \dot{x}_1 = V_+, & t > 0, \\ \dot{x}_i = v\left(\frac{\ell}{x_{i-1} - x_i}, x_i\right), & t > 0, \ i \in \{2, \ldots, n+1\}, \\ x_i(0) = \bar{x}_i, & i \in \{1, \ldots, n+1\}. \end{cases} \qquad (11)$$

Above $\bar{x}_1 \doteq x_{\max}$, $\bar{x}_{n+1} \doteq x_{\min}$ and for $i \in \{2, \ldots, n\}$ we define

$$\rho_L \neq 0 \implies x_{i+1}^0 = -\delta + (n-i)\frac{\ell}{\rho_L} \leq 0, \qquad i \geq \left\lceil \frac{\rho_R}{\rho_L + \rho_R} n \right\rceil,$$

$$\rho_R \neq 0 \implies x_{i+1}^0 = \delta - i\frac{\ell}{\rho_R} \geq 0, \qquad i \leq \left\lfloor \frac{\rho_R}{\rho_L + \rho_R} n \right\rfloor.$$

Notice that by definition $\int_{\bar{x}_i}^{\bar{x}_{i-1}} \bar{\rho}(x)\,dx = \ell$ and for any $t \geq 0$ we have

$$\|r(t)\|_{L^1(\mathbb{R})} = \delta(\rho_L + \rho_R), \qquad x_1(t) = x_{\max} + V_+ t.$$

We recall that (11) can be interpreted as a Lagrangian many particle approximation of (9). Moreover (11) corresponds to the following basic rules:

- Cars respect the speed limits.
- Cars do not overtake each other.
- Each car adjusts instantaneously its velocity according to the distance from its predecessor.
- Higher velocities correspond to higher distances from the predecessor.

In [29] it is proved that in the case $V_- = V_+ = V_{\max} > 0$ the discretized density r given in (10) converges a.e. in $L^1_{\text{loc}}(\mathbb{R}_+ \times \mathbb{R})$ to the unique Kruzhkov [41] solution ρ to (9). The rigorous proof for the convergence of the discretized density r in the general case $V_- \neq V_+$ is beyond the purposes of the present paper and is left to a future work. Here we are just interested in empirically deducing a Riemann solver from *ad hoc* computer assisted numerical simulations of (10), (11). It should be stressed that the simulations appear indeed as stable and convergent in all simulations we have run.

We are not interested in the solutions of (9) in the whole of $\mathbb{R}$, but only in the vicinity of $x = 0$. For this reason we do not investigate the nature of the waves created at $x = \pm\delta$ and their interactions with the waves starting from $x = 0$.

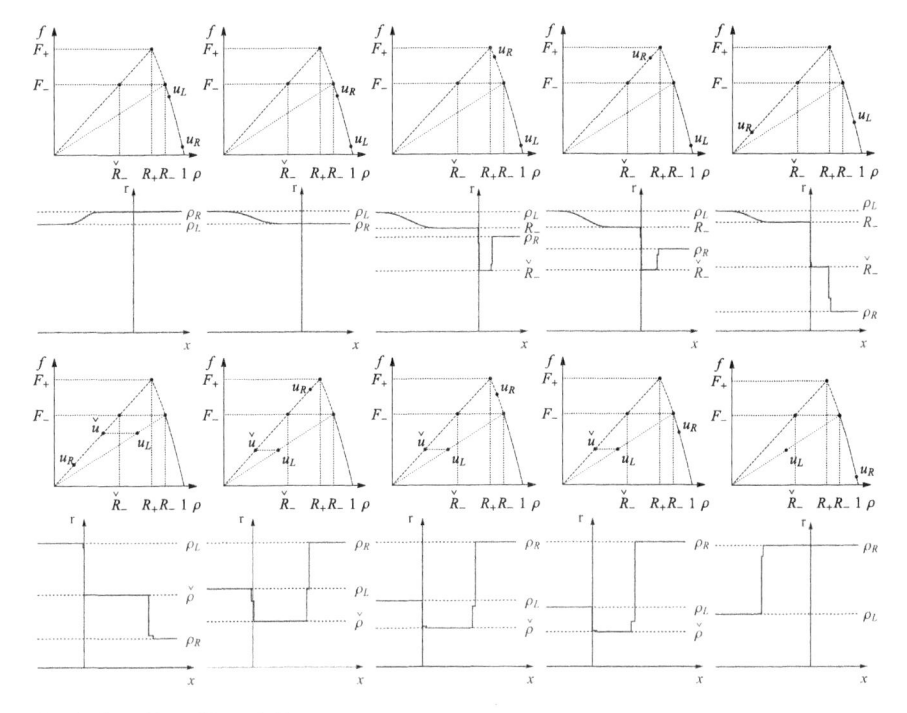

Fig. 1 Case $V_- < V_+ < 1/2$

Computer assisted numerical simulations allow to describe the limit solutions for any Riemann problem (2), (6). Our choice for the initial data $(\rho_L, \rho_R) \in [0, 1]^2$ used for the simulations allows to encompass all the cases necessary to deduce a Riemann solver, which turns out to coincide with RS_M.

In Figs. 1 and 2 we present the outputs of the numerical simulations in the case $V_-, V_+ \in (0, 1/2)$; the remaining cases are analogous and are not furnished here not to overload the paper. In Figs. 1 and 2 we show $u_L \doteq (\rho_L, f_-(\rho_L))$ and $u_R \doteq (\rho_R, f_+(\rho_R))$ in the fundamental diagram and below it we show the profile of the corresponding approximate discrete density r at a time $t > 0$ of interest. For each simulation we take $n = 1000$, $V_{\max} = 1$ and v_-, v_+ given by (3) with $V_-, V_+ \in \{\frac{2-\sqrt{2}}{4}, \frac{1}{4}\}$.

4 Qualitative Comparison of LWR and FTL Models

In this section we consider a homogeneous road and assume $V_- = V_+ = V_{\max}$. In [29] it is proved that in this case the unique entropy solution to the LWR model is the limit of discrete densities corresponding to the FTL model. Despite this link between the two models, there are deep differences between them. We show below that the

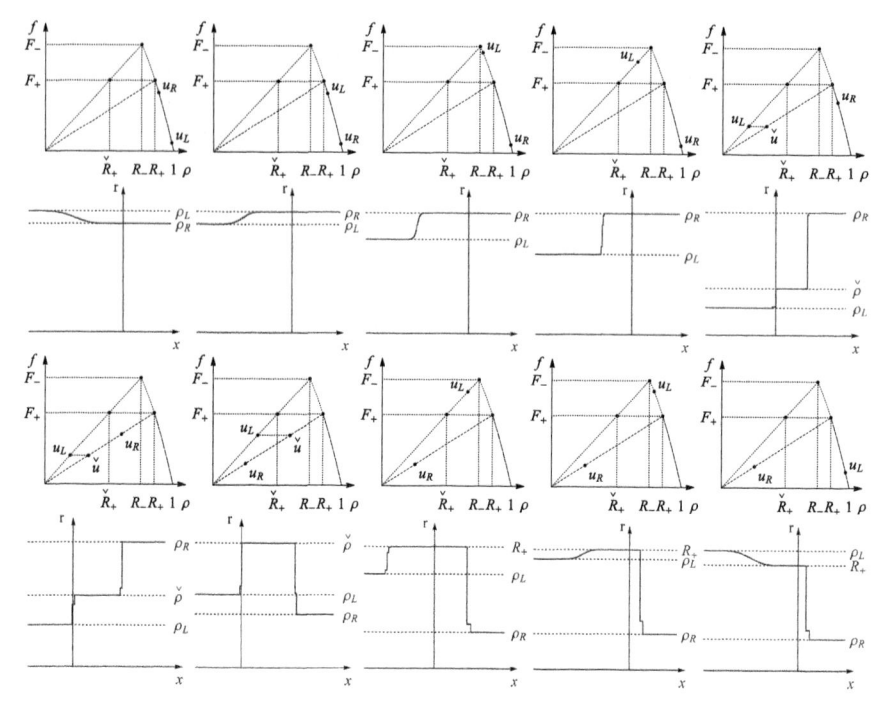

Fig. 2 Case $V_+ < V_- < 1/2$

order-preservation property and the finite speed of propagation, well known for the LWR model, are not shared by the FTL model. In a sense, these properties are gained only at the limit $n \to +\infty$.

We stress that the examples given below can be easily adapted to the case of inhomogeneous roads: it is enough to place all the cars on the right of $x = 0$. In fact, properties of the discontinuous-flux semigroup cannot be better than those of the continuous-flux one.

For simplicity consider a single lane road with neither entrances nor exits and parametrized by $x \in \mathbb{R}$. Let $\ell > 0$ be the length of each car. The evolution of the traffic can be then described at the macroscopic level by the LWR model and at the microscopic level by the FTL model. The LWR model is expressed by the conservation law

$$\partial_t \rho + \partial_x f(\rho) = 0, \quad x \in \mathbb{R}, \ t > 0, \tag{12}$$

where $\rho = \rho(t, x) \in [0, 1]$ is the (normalized) density at time t and position x, and $f(\rho) \doteq \rho \, v(\rho)$ is the flux corresponding to the velocity

$$v(\rho) \doteq V_{\max} (1 - \rho).$$

The FTL model is given by the system of ordinary differential equations

$$\begin{cases} \dot{x}_1 = V_{\max}, & t > 0, \\ \dot{x}_i = v\left(\dfrac{\ell}{x_{i-1}-x_i}\right), & t > 0, \ i \in \{2,\ldots,n\}, \end{cases} \quad (13)$$

where $x_i = x_i(t) \in \mathbb{R}$ is the position at time t of the front bumper of the ith car labeled starting from the right and n is the total number of cars. The corresponding discrete density is

$$r(t,x) \doteq \sum_{i=1}^{n-1} \frac{\ell}{x_i(t) - x_{i+1}(t)} \cdot \mathbb{1}_{[x_{i+1}(t), x_i(t))}(x). \quad (14)$$

Notice that the total length of the traffic is ℓn and is linked to the discrete density r and the solution ρ to LWR model (12) as follows:

$$\int_\mathbb{R} r(t,x)\,dx + \ell = \ell n = \int_\mathbb{R} \rho(t,x)\,dx, \quad t \geq 0.$$

4.1 Finite Speed of Propagation

It is well known that in the LWR model, information propagates with speed bounded by $V_{\max}$. On the contrary, according to the FTL model the propagation of information can be instantaneous. The next example shows that this drawback of the FTL model is responsible for an unrealistic behavior of the cars.

Example 1 Consider a traffic light placed at $x = 0$, which at time $t = 0$ turns from red to green. Assume that at time $t = 0$ the cars in $x \leq 0$ are bumper to bumper while in $x > 0$ no car is present. The corresponding FTL model (13) is

$$\begin{cases} \dot{x}_1 = V_{\max}, & t > 0, \\ \dot{x}_i = v\left(\dfrac{\ell}{x_{i-1}-x_i}\right), & t > 0, \ i \in \{2,\ldots,n\}, \\ x_i(0) = -(i-1)\ell, & i \in \{1,\ldots,n\}. \end{cases} \quad (15)$$

We obviously have

$$\begin{cases} x_1(t) = V_{\max}\, t, & t > 0, \\ x_i(t) > -(i-1)\ell, & t > 0, \ i \in \{2,\ldots,n\}. \end{cases} \quad (16)$$

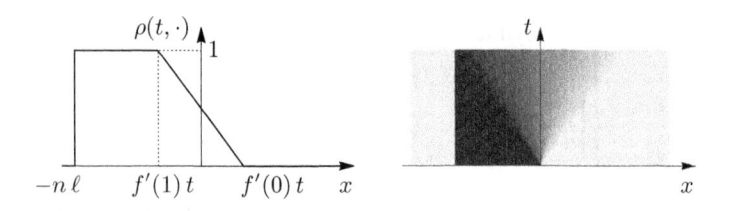

Fig. 3 Solution (18) for $t \in \bigl(0, -n\,\ell/f'(1)\bigr)$. On the right, darker (x, t)-regions correspond to higher densities

The corresponding LWR model (12) is

$$\begin{cases} \partial_t \rho + \partial_x f(\rho) = 0, & x \in \mathbb{R},\ t > 0, \\ \rho(0, x) = \begin{cases} 1 & \text{if } x \in [-n\,\ell, 0), \\ 0 & \text{otherwise}, \end{cases} & x \in \mathbb{R}. \end{cases} \quad (17)$$

In the time interval $t \in \bigl(0, -n\,\ell/f'(1)\bigr)$, its entropy solution ρ is the juxtaposition of a stationary shock $\mathsf{S}(0, 1)$ starting from $x = -n\,\ell$ and a rarefaction $\mathsf{R}(1, 0)$ starting from $x = 0$, namely

$$\rho(t, x) = \begin{cases} 1 & \text{if } -n\,\ell \leqslant x/t < f'(1) \\ (f')^{-1}(x/t) & \text{if } f'(1) \leqslant x/t < f'(0) \\ 0 & \text{otherwise} \end{cases} \quad (18)$$

$$= \begin{cases} 1 & \text{if } -n\,\ell \leqslant x/t < -V_{\max}, \\ \frac{1}{2}\left(1 - \frac{\xi}{V_{\max}}\right) & \text{if } -V_{\max} \leqslant x/t < V_{\max}, \\ 0 & \text{otherwise}, \end{cases}$$

see Fig. 3.

From (18) we can deduce the corresponding trajectories of the cars. Indeed, if $x = X_i(t)$ is the trajectory of the ith car, then

$$\begin{cases} \dot{X}_i = v\bigl(\rho(t, X_i)\bigr), & t > 0,\ i \in \{1, \ldots, n\}, \\ X_i(0) = -(i-1)\,\ell, & t > 0,\ i \in \{1, \ldots, n\}. \end{cases} \quad (19)$$

We obviously have

$$\begin{cases} X_1(t) = V_{\max}\, t, & t > 0, \\ X_i(t) = -i\,\ell, & t \in \bigl(0, -(i-1)\,\ell/f'(1)\bigr),\ i \in \{2, \ldots, n\}. \end{cases} \quad (20)$$

By comparing (16) and (20) it is clear that the trajectories corresponding to the microscopic (15) and macroscopic (17) descriptions do not coincide. Indeed, accord-

Fig. 4 Comparison of microscopic (15) (dotted) and macroscopic (19) (dashed) trajectories

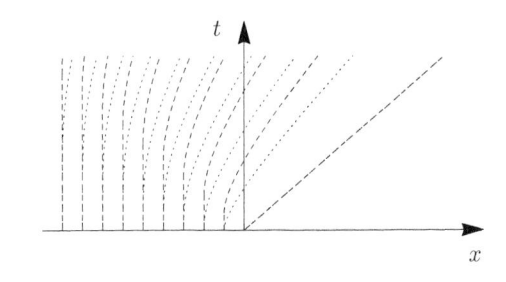

Fig. 5 The graphs of $r^1(0, \cdot)$ (dotted) and $r^2(0, \cdot)$ (dashed) considered in Example 2

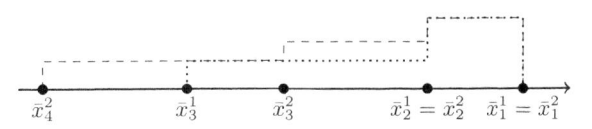

ing to the microscopic description each car start to move at time $t = 0^+$ with strictly positive speed. On the contrary, according to the macroscopic description, with exception for the rightmost car, all the other cars need some time before starting to move. In Fig. 4 we plot the trajectories corresponding to the two levels of descriptions.

Remark 1 We stress that real life experience shows that, at least in the case considered in Example 1, the macroscopic description is more appropriate than the microscopic one. Indeed, everyday experience shows that the cars don't start to move all together immediately after $t = 0$.

4.2 Order-Preservation Property

In this subsection we consider the order-preservation property. It is well known that the LWR model satisfies it: if ρ^1 and ρ^2 are solutions to (12) corresponding to the initial data $\bar\rho^1$ and $\bar\rho^2$, and $\bar\rho^1 \leqslant \bar\rho^2$, then also $\rho^1 \leqslant \rho^2$. The next example shows that FTL model (13) is not order-preserving.

Example 2 Fix $0 < \bar{x}_3^1 < \bar{x}_2^1 < \bar{x}_1^1$ and $0 < \bar{x}_4^2 < \ldots < \bar{x}_1^2$ as in Fig. 5 with

$$\bar{x}_3^2 - \bar{x}_4^2 = \bar{x}_2^1 - \bar{x}_3^1 > \bar{x}_2^2 - \bar{x}_3^2 > \bar{x}_1^2 - \bar{x}_2^2 = \bar{x}_1^1 - \bar{x}_2^1 = \ell.$$

By considering $\bar{x}_3^1, \bar{x}_2^1, \bar{x}_1^1$ and $\bar{x}_4^2, \ldots, \bar{x}_1^2$ as initial positions for FTL model (13), we obtain the discrete densities r^1 and r^2 via (14). By construction, at the initial time we have $r^1(0, \cdot) \leqslant r^2(0, \cdot)$. Clearly $\dot{x}_3^1(0) = \dot{x}_4^2(0)$ and $\dot{x}_2^1(0) = 0 < \dot{x}_3^2(0)$; thus, at least for $t > 0$ sufficiently small, we have $x_3^2(t) - x_4^2(t) > x_2^1(t) - x_3^1(t)$ and therefore $r^1(t, \cdot) > r^2(t, \cdot)$ in the space interval $\left(x_3^1(t), x_3^2(t)\right)$.

Remark 2 It should be stressed that order-preservation implies $\mathbf{L}^1$-contractivity by virtue of the classical Crandall-Tartar lemma. The semigroups of solutions of (1) considered in the present paper are $\mathbf{L}^1$-contractive (cf. [3]), governed by accretive operators on $\mathbf{L}^1$. As we can now see, the FTL semigroup (defined on densities) is conservative but not order-preserving: the FTL semigroup is *not* governed by an accretive operator on $\mathbf{L}^1$. However, since the above example concerns the case with space-homogeneous flux, the convergence results of, e.g., [29] apply. Thus we observe that the approximation of (1), although convergent and motivated by modeling considerations, fails to share the key mathematical structure of the limit problem.

It is easily seen that the lack of order-preservation for the homogeneous FTL model underlying (12) is inherited by the FTL model underlying the discontinuous-flux équation (1): indeed, its is enough to consider initial data supported in $\mathfrak{R}_+$.

Remark 3 In the context of (1), preservation by various approximation schemes of specific steady states is the cornerstone of the proofs of convergence of vanishing (or adapted) viscosity approximations, of Godunov Finite Volume approximations, or of the Wave-Front Tracking approximations to (1), see [4, 10]. In the context of fluxes of kind (3), it is not difficult to show that the FTL approximation does preserve the key steady state corresponding to the maximization of the interface flux. Yet, the absence of $\mathbf{L}^1$-contraction at the FTL level precludes us from exploiting the convergence paradigm suitable for above mentioned order-preserving approximations. This leaves space to future work on rigorous justification of the convergence behavior reported in the present paper.

5 Conservation Laws with Point Constraint on the Flow

In this section we show how the above arguments justify the Riemann solver RS_F introduced in [21] for the scalar conservation law with point constraint on the flow

$$\begin{cases} \partial_t \rho + \partial_x f(\rho) = 0, & x \in \mathbb{R}, \, t > 0, \\ \rho(0, x) = \bar{\rho}(x), & x \in \mathbb{R}, \\ f\bigl(\rho(t, 0^\pm)\bigr) \leqslant F, & t > 0. \end{cases} \qquad (21)$$

Above $\rho \in [0, 1]$ is again the (normalized) density of cars, $f \colon [0, 1] \to [0, f_{\max}]$ is the flux and $F \in (0, f_{\max})$ is the capacity of a pointwise bottleneck localized at $x = 0$, e.g., a toll gate. We assume that the flux is bell shaped, attains its maximal value $f_{\max} > 0$ at $R_{\max} \in (0, 1)$ and takes the form $f(\rho) = \rho \, v(\rho)$, where the velocity function $v \colon [0, 1] \to [0, V_{\max}]$ is a continuous non-increasing function such that $v(0) = V_{\max}$ and $v(1) = 0$.

Since we are interested only in deducing a Riemann solver, we consider only Riemann initial conditions

$$\bar{\rho}(x) \doteq \rho_L \, \mathbb{1}_{x<0}(x) + \rho_R \, \mathbb{1}_{x \geqslant 0}(x). \qquad (22)$$

As in [21] we introduce an approximate Riemann problem

$$\begin{cases} \partial_t \rho + \partial_x f_\varepsilon(\rho, x) = 0, & x \in \mathbb{R}, t > 0, \\ \rho(0, x) = \bar{\rho}(x), & x \in \mathbb{R}, \end{cases} \quad (23)$$

where $f_\varepsilon : [0, 1] \times \mathbb{R} \to [0, f_{\max}]$, $\varepsilon > 0$, is defined by

$$f_\varepsilon(\rho, x) \doteq \begin{cases} f(\rho) & \text{if } |x| \geq \varepsilon, \\ f_F(\rho) & \text{if } |x| < \varepsilon, \end{cases} \quad (24)$$

where $f_F(\rho) \doteq \rho \min\{V_F, v(\rho)\}$, with $V_F \in (0, V_{\max})$ implicitly given by

$$V_F = v(R_F), \quad R_F V_F = F, \quad R_F > R_{\max}.$$

In words, we zoom at $x = 0$, extend the location of the bottleneck to the interval $(-\varepsilon, \varepsilon)$, consider there a flux f_F having F as maximum value and corresponding to a velocity function of the form (3).

We then apply RS_M at the discontinuity points $x = \pm \varepsilon$ of f_ε, construct the approximate solution ρ_ε, let ε go to zero and obtain the limit function ρ. It is easy to seen then that $\rho \equiv \mathsf{RS}_F[\rho_L, \rho_R]$. For clarity, we consider in the next example one case.

Example 3 Fix $0 < V < V_{\max}$ and consider the flux $f(r) \doteq r \, v(r)$ with

$$v(r) \doteq \min\{V, V_{\max}(1 - r)\},$$

see Fig. 6. By definition $R_{\max} = 1 - \frac{V}{V_{\max}}$. Fix $\rho_L, \rho_R \in (0, R_{\max})$ and assume that $f(\rho_L) > F$. The solution to (22), (23) can be constructed as follows. Introduce the following notation

$$\begin{aligned} u_L &\doteq (\rho_L, f(\rho_L)), & u_R &\doteq (\rho_R, f(\rho_R)), & U_F &\doteq (R_F, F), \\ u_L^\varepsilon &\doteq (\rho_L, f_F(\rho_L)), & u_R^\varepsilon &\doteq (\rho_R, f_F(\rho_R)), & & \\ \check{u}_L &\doteq (\check{\rho}_L, f(\check{\rho}_L)), & \check{u}_R &\doteq (\check{\rho}_R, f(\check{\rho}_R)), & \check{U}_F &\doteq (\check{R}_F, f(\check{R}_F)), \end{aligned}$$

where $\check{\rho}_L, \check{\rho}_R, \check{R}_F \in (0, R_{\max})$ are implicitly defined by

$$f(\check{\rho}_L) = f_F(\rho_L), \quad f(\check{\rho}_R) = f_F(\rho_R), \quad f(\check{R}_F) = f(R_F) = F.$$

We first apply RS_M at the discontinuity points $x = \pm \varepsilon$ of the flux f_ε. At $x = -\varepsilon$ we obtain the shock $\mathsf{S}(u_L, U_F)$ with negative speed (of propagation) and the contact discontinuity $\mathsf{CD}(U_F, u_L^\varepsilon)$ with speed V_F. At $x = \varepsilon$ we obtain the (stationary) nonclassical shock $\mathsf{NS}(u_R^\varepsilon, \check{u}_R)$ and $\mathsf{CD}(\check{u}_R, u_r)$ with speed V. At $x = 0$ we apply the Lax Riemann solver associated to the flux f_F and obtain $\mathsf{CD}(u_L^\varepsilon, u_R^\varepsilon)$ with speed V_F. Once $\mathsf{CD}(u_L^\varepsilon, u_R^\varepsilon)$ reaches $x = \varepsilon$, we apply RS_M and obtain $\mathsf{NS}(u_L^\varepsilon, \check{u}_L)$ and

Fig. 6 The approximate solution constructed in Example 3

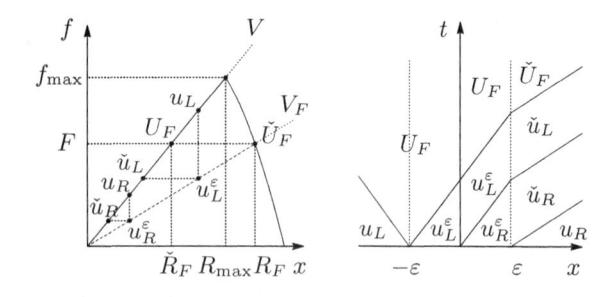

$\mathrm{CD}(\check{u}_L, \check{u}_R)$ with speed V. Analogously, when $\mathrm{CD}(U_F, u_L^\varepsilon)$ reaches $x = \varepsilon$, we apply again RS_M and obtain $\mathrm{NS}(U_F, \check{U}_F)$ and $\mathrm{CD}(\check{U}_F, \check{u}_L)$ with speed V. At last, by letting $\varepsilon \to 0^+$ we obtain

$$\mathrm{RS}_F[\rho_L, \rho_R](\xi) = \begin{cases} \rho_L & \text{if } \xi < \frac{f(\rho_L)-F}{\rho_L-R_F}, \\ R_F & \text{if } \frac{f(\rho_L)-F}{\rho_L-R_F} \leqslant \xi < 0, \\ \check{R}_F & \text{if } 0 \leqslant \xi < V, \\ \rho_R & \text{if } \xi \geqslant V. \end{cases}$$

Remark 4 Another possible way to deduce RS_F is to consider the microscopic FTL model corresponding to the Cauchy problem for (23) with truncated Riemann datum

$$\bar{\rho}(x) \doteq \rho_L \, \mathbb{1}_{[-\delta,0)}(x) + \rho_R \, \mathbb{1}_{[0,\delta]}(x).$$

By letting then $n \to +\infty$, $\delta \to +\infty$ and $\varepsilon \to 0^+$ we obtain the solution selected by RS_F for Riemann problem (21), (22).

Remark 5 We underline that in [21] it is considered the discontinuous flux (24) with $f_F(\rho) \doteq \frac{F}{f_{\max}} f(\rho)$. Even if it differs from our choice, at the limit we obtain the same result. The reason is simply because in any cases we obtain at the limit the Riemann solver RS_M which maximizes the flow.

Remark 6 Note that a yet different justification of the model of [21] can be given in terms of a velocity constraint; roughly speaking, prescribing limited velocity results in prescribing a specific flux limitation, due to the boundary layer phenomena proper to hyperbolic scalar conservation laws [2, 4, 12, 13, 23]. However this line of justification is beyond the scope of the present paper.

Acknowledgements The publication has been prepared with the support of the RUDN University Program 5-100.

References

1. Adimurthi, Mishra, S., Gowda, G.D.V.: Optimal entropy solutions for conservation laws with discontinuous flux-functions. J. Hyperbolic Differ. Equ. **2**(4), 783–837 (2005)
2. Amadori, D.: Initial-boundary value problems for nonlinear systems of conservation laws. NoDEA Nonlinear Differ. Equ. Appl. **4**(1), 1–42 (1997)
3. Andreianov, B.: The semigroup approach to conservation laws with discontinuous flux. In: Hyperbolic Conservation Laws and Related Analysis with Applications, Springer Proceedings in Mathematics and Statistics, vol. 49, pp. 1–22. Springer, Heidelberg (2014)
4. Andreianov, B.: New approaches to describing admissibility of solutions of scalar conservation laws with discontinuous flux. In: CANUM 2014–42e Congrès National d'Analyse Numérique, ESAIM Proceedings Surveys, vol. 50, pp. 40–65. EDP Sci, Les Ulis (2015)
5. Andreianov, B., Cancès, C.: Vanishing capillarity solutions of Buckley-Leverett equation with gravity in two-rocks' medium. Comput. Geosci. **17**(3), 551–572 (2013)
6. Andreianov, B., Cancès, C.: On interface transmission conditions for conservation laws with discontinuous flux of general shape. J. Hyperbolic Differ. Equ. **12**(2), 343–384 (2015)
7. Andreianov, B., Donadello, C., Razafison, U., Rosini, M.D.: Qualitative behaviour and numerical approximation of solutions to conservation laws with non-local point constraints on the flux and modeling of crowd dynamics at the bottlenecks. ESAIM Math. Model. Numer. Anal. **50**(5), 1269–1287 (2016)
8. Andreianov, B., Donadello, C., Razafison, U., Rosini, M.D.: Analysis and approximation of one-dimensional scalar conservation laws with general point constraints on the flux. J. Math. Pures Appl. **9**(116), 309–346 (2018)
9. Andreianov, B., Donadello, C., Rosini, M.D.: Crowd dynamics and conservation laws with nonlocal constraints and capacity drop. Math. Models Methods Appl. Sci. **24**(13), 2685–2722 (2014)
10. Andreianov, B., Karlsen, K.H., Risebro, N.H.: A theory of L^1-dissipative solvers for scalar conservation laws with discontinuous flux. Arch. Ration. Mech. Anal. **201**(1), 27–86 (2011)
11. Andreianov, B., Mitrović, D.: Entropy conditions for scalar conservation laws with discontinuous flux revisited. Ann. Inst. H. Poincaré Anal. Non Linéaire **32**(6), 1307–1335 (2015)
12. Andreianov, B., Sbihi, K.: Well-posedness of general boundary-value problems for scalar conservation laws. Trans. Am. Math. Soc. **367**(6), 3763–3806 (2015)
13. Bardos, C., le Roux, A.Y., Nédélec, J.C.: First order quasilinear equations with boundary conditions. Commun. Partial Differ. Equ. **4**(9), 1017–1034 (1979)
14. Benyahia, M., Rosini, M.D.: A macroscopic traffic model with phase transitions and local point constraints on the flow. Netw. Heterog. Media **12**(2), 297–317 (2017)
15. Braess, D.: Über ein paradoxon aus der verkehrsplanung. Unternehmensforschung **12**(1), 258–268 (1968)
16. Bürger, R., Karlsen, K., Risebro, N., Towers, J.: Monotone difference approximations for the simulation of clarifier-thickener units. Comput. Vis. Sci. **6**(2), 83–91 (2004)
17. Bürger, R., Karlsen, K.H., Towers, J.D.: An Engquist-Osher-type scheme for conservation laws with discontinuous flux adapted to flux connections. SIAM J. Numer. Anal. **47**(3), 1684–1712 (2009)
18. Bürger, R., Karlsen, K.H., Towers, J.D.: On some difference schemes and entropy conditions for a class of multi-species kinematic flow models with discontinuous flux. Netw. Heterog. Media **5**(3), 461–485 (2010)
19. Cancès, C.: Asymptotic behavior of two-phase flows in heterogeneous porous media for capillarity depending only on space. I. Convergence to the optimal entropy solution. SIAM J. Math. Anal. **42**(2), 946–971 (2010)

20. Colombo, R.M., Garavello, M.: Phase transition model for traffic at a junction. J. Math. Sci. (N.Y.) **196**(1), 30–36 (2014)
21. Colombo, R.M., Goatin, P.: A well posed conservation law with a variable unilateral constraint. J. Differ. Equ. **234**(2), 654–675 (2007)
22. Colombo, R.M., Rosini, M.D.: Pedestrian flows and non-classical shocks. Math. Methods Appl. Sci. **28**(13), 1553–1567 (2005)
23. Colombo, R.M., Rosini, M.D.: Well posedness of balance laws with boundary. J. Math. Anal. Appl. **311**(2), 683–702 (2005)
24. Corli, A., Rosini, M.D.: Coherence and chattering of a one-way valve. ZAMM—J. Appl. Math. Mech./Zeitschrift für Angewandte Mathematik und Mechanik (2019)
25. Dal Santo, E., Donadello, C., Pellegrino, S.F., Rosini, M.D.: Representation of capacity drop at a road merge via point constraints in a first order traffic model. ESAIM Math. Model. Numer. Anal. **53**(1), 1–34 (2019)
26. Delle Monache, M.L., Goatin, P., Piccoli, B.: Priority-based Riemann solver for traffic flow on networks. Commun. Math. Sci. **16**(1), 185–211 (2018)
27. Di Francesco, M., Fagioli, S., Rosini, M.D.: Deterministic particle approximation of scalar conservation laws. Boll. Unione Mat. Ital. **10**(3), 487–501 (2017)
28. Di Francesco, M., Fagioli, S., Rosini, M.D., Russo, G.: Follow-the-leader approximations of macroscopic models for vehicular and pedestrian flows. In: Active Particles. vol. 1. Advances in Theory, Models, and Applications, pp. 333–378. Model. Simul. Sci. Eng. Technol. Birkhäuser/Springer, Cham (2017)
29. Di Francesco, M., Rosini, M.: Rigorous derivation of nonlinear scalar conservation laws from follow-the-leader type models via many particle limit. Arch. Rat. Mech. Anal. **217**(3), 831–871 (2015)
30. Di Francesco, M., Fagioli, S., Rosini, M.D.: Many particle approximation of the Aw-Rascle-Zhang second order model for vehicular traffic. Math. Biosci. Eng. **14**(1), 127–141 (2017)
31. Di Francesco, M., Fagioli, S., Rosini, M.D., Russo, G.: Deterministic particle approximation of the Hughes model in one space dimension. Kinet. Relat. Models **10**(1), 215–237 (2017)
32. Diehl, S.: Continuous sedimentation of multi-component particles. Math. Methods Appl. Sci. **20**(15), 1345–1364 (1997)
33. Diehl, S.: A uniqueness condition for nonlinear convection-diffusion equations with discontinuous coefficients. J. Hyperbolic Differ. Equ. **6**(1), 127–159 (2009)
34. Garavello, M., Goatin, P.: The Aw-Rascle traffic model with locally constrained flow. J. Math. Anal. Appl. **378**(2), 634–648 (2011)
35. Garavello, M., Natalini, R., Piccoli, B., Terracina, A.: Conservation laws with discontinuous flux. Netw. Heterog. Media **2**(1), 159–179 (2007)
36. Gimse, T., Risebro, N.H.: Solution of the Cauchy problem for a conservation law with a discontinuous flux function. SIAM J. Math. Anal. **23**(3), 635–648 (1992)
37. Hopf, E.: The partial differential equation $u_t + uu_x = \mu u_{xx}$. Commun. Pure Appl. Math. **3**, 201–230 (1950)
38. Kaasschieter, E.F.: Solving the Buckley-Leverett equation with gravity in a heterogeneous porous medium. Comput. Geosci. **3**(1), 23–48 (1999)
39. Karlsen, K.H., Risebro, N.H., Towers, J.D.: L^1 stability for entropy solutions of nonlinear degenerate parabolic convection-diffusion equations with discontinuous coefficients. Skr. K. Nor. Vidensk. Selsk. **3**, 1–49 (2003)
40. Kolb, O., Costeseque, G., Goatin, P., Göttlich, S.: Pareto-optimal coupling conditions for the Aw-Rascle-Zhang traffic flow model at junctions. SIAM J. Appl. Math. **78**(4), 1981–2002 (2018)
41. Kruzhkov, S.N.: First order quasilinear equations with several independent variables. Mat. Sb. (N.S.) **81 (123)**, 228–255 (1970)
42. Lax, P.: Shock Waves and Entropy, pp. 603–634 (1971)
43. LeFloch, P.G.: Hyperbolic Systems of Conservation Laws. Lectures in Mathematics ETH Zürich, Birkhäuser Verlag, Basel (2002)

44. Lighthill, M., Whitham, G.: On kinematic waves. II. A theory of traffic flow on long crowded roads. In: Royal Society of London. Series A, Mathematical and Physical Sciences. vol. 229, pp. 317–345 (1955)
45. Moutari, S., Herty, M., Klein, A., Oeser, M., Steinauer, B., Schleper, V.: Modelling road traffic accidents using macroscopic second-order models of traffic flow. IMA J. Appl. Math. **78**(5), 1087–1108 (2013)
46. Oleĭnik, O.A.: Uniqueness and stability of the generalized solution of the Cauchy problem for a quasi-linear equation. Uspehi Mat. Nauk **14**(2(86)), 165–170 (1959)
47. Rayleigh, L.: Aerial plane waves of finite amplitude [Proc. Roy. Soc. London Ser. A **84** (1910), 247–284]. In: Classic papers in shock compression science, pp. 361–404. High-press. Shock Compression Condens. Matter, Springer, New York (1998)
48. Richards, P.I.: Shock waves on the highway. Oper. Res. **4**(1), 42–51 (1956)
49. Ross, D.S.: Two new moving boundary problems for scalar conservation laws. Commun. Pure Appl. Math. **41**(5), 725–737 (1988)
50. Seguin, N., Vovelle, J.: Analysis and approximation of a scalar conservation law with a flux function with discontinuous coefficients. Math. Models Methods Appl. Sci. **13**(2), 221–257 (2003)
51. Shen, W.: Traveling wave profiles for a follow-the-leader model for traffic flow with rough road condition. Netw. Heterog. Media **13**(3), 449–478 (2018)
52. Srivastava, A., Geroliminis, N.: Empirical observations of capacity drop in freeway merges with ramp control and integration in a first-order model. Transp. Res. Part C: Emerg. Technol. **30**, 161–177 (2013)
53. Towers, J.D.: Convergence of a difference scheme for conservation laws with a discontinuous flux. SIAM J. Numer. Anal. **38**(2), 681–698 (2000)
54. Villa, S., Goatin, P., Chalons, C.: Moving bottlenecks for the Aw-Rascle-Zhang traffic flow model. Discrete Contin. Dyn. Syst. Ser. B **22**(10), 3921–3952 (2017)
55. Vol'pert, A.I.: Spaces BV and quasilinear equations. Mat. Sb. (N.S.) **73 (115)**, 255–302 (1967)

Newton's Method for the McKendrick-von Foerster Equation

Agnieszka Bartłomiejczyk and Monika Wrzosek

Abstract In the paper we study an age-structured model which describes the dynamics of one population with growth, reproduction and mortality rates. We apply Newton's method to the McKendrick-von Foerster equation in the semigroup setting. We prove its first- and second-order convergence.

Keywords Newton's method · McKendrick-von Foerster equation · Semigroup

2010 Mathematics Subject Classifications 35L04, 49M15

1 Introduction

Analyzis of the variability of the distribution of individuals, with particular consideration of models in which a structural variable associated with the physiological properties of individuals appears, is the direction of research of many scientists [4, 17, 19]. In [2] it has been shown how the variability of an individual trait (e.g. the size of the individual) affects the dynamics of the entire population. Such a feature may be not only the size of the individual, but also the age, maturity or location in space. The oldest structural models include a system describing the age distribution of the population. Integral equation models for age-structured populations were first posed in 1911 by Sharpe and Lotka [18]. Five years later McKendrick [14] developed a partial differential equation model, which was then formulated independently by von Foerster in his seminal 1959 paper [5]. Hence it is often referenced as

---

A. Bartłomiejczyk (✉)
Faculty of Applied Physics and Mathematics, Gdańsk University of Technology, Gabriela Narutowicza 11/12, 80-233 Gdańsk, Poland
e-mail: agnbartl@pg.edu.pl

M. Wrzosek
Institute of Mathematics, University of Gdańsk, Wita Stwosza 57, 80-952 Gdańsk, Poland
e-mail: mwrzosek@mat.ug.edu.pl

© Springer Nature Switzerland AG 2020
J. Banasiak et al. (eds.), *Semigroups of Operators – Theory and Applications*,
Springer Proceedings in Mathematics & Statistics 325,
https://doi.org/10.1007/978-3-030-46079-2_8

Lotka-McKendrick, McKendrick-von Foerster or von Foerster equation (see [8] for more details).

The McKendrick-von Foerster equation is a basic tool of mathematical modelling of phenomena in demography and epidemiology. In this equation a one-sex female population is described by the density $u(t, x)$ of individuals of age x at time t. The model is often written as

$$\frac{\partial u}{\partial t} + c\frac{\partial u}{\partial x} = \lambda u, \quad (t, x) \in [0, T] \times \mathbb{R}_+ \tag{1.1}$$

with an initial density $u_0(x)$ and the renewal condition

$$u(t, 0) = \int_0^\infty k(y) u(t, y) \, dy, \quad t \in [0, T],$$

which is interpreted as giving birth to young individuals of age 0 by mature individuals. The function λ is called the mortality function or death modulus and the function k, governing the birth process, is called the birth modulus. The function c is the growth rate of e.g. the age of the individual, its size or cell maturity. Thus, the individual's maturity (or more generally aging of the population) changes according to the equation $dx/dt = c$. So, if the growth rate c is constant, then the characteristics (i.e. the solutions of $dx/dt = const.$) are straight lines in the (t, x)–plot with slope equal to c. A more realistic approach enforces the dependence of the function c on age x of the individual. In this context the solution propagates along characteristics, which are the curves $t - c(x) = const.$

In this paper we are interested in a more general situation when a function c depends on time and age of population, i.e. $c = c(t, x)$. Note that, if we treat the Eq. (1.1) as the advection equation (in fluid mechanics), then the function $c(t, x)$ is interpreted as velocity of the fluid at time t in position x. In addition, the generalization of the model (1.1) presented here concerns also the function λ. Namely, in our paper the death modulus λ is the function of time t, age x, the population density u and the total population size $\int_0^\infty u(t, x) dx$. Existence and uniqueness results for such forms of McKendrick-von Foerster equation are established in [3, 6, 10].

Since our paper concerns numerical methods we give a brief review of the problems being discussed in the literature. An interesting review of numerical methods applied to age-structured models, their stability and convergence results can be found in [1]. More recent studies include numerical methods for McKendrick-von Foerster equation with functional dependence: the method of lines [11], the Rothe method [12], the finite difference method [13]. In [7] authors consider the case of a finite lifespan of the population and prove the convergence of finite difference methods with the unbounded mortality rate. Some algorithms inspired by the approach based on Volterra integral equations are designed in [16]. In [21] a numerical method based on Bernstein polynomials approximation is presented. A discontinuous Galerkin finite element method is applied in [9]. The probabilistic second-order convergence of

Newton's method for first-order hyperbolic partial differential equations subject to random perturbations is proved in [20]. The Newton method can be combined with numerical discretization methods to obtain a significant acceleration of convergence.

The paper is organized as follows. In Sect. 2 we introduce the basic notation and reformulate a generalized form of the McKendrick-von Foerster equation in the semigroup setting. In Sect. 3 we apply Newton's method to the McKendrick-von Foerster equation. Its first- and second-order convergence is proved in Sect. 4.

2 Formulation of the Problem

Let $T > 0$ and denote $\mathbb{R}_+ = [0, \infty)$. Suppose that $c : [0, T] \times \mathbb{R}_+ \to (0, \infty]$, $\lambda : [0, T] \times \mathbb{R}_+^3 \to \mathbb{R}_+$, $u_0 : \mathbb{R}_+ \to \mathbb{R}_+$ and $k : \mathbb{R}_+ \to \mathbb{R}_+$ are given functions. We consider a generalized form of the McKendrick-von Foerster equation (1.1), i.e.

$$\frac{\partial u}{\partial t}(t, x) + c(t, x) \frac{\partial u}{\partial x}(t, x) = u(t, x) \lambda \left(t, x, u(t, x), \int_0^\infty |u(t, y)| \, dy \right) \quad (2.1)$$

for $(t, x) \in [0, T] \times \mathbb{R}_+$ with the initial condition

$$u(0, x) = u_0(x), \quad x \in \mathbb{R}_+ \quad (2.2)$$

and the renewal condition

$$u(t, 0) = \int_0^\infty k(y) u(t, y) \, dy, \quad t \in [0, T]. \quad (2.3)$$

This model is nonlocal as it contains the total size of population $\int_0^\infty |u(t, y)| \, dy$.

We make the following assumptions.

(A1) $c : [0, T] \times \mathbb{R}_+ \to \mathbb{R}_+$ is positive, bounded, continuous in (t, x) and satisfies

$$|c(t, x) - c(t, \bar{x})| \leq L_c |x - \bar{x}|$$

for some $L_c > 0$.

(A2) $\lambda : [0, T] \times \mathbb{R}_+^3 \to \mathbb{R}$ is continuous in (t, x, p, q) and satisfies

$$|\lambda(t, x, p, q) - \lambda(t, \bar{x}, \bar{p}, \bar{q})| \leq L_\lambda (|x - \bar{x}| + |p - \bar{p}| + |q - \bar{q}|)$$

for some $L_\lambda > 0$.

(A3) $|\lambda(t, x, p, q)| \leq C_\lambda$ for some $C_\lambda > 0$.

(A4) Derivatives $\partial_p \lambda, \partial_q \lambda$ are continuous and satisfy $|\partial_p \lambda| \leq L_\lambda$, $|\partial_q \lambda| \leq L_\lambda$.

(A5) $k : \mathbb{R}_+ \to \mathbb{R}_+$ is a non-negative, continuous and bounded function.
(A6) $u_0 : \mathbb{R}_+ \to \mathbb{R}_+$ is non-negative, continuous and bounded with some $M_0 > 0$.

Under assumptions (A1)–(A6) there exists a unique solution to problem (2.1)–(2.3), see [6].

We denote by $\mathcal{X}$ the space of all functions $f : \mathbb{R}_+ \to \mathbb{R}_+$ such that $f \in L^1 \cap L^\infty \cap C$ and with the norm

$$|f|_{\mathcal{X}} = \max\{|f|_{L^1}, |f|_{L^\infty}\}, \qquad (2.4)$$

where

$$|f|_{L^1} = \int_0^\infty |f(x)|\,dx < \infty \quad \text{and} \quad |f|_{L^\infty} = \operatorname{ess\,sup}_{x \in \mathbb{R}_+} |f(x)| < \infty.$$

Let $\mathcal{X}_{ren}$ stand for these functions $f \in \mathcal{X}$ which additionally satisfy the renewal condition, i.e.

$$f(0) = \int_0^\infty k(y)\,f(y)\,dy.$$

Note that since $\mathcal{X}_{ren}$ is a closed subset of $\mathcal{X}$, the norms are the same in both spaces. So we use the same symbol $|\cdot|_{\mathcal{X}}$ to denote the norm (2.4) for functions $f \in \mathcal{X}_{ren} \subset \mathcal{X}$.

Let us apply the semigroup approach to the Eq. (2.1). For $u(t,\cdot) \in \mathcal{X}_{ren}$ denote $\mathbf{u}(t) = u(t,\cdot)$, $\mathbf{u} : [0, T] \to \mathcal{X}_{ren}$. For each $t \in [0, T]$ let $\mathcal{A}(t) : \mathcal{D}(\mathcal{A}(t)) \subset \mathcal{X}_{ren} \to \mathcal{X}$ with

$$\mathcal{D}(\mathcal{A}(t)) := \{\mathbf{u} \in C^1 : \mathbf{u} \in \mathcal{X}_{ren}\}$$

be the operator given by

$$\mathcal{A}(t) = c(t, x)\frac{\partial}{\partial x}, \quad x \in \mathbb{R}_+.$$

Problem (2.1)–(2.3) can be rewritten as follows

$$\frac{d\mathbf{u}}{dt} + \mathcal{A}(t)\mathbf{u} = \mathbf{u}\,\Lambda(t, \mathbf{u}), \quad t \in [0, T], \qquad (2.5)$$

$$\mathbf{u}(0) = u_0,$$

where $\Lambda : [0, T] \times \mathcal{X}_{ren} \to L^\infty \cap C$, $\Lambda(t, \mathbf{u}) = \lambda(t, x, u, \int_0^\infty |u(t, y)|\,dy)$ and $u_0 \in \mathcal{D}(\mathcal{A}(t))$. Assumptions (A2)–(A4) take the following form

(A2') $|\Lambda(t, \mathbf{u}) - \Lambda(t, \bar{\mathbf{u}})|_{L^\infty} \le L_\lambda |\mathbf{u} - \bar{\mathbf{u}}|_{\mathcal{X}}$,
(A3') $|\Lambda(t, \mathbf{u})|_{L^\infty} \le C_\lambda$,
(A4') $|D\Lambda(t, \mathbf{u})|_{\mathcal{L}(\mathcal{X}_{ren}, L^\infty)} \le L_\lambda$,

where $\mathcal{L}(\mathcal{X}_{ren}, L^\infty)$ denotes the space of all linear bounded operators from $\mathcal{X}_{ren}$ to L^∞ and

$$D\Lambda(t, \mathbf{u}(t))\mathbf{h}(t) := \partial_p \lambda(t, x, u(t, x), |u(t, \cdot)|_{L^1})h(t, x)$$
$$+ \partial_q \lambda(t, x, u(t, x), |u(t, \cdot)|_{L^1}) \int_0^\infty h(t, y)dy,$$

$$|D\Lambda(t, \mathbf{u})|_{\mathcal{L}(\mathcal{X}_{ren}, L^\infty)} := \sup_{|\mathbf{h}(t)|_\mathcal{X} \leq 1} |D\Lambda(t, \mathbf{u}(t))\mathbf{h}(t)|_{L^\infty}$$

for $\mathbf{h}(t) = h(t, \cdot) \in \mathcal{X}_{ren}$. It can be shown that the family of operators $\{\mathcal{A}(t)\}_{t \in [0,T]}$ satisfies the "hyperbolic" conditions (cf. [15], p.135), so that there exists a unique evolution system $U(t, s), 0 \leq s \leq t \leq T$ i.e.

(1) $U(t, t) = I, U(t, s) = U(t, r)U(r, s)$ for $0 \leq s \leq r \leq t \leq T$,
(2) $(t, s) \mapsto U(t, s)$ is strongly continuous for $0 \leq s \leq r \leq t \leq T$.

The evolution system $U(t, s), 0 \leq s \leq t \leq T$ is called the fundamental solution to homogeneous problem associated with (2.5) (see Theorem 3.1, [15]). Applying the Duhamel principle (or variation of constants formula) we obtain the mild solution to inhomogeneous problem (2.5)

$$\mathbf{u}(t) = U(t, 0)u_0 + \int_0^t U(t, s)\mathbf{u}(s)\Lambda(s, \mathbf{u}(s))ds, \quad t \in [0, T]. \quad (2.6)$$

Notice that by (2.6), $U(t, s) : \mathcal{X} \to \mathcal{X}$ for $0 \leq s \leq t \leq T$. Moreover, we assume that there exists $N > 0$ such that

$$|U(t, s)|_{\mathcal{L}(\mathcal{X}, \mathcal{X})} \leq N \quad \text{for} \quad 0 \leq s \leq t \leq T. \quad (2.7)$$

3 Newton's Method

Now we formulate the Newton scheme to solve the initial value problem (2.5). Let $\mathbf{u}^{(0)} = u_0$ and

$$\begin{cases} \dfrac{d}{dt}\mathbf{u}^{(k+1)} + \mathcal{A}(t)\mathbf{u}^{(k+1)} = \mathbf{u}^{(k)}\Lambda(t, \mathbf{u}^{(k)}) + \mathbf{u}^{(k)}D\Lambda(t, \mathbf{u}^{(k)})\varepsilon^{(k)} \\ \qquad\qquad\qquad\qquad + \Lambda(t, \mathbf{u}^{(k)})\varepsilon^{(k)}, \quad t \in [0, T] \\ \mathbf{u}^{(k+1)}(0) = u_0, \end{cases} \quad (3.1)$$

where $\varepsilon^{(k)} := \mathbf{u}^{(k+1)} - \mathbf{u}^{(k)}$, $k = 0, 1, \ldots$ We have the following differential equations for increments

$$\begin{cases} \dfrac{d}{dt}\varepsilon^{(k+1)} + \mathcal{A}(t)\varepsilon^{(k+1)} = \mathbf{u}^{(k+1)}\Lambda(t,\mathbf{u}^{(k+1)}) - \mathbf{u}^{(k)}\Lambda(t,\mathbf{u}^{(k)}) \\ \qquad\qquad\qquad\qquad + \mathbf{u}^{(k+1)}D\Lambda(t,\mathbf{u}^{(k+1)})\varepsilon^{(k+1)} - \mathbf{u}^{(k)}D\Lambda(t,\mathbf{u}^{(k)})\varepsilon^{(k)} \\ \qquad\qquad\qquad\qquad + \Lambda(t,\mathbf{u}^{(k+1)})\varepsilon^{(k+1)} - \Lambda(t,\mathbf{u}^{(k)})\varepsilon^{(k)}, \quad t \in [0,T] \\ \varepsilon^{(k+1)}(0) = 0 \end{cases}$$

(3.2)

for $k = 0, 1, \ldots$

Lemma 3.1 *The Newton sequence* $(\mathbf{u}^{(k)})_{k\in\mathbb{N}}$ *defined by* (3.1) *preserves the class* $\mathcal{X}_{ren}$ *i.e.* $\mathbf{u}^{(k)}(t) \in \mathcal{X}_{ren}$ *for* $t \in [0,T]$, $k = 0, 1, \ldots$ *Moreover, there exists* $M_u > 0$ *such that* $|\mathbf{u}^{(k)}(t)|_{\mathcal{X}} < M_u$ *for* $t \in [0,T]$, $k = 0, 1, \ldots$

Proof It suffices to consider the following differential equation

$$\begin{cases} \dfrac{d}{dt}\mathbf{z} = C\left(\mathbf{z} + \mathbf{z}^2\right), \\ \mathbf{z}(0) = |u_0|_{\mathcal{X}}, \end{cases}$$

which solution

$$\mathbf{z}(t) = \dfrac{|u_0|_{\mathcal{X}} e^{Ct}}{1 + |u_0|_{\mathcal{X}} - |u_0|_{\mathcal{X}} e^{Ct}}$$

is bounded for $t \in [0, T]$.

4 Main Results

In the following theorem we establish the first-order convergence of the Newton method applied to (2.5).

Theorem 1 *Suppose that assumptions* (A2')–(A4') *and* (2.7) *hold. Then the Newton sequence* $(\mathbf{u}^{(k)})_{k\in\mathbb{N}}$ *defined by* (3.1) *converges to the unique solution* $\mathbf{u}$ *of problem* (2.1)–(2.3) *in the following sense*

$$\lim_{k\to\infty} |\mathbf{u}^{(k)}(t) - \mathbf{u}(t)|_{\mathcal{X}} = 0, \quad t \in [0,T].$$

Moreover, for some $C_0 > 0$:

$$|\mathbf{u}^{(k+1)}(t) - \mathbf{u}^{(k)}(t)|_{\mathcal{X}} \leq C_0 \dfrac{C^k t^k}{k!} |\mathbf{u}^{(k)}(t) - \mathbf{u}^{(k-1)}(t)|_{\mathcal{X}}, \quad k = 0, 1, \ldots$$

Proof We show that $(\mathbf{u}^{(k)})_{k\in\mathbb{N}}$ satisfies the Cauchy condition with respect to the norm $|\cdot|_{\mathcal{X}}$. Rewrite (3.2) into the following form

$$\begin{cases} \dfrac{d}{dt}\varepsilon^{(k+1)} + \mathcal{A}(t)\varepsilon^{(k+1)} = \mathbf{u}^{(k+1)}\left(\Lambda(t,\mathbf{u}^{(k+1)}) - \Lambda(t,\mathbf{u}^{(k)})\right) - \mathbf{u}^{(k)}D\Lambda(t,\mathbf{u}^{(k)})\varepsilon^{(k)} \\ \qquad\qquad + \Lambda(t,\mathbf{u}^{(k+1)})\varepsilon^{(k+1)} + \mathbf{u}^{(k+1)}D\Lambda(t,\mathbf{u}^{(k+1)})\varepsilon^{(k+1)}, \\ \varepsilon^{(k+1)}(0) = 0 \end{cases}$$

(4.1)

for $k = 0, 1, \ldots$ First, we estimate L^∞ norm of $\varepsilon^{(k+1)}$:

$$|\varepsilon^{(k+1)}(t)|_{L^\infty} \leq \left|\int_0^t U(t,s)\left[\mathbf{u}^{(k+1)}(s)\left(\Lambda(t,\mathbf{u}^{(k+1)}(s)) - \Lambda(s,\mathbf{u}^{(k)}(s))\right)\right]ds\right|_{L^\infty}$$

$$+ \left|\int_0^t U(t,s)\left[\mathbf{u}^{(k)}(s)\left(D\Lambda(s,\mathbf{u}^{(k)}(s))\right)\left[\varepsilon^{(k)}(s)\right]\right)\right]ds\right|_{L^\infty}$$

$$+ \left|\int_0^t U(t,s)\left[\Lambda(s,\mathbf{u}^{(k+1)}(s))\varepsilon^{(k+1)}(s)\right.\right.$$

$$\left.\left. + \mathbf{u}^{(k+1)}(s)\left(D\Lambda(s,\mathbf{u}^{(k+1)}(s))\right)\left[\varepsilon^{(k+1)}(s)\right]\right)\right]ds\right|_{L^\infty}$$

$$= I_1 + I_2 + I_3.$$

By (2.7), Lemma 3.1 and the Lipschitz condition (A2') we have

$$I_1 \leq \int_0^t |U(t,s)|_{\mathcal{L}(\mathcal{X},\mathcal{X})}\left|\mathbf{u}^{(k+1)}(s)\cdot\left(\Lambda(t,\mathbf{u}^{(k+1)}(s)) - \Lambda(s,\mathbf{u}^{(k)}(s))\right)\right|_{L^\infty}ds$$

$$\leq N \int_0^t |\mathbf{u}^{(k+1)}(s)|_{L^\infty}\left|\Lambda(t,\mathbf{u}^{(k+1)}(s)) - \Lambda(s,\mathbf{u}^{(k)}(s))\right|_{L^\infty}ds$$

$$\leq L_\lambda M_u N \int_0^t |\varepsilon^{(k)}(s)|_{\mathcal{X}}ds.$$

Applying (2.7), Lemma 3.1 and (A4') we obtain

$$I_2 \leq N \int_0^t |\mathbf{u}^{(k)}(s)|_{\mathcal{X}}|D\Lambda(s,\mathbf{u}^{(k)}(s))|_{\mathcal{L}(\mathcal{X},L^\infty)}|\varepsilon^{(k)}(s)|_{L^\infty}ds$$

$$\leq L_\lambda M_u N \int_0^t |\varepsilon^{(k)}(s)|_{\mathcal{X}}ds.$$

From (2.7), (A3'), Lemma 3.1 and (A4') we get

$$I_3 \leq N \int_0^t \left\{ |\Lambda(s, \mathbf{u}^{(k+1)}(s))|_{L^\infty} + |\mathbf{u}^{(k+1)}(s)|_{\mathcal{X}} |D\Lambda(s, \mathbf{u}^{(k+1)}(s))|_{\mathcal{L}(\mathcal{X}_{ren}, L^\infty)} \right\}$$

$$\times |\varepsilon^{(k+1)}(s)|_{\mathcal{X}} ds \leq N(C_\lambda + L_\lambda M_u) \int_0^t |\varepsilon^{(k+1)}(s)|_{\mathcal{X}} ds.$$

Hence

$$|\varepsilon^{(k+1)}(t)|_{L^\infty} \leq I_1 + I_2 + I_3$$

$$= 2L_\lambda M_u N \int_0^t |\varepsilon^{(k)}(s)|_{\mathcal{X}} ds + N(C_\lambda + L_\lambda M_u) \int_0^t |\varepsilon^{(k+1)}(s)|_{\mathcal{X}} ds.$$

Similarly, we have L^1 norm estimate:

$$|\varepsilon^{(k+1)}(t)|_{L^1} \leq 2L_\lambda M_u N \int_0^t |\varepsilon^{(k)}(s)|_{\mathcal{X}} ds + N(C_\lambda + L_\lambda M_u) \int_0^t |\varepsilon^{(k+1)}(s)|_{\mathcal{X}} ds.$$

Hence

$$|\varepsilon^{(k+1)}(t)|_{\mathcal{X}} \leq 2L_\lambda M_u N \int_0^t |\varepsilon^{(k)}(s)|_{\mathcal{X}} ds + N(C_\lambda + L_\lambda M_u) \int_0^t |\varepsilon^{(k+1)}(s)|_{\mathcal{X}} ds$$

for $t \in [0, T]$, $k = 0, 1, \ldots$ Applying the Gronwall inequality, we obtain

$$|\varepsilon^{(k+1)}(t)|_{\mathcal{X}} \leq C \int_0^t |\varepsilon^{(k)}(s)|_{\mathcal{X}} ds, \quad t \in [0, T], \ k = 0, 1, \ldots, \tag{4.2}$$

where

$$C = 2L_\lambda M_u N e^{N(C_\lambda + M_u L_\lambda)T}.$$

The recursive use of (4.2) leads to

$$|\varepsilon^{(k)}(t)|_{\mathcal{X}} \leq \frac{C^k t^k}{k!} |\varepsilon^{(0)}(t)|_{\mathcal{X}} \leq 2M_u \frac{C^k t^k}{k!}, \quad k = 0, 1, \ldots$$

which completes the proof.

The additional assumption on the derivative of function Λ to satisfy the Lipschitz condition results in the higher order of convergence of the Newton method applied to (2.5).

Theorem 2 *Suppose (A2')–(A4'), (2.7) hold and assume additionally that there exists a non-negative constant L_D such that*

$$|D\Lambda(t, \boldsymbol{u}) - D\Lambda(t, \bar{\boldsymbol{u}})|_{\mathcal{L}(\mathcal{X}_{ren}, L^\infty)} \leq L_D |\boldsymbol{u}(t) - \bar{\boldsymbol{u}}(t)|_{\mathcal{X}}, \quad t \in [0, T]. \qquad (4.3)$$

Then the Newton method (3.1) is of the second order:

$$|\boldsymbol{u}^{(k+1)}(t) - \boldsymbol{u}^{(k)}(t)|_{\mathcal{X}} \leq C_1 |\boldsymbol{u}^{(k)}(t) - \boldsymbol{u}^{(k-1)}(t)|_{\mathcal{X}}^2, \quad k = 0, 1, \ldots$$

for some $C_1 = C_1(T) > 0$.

Proof We show that there exists some constant $\widetilde{C} > 0$ such that

$$|\varepsilon^{(k+1)}(t)|_{\mathcal{X}} \leq \widetilde{C} \int_0^t |\varepsilon^{(k)}(s)|_{\mathcal{X}}^2 ds, \quad t \in [0, T], \ k = 0, 1, \ldots \qquad (4.4)$$

By the Taylor formula, we have

$$\Lambda(t, \boldsymbol{u}^{(k+1)}) - \Lambda(t, \boldsymbol{u}^{(k)}) = D\Lambda(t, \bar{\boldsymbol{u}}^{(k)})\varepsilon^{(k)}.$$

Hence (4.1) takes the form

$$\begin{cases} \dfrac{d}{dt}\varepsilon^{(k+1)} + \mathcal{A}(t)\varepsilon^{(k+1)} = \boldsymbol{u}^{(k+1)} D\Lambda(t, \bar{\boldsymbol{u}}^{(k)})\varepsilon^{(k)} - \boldsymbol{u}^{(k)} D\Lambda(t, \boldsymbol{u}^{(k)})\varepsilon^{(k)} \\ \qquad\qquad\qquad + \Lambda(t, \boldsymbol{u}^{(k+1)})\varepsilon^{(k+1)} + \boldsymbol{u}^{(k+1)} D\Lambda(t, \boldsymbol{u}^{(k+1)})\varepsilon^{(k+1)}, \\ \varepsilon^{(k+1)}(0) = 0 \end{cases}$$

for $k = 0, 1, \ldots$ Moreover, we have

$$\boldsymbol{u}^{(k+1)} D\Lambda(t, \bar{\boldsymbol{u}}^{(k)})\varepsilon^{(k)} - \boldsymbol{u}^{(k)} D\Lambda(t, \boldsymbol{u}^{(k)})\varepsilon^{(k)}$$
$$= \boldsymbol{u}^{(k+1)} \left(D\Lambda(t, \bar{\boldsymbol{u}}^{(k)}) - D\Lambda(t, \boldsymbol{u}^{(k)}) \right) \varepsilon^{(k)} - \varepsilon^{(k)} D\Lambda(t, \boldsymbol{u}^{(k)})\varepsilon^{(k)}, \quad k = 0, 1, \ldots$$

Lemma 3.1 and assumptions (A2')–(A4'), (2.7), (4.3) yield

$$|\varepsilon^{(k+1)}(t)|_{\mathcal{X}} \leq N(M_u L_D + L_\lambda) \int_0^t |\varepsilon^{(k)}(s)|_{\mathcal{X}}^2 ds + (C_\lambda + M_u L_\lambda) N \int_0^t |\varepsilon^{(k+1)}(s)|_{\mathcal{X}} ds$$

for $t \in [0, T], k = 0, 1, \ldots$ Applying the Gronwall inequality we obtain (4.4) with

$$\widetilde{C} = N(M_u L_D + L_\lambda)e^{N(C_\lambda + M_u L_\lambda)T},$$

which completes the proof.

Acknowledgements We would like to thank Karolina Lademann for discussions on this topic.

References

1. Abia, L.M., Angulo, O., López-Marcos, J.C.: Age-structured population models and their numerical solution. Ecol. Model. **188**, 112–136 (2005)
2. Bartłomiejczyk, A., Leszczyński, H.: Structured populations with diffusion and Feller conditions. Math. Biosci. Eng. **13**(2), 261–279 (2016)
3. Bartłomiejczyk, A., Leszczyński, H., Zwierkowski, P.: Existence and uniqueness of solutions for single-population McKendrick-von Foerster models with renewal. Rocky Mt. J. Math. **45**, 401–426 (2015)
4. Brauer, F., Castillo-Chávez, C.: Mathematical Models in Population Biology and Epidemiology. Springer, New York (2012)
5. von Foerster, H.: Some remarks on changing populations. In: The Kinetics of Cellular Proliferation, Grune and Stratton. New York (1959)
6. Gurtin, M.E., MacCamy, R.C.: Non-linear age-dependent population dynamics. Arch. Ration. Mech. Anal. **54**, 281–300 (1974)
7. Iannelli, M., Milner, F.A.: On the approximation of the Lotka-McKendrick equation with finite life-span. J. Comput. Appl. Math. **136**, 245–254 (2001)
8. Keyfitz, B.L., Keyfitz, N.: The McKendrick partial differential equation and its uses in epidemiology and population study. Math. Comput. Model. **26**, 1–9 (1997)
9. Kim, M.Y., Selenge, T.: Age-time discontinuous Galerkin method for the Lotka-McKendrick equation. Commun. Korean Math. Soc. **18**, 569–580 (2003)
10. Leszczyński, H.: Differential functional von Foerster equations with renewal. Cond. Mat. Phys. **54**, 361–370 (2008)
11. Leszczyński, H., Zwierkowski, P.: Stability of the method of lines for the McKendrick-von Foerster equation. Funct. Differ. Equ. **20**, 201–225 (2013)
12. Leszczyński, H., Zwierkowski, P.: The Rothe method for the McKendrick-von Foerster equation. Czechoslovak Math. J. **63**, 589–602 (2013)
13. Leszczyński, H., Zwierkowski, P.: Stability of finite difference schemes for generalized von Foerster equations with renewal. Opuscula Math. **34**, 375–386 (2014)
14. McKendrick, A.G.: Applications of mathematics to medical problems. Proc. Edinb. Math. Soc. **44**, 98–130 (1926)
15. Pazy, A.: Semigroups of Linear Operators and Applications to Partial Differential Equations. Springer, Berlin (1983)
16. Pelovska, G., Iannelli, M.: Numerical methods for the Lotka-McKendrick's equation. J. Comput. Appl. Math. **197**, 534–557 (2006)
17. Polański, A., Bobrowski, A., Kimmel, M.: A note on distributions of times to coalescence, under time-dependent population size. Theor. Popul. Biol. **63**, 33–40 (2003)
18. Sharpe, F.R., Lotka, A.J.: A problem in age-distribution. Philos. Mag. **21**, 435–438 (1911)
19. Ważewska-Czyżewska, M., Lasota, A.: Mathematical problems of the dynamics of the red blood cells. Mathematica Applicanda **4**, 23–40 (1976)
20. Wrzosek, M.: Newton's method for first-order stochastic functional partial differential equations. Comment. Math. **54**, 51–64 (2014)
21. Yousefi, S.A., Behroozifar, M., Dehghan, M.: Numerical solution of the nonlinear age-structured population models by using the operational matrices of Bernstein polynomials. Appl. Math. Model. **36**, 945–963 (2012)

# Singular Thermal Relaxation Limit for the Moore-Gibson-Thompson Equation Arising in Propagation of Acoustic Waves

Marcelo Bongarti, Sutthirut Charoenphon, and Irena Lasiecka

Abstract Moore-Gibson-Thompson (MGT) equations, which describe acoustic waves in a heterogeneous medium, are considered. These are the third order in time evolutions of a predominantly hyperbolic type. MGT models account for a finite speed propagation due to the appearance of thermal relaxation coefficient $\tau > 0$ in front of the third order time derivative. Since the values of τ are relatively small and often negligible, it is important to understand the asymptotic behavior and characteristics of the model when $\tau \to 0$. This is a particularly delicate issue since the $\tau-$dynamics is governed by a generator which is singular as $\tau \to 0$. It turns out that the limit dynamics corresponds to the linearized Westervelt equation which is of a parabolic type. In this paper, we provide a rigorous analysis of the asymptotics which includes strong convergence of the corresponding evolutions over infinite horizon. This is obtained by studying convergence rates along with the uniform exponential stability of the third order evolutions. Spectral analysis for the MGT-equation along with a discussion of spectral uppersemicontinuity for both equations (MGT and linearized Westervelt) will also be provided.

Keywords Moore-Gibson-Thompson equation · Third-order evolutions, singular limit · Strong convergence of semigroup · Uniform exponential decays · Acoustic waves · Spectral analysis

1 Introduction

In this paper, we consider PDE system describing the propagation of acoustic waves in a heterogeneous medium. The corresponding models, referred as Westervelt, Kuznetsov or Moore-Gibson-Thompson equation (MGT), have attracted considerable attention triggered by important applications in medicine, engineering and

---

M. Bongarti · S. Charoenphon · I. Lasiecka (✉)
Department of Mathematical Sciences, University of Memphis, Memphis, TN 38152, USA
e-mail: lasiecka@memphis.edu

IBS, Polish Academy of Sciences, Warsaw, Poland

© Springer Nature Switzerland AG 2020
J. Banasiak et al. (eds.), *Semigroups of Operators – Theory and Applications*,
Springer Proceedings in Mathematics & Statistics 325,
https://doi.org/10.1007/978-3-030-46079-2_9

life sciences (see [10, 11, 15–17, 29]). Processes such as welding, lithotripsy or high frequency focused ultrasound depend on accurate modeling involving acoustic equations. From a mathematical point of view, these are either second-order-in-time equations with strong diffusion or third-order-in-time dynamics. While in the first case the equation is of strongly parabolic type (diffusive effects are dominant), in the second case the system displays partial hyperbolic effect which can be easily attested by spectral analysis. From a physical point of view, the difference between two types manifests itself by accounting for finite speed of propagation for the MGT equation vs infinite speed of propagation for diffusive phenomena. By accounting for thermal relaxation in the process, MGT equation resolves infinite speed of propagation paradox associated with Westervelt-Kuznetsov equation. The goal of this work is a careful asymptotic analysis of MGT equation with respect to vanishing relaxation parameter. We will show that the Westervelt-Kuznetsov equation is a limit (in terms of projected semigroups) of MGT equation, when the relaxation parameter vanishes. A quantitative rate of convergence of the corresponding solutions will be derived as well.

1.1 Physical Motivation, Modeling and Thermal Relaxation Parameter

Physical models for nonlinear acoustics depend on what constitutive law we choose to describe the dynamics of the heat conduction. According to Fourier (classical continuum mechanics), the dynamics of the thermal flux in a homogeneous and isotropic thermally conducting medium obeys the relation

$$\vec{q} = -K\nabla\theta, \tag{1.1}$$

where $\vec{q}$ is the flux vector, $\theta = \theta(t, x)$ is the absolute temperature and the constant $K > 0$ is the thermal conductivity, see [8] for more details.

Along with the conservation of mass, momentum and energy, the use of Fourier's law for the heat flux lead us to obtain a number of known equations among which we find the classic second order (in time) nonlinear Westervelt's equation for the acoustic pressure $u = u(t, x)$ which can be written as

$$(1 - 2ku)u_{tt} - c^2 \Delta u - \delta \Delta u_t = 2k(u_t)^2, \tag{1.2}$$

where k is a parameter that depends on the mass density and the constant of the nonlinearity of the medium and c and δ denote the speed and diffusivity of the sound, respectively. There are many references addressing various modeling aspects. Within the context of this paper, we refer to [8, 16–18] and references therein.

Unfortunately, Fourier's law does not fully describe the heat diffusion process. Physically, Fourier's law predicts the propagation of the thermal signals at infinite speed, which is unrealistic (see [13]). Mathematically, the so-called *paradox of infi-*

nite speed of propagation (or *paradox of heat conduction*, in physics) intuitively means that initial data has an instantaneous effect on the entire space. Quantitatively we translate this notion in terms of support.

In order to make the notion clear, consider the linearized homogeneous Westervelt's equation

$$\alpha u_{tt} - c^2 \Delta u - \delta \Delta u_t = 0, \tag{1.3}$$

(α being a real constant) with initial conditions $u(0, x) = u_0(x)$ and $u_t(0, x) = u_1(x)$. We can simply assume, for the time being, that $x \in \mathbb{R}^n$.

Suppose that $\operatorname{supp}(u_0) \cup \operatorname{supp}(u_1) \subset B(z, R)$, that is, u_0 and u_1 have supports inside the ball of radius R and center $z \in \mathbb{R}^n$. We say that the Partial Differential Equation (PDE) above has *finite speed of propagation* if the solution u is such that $u(t, \cdot)$ has compact support for every $t > 0$. More precisely we call *speed of propagation* the number C defined as the infimum of the values $c > 0$ such that $\operatorname{supp}(u(t, \cdot)) \subset B(z, R + ct)$. If there is no finite c with the above property, we say that the PDE has infinite speed of propagation.

One can fairly easily see why Fourier's law leads to infinite speed of propagation. In general lines, neglecting internal dissipation and all sort of thermal sources, the authors of [8, 22] used the equation of continuity

$$\frac{\partial \rho}{\partial t} + \nabla \cdot (\rho \vec{u}) = 0,$$

(here $\vec{u}$ is the velocity vector of the material point, t is time and ρ is the mass density) to write the balance law for internal heat energy as

$$\rho C_p \frac{D\theta}{Dt} + \nabla \cdot \vec{q} = 0, \tag{1.4}$$

where C_p is the specific heat at constant pressure and the operator

$$\frac{D}{Dt} \equiv \frac{\partial}{\partial t} + \vec{u} \cdot \nabla$$

represents the material derivative.

Therefore, by simply replacing the flux vector in (1.4) by the Fourier's law and using the definition of the material derivative we end up with what is called *heat transport equation*

$$\theta_t + \vec{u} \cdot \nabla \theta - \eta \nabla^2 \theta = 0, \tag{1.5}$$

where $\eta = \frac{K}{\rho C_p}$ is the constant of thermal diffusivity. Assuming for one moment that the material point does not move (i.e., $\vec{u} = 0$) the heat transport equation (1.5) reduces to the classical diffusion equation which is a PDE with parabolic behavior. The solution for the heat equation, as we know, is given by a convolution $u = \phi \star u_0$ where ϕ is the fundamental solution of the Laplace equation and u_0 is the initial data. From this structure, we can see that small disturbance on the initial data has the potential to affect the whole solution in the entire space.

In order to address this *defect* and account for a finite speed to the heat conduction, several improvements and modifications to the Fourier's law were studied (see [32]). Although different, all the modifications agree with the fact that it is unrealistic to consider that any change of temperature is immediately felt regardless of position. It is interesting to note that the first work to notice this phenomena with a derivation of a new third order in time model is [31] by Professor G. G. Stokes. After 97 years, in [3], C. Cattaneo derived what today is known as the Maxwell-Cattaneo law (see also [4, 13, 32]).

The Maxwell-Cattaneo Law is given by

$$\vec{q} + \tau \frac{\partial \vec{q}}{\partial t} = -K\nabla\theta, \tag{1.6}$$

and managed to remove the infinite speed paradox by adding the so called *thermal inertia* term which is proportional to the time derivative of the flux vector.

It is important to observe that Maxwell-Cattaneo law as we presented in (1.6) resolves the paradox of infinite speed of propagation, but the diffusion process is only free of paradoxes in the case where the body (or the object) of the dynamic is resting. In the moving frame, this same constitutive law gives rise to another paradox related to the Galilean relativity regarding the invariance of physical laws in all frames. This latter and last paradox can be resolved by replacing the time derivative in (1.6) by the material derivative. More details about this issue can be found in [8].

The material-dependent constant τ is known as the *thermal relaxation parameter* or *time relaxation parameter* and is the center of this paper. Physically τ represents the time necessary to achieve steady heat conduction once a temperature gradient is imposed to a volume element. This time lag can be (and in fact it is) translated to different phenomena and contexts, as is the case where the models are used to study problems of High-Frequency Ultrasound (HFU) in lithotripsy, thermotherapy, ultrasound cleaning and sonochemistry. See [8, 17, 20].

The goal of this paper is to *quantify* the sensitivity the thermal relaxation parameter τ on a variety of materials by studying a singular perturbation problem, which makes sense since a number of experiments found this parameter to be small in several mediums, although not all. Among the ones where τ is not small we find biological tissue (1–100 s), sand (21 s), H acid (25 s) and $NaHCO_3$ (29 s) (see [8]). Among the ones with τ small we find cells and melanosome (order of milliseconds), blood vessels (order of microseconds, depending on the diameter) (see [34]) and most metals (order of picoseconds) (see [8]).

The same procedure as to obtain the Westervelt's equation leads us now to the third-order (in time) nonlinear Moore-Gibson-Thompson (MGT) equation

$$\tau u_{ttt} + (1 - 2ku)u_{tt} - c^2\Delta u - b\Delta u_t = 2k(u_t)^2, \tag{1.7}$$

where k and c has the same meaning as the ones in the Westervelt's equation but the diffusivity of the sound δ also suffers a change due to the presence of the thermal relaxation parameter τ and gives place to a new parameter $b = \delta + \tau c^2$. The operator Δ is understood as the Laplacian subject to suitable boundary conditions-Dirichlet, Neumann or Robin. It should be noted that the original version of this model dates back to Stokes paper [31]. A typical JMGT equation is equipped with the additional more precise physical parameters [20], however, for the sake of transparency only the canonical abstract form is retained.

1.2 Past Literature and Introduction of the Problem

This section collects the relevant past results pertinent to the model under consideration.

Let Ω be a bounded domain in $\mathbb{R}^n$ ($n = 2, 3$) with a C^2-boundary $\Gamma = \partial\Omega$ immersed in a resting medium. We work with $L^2(\Omega)$ but the treatment could be similarly carried out to any (separable) Hilbert space H. Consider $A : \mathcal{D}(A) \subset L^2(\Omega) \to L^2(\Omega)$ defined as the Dirichlet Laplacian, i.e., $A = -\Delta$ with $\mathcal{D}(A) = H_0^1(\Omega) \cap H^2(\Omega)$. All the results remain true if we assume A to be any unbounded positive self-adjoint operator with compact resolvent defined on H.

Consider the nonlinear third order evolution

$$\begin{cases} \tau u_{ttt} + (1 + 2ku)u_{tt} + c^2 Au + bAu_t = 0, \\ u(0, \cdot) = u_0, u_t(0, \cdot) = u_1, u_{tt}(0, \cdot) = u_2, \end{cases} \quad (1.8)$$

and its linearization

$$\begin{cases} \tau u_{ttt} + \alpha u_{tt} + c^2 Au + bAu_t = 0, \\ u(0, \cdot) = u_0, u_t(0, \cdot) = u_1, u_{tt}(0, \cdot) = u_2. \end{cases} \quad (1.9)$$

The natural phase spaces associated with these evolutions are the following:

$$\mathbb{H}_0 \equiv \mathcal{D}(A^{1/2}) \times \mathcal{D}(A^{1/2}) \times L^2(\Omega) \ and \ \mathbb{H}_1 \equiv \mathcal{D}(A) \times \mathcal{D}(A^{1/2}) \times L^2(\Omega). \quad (1.10)$$

Generation of linear semigroups associated with (1.9) has been studied in [19, 28] where it was shown that for any $\tau > 0$, $b > 0$ (1.9) generates a strongly continuous group on either $\mathbb{H}_0$ or $\mathbb{H}_1$. This result depends on $b > 0$. When $b = 0$ the generation of semigroups fails [14].

The nonlinear (quasilinear) model (1.7) has been treated in [20] where it was shown that for the initial data sufficiently small in $\mathbb{H}_1$, i.e., in a ball $B_{\mathbb{H}_1}(r)$ there exists nonlinear semigroup operator defined on $\mathbb{H}_1$ for all $t > 0$. The value of r depends only on the values of the physical parameters in the equation and not on $t > 0$. The aforementioned result depends on uniform stability of the dynamics of (1.9) and this holds for $\gamma = \alpha - \tau c^2 b^{-1} > 0$.

Subsequently, the authors in [28] showed that the linear equation generates a C_0-group in four different spaces with exponential stability provided $\gamma = \alpha - \tau c^2 b^{-1} > 0$. In case $\gamma = 0$ the system is conservative and in case $\gamma < 0$, by assuming very regular energy spaces the authors in [9] showed that (1.9) generates a chaotic semigroup.

Spectral analysis of the linear problem was also studied [19, 20, 28]. The spectrum consists of continuous spectrum and point spectrum. The location of the eigenvalues confirms partially hyperbolic character of the dynamics.

The same model with added memory, where the latter accounts for molecular relaxation, was considered in [1, 2, 12, 25, 26] for linear case and in [23] for the nonlinear case.

All the results obtained and mentioned above pertain to the situation when $\tau > \tau_0 > 0$. Since the parameter τ in many applications is typically very small, it is essential to understand the effects of diminishing values of relaxation parameter on quantitative properties of the underlined dynamics. This will provide important information on sensitivity of the model with respect to time relaxation. The goal of this paper is precisely to consider the vanishing parameter $\tau \to 0$ and its consequences on the resulting evolution. Specific questions we ask are the following:

- Convergence of semigroups with respect to vanishing relaxation parameter $\tau \geq 0$.

- Uniform (with respect to $\tau > 0$) asymptotic stability properties of the "relaxed" groups.

- Asymptotic (in τ) behavior of the spectrum for the family of the generators.

To our best knowledge, this is the first work that takes into consideration asymptotic properties of the MGT dynamics with respect to the vanishing relaxation parameter. The limiting evolution changes the character from a hyperbolic group to a parabolic semigroup. This change is expected to be reflected by the asymptotic properties of the spectrum and quantitative estimates for the corresponding evolutions. It should also be noted that the problem under consideration does not fit the usual Trotter-Kato type of the framework. This is due to the fact that the limit problem corresponds formally to degenerated structure. Thus, convergence of the resolvents (condition required by Trotter Kato framework) does not have a natural interpretation.

2 Main Results

2.1 Convergence of the Projected Semigroup Solutions

As before, let Ω be a bounded domain in $\mathbb{R}^n$ ($n = 2, 3$) with a C^2–boundary $\Gamma = \partial\Omega$ immersed in a resting medium and $A : \mathcal{D}(A) \subset L^2(\Omega) \to L^2(\Omega)$ defined as the Dirichlet Laplacian, i.e., $A = -\Delta$ with $\mathcal{D}(A) = H_0^1(\Omega) \cap H^2(\Omega)$.

Let $T > 0$. We consider a family of "hyperbolic" abstract third order problems

$$\begin{cases} \tau u_{ttt}^\tau + \alpha u_{tt}^\tau + c^2 A u^\tau + b^\tau A u_t^\tau = 0, \ t > 0, \\ u^\tau(0, \cdot) = u_0, u_t^\tau(0, \cdot) = u_1, u_{tt}^\tau(0, \cdot) = u_2, \end{cases} \quad (2.1)$$

where $b^\tau = \delta + \tau c^2$ and $\alpha, c, \delta, \tau > 0$.

We rewrite (2.1) abstractly by using a mass operator M_τ as below:

$$\begin{cases} M_\tau U_t^\tau(t) = \mathcal{A}_0^\tau U^\tau(t), \ t > 0, \\ U^\tau(0) = U_0 = (u_0, u_1, u_2)^T, \end{cases} \quad (2.2)$$

or equivalently with $\mathcal{A}^\tau = M_\tau^{-1} \mathcal{A}_0^\tau$

$$\begin{cases} U_t^\tau(t) = \mathcal{A}^\tau U^\tau(t), \ t > 0, \\ U^\tau(0) = U_0 = (u_0, u_1, u_2)^T, \end{cases} \quad (2.3)$$

where

$$U^\tau \equiv \begin{pmatrix} u^\tau \\ u_t^\tau \\ u_{tt}^\tau \end{pmatrix}; \ \mathcal{A}^\tau \equiv \begin{pmatrix} 0 & 1 & 0 \\ 0 & 0 & 1 \\ -\tau^{-1} c^2 A & -\tau^{-1} b^\tau A & -\tau^{-1} \alpha \end{pmatrix}; \ M_\tau \equiv \begin{pmatrix} 1 & 0 & 0 \\ 0 & 1 & 0 \\ 0 & 0 & \tau \end{pmatrix}. \quad (2.4)$$

The evolution described in (2.3) can be considered on several product spaces with the results depending on the space and the domain where $\mathcal{A}^\tau$ is defined.

Remark 2.1 The generator $\mathcal{A}^\tau$ "blows up" when $\tau \to 0$.

The following three spaces are important for the development for our result. We define $\mathbb{H}_0, \mathbb{H}_1, \mathbb{H}_2$ as

$$\mathbb{H}_0 \equiv \mathcal{D}(A^{1/2}) \times \mathcal{D}(A^{1/2}) \times L^2(\Omega);$$

$$\mathbb{H}_1 \equiv \mathcal{D}(A) \times \mathcal{D}(A^{1/2}) \times L^2(\Omega);$$

$$\mathbb{H}_2 \equiv \mathcal{D}(A) \times \mathcal{D}(A) \times \mathcal{D}(A^{1/2}).$$

The operators $\mathcal{A}^\tau$ are considered on each of these spaces with natural domains induced by the given topology. For instance, $\mathcal{A}^\tau : \mathcal{D}(\mathcal{A}^\tau) \subset \mathbb{H}_0 \to \mathbb{H}_0$ has the domain defined by

$$\mathcal{D}(\mathcal{A}^\tau) = \{(u, v, w) \in \mathbb{H}_0; c^2 u + b^\tau v \in \mathcal{D}(A)\}.$$

Clearly, the domains are not compact in $\mathbb{H}_0$. Analogous setups are made for $\mathbb{H}_1$ and $\mathbb{H}_2$.

For each $\tau > 0$, we will consider weighted norms defined by the means of the mass operator M_τ.

$$\|M_\tau^{1/2} U\|_{\mathbb{H}_0}^2, \quad \|M_\tau^{1/2} U\|_{\mathbb{H}_1}^2, \quad \|M_\tau^{1/2} U\|_{\mathbb{H}_2}^2,$$

that is,

$$\|(u, v, w)\|_{\tau,0}^2 = \|u\|_{\mathcal{D}(A^{1/2})}^2 + \|v\|_{\mathcal{D}(A^{1/2})}^2 + \tau\|w\|_2^2 = \|M_\tau^{1/2} U\|_{\mathbb{H}_0}^2;$$

$$\|(u, v, w)\|_{\tau,1}^2 = \|u\|_{\mathcal{D}(A)}^2 + \|v\|_{\mathcal{D}(A^{1/2})}^2 + \tau\|w\|_2^2 = \|M_\tau^{1/2} U\|_{\mathbb{H}_1}^2;$$

$$\|(u, v, w)\|_{\tau,2}^2 = \|u\|_{\mathcal{D}(A)}^2 + \|v\|_{\mathcal{D}(A)}^2 + \tau\|w\|_{\mathcal{D}(A^{1/2})}^2 = \|M_\tau^{1/2} U\|_{\mathbb{H}_2}^2$$

with $\|\cdot\|_2$ representing the standard L^2–norm. We shall also use the rescaled notation: $\mathbb{H}_0^\tau = M_\tau^{1/2} \mathbb{H}_0$, $\mathbb{H}_1^\tau = M_\tau^{1/2} \mathbb{H}_1$, $\mathbb{H}_2^\tau = M_\tau^{1/2} \mathbb{H}_2$ with an obvious interpretation for the composition where the elements of $\mathbb{H}_0^\tau$ coincide with the elements of $\mathbb{H}_0$ and induced topology given by $\|(u, v, w)\|_{\tau,0}$.

Theorem 2.1 (Generation of a group on $\mathbb{H}_0$ and $\mathbb{H}_2$) *Let $\alpha, c, \delta > 0$. Then, for each $\tau > 0$ the operator $\mathcal{A}^\tau$ generates a C_0–group $\{T^\tau(t)\}_{t \geq 0}$ on $\mathbb{H}_0$ and also on $\mathbb{H}_2$.*

Theorem 2.1 follows from [28] applied to $\mathbb{H}_0$ space. The invariance of the generator under the multiplication by fractional powers of A leads to the result stated for $\mathbb{H}_2$.

Theorem 2.2 (Equi-boundedness and uniform (in τ) exponential stability in $\mathbb{H}_0^\tau$) *Consider the family $\mathcal{F} = \{T^\tau(t)\}_{\tau > 0}$ of groups generated by $\mathcal{A}^\tau$ on $\mathbb{H}_0$. Assume that $\gamma^\tau \equiv \alpha - c^2 \tau (b^\tau)^{-1} \geq \gamma_0 > 0$. Then, there exists $\tau_0 > 0$ and constants $M = M(\tau_0), \omega = \omega(\tau_0) > 0$ (both independent on τ) such that*

$$\|T^\tau(t)\|_{\mathcal{L}(\mathbb{H}_0^\tau)} \leq M e^{-\omega t} \text{ for all } \tau \in (0, \tau_0] \text{ and } t \geq 0.$$

Theorem 2.3 *Let $\alpha, c, \delta > 0$. Then*

(a) (**generation on $\mathbb{H}_1$**) *For each $\tau > 0$ the operator $\mathcal{A}^\tau$ generates a C_0–group $\{T^\tau(t)\}_{t \geq 0}$ on $\mathbb{H}_1$.*

(b) (**equi-boundedness and uniform (in τ) exponential stability**) *Consider the family $\mathcal{F}_1 = \{T^\tau(t)\}_{\tau > 0}$ of groups generated by $\mathcal{A}^\tau$ on $\mathbb{H}_1$. Assume $\gamma^\tau > \gamma_0 > 0$. Then, there exists $\tau_0 > 0$ and constants $\overline{M}_1 = \overline{M}_1(\tau_0)$, $\overline{\omega}_1 = \overline{\omega}_1(\tau_0) > 0$, both independent on τ such that*

$$\|T^\tau(t)\|_{\mathcal{L}(\mathbb{H}_1^\tau)} \leq \overline{M}_1 e^{-\overline{\omega}_1 t} \text{ for all } \tau \in (0, \tau_0] \text{ and } t \geq 0.$$

Remark 2.2 Notice that the space $\mathbb{H}_2$ is obtained by multiplication of elements in $\mathbb{H}_0$ by $A^{1/2}$ (componentwise), therefore, if we assume initial data in $\mathbb{H}_2$, it follows that uniform (in τ) boundedness and stability of the dynamics remain true.

In order to characterize asymptotic behavior of the family $\mathcal{F}$ of the groups $T^\tau(t)$ we introduce the space $\mathbb{H}_0^0 \equiv \mathcal{D}(A^{1/2}) \times \mathcal{D}(A^{1/2})$ and the projection operator $P : \mathbb{H}_0 \to \mathbb{H}_0^0$ defined as

$$\mathbb{H}_0 \ni (u, v, w) \mapsto (u, v) \in \mathbb{H}_0^0.$$

With this notation Theorem 2.2 implies uniform boundedness of the sequence

$$\|PT^\tau(t)E\|_{L(\mathbb{H}_0^0)} \leq Me^{-\omega t}, t > 0, \tag{2.5}$$

where E denotes the extension operator from $\mathbb{H}_0^0 \to \mathbb{H}_0^\tau$ defined by $E(u, v) \equiv (u, v, 0)$. From (2.5) we deduce that for every $U^0 = (u_0, u_1) \in \mathbb{H}_0^0$ the corresponding projected solutions $PT^\tau(t)EU^0$ have a weakly convergent subsequence in $\mathbb{H}_0^0$ and weakly star in $L^\infty(0, \infty; \mathbb{H}_0^0)$. By standard distributional calculus one shows that such subsequence converges *weakly* to $U^0(t) = (u^0(t), u_t^0(t))$ which satisfies (distributionally) the following **limit** equation

$$\begin{cases} \alpha u_{tt}^0 + c^2 A u^0 + \delta A u_t^0 = 0, \\ u^0(0, \cdot) = u_0, u_t^0(0, \cdot) = u_1, \end{cases} \tag{2.6}$$

which rewritten as first order system becomes

$$\begin{cases} U_t^0(t) = \mathcal{A}U^0(t), \ t > 0, \\ U^0(0) = U_0^0 = (u_0, u_1)^T, \end{cases} \tag{2.7}$$

where

$$U^0 \equiv \begin{pmatrix} u^0 \\ u_t^0 \end{pmatrix}; \ \mathcal{A} = \begin{pmatrix} 0 & I \\ -c^2\alpha^{-1}A & -\delta\alpha^{-1}A \end{pmatrix} \tag{2.8}$$

and

$$\mathcal{A} : \mathcal{D}(\mathcal{A}) \subset \mathbb{H}_0^0 \to \mathbb{H}_0^0$$

with

$$\mathcal{D}(\mathcal{A}) = \{(u, v) \in \mathcal{D}(A^{1/2}) \times \mathcal{D}(A^{1/2}); c^2 u + \delta v \in \mathcal{D}(A^{3/2})\}.$$

Equation (2.6) is a known and well studied in the literature strongly damped wave equations. In fact, generation of an analytic and exponentially decaying semigroup on the space $\mathcal{D}(A^{1/2}) \times L^2(\Omega)$ is a standard by now result [6, 7, 27]. Less standard is the analysis on $\mathbb{H}_0^0 \equiv \mathcal{D}(A^{1/2}) \times \mathcal{D}(A^{1/2})$, where contractivity and dissipativity are no longer valid. This latter is the framework relevant to our analysis.

Proposition 2.1 (a) (**generation of a semigroup on** $\mathbb{H}_0^0$) Let $\mathbb{H}_0^0 \equiv \mathcal{D}(A^{1/2}) \times \mathcal{D}(A^{1/2})$ and $\alpha, \delta, c > 0$. Then the operator $\mathcal{A}$ generates an (noncontractive) analytic semigroup $\{T(t)\}_{t \geq 0}$ in $\mathbb{H}_0^0$.

(b) (**exponential stability**) There exist constants $M_0, \omega_0 > 0$ such that

$$\|T(t)\|_{\mathcal{L}(\mathbb{H}_0^0)} \leq M_0 e^{-\omega_0 t}, \ t \geq 0.$$

Proof The well-posedness and analyticity of the associated generator on the space $L^2(\Omega) \times L^2(\Omega)$ is a direct consequence of [24] (Theorem 3B.6, p. 293) and [33] (Proposition 2.2, p. 387). Invariance of the semigroup under the action of $A^{1/2}$ implies the same result in $\mathbb{H}_0^0$, hence justifying the part (a) of Proposition 2.1. As to the exponential stability, while this is a well known fact proved by energy methods on the space $\mathcal{D}(A^{1/2}) \times L^2(\Omega)$, the decay rates on $\mathbb{H}_0^0$ (nondissipative case) need a justification. In our case, this follows from the estimate in (2.5) along with weak lower semicontinuity of $\|PT^\tau(t)EU^0\|_{\mathbb{H}_0^0}^2$. The conclusion on exponential stability can also be derived independently of the family $\mathcal{F}$, by evoking analyticity of the generator [7] along with the spectrum growth determined condition and the analysis of the location of the spectrum (see Sect. 2.2 below). □

Remark 2.3 Proposition 2.1 also holds with $\mathbb{H}_0^0$ replaced by $\mathbb{H}_1^0 \equiv \mathcal{D}(A) \times \mathcal{D}(A)$.

Our main interest and goal of this work is to provide a quantitative description of strong convergence, when $\tau \to 0$, of hyperbolic groups $T^\tau(t)$ to the parabolic like semigroup $T(t)$. Our result is formulated below.

Theorem 2.4 (a) (**Rate of convergence**) Let $U_0 \in \mathbb{H}_2$. Then there exists $C = C(T, \tau_0)$ such that

$$\|PT^\tau(t)U_0 - T(t)PU_0\|_{\mathbb{H}_0^0}^2 \leq C\tau \|U_0\|_{\mathbb{H}_2}^2$$

uniformly for $t \in [0, T]$.

(b) (**Strong convergence**) Let $U_0 \in \mathbb{H}_0$. Then the following strong convergence holds

$$\|PT^\tau(t)U_0 - T(t)PU_0\|_{\mathbb{H}_0^0} \to 0 \text{ as } \tau \to 0 \tag{2.9}$$

uniformly for all $t \geq 0$.

Remark 2.4 Note that Theorem 2.4 pertains to uniform (in time) strong convergence on infinite time horizon. This fact is essential in infinite horizon optimal control theory [24].

Remark 2.5 A standard tool for proving strong convergence of semigroups is Trotter-Kato Theorem [21]. However, this approach does not apply to the problem under consideration due to the singularity of the family of generators. A consistency requirement (convergence of the resolvents) is problematic due to specific framework where the family of $\mathcal{A}^\tau$ becomes singular when $\tau = 0$. More refined approach applicable to this particular framework will be developed.

The Theorem 2.4 provides the information about strong convergence of the solution u^τ and its first derivative in time. Regarding the second time derivative, we have the following.

Proposition 2.2 *Let $U_0 \in \mathbb{H}_0$, then we have*

$$\tau^{1/2} u^\tau_{tt} \to 0 \text{ weakly* in } L^\infty(0, \infty; L^2(\Omega)).$$

Proof. It is shown in [6].

2.2 Spectral Analysis and Comparison Between $\sigma(\mathcal{A}^\tau)$ and $\sigma(\mathcal{A})$

Recall that A is assumed to be a positive self-adjoint operator with compact resolvent defined on a infinite-dimensional Hilbert space H ($L^2(\Omega)$ for instance). This allow us to infer that the spectrum of A consists purely of the point spectrum. Moreover, it is countable and positive. In other words:

$$\sigma(A) = \sigma_p(A) = \{\mu_n\}_{n \in \mathbb{N}} \subset \mathbb{R}^*_+$$

and $\mu_n \to \infty$ as $n \to \infty$.

We begin with the characterization of the spectrum of $\mathcal{A}$, the operator corresponding to the limit problem. See Fig. 1.

Proposition 2.3 (a) *The residual spectrum is empty: $\sigma_r(\mathcal{A}) = \emptyset$.*

(b) *The continuous spectrum consists of one single real value: $\sigma_c(\mathcal{A}) = \left\{-\dfrac{c^2}{\delta}\right\}$.*

(c) *The point spectrum is given by*

$$\sigma_p(\mathcal{A}) = \left\{\lambda \in \mathbb{C}; \alpha \lambda^2 + \delta \mu_n \lambda + c^2 \mu_n = 0, \ n \in \mathbb{N}\right\} = \left\{\lambda^0_n, \lambda^1_n, n \in \mathbb{N}\right\},$$

where $\text{Re}(\lambda^i_n) \in \mathbb{R}^*_-$ for all $n \in \mathbb{N}$ and $i = 1, 2$. Moreover, both branches are eventually real and the following limits hold:

$$\lim_{n \to \infty} \lambda^0_n = -\frac{c^2}{\delta} \text{ and } \lim_{n \to \infty} \lambda^1_n = -\infty.$$

Regarding the spectrum of $\mathcal{A}^\tau$ we have, see Fig. 2 for each $\tau > 0$.

Proposition 2.4 (a) *The residual spectrum is empty $\sigma_r(\mathcal{A}^\tau) = \emptyset$ for all $\tau > 0$.*

(b) *The continuous spectrum is either empty or consists of a single real value:*

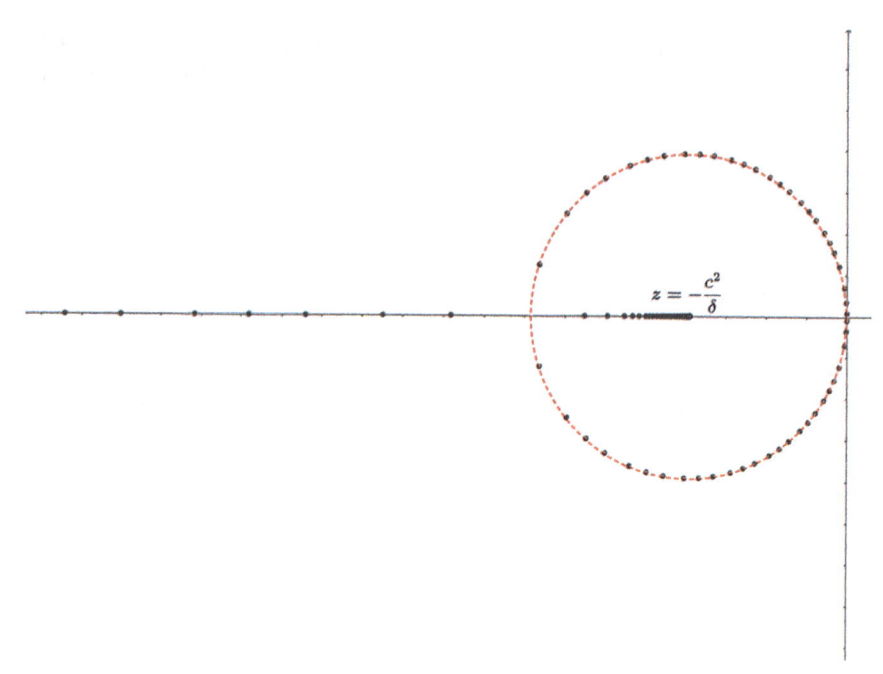

Fig. 1 Graphical representation of $\sigma_p(\mathcal{A})$ and $\sigma_c(\mathcal{A})$: the circle in red is centered at $\left(-\frac{c^2}{\delta}, 0\right)$ and has radius $r = \frac{c^2}{\delta}$. The shape of the eigenvalues is represented by the curve of black dots. Clearly, the bigger is the radius, the bigger is number of complex roots. Nevertheless, the asymptotic behavior of the eigenvalues are the same: when n becomes sufficiently large, all the eigenvalues are real and we have one branch converging to $-\infty$ while the other converge to the point in the continuous spectrum $z = -\frac{c^2}{\delta}$

$$\sigma_c(\mathcal{A}^\tau) = \begin{cases} \left\{-\frac{c^2}{b^\tau}\right\} & \text{if } \gamma^\tau > 0, \\ \emptyset & \text{if } \gamma^\tau = 0, \end{cases}$$

where $\gamma^\tau \equiv \alpha - c^2 \tau (b^\tau)^{-1}$.

(c) *The point spectrum is given by*

$$\sigma_p(\mathcal{A}^\tau) = \{\lambda \in \mathbb{C};\ \tau \lambda^3 + \alpha \lambda^2 + b^\tau \mu_n \lambda + c^2 \mu_n = 0,\ n \in \mathbb{N}\} = \left\{\lambda_n^{0,\tau}, \lambda_n^{1,\tau}, \lambda_n^{2,\tau}, n \in \mathbb{N}\right\}.$$

One of the branches, say $\lambda_n^{0,\tau}$, is eventually real while the other two branches are eventually complex, conjugate of each other and the following limits hold:

$$\lim_{n \to \infty} \lambda_n^{0,\tau} = -\frac{c^2}{b^\tau}, \quad \lim_{n \to \infty} \text{Re}(\lambda_n^{1,\tau}) = -\frac{\gamma^\tau}{2\tau} \text{ and } \lim_{n \to \infty} |\text{Img}(\lambda_n^{1,\tau})| = \infty.$$

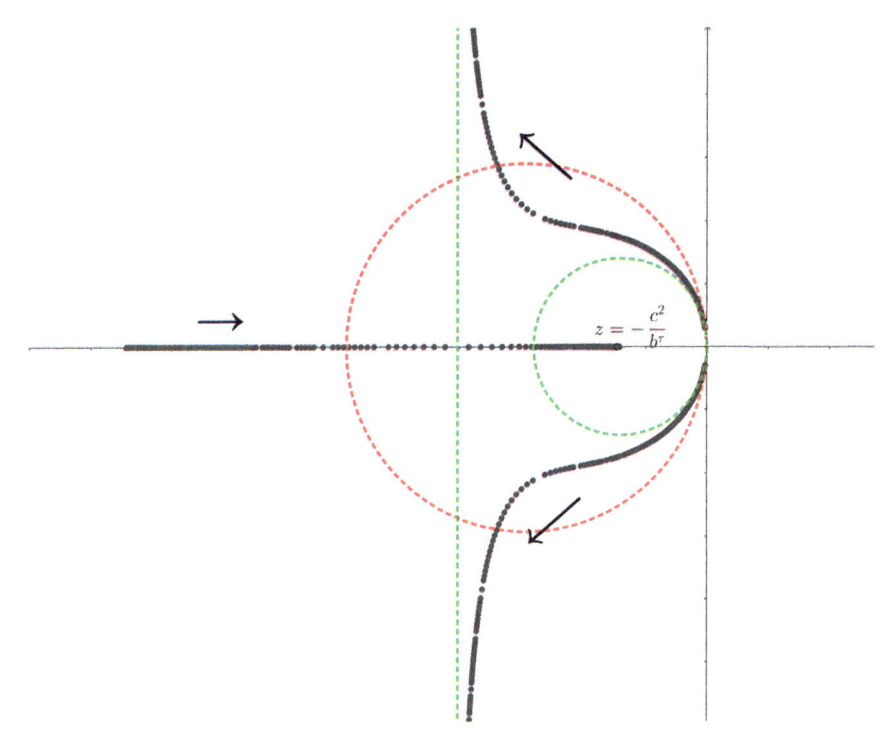

Fig. 2 Graphical representation of $\sigma_p(\mathcal{A}^\tau)$ and $\sigma_c(\mathcal{A}^\tau)$ for a $\tau = \tau_0 \in (0, 1]$ **fixed**: the circle in red is the same as in Fig. 1 while the circle in green is centered at $\left(-\frac{c^2}{b^\tau}, 0\right)$ and has radius $r = \frac{c^2}{b^\tau}$. The shape of the eigenvalues is represented by the curve of black dots. The green vertical line is given by $x = -\frac{\gamma^\tau}{2\tau}$ (see Theorem 2.4). The asymptotic behavior of the eigenvalues is exactly as we described in Theorem 2.4: as n becomes large, two branches of the eigenvalues have their respective real parts converging to $-\frac{\gamma^\tau}{2\tau}$ while their imaginary parts split into $\pm\infty$, and the other branch converges to the continuous spectrum, which in this case is given by the single point $z = -\frac{c^2}{b^\tau}$

The next lemma allows us to establish quantitative relation between $\sigma(\mathcal{A})$ and $\sigma(\mathcal{A}^\tau)$ for small τ.

Lemma 2.5 *Let $n \in \mathbb{N}$ fixed. Then, among the three roots $\{\lambda_n^{0,\tau}, \lambda_n^{1,\tau}, \lambda_n^{2,\tau}\}_{\tau > 0}$ of the equation*

$$\tau\lambda^3 + \alpha\lambda^2 + b^\tau \mu_n \lambda + c^2 \mu_n = 0,$$

there are two converging, as $\tau \to 0$, to the two roots $\{\lambda_n^1, \lambda_n^2\}$ of the equation

$$\alpha\lambda^2 + \delta\mu_n\lambda + c^2\mu_n = 0.$$

Proof Write

$$\tau\lambda^3 + \alpha\lambda^2 + b^T\mu_n\lambda + c^2\mu_n = p(\lambda)(\alpha\lambda^2 + \delta\mu_n\lambda + c^2\mu_n) + q_\tau(\lambda)$$

with

$$p(\lambda) = \frac{1}{\alpha^2}\left[\tau\alpha\lambda + \alpha^2 - \tau\delta\mu_n\right]$$

$$q_\tau(\lambda) = \frac{\tau\mu_n}{\alpha^2}\left[\alpha(\alpha-1)c^2 - \delta^2\mu_n\right]\lambda + \frac{\tau\mu_n^2 c^2\delta}{\alpha^2}$$

and notice that

$$\lim_{\tau\to 0} q_\tau(\lambda) = 0 \text{ and } \lim_{\tau\to 0} p_\tau(\lambda) = 1$$

for every λ. □

The above statement implies the following corollary.

Corollary 2.6 (**Uppersemicontinuity of the spectrum**) *Let $\varepsilon > 0$ given. Then, for each $z^0 \in \sigma_p(\mathcal{A})$ there exists $\delta = \delta_\varepsilon > 0$ and $\tau < \delta$ such that the set*

$$\{z^\tau \in \sigma(\mathcal{A}^\tau); |z^\tau - z^0| < \varepsilon\}$$

is nonempty.

Remark 2.6 The goal of Proposititon 2.4 is to localize the vertical asymptote in the spectrum explicitly and support the later claim that, as τ vanishes, it becomes arbitrarily far from the imaginary axis. Notice that the proofs of Lemma 2.5 and part (c) of Proposition 2.4 (see page 178 [Sect. 3.6]) have some similarities of algebraic manipulation but have different meaning. The proof of part (c) of Proposition 2.4 takes advantage of the known single-point continuous spectrum to conclude that for n very large the third degree polynomial must have no more then one real root and then quantify the imaginary and real parts of the complex roots. However, the proof of Lemma 2.5 makes use of the quadratic structure of the point spectrum of $\mathcal{A}$ in order to conclude the expected approximation.

The remaining part of the paper is devoted to the proofs of the main results (Figs. 3 and 4).

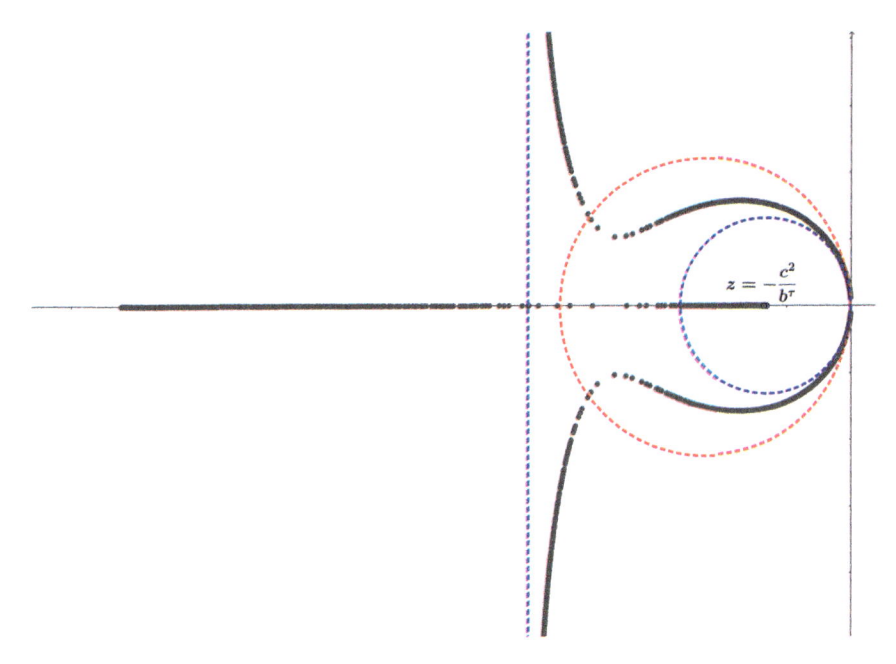

Singular Thermal Relaxation Limit for the Moore-Gibson-Thompson …

Fig. 3 Graphical representation of $\sigma_p(\mathcal{A}^\tau)$ and $\sigma_c(\mathcal{A}^\tau)$ for a $\tau = \tau_0/10$

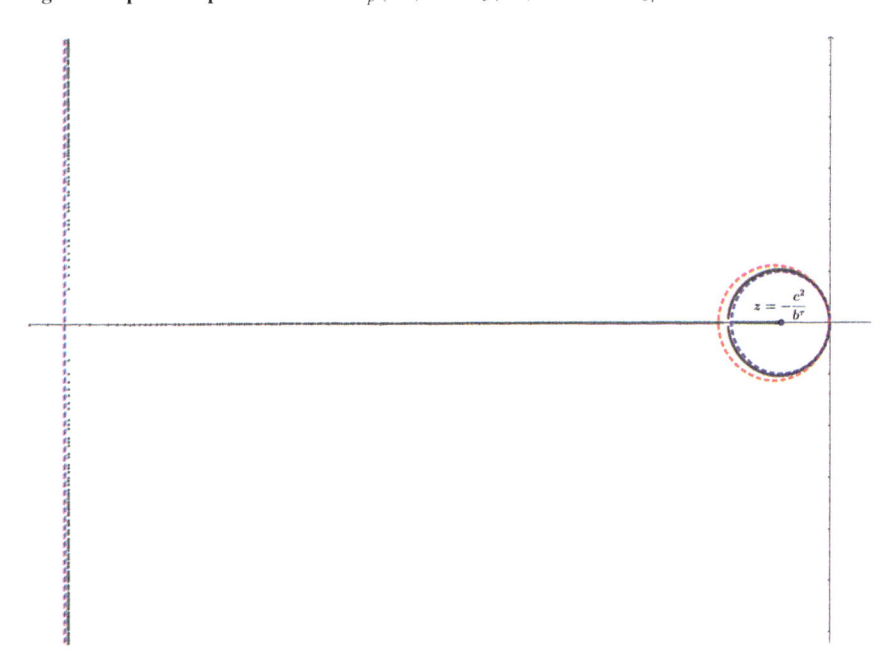

Fig. 4 Graphical representation of $\sigma_p(\mathcal{A}^\tau)$ and $\sigma_c(\mathcal{A}^\tau)$ for a $\tau = \tau_0/100$: It is important to observe how the "vertical" spectrum escapes to $-\infty$ as τ becomes small

3 Proofs

3.1 Proof of Theorem 2.2

Part I: Equi-boundedness of the groups

Let u^τ be the solution for (2.1) and consider the energy functional $E^\tau(t)$ defined as

$$E^\tau(t) = E_0^\tau(t) + E_1^\tau(t),$$

where

$$E_0^\tau(t) = \frac{\alpha}{2}\|u_t^\tau(t)\|_2^2 + \frac{c^2}{2}\|A^{1/2}u^\tau(t)\|_2^2 \qquad (3.1)$$

and

$$E_1^\tau(t) = \frac{b\tau}{2}\left\|A^{1/2}\left(u_t^\tau(t) + \frac{c^2}{b\tau}u^\tau(t)\right)\right\|_2^2 + \frac{\tau}{2}\left\|u_{tt}^\tau(t) + \frac{c^2}{b\tau}u_t^\tau(t)\right\|_2^2 + \frac{c^2\gamma^\tau}{2b\tau}\|u_t^\tau(t)\|_2^2. \qquad (3.2)$$

The following differential identity can be derived for the functional $E_1^\tau(t)$:

$$\frac{d}{dt}E_1^\tau(t) + \gamma^\tau \|u_{tt}^\tau(t)\|_2^2 = 0, \qquad (3.3)$$

where $\gamma^\tau = \alpha - c^2\tau(b^\tau)^{-1}$. The proof of (3.3) follows along the same lines as in Lemma 3.1 [20]. The equality is derived first for smooth solution and then extended by density to the "energy" level solutions. For readers convenience the details of the derivation are given in the Appendix. Moreover, for each fixed value of $\tau > 0$, the authors in [20] establish exponential decay (with decay rates depending on τ) provided $\gamma^\tau > 0$.

We aim to prove that the family of semigroups $\mathcal{F}$ is equi-bounded in τ. In other words, there exists τ_0 small enough such that if we consider $\tau \in (0, \tau_0]$, we can provide an uniform (in τ) bound for the norm of the solutions in $(\mathbb{H}_0, \|\cdot\|_{\tau,0}) = \mathbb{H}_0^\tau$. To achieve this, we establish first topological equivalence of energy function with respect to the topology defined on $\mathbb{H}_0^\tau$. This is given in the lemma to follow.

Lemma 3.1 *Let $U_0 \in \mathbb{H}_0$ and define $\tau_0 \equiv \inf\{c > 0; \gamma^\tau > 0 \text{ for all } \tau \in (0, c]\} > 0$. Then there exist $k = k(\tau_0), K = K(\tau_0) > 0$ such that*

$$k\|T^\tau(t)U_0\|_{\tau,0}^2 \leq E^\tau(t) \leq K\|T^\tau(t)U_0\|_{\tau,0}^2, \quad \tau \in (0, \tau_0] \qquad (3.4)$$

for all $t \geq 0$.

Proof We begin with the second inequality-as an easier one. In order to get it, we observe that from (3.2) we have[1]:

---
[1] We have omitted the obvious dependence on t.

$$E^\tau(t) = \frac{b^\tau}{2} \left\| A^{1/2} \left(u_t^\tau + \frac{c^2}{b^\tau} u^\tau \right) \right\|_2^2 + \frac{\tau}{2} \left\| u_{tt}^\tau + \frac{c^2}{b^\tau} u_t^\tau \right\|_2^2 + \left(\frac{c^2 \gamma^\tau}{2b^\tau} + \frac{\alpha}{2} \right) \|u_t^\tau\|_2^2 + \frac{c^2}{2} \|A^{1/2} u^\tau\|_2^2$$

$$\leqslant \left(\frac{b^\tau + c^2}{2} \right) \|A^{1/2} u_t^\tau\|_2^2 + \frac{c^2}{2} \left(2 + \frac{c^2}{b^\tau} \right) \|A^{1/2} u^\tau\|_2^2 + \frac{c^2}{2b^\tau} \left(\tau + \frac{c^2 \tau}{b^\tau} + \frac{\alpha b^\tau}{c^2} + \gamma^\tau \right) \|u_t^\tau\|_2^2 + \frac{\tau}{2} \left(1 + \frac{c^2}{b^\tau} \right) \|u_{tt}^\tau\|_2^2$$

$$\leqslant \frac{c^2}{2} \left(2 + \frac{c^2}{b^\tau} \right) \|A^{1/2} u^\tau\|_2^2 + \frac{c^2}{2b^\tau} \left[C^* \left(\tau + \frac{c^2 \tau}{b^\tau} + \gamma^\tau \right) + \frac{(\alpha C^* + b^\tau) b^\tau}{c^2} + b^\tau \right] \|A^{1/2} u_t^\tau\|_2^2 + \frac{\tau}{2} \left(1 + \frac{c^2}{b^\tau} \right) \|u_{tt}^\tau\|_2^2$$

$$\leqslant \frac{c^2}{2} \left(2 + \frac{c^2}{\delta} \right) \|A^{1/2} u^\tau\|_2^2 + \frac{c^2}{2\delta} \left[C^* \left(\tau_0 + \frac{c^2 \tau_0}{\delta} + \gamma^{\tau_0} \right) + \frac{(\alpha C^* + \delta) \delta}{c^2} + \delta \right] \|A^{1/2} u_t^\tau\|_2^2 + \frac{\tau}{2} \left(1 + \frac{c^2}{\delta} \right) \|u_{tt}^\tau\|_2^2$$

$$\leqslant \max \left\{ \frac{c^2}{2} \left(2 + \frac{c^2}{\delta} \right); \frac{c^2}{2\delta} \left[C^* \left(\tau_0 + \frac{c^2 \tau_0}{\delta} + \gamma^{\tau_0} \right) + \frac{(\alpha C^* + \delta) \delta}{c^2} + \delta \right]; \frac{1}{2} + \frac{c^2}{2\delta} \right\} \|U^\tau(t)\|_{\tau,0}^2.$$

[2]where $C^* = C^*(n, \Omega)$ is the constant that appears in Poincaré's Inequality.

Now the second inequality in (3.4) follows after we define

$$K(\tau_0) \equiv \max \left\{ \frac{c^2}{2} \left(2 + \frac{c^2}{\delta} \right); \frac{c^2}{2\delta} \left[C^* \left(\tau_0 + \frac{c^2 \tau_0}{\delta} + \gamma^{\tau_0} \right) + \frac{(\alpha C^* + \delta) \delta}{c^2} + \delta \right]; \frac{1}{2} + \frac{c^2}{2\delta} \right\}. \tag{3.5}$$

For the first inequality, fix $\varepsilon > 0$ to be determined later. Peter-Paul inequality then implies that

$$-\frac{c^2 \varepsilon \|A^{1/2} u^\tau\|_2^2}{2} - \frac{c^2 \|A^{1/2} u_t^\tau\|_2^2}{2\varepsilon} \leqslant c^2 (A^{1/2} u^\tau, A^{1/2} u_t^\tau) \text{ and } -\frac{\varepsilon \tau c^2}{2b^\tau} \|u_{tt}^\tau\|_2^2 - \frac{\tau c^2}{2\varepsilon b^\tau} \|u_t^\tau\|_2^2 \leqslant \frac{\tau c^2}{b^\tau} (u_{tt}^\tau, u_t^\tau). \tag{3.6}$$

Then we have

$$E^\tau(t) = \frac{b^\tau}{2} \left\| A^{1/2} \left(u_t^\tau + \frac{c^2}{b^\tau} u^\tau \right) \right\|_2^2 + \frac{\tau}{2} \left\| u_{tt}^\tau + \frac{c^2}{b^\tau} u_t^\tau \right\|_2^2 + \left(\frac{c^2 \gamma^\tau}{2b^\tau} + \frac{\alpha}{2} \right) \|u_t^\tau\|_2^2 + \frac{c^2}{2} \|A^{1/2} u^\tau\|_2^2$$

$$= \frac{c^2}{2} \left(1 + \frac{c^2}{b^\tau} \right) \|A^{1/2} u^\tau\|_2^2 + \frac{b^\tau}{2} \|A^{1/2} u_t^\tau\|_2^2 + c^2 (A^{1/2} u^\tau, A^{1/2} u_t^\tau) + \frac{\tau}{2} \|u_{tt}^\tau\|_2^2 + \frac{\alpha}{2} \left(1 + \frac{c^2}{b^\tau} \right) \|u_t^\tau\|_2^2 + \frac{\tau c^2}{b^\tau} (u_{tt}^\tau, u_t^\tau)$$

$$\geqslant \frac{c^2}{2} \left(1 + \frac{c^2}{b^\tau} - \frac{1}{\varepsilon} \right) \|A^{1/2} u^\tau\|_2^2 + \left(\frac{b^\tau}{2} - \frac{c^2 \varepsilon}{2} \right) \|A^{1/2} u_t^\tau\|_2^2 + \frac{\tau}{2} \left(1 - \frac{c^2 \varepsilon}{b^\tau} \right) \|u_{tt}^\tau\|_2^2 + \left(\frac{\alpha}{2} \left(1 + \frac{c^2}{b^\tau} \right) - \frac{\tau c^2}{2\varepsilon b^\tau} \right) \|u_t^\tau\|_2^2,$$

Pick an $\varepsilon > 0$ such that $\frac{b^\tau}{b^\tau + c^2} < \varepsilon < \frac{b^\tau}{c^2}$ and observe that

$$\frac{b^\tau}{2} - \frac{c^2 \varepsilon}{2} > 0 \text{ and } \frac{c^2}{2} \left(\frac{c^2}{2} \right)_1 - \frac{c^2}{2\varepsilon} = \frac{c^2}{2} \left(\frac{c^2}{b^\tau} + 1 - \frac{1}{\varepsilon} \right) > 0$$

and similarly

$$\frac{\tau}{2} - \frac{\varepsilon \tau c^2}{2b^\tau} > 0 \text{ and } \frac{\alpha}{2} \left(\frac{c^2}{b^\tau} + 1 \right) - \frac{\tau c^2}{2\varepsilon b^\tau} = \frac{\alpha}{2} \left(\frac{c^2}{b^\tau} + 1 - \frac{1}{\varepsilon} \right) = \frac{\gamma^\tau}{2} \left(\frac{c^2}{b^\tau} + 1 \right) + \frac{\tau c^2}{2b^\tau} \left(\frac{c^2}{b^\tau} + 1 - \frac{1}{\varepsilon} \right) > 0.$$

Hence, picking

$$\varepsilon = \frac{2b^\tau}{b^\tau + 2c^2}$$

and continuing the lower bound estimate of $E^\tau(t)$ we have

---

[2]We have omitted the obvious dependence on t.

$$E^\tau(t) \geq \frac{c^2}{2}\left(1+\frac{c^2}{b^\tau}-\frac{1}{\varepsilon}\right)\|A^{1/2}u^\tau\|_2^2 + \left(\frac{b^\tau}{2}-\frac{c^2\varepsilon}{2}\right)\|A^{1/2}u_t^\tau\|_2^2 + \frac{\tau}{2}\left(1-\frac{c^2\varepsilon}{b^\tau}\right)\|u_{tt}^\tau\|_2^2$$

$$\geq \frac{c^2}{4}\|A^{1/2}u^\tau\|_2^2 + \frac{\delta^2}{(4+\tau_0)c^2+2\delta}\|A^{1/2}u_t^\tau\|_2^2 + \frac{\tau\delta}{2\delta+(4+\tau_0)c^2}\|u_{tt}^\tau\|_2^2$$

$$\geq \min\left\{\frac{c^2}{4}; \frac{\delta^2}{(4+\tau_0)c^2+2\delta}; \frac{\delta}{2\delta+(4+\tau_0)c^2}\right\}\|U^\tau(t)\|_{\tau,0}^2.$$

where we have used

$$\frac{c^2}{2}\left(1+\frac{c^2}{b^\tau}-\frac{1}{\varepsilon}\right) = \frac{c^2}{4},$$

$$\frac{b^\tau}{2}-\frac{c^2\varepsilon}{2} = \frac{(b^\tau)^2}{4c^2+2b^\tau} \geq \frac{\delta^2}{(4+\tau_0)c^2+2\delta}$$

and similarly

$$\frac{\tau}{2}\left(1-\frac{c^2\varepsilon}{b^\tau}\right) = \frac{\tau b^\tau}{2b^\tau+4c^2} \geq \frac{\tau\delta}{2\delta+(4+\tau_0)c^2}.$$

Setting

$$k(\tau_0) \equiv \min\left\{\frac{c^2}{4}; \frac{\delta^2}{(4+\tau_0)c^2+2\delta}; \frac{\delta}{2\delta+(4+\tau_0)c^2}\right\} \tag{3.7}$$

gives the first part of the inequality in Lemma 3.1. This completes the proof of the Lemma. □

Lemma 3.1 along with the identity (3.3) imply the equi-boundedness of the family $\mathcal{F}$. Indeed,

Lemma 3.2 *Let $U_0 \in \mathbb{H}_0$. For given k in (3.7) and K in (3.5), there exists a constant $L_1 > 0$ (independent on $\tau \in (0, \tau_0)$) such that*

$$\|T^\tau(t)U_0\|_{\tau,0}^2 \leq \frac{1}{k}E^\tau(t) \leq \frac{L_1 K}{k}\|U_0\|_{\tau,0}^2.$$

Proof We work with sufficiently smooth solutions guaranteed by the well-posedness-regularity theory. The final estimates are obtained via density.

Taking the L^2-inner product of (2.1) with u_t^τ we obtain

$$b^\tau\|A^{1/2}u_t^\tau(t)\|_2^2 = \tau\|u_{tt}^\tau(t)\|_2^2 - \frac{d}{dt}\left[\frac{\alpha}{2}\|u_t^\tau\|_2^2 + \frac{c^2}{2}\|A^{1/2}u^\tau\|_2^2 + \tau(u_{tt}^\tau, u_t^\tau)\right]. \tag{3.8}$$

Multiplying (3.8) by γ^τ with using (3.1) gives

$$\gamma^\tau b^\tau\|A^{1/2}u_t^\tau(t)\|_2^2 = \gamma^\tau\tau\|u_{tt}^\tau(t)\|_2^2 - \gamma^\tau\frac{d}{dt}E_0^\tau(t) - \gamma^\tau\tau\frac{d}{dt}(u_{tt}^\tau, u_t^\tau) \tag{3.9}$$

Combine (3.9) with the identity (3.3) we obtain

$$\frac{d}{dt}E_1^\tau(t) + \gamma^\tau \frac{d}{dt}E_0^\tau(t) + b^\tau \gamma^\tau \|A^{1/2}u_t^\tau(t)\|_2^2 = \gamma^\tau(\tau-1)\|u_{tt}^\tau(t)\|_2^2 - \gamma^\tau \tau \frac{d}{dt}(u_{tt}^\tau, u_t^\tau). \tag{3.10}$$

Since τ is very small and we assume $0 < \tau < 1$, we have $\gamma^\tau(\tau-1)\|u_{tt}^\tau(t)\|_2^2 < 0$ for all $t \in [0, T]$.

Then integrating w.r.t. time from 0 to t we have

$$E_1^\tau(t) + \gamma^\tau E_0^\tau(t) + \gamma^\tau b^\tau \int_0^t \|A^{1/2}u_t^\tau(s)\|_2^2 ds \leqslant E_1^\tau(0) + \gamma^\tau E_0^\tau(0) + \gamma^\tau \tau (u_{tt}^\tau, u_t^\tau)|_0^t. \tag{3.11}$$

From (3.3), we have $E_1^\tau(t) \leqslant E_1^\tau(0)$ and notice that

$$\tau(u_{tt}^\tau, u_t^\tau)|_0^t = \tau(u_{tt}^\tau(t), u_t^\tau(t)) - \tau(u_2, u_1)$$
$$\leqslant \frac{\tau}{2}\|u_{tt}^\tau(t)\|_2^2 + \frac{\tau}{2}\|u_t^\tau(t)\|_2^2 + \frac{\tau}{2}\|u_2\|_2^2 + \frac{\tau}{2}\|u_1\|_2^2$$
$$\leqslant \left(1 + \frac{\tau b}{\alpha c^2}\right)\left[E_1^\tau(t) + E_1^\tau(0)\right] \leqslant 2\left(\frac{\alpha c^2 + \tau b}{\alpha c^2}\right)E_1^\tau(0).$$

Then

$$E_1^\tau(t) + \gamma^\tau E_0^\tau(t) + \gamma^\tau b^\tau \int_0^t \|A^{1/2}u_t^\tau(s)\|_2^2 ds \leqslant E_1^\tau(0) + \gamma^\tau E_0^\tau(0) + 2\gamma^\tau\left(\frac{\alpha c^2 + \tau b^\tau}{\alpha c^2}\right)E_1^\tau(0).$$

$$E_1^\tau(t) + \gamma^\tau E_0^\tau(t) \leqslant \max\left\{\frac{\alpha c^2 + 2\gamma^\tau(\alpha c^2 + \tau b^\tau)}{\alpha c^2}, \gamma^\tau\right\} E^\tau(0)$$

Therefore

$$E^\tau(t) \leqslant \frac{\max\left\{\frac{\alpha c^2 + 2\gamma^{\tau_0}\alpha c^2 + \tau b^\tau}{\alpha c^2}, \gamma^{\tau_0}\right\}}{\min\{1, \gamma^{\tau_0}\}} E^\tau(0) = L_1 E^\tau(0) \tag{3.12}$$

This means that the *total* energy $E^\tau(t)$ is bounded in time by the initial *total* energy. Thus the proof is obtained. □

From Lemma 3.2 we conclude

Corollary 3.3 *There exists a constant $M > 0$ (independent on $\tau \in (0, \tau_0)$) such that*

$$\|T^\tau(t)\|_{\mathcal{L}(\mathbb{H}_0^\tau)} \leqslant M, \forall t > 0. \tag{3.13}$$

Part II: Uniform (in τ) decay rates

In order to prove the uniformity of the decay rates we use the Pazy-Datko Theorem. The first step consists of showing the map $t \mapsto \|T^\tau(t)U_0\|_{\tau,0}^2$ belongs to $L^1(0, \infty; \mathbb{H}_0)$. This is the statement of the next Lemma.

Lemma 3.4 *There exists $\overline{K} > 0$ independent on τ such that*

$$\int_0^\infty \|T^\tau(s)U_0\|_{\tau,0}^2 ds \leq \overline{K}\|U_0\|_{\tau,0}^2 < \infty. \tag{3.14}$$

Proof Multiplying the identity (3.3) by 2τ gives

$$2\tau \frac{d}{dt}E_1^\tau(t) + 2\tau\gamma^\tau \|u_{tt}^\tau(t)\|_2^2 = 0 \tag{3.15}$$

Multiply (3.8) by γ^τ we have

$$\gamma^\tau b^\tau \|A^{1/2}u_t^\tau(t)\|_2^2 = \gamma^\tau \tau \|u_{tt}^\tau(t)\|_2^2 - \gamma^\tau \frac{d}{dt}E_0^\tau(t) + \gamma^\tau \tau \frac{d}{dt}(u_{tt}^\tau, u_t^\tau) \tag{3.16}$$

With (3.15) and (3.16) we get

$$2\tau \frac{d}{dt}E_1^\tau(t) + \gamma^\tau \frac{d}{dt}E_0^\tau(t) + \tau\gamma^\tau \|u_{tt}^\tau(t)\|_2^2 + b^\tau \gamma^\tau \|A^{1/2}u_t^\tau(t)\|_2^2 = -\gamma^\tau \tau \frac{d}{dt}(u_{tt}^\tau, u_t^\tau). \tag{3.17}$$

Then integrating w.r.t. time from 0 to t we have

$$2\tau E_1^\tau(t) + \gamma^\tau E_0^\tau(t) + \gamma^\tau \int_0^t \left[\tau\|u_{tt}^\tau(s)\|_2^2 + b^\tau\|A^{1/2}u_t^\tau(s)\|_2^2\right] ds = 2\tau E_1^\tau(0) + \gamma^\tau E_0^\tau(0) - \gamma^\tau \tau(u_{tt}^\tau, u_t^\tau)\big|_0^t. \tag{3.18}$$

Notice that from Lemma 3.2 we have

$$\tau(u_{tt}^\tau, u_t^\tau)\big|_0^t \leq (1+\tau_0)\left(\|T^\tau(t)U_0\|_\tau^2 + \|U_0\|_\tau^2\right) \leq \frac{(1+\tau_0)(L_1K+k)}{k}\|U_0\|_\tau^2.$$

Therefore,

$$\int_0^t \left[\tau\|u_{tt}^\tau(s)\|_2^2 + b^\tau\|A^{1/2}u_t^\tau(s)\|_2^2\right] ds \leq \frac{\gamma^\tau(1+\tau_0)(L_1K+k) + (2\tau_0+\gamma^\tau)L_1Kk}{k\gamma^\tau}\|U_0\|_{\tau,0}^2$$

$$\leq \frac{\gamma^{\tau_0}(1+\tau_0)(L_1K+k) + (2\tau_0+\gamma^{\tau_0})L_1Kk}{k\gamma^{\tau_0}}\|U_0\|_{\tau,0}^2 = M_1\|U_0\|_{\tau,0}^2. \tag{3.19}$$

Similarly, taking the L^2-inner product of (2.1) with u^τ we have

$$\frac{b^\tau}{2}\frac{d}{dt}\|A^{1/2}u^\tau\|_2^2 + c^2\|A^{1/2}u^\tau(t)\|_2^2 = \alpha\|u_t^\tau(t)\|_2^2 + \frac{d}{dt}\left[\frac{\tau}{2}\|u_t^\tau\|_2^2 - \tau(u_{tt}^\tau, u^\tau) - \alpha(u_t^\tau, u^\tau)\right] \tag{3.20}$$

and integrating (3.20) w.r.t. time from 0 to t we have

$$\frac{b^\tau}{2}\|A^{1/2}u^\tau(t)\|_2^2 + c^2 \int_0^t \|A^{1/2}u^\tau(s)\|_2^2 ds$$

$$= \frac{b^\tau}{2}\|A^{1/2}u_0\|_2^2 + \alpha \int_0^t \|u_t^\tau(s)\|_2^2 ds + \left[\frac{\tau}{2}\|u_t^\tau\|_2^2 - \tau(u_{tt}^\tau, u^\tau) - \alpha(u_t^\tau, u^\tau)\right]\Big|_0^t$$

$$\leqslant b^\tau\|U_0\|_{\tau,0}^2 + \frac{\alpha M_1}{b^\tau}\|U_0\|_{\tau,0}^2 + \frac{[(\alpha+\tau_0)C^*+1](L_1K+k)}{k}\|U_0\|_{\tau,0}^2$$

$$= \frac{kb^\tau + \alpha k M_1 + b^\tau[(\alpha+\tau_0)C^*+1](L_1K+k)}{b^\tau k}\|U_0\|_{\tau,0}^2$$

$$\leqslant \frac{k\delta + \alpha k M_1 + \delta[(\alpha+\tau_0)C^*+1](L_1K+k)}{\delta k}\|U_0\|_{\tau,0}^2 = M_2\|U_0\|_{\tau,0}^2. \quad (3.21)$$

Hence, by (3.19) and (3.21) we get

$$\int_0^\infty \|T^\tau(s)U_0\|_\tau^2 ds \leqslant \frac{M_1+M_2}{1+\delta+c^2}\|U_0\|_\tau^2 = M_3\|U_0\|_\tau^2 < \infty. \quad (3.22)$$

Therefore, according to Theorem 4.1 ([30], p. 116) the rate ω can be determined as follows: We first chose a number ρ such that $0 < \rho < M_3^{-1}$, then we define a number $\eta_0 = M_3\rho^{-1}$ and choose another number η such that $\eta > \eta_0$. The rate is then given by

$$\omega = -\frac{1}{\eta}\log(M_3\rho) > 0,$$

and is clearly independent on τ. The proof is thus completed. $\square$

3.2 Proof of Theorem 2.3

(a) Well-posedness
The well-posedness follows directly from Theorem 1.4 in [19] and the fact that on the space $\mathbb{H}_1$ the standard sum norm $\|\cdot\|$ and $\|\cdot\|_{\tau,1}$ are equivalent for each $\tau \in (0,1]$.
(b) Equi-boundedness on $\mathbb{H}_1^\tau$ and uniform (in τ) exponential stability
Let u^τ be the solution for (2.1) and consider the energy functional $\mathcal{E}^\tau(t)$ defined as

$$\mathcal{E}^\tau(t) = E^\tau(t) + \|Au^\tau(t)\|_2^2. \quad (3.23)$$

By Theorem 1.3 [19] the energy functional above (which is equivalent to the $\|(u^\tau, u_t^\tau, u_{tt}^\tau)\|$ for all $t \geqslant 0$) decays exponentially with time [for each fixed $\tau > 0$, provided $\gamma^\tau \equiv \alpha - c^2\tau(b^\tau)^{-1} > 0$.

As in the previous theorem, the equi-boundedness of the family $\mathcal{F}_1$ follows from the lemma:

Lemma 3.5 *Let $U_0 \in \mathbb{H}_1$ and define $\tau_0 \equiv \inf\{c > 0; \gamma^\tau > 0 \text{ for all } \tau \in (0, c]\} > 0$. Then there exist $k_1 = k_1(\tau_0)$ and $K_1 = K_1(\tau_0)$ such that*

$$k_1\|T^\tau(t)U_0\|_{\tau,1}^2 \leqslant \mathcal{E}^\tau(t) \leqslant K_1\|T^\tau(t)U_0\|_{\tau,1}^2, \quad \tau \in (0, \tau_0]$$

for all $t \geqslant 0$.

The proof of Lemma 3.5 as well as the conclusion of the equi-boundedness in $\mathbb{H}_1$ capitalizes on the estimates already derived for $\mathbb{H}_0$. We shall focus on additional terms which need to be estimated additionally.

Lemma 3.5 is obtained from Lemma 3.1 and the estimates already derived for the space $\mathbb{H}_0^\tau$ by adding the term $\|Au^\tau(t)\|_2^2$. This gives the inequality stated in Lemma 3.5. Recall (3.12), we will obtain the apriori bound for $\mathcal{E}^\tau(t)$ from the relation

$$\mathcal{E}^\tau(t) = E^\tau(t) + \|Au^\tau(t)\|_2^2 \leq L_1 E^\tau(0) + \sup_{t \geq 0} \|Au^\tau(t)\|_2^2.$$

To achieve the goal, the second term needs to be accounted for. To estimate the second term we employ the equality by taking the L^2-inner product of (2.1) with the multiplier Au^τ and integrating w.r.t. time from 0 to t

$$\frac{b^\tau}{2}\|Au^\tau(t)\|_2^2 + c^2\int_0^t \|Au^\tau(s)\|_2^2 ds = \frac{b^\tau}{2}\|Au_0\|_2^2$$
$$+ \alpha\int_0^t \|A^{1/2}u_t^\tau(s)\|_2^2 ds + \left[\frac{\tau}{2}\|A^{1/2}u_t^\tau\|_2^2 - \tau(u_{tt}^\tau, Au^\tau) - \alpha(u_t^\tau, Au^\tau)\right]\Big|_0^t \tag{3.24}$$

and by using the already obtained estimates in $\mathbb{H}_0^\tau$

$$\frac{(b^\tau - \varepsilon)}{2}\|Au^\tau(t)\|_2^2 + c^2\int_0^t \|Au^\tau(s)\|_2^2 ds \leqslant b^\tau\|U_0\|_{\tau,1}^2 + \alpha M_3\|U_0\|_{\tau,1}^2 + \frac{(\frac{\tau}{2} + \tau C_\varepsilon + \alpha)(K_1 + k_1)}{k_1}\|U_0\|_{\tau,1}^2$$
$$= \frac{2k_1 b^\tau + 2k_1\alpha M_3 + [\tau(1 + 2C_\varepsilon) + 2\alpha](K_1 + k_1)}{2k_1}\|U_0\|_{\tau,1}^2$$
$$\leqslant \frac{2k_1\delta + 2k_1\alpha M_3 + [\tau_0(1 + 2C_\varepsilon) + 2\alpha](K_1 + k_1)}{2k_1}\|U_0\|_{\tau,1}^2$$
$$= \hat{C}\|U_0\|_{\tau,1}^2. \tag{3.25}$$

Rescaling ε allows to estimate $\sup_t \|Au^\tau(t)\|$, hence $\mathcal{E}^\tau(t)$. Then we have

$$\|T^\tau(t)U_0\|_{\tau,1}^2 \leqslant \frac{1}{k_1}\mathcal{E}^\tau(t) \leqslant \left(\frac{\hat{C}K_1}{k_1}\right)\|U_0\|_{\tau,1}^2,$$

from where it follows that the groups are equibounded also on $\mathbb{H}_1^\tau$, i.e.,

Corollary 3.6 *There exists a constant $N > 0$ [independent on $\tau \in (0, \tau_0]$] such that*

$$\|T^\tau(t)\|_{\mathcal{L}(\mathbb{H}_1^\tau)} \leqslant \left(\frac{\hat{C}K_1}{k_1}\right)^{1/2} = N. \tag{3.26}$$

As for exponential uniform decays, we shall evoke again the Pazy-Datko Theorem. This shows the existence of uniform (in τ) decay rate with existence of $\overline{K_1} > 0$ independent on τ such that

Lemma 3.7
$$\int_0^\infty \|T^\tau(s)U_0\|_{\tau,1}^2 ds \leqslant \overline{K_1}\|U_0\|_{\tau,1}^2 < \infty. \tag{3.27}$$

Proof Direct from (3.25), we have

$$c^2 \int_0^t \|Au^\tau(s)\|_2^2 ds \leqslant N_1 \|U_0\|_{\tau,1}^2. \tag{3.28}$$

Hence, by (3.22) and (3.24) we have

$$\int_0^\infty \|T^\tau(s)U_0\|_{\tau,1}^2 ds \leqslant \frac{M_3 + N_1}{c^2}\|U_0\|_{\tau,1}^2 = N_2\|U_0\|_{\tau,1}^2 < \infty. \tag{3.29}$$

Thus by Theorem 4.1 ([30], p. 116) the rate ω can be taken as we first chose a number ρ such that $0 < \rho < N_2^{-1}$, then we define a number $\eta_0 = N_2 \rho^{-1}$ and choose another number η such that $\eta > \eta_0$. The rate is then given by

$$\omega = -\frac{1}{\eta}\log(N_2 \rho) > 0,$$

and is clearly independent on τ. Then the proof is completed. □

3.3 Proof of Theorem 2.4

Proof of part (a)-convergence rates.
Let $x^\tau = u^\tau - u^0$ where u^τ and u^0 are the solutions for the problems (2.1) and (2.6) respectively with the same initial values for $u(t=0)$ and $u_t(t=0)$. By taking the difference of the two problems we can write a x^τ-problem given by

$$\begin{cases} \alpha x_{tt}^\tau + c^2 A x^\tau + \delta A x_t^\tau = -\tau u_{ttt}^\tau - \tau c^2 A u_t^\tau & \text{in } (0, T) \times \Omega, \\ x^\tau(0) = 0, \ x_t^\tau(0) = 0. \end{cases} \tag{3.30}$$

Observe that since $\mathbb{H}_2$ is $\mathbb{H}_0$ subject to the multiplication by $A^{1/2}$ where the latter leaves the dynamics invariant and $\mathcal{A}^\tau$ generates a C_0-group in $\mathbb{H}_0$, we also have $\mathcal{A}^\tau$ generating a C_0-group in $\mathbb{H}_2$.

We aim to prove that

$$\|PT^\tau(t)U_0 - T(t)PU_0\|_{\mathbb{H}_0^0}^2 \leqslant \tau C\|U_0\|_{\mathbb{H}_2}^2$$

which is the same as showing that

$$\|A^{1/2}x^\tau(t)\|_2^2 + \|A^{1/2}x_t^\tau(t)\|_2^2 \leqslant \tau C \|U_0\|_{\mathbb{H}_2}^2$$

for all $t \in [0, T]$.

Step 1: Reconstruction of $\|A^{1/2}x^\tau(t)\|_2^2$

We start by taking the L^2-inner product of x^τ-equation (3.30) with x_t^τ. This gives

$$\alpha(x_{tt}^\tau(t), x_t^\tau(t)) + c^2(Ax^\tau(t), x_t^\tau(t)) + \delta(Ax^\tau(t)_t, x_t^\tau(t)) = -\tau(u_{ttt}^\tau(t), x_t^\tau(t)) - \tau c^2(Au_t^\tau(t), x_t^\tau(t))$$

which can be rewritten as

$$\frac{\alpha}{2}\frac{d}{dt}\|x_t^\tau\|_2^2 + \frac{c^2}{2}\frac{d}{dt}\|A^{1/2}x^\tau\|_2^2 + \delta\|A^{1/2}x_t^\tau(t)\|_2^2 = -\tau(u_{ttt}^\tau(t), x_t^\tau(t)) - \tau c^2(A^{1/2}u_t^\tau(t), A^{1/2}x_t^\tau(t)). \tag{3.31}$$

We now integrate (3.31) with respect to time from 0 to $t \in (0, T]$. This gives

$$\frac{\alpha}{2}\|x_t^\tau(t)\|_2^2 + \frac{c^2}{2}\|A^{1/2}x^\tau(t)\|_2^2 + \delta\int_0^t \|A^{1/2}x_t^\tau(s)\|_2^2 ds$$

$$\leqslant \frac{\delta}{2}\int_0^t \|A^{1/2}x_t^\tau(s)\|_2^2 ds + \frac{\tau C^*}{\delta}\int_0^t \|\sqrt{\tau}u_{ttt}^\tau(s)\|_2^2 ds + \frac{\tau^2 c^4 T}{\delta}\sup_{t\in[0,T]}\|A^{1/2}u_t^\tau(t)\|_2^2$$

where we have used the zero initial conditions of the x^τ-equation and C^* is the Poincaré's constant. Then

$$\frac{\alpha}{2}\|x_t^\tau(t)\|_2^2 + \frac{c^2}{2}\|A^{1/2}x^\tau(t)\|_2^2 + \frac{\delta}{2}\int_0^t \|A^{1/2}x_t^\tau(s)\|_2^2 ds$$

$$\leqslant \frac{\tau C^*}{\delta}\int_0^t \|\sqrt{\tau}u_{ttt}^\tau(s)\|_2^2 ds + \frac{\tau^2 c^4 T}{\delta}\sup_{t\in[0,T]}\|A^{1/2}u_t^\tau(t)\|_2^2 \tag{3.32}$$

This was the reconstruction we needed.

Step 2: Reconstruction of $\|A^{1/2}x_t^\tau(t)\|_2^2$

We start by taking the L^2-inner product of x^τ-equation (3.30) with Ax_t^τ. This gives

$$\alpha(x_{tt}^\tau(t), Ax_t^\tau(t)) + c^2(Ax^\tau(t), Ax_t^\tau(t)) + \delta(Ax_t^\tau(t), Ax_t^\tau(t)) = -\tau(u_{ttt}^\tau(t), Ax_t^\tau(t)) - \tau c^2(Au_t^\tau(t), Ax_t^\tau(t))$$

which can be rewritten as

$$\frac{\alpha}{2}\frac{d}{dt}\|A^{1/2}x_t^\tau\|_2^2 + \frac{c^2}{2}\frac{d}{dt}\|Ax^\tau\|_2^2 + \delta\|Ax_t^\tau(t)\|_2^2 = -\tau(u_{ttt}(t), Ax_t^\tau(t)) - \tau c^2(Au_t^\tau(t), Ax_t^\tau(t)). \tag{3.33}$$

We now integrate (3.33) with respect to time from 0 to $t \in (0, T]$. This gives

$$\frac{\alpha}{2}\|A^{1/2}x_t^\tau(t)\|_2^2 + \frac{c^2}{2}\|Ax^\tau(t)\|_2^2 + \delta\int_0^t \|Ax_t^\tau(s)\|_2^2 ds$$
$$\leqslant \frac{\delta}{2}\int_0^t \|Ax_t^\tau(s)\|_2^2 ds + \frac{\tau}{\delta}\int_0^t \|\sqrt{\tau}u_{ttt}^\tau(s)\|_2^2 ds + \frac{\tau^2 c^4 T}{\delta}\sup_{t\in[0,T]}\|Au_t^\tau(t)\|_2^2 \quad (3.34)$$

Then

$$\frac{\alpha}{2}\|A^{1/2}x_t^\tau(t)\|_2^2 + \frac{c^2}{2}\|Ax^\tau(t)\|_2^2 + \frac{\delta}{2}\int_0^t \|Ax_t^\tau(s)\|_2^2 ds \leqslant \frac{\tau}{\delta}\int_0^t \|\sqrt{\tau}u_{ttt}^\tau(s)\|_2^2 ds + \frac{\tau^2 c^4 T}{\delta}\sup_{t\in[0,T]}\|Au_t^\tau(t)\|_2^2, \quad (3.35)$$

where we have used the zero initial conditions of the x^τ-equation.

This was the reconstruction we needed.

Step 3: Uniform (in τ) bound for $\|\sqrt{\tau}u_{ttt}^\tau(t)\|_2^2$

Recall that the problem (2.1) is written abstractly as (see (2.4))

$$\begin{cases} M_\tau U_t^\tau(t) = M_\tau \mathcal{A}^\tau U^\tau(t), \ t > 0, \\ U^\tau(0) = U_0 = (u_0, u_1, u_2)^T. \end{cases} \quad (3.36)$$

In order to estimate $\|\sqrt{\tau}u_{ttt}^\tau(t)\|_2^2$, we differentiate (3.36) in time, which leads us to

$$\begin{cases} M_\tau U_{tt}^\tau(t) = M_\tau \mathcal{A}^\tau U_t^\tau(t), \ t > 0, \\ U_t^\tau(0) = \mathcal{A}^\tau U_0. \end{cases} \quad (3.37)$$

and by relabeling $V^\tau = U_t^\tau$ we can further rewrite

$$\begin{cases} M_\tau V_t^\tau(t) = M_\tau \mathcal{A}^\tau V^\tau(t), \ t > 0, \\ V^\tau(0) = V_0 = \mathcal{A}^\tau U_0. \end{cases} \quad (3.38)$$

Now, since we are considering $U_0 \in \mathbb{H}_2$, which means $u_0, u_1 \in \mathcal{D}(A)$ and $u_2 \in \mathcal{D}(A^{1/2})$, we have $\mathcal{A}^\tau U_0 \in \mathbb{H}_1$. Therefore, by Theorem 2.2 and Remark 2.2 we get

$$\|\sqrt{\tau}u_{ttt}^\tau(t)\|_2^2 \leqslant \|T^\tau(t)V_0\|_{\tau,1}^2 \leqslant \overline{M_1}^2 \|V_0\|_{\tau,1}^2 \leqslant \frac{\overline{K}}{\tau}\|U_0\|_{\mathbb{H}_2}^2, \quad (3.39)$$

for all $t \in [0, T]$, where $\overline{K}$ does not depend on τ.

We also obtain

$$\gamma^\tau \tau \int_0^t \|u_{ttt}^\tau(s)\|_2^2 ds \leqslant C\|U_0\|_{\mathbb{H}_2}^2, \quad (3.40)$$

where C does not depend on τ.

Proof. Apply (3.3) to time derivatives. This gives

$$\gamma \int_0^T |u_{ttt}|^2 \leqslant C|A^{1/2}u_{tt}(0)|^2 + |A^{1/2}u_t(0)|^2 + \tau |u_{ttt}(0)|^2$$

From the equation read of $\tau u_{ttt}(0)$

$$\tau|u_{ttt}(0)|^2 \leqslant \frac{C}{\tau}[|u_2|^2 + |Au_0|^2 + |Au_1|^2]$$

$\square$

This gives the conclusion in (3.40).
Step 4: Collecting the estimates
By adding (3.32) and (3.35) and using (3.39) we conclude

$$\|A^{1/2}x^\tau(t)\|_2^2 + \|A^{1/2}x_t^\tau(t)\|_2^2 \leqslant \tau C\|U_0\|_{\mathbb{H}_2}^2,$$

This finishes the proof of part (a) of Theorem 2.4.
Proof of part (b)-strong convergence. This amounts to showing that given $U_0 \in \mathbb{H}_0^\tau$ and given $\varepsilon > 0$ there exist $\delta > 0$ such that if $\tau < \delta$ then

$$\|PT^\tau(t)U_0 - T(t)PU_0\|_{\mathbb{H}_0^0}^2 < \varepsilon$$

for all times $t > 0$. The strategy is prove that for any given fixed time T the above inequality is true and then to choose a suitable T such that for $t > T$ the energy is still bounded above by ε.
Step 1: Finite time. Let $U_0 \in \mathbb{H}_0$ and $T > 0$. Let $\varepsilon > 0$ be arbitrary.
Since $\mathbb{H}_2$ is dense in $\mathbb{H}_0$, if we define

$$\varepsilon' = \frac{\varepsilon}{2(M + M_0)},$$

we can find $U_{\varepsilon'} \in \mathbb{H}_2$ such that

$$\|U_0 - U_{\varepsilon'}\|_{\tau,0}^2 < \varepsilon'.$$

Define

$$\delta = \delta_\varepsilon = \frac{\varepsilon}{2C\|U_{\varepsilon'}\|_{\mathbb{H}_2}^2}$$

(where C comes from Step 4 in the proof of part (a) considering $U_{\varepsilon'}$ as the initial condition) and notice that $\delta \to 0$ as $\varepsilon \to 0$.
Then, for $\tau < \delta$ we estimate by using (3.13),

$$\|PT^\tau(t)U_0 - T(t)PU_0\|_{P(\mathbb{H}_0)}^2 \leqslant \|PT^\tau(t)(U_0 - U_{\varepsilon'}) - T(t)P(U_0 - U_{\varepsilon'})\|_{\mathbb{H}_0^0}^2 + \|PT^\tau(t)U_{\varepsilon'} - T(t)PU_{\varepsilon'}\|_{\mathbb{H}_0^0}^2$$
$$\leqslant \|T^\tau(t)\|_{\mathcal{L}(\mathbb{H}_0^\tau)}^2 \|U_0 - U_{\varepsilon'}\|_{\tau,0}^2 + \|T(t)\|_{\mathcal{L}(\mathbb{H}_0^0)}^2 \|U_0 - U_{\varepsilon'}\|_{\tau,0}^2 + C\tau\|U_{\varepsilon'}\|_{\tau,2}^2$$
$$< \varepsilon'(M + M_0) + C\delta\|U_{\varepsilon'}\|_{\tau,2}^2 < \varepsilon.$$

Step 2: Infinite time. Integrating Equation (3.31) in time from s to t we have

Singular Thermal Relaxation Limit for the Moore-Gibson-Thompson ...

$$I(x) = \frac{\alpha}{2}\|x_t^T(t)\|_2^2 + \frac{c^2}{2}\|A^{1/2}x^T(t)\|_2^2 = \left[\frac{\alpha}{2}\|x_t^T(s)\|_2^2 + \frac{c^2}{2}\|A^{1/2}x^T(s)\|_2^2\right]$$

$$-\delta\int_s^t \|A^{1/2}x_t^T(\sigma)\|_2^2 d\sigma - \tau\int_s^t (u_{ttt}^T(\sigma), x_t^T(\sigma))d\sigma - \tau c^2 \int_s^t (A^{1/2}u_t^T(\sigma), A^{1/2}x_t(\sigma))d\sigma$$

$$\leq \left[\frac{\alpha}{2}\|x_t^T(s)\|_2^2 + \frac{c^2}{2}\|A^{1/2}x^T(s)\|_2^2\right] + \tau\int_s^t (u_{ttt}^T(\sigma), x_t^T(\sigma))d\sigma + \tau c^2 \int_s^t (A^{1/2}u_t^T(\sigma), A^{1/2}x_t(\sigma))d\sigma$$

$$\leq \left[\frac{\alpha}{2}\|x_t^T(s)\|_2^2 + \frac{c^2}{2}\|A^{1/2}x^T(s)\|_2^2\right] + \tau\int_s^t [\|\sqrt{\tau}u_{ttt}^T(\sigma)\|_2^2 + \|x_t^T(\sigma)\|_2^2]d\sigma$$

$$+ \tau c^2 \int_s^t [\|A^{1/2}u_t^T(\sigma)\|_2^2 + \|A^{1/2}x_t(\sigma)\|_2^2]d\sigma.$$

Now observe that all the terms on the right hand side above (both inside and outside the integral) are uniformly exponentially stable. Therefore, there exist positive constants L_1, L_2, a, b such that

$$I(x) \leq \left[\frac{\alpha}{2}\|x_t^T(s)\|_2^2 + \frac{c^2}{2}\|A^{1/2}x^T(s)\|_2^2\right] + \tau\int_s^\infty [\|\sqrt{\tau}u_{ttt}^T(\sigma)\|_2^2 + \|x_t^T(\sigma)\|_2^2]d\sigma$$

$$+ \tau c^2 \int_s^\infty [\|A^{1/2}u_t^T(\sigma)\|_2^2 + \|A^{1/2}x_t(\sigma)\|_2^2]d\sigma \leq L_1 e^{-as} + \frac{L_2}{b}e^{-bs} < \varepsilon,$$

as long as

$$s > T_{1,\varepsilon} \equiv \max\left\{-\frac{1}{a}\ln\left(\frac{\varepsilon}{2L_1}\right), -\frac{1}{b}\ln\left(\frac{\varepsilon b}{2L_2}\right)\right\}.$$

Similar estimates are valid when one integrates (3.33) in time from s to t. This will then gives rise to an $T_{2,\varepsilon}$.

Thus taking $T = T_\varepsilon \equiv \max\{T_{1,\varepsilon}, T_{2,\varepsilon}\}$ in Step 1 and combining it with control of the tail of the integral leads to the convergence in (2.9) uniformly for all $t \geq 0$. This completes the proof of part (b) of Theorem 2.4.

3.4 Proof of Proposition 2.2

The uniform bounds imply, among other things, that there exist z_1 and z_2 such that

$$\begin{cases} u_t^T \to z_1 & \text{weakly* in } L^\infty(0, T; \mathcal{D}(A^{1/2})), \\ \tau^{1/2}u_{tt}^T \to z_2 & \text{weakly* in } L^\infty(0, T; L^2(\Omega)). \end{cases} \quad (3.41)$$

An argument of Distributional Calculus shows that $u_{tt}^T \to z_{tt}$ in $H^{-1}(0, T; L^2(\Omega))$ and therefore $\tau^{1/2}u_{tt}^T \to \tau^{1/2}z_{tt} \to 0$ in $H^{-1}(0, T; L^2(\Omega))$. Uniqueness of the limit then leads to the conclusion.

3.5 Proof of Proposition 2.3

Recall that $\{\mu_n\}_{n\in\mathbb{N}}$ is the set of eigenvalues of A and since A is unbounded we can assume $\mu_n \to \infty$ as $n \to \infty$. Proving the Proposition 2.3 amounts to the study of spectrum of $\mathcal{A}$ on the space $\mathbb{H}_0^0$.

Lemma 3.8

$$\sigma_p(\mathcal{A}) = \sigma_p(\mathcal{A}^*) = \{\lambda \in \mathbb{C};\ \alpha\lambda^2 + \delta\mu_n\lambda + c^2\mu_n = 0,\ n \in \mathbb{N}\} = \left\{ \frac{-\delta\mu_n \pm \sqrt{\delta^2\mu_n^2 - 4\alpha c^2 \mu_n}}{2\alpha},\ n \in \mathbb{N} \right\}.$$

Proof Since A is a positive self-adjoint operator with compact resolvent,

$$\sigma(A) = \sigma_p(A) \subset \mathbb{R}_+^*.$$

The spectrum of A is countable and positive. So we assume the point spectrum is then a sequence (μ_n) such that $\mu_n \to +\infty$ as $n \to +\infty$. We shall consider the operator $\mathcal{A}$ acting on $\mathcal{D}(A^{1/2}) \times \mathcal{D}(A^{1/2})$ with the domain

$$\mathcal{D}(\mathcal{A}) = \{(u, v) \in \mathcal{D}(A^{1/2}) \times \mathcal{D}(A^{1/2}); c^2 u + \delta v \in \mathcal{D}(A^{3/2})\}.$$

Let $\varphi = (\varphi_1, \varphi_2)^T \in \mathcal{D}(\mathcal{A})$. We seek to the describe the values of $\lambda \in \mathbb{C}$ such that

$$\mathcal{A}\varphi = \lambda\varphi. \tag{3.42}$$

We compute:

$$\mathcal{A}\varphi = \begin{pmatrix} 0 & I \\ -c^2\alpha^{-1}A & -\delta\alpha^{-1}A \end{pmatrix} \begin{pmatrix} \varphi_1 \\ \varphi_2 \end{pmatrix} = \begin{pmatrix} \varphi_2 \\ -c^2\alpha^{-1}A\varphi_1 - \delta\alpha^{-1}A\varphi_2 \end{pmatrix}.$$

Therefore the Eq. (3.42) will be satisfied if and only if

$$\varphi_2 = \lambda\varphi_1$$

and

$$c^2 A\varphi_1 + \delta A\varphi_2 = -\alpha\lambda\varphi_2$$

which is the same as

$$c^2 A\varphi_1 + \delta\lambda A\varphi_1 = -\alpha\lambda^2\varphi_1$$

or further

$$A\varphi_1 = \frac{-\alpha\lambda^2}{c^2 + \delta\lambda}\varphi_1.$$

The last equation means that φ_1 is an eigenvector of A and because of that must be associated with some eigenvalue μ_n. Therefore, the relation between λ and μ_n can be easily derived to be the quadratic equation

$$\alpha\lambda^2 + \delta\mu_n\lambda + c^2\mu_n = 0$$

from where follows that
$$\lambda_n = -\frac{\delta\mu_n}{2\alpha} \pm \frac{\sqrt{\delta^2\mu_n^2 - 4\alpha c^2 \mu_n}}{2\alpha}. \tag{3.43}$$

We now characterize the point spectrum of $\mathcal{A}^*$. Keeping in mind the following facts:
(i) A is self-adjoint in $H = L^2(\Omega)$.
(ii) Fractional powers preserve self-adjointness.
We begin by computing $\mathcal{A}^*$.
Let $\varphi, \psi \in \mathcal{D}(A^{1/2}) \times \mathcal{D}(A^{1/2})$, $\varphi = (\varphi_1, \varphi_2)^T$ and $\psi = (\psi_1, \psi_2)^T$. We have

$$(\mathcal{A}\varphi, \psi)_{\mathcal{D}(A^{1/2}) \times \mathcal{D}(A^{1/2})} = \left(\begin{pmatrix} 0 & I \\ -c^2\alpha^{-1}A & -\delta\alpha^{-1}A \end{pmatrix} \begin{pmatrix} \varphi_1 \\ \varphi_2 \end{pmatrix}, \begin{pmatrix} \psi_1 \\ \psi_2 \end{pmatrix} \right)_{\mathcal{D}(A^{1/2}) \times \mathcal{D}(A^{1/2})}$$

$$= \left(\begin{pmatrix} \varphi_2 \\ -c^2\alpha^{-1}A\varphi_1 - \delta\alpha^{-1}A\varphi_2 \end{pmatrix}, \begin{pmatrix} \psi_1 \\ \psi_2 \end{pmatrix} \right)_{\mathcal{D}(A^{1/2}) \times \mathcal{D}(A^{1/2})}$$

$$= (\varphi_2, \psi_1)_{\mathcal{D}(A^{1/2})} + (-c^2\alpha^{-1}A\varphi_1 - \delta\alpha^{-1}A\varphi_2, \psi_2)_{\mathcal{D}(A^{1/2})}$$

$$= (\varphi_2, \psi_1)_{\mathcal{D}(A^{1/2})} + (\varphi_1, -c^2\alpha^{-1}A\psi_2)_{\mathcal{D}(A^{1/2})} + (\varphi_2, -\delta\alpha^{-1}A\psi_2)_{\mathcal{D}(A^{1/2})}$$

$$= (\varphi_1, -c^2\alpha^{-1}A\psi_2)_{\mathcal{D}(A^{1/2})} + (\varphi_2, \psi_1 - \delta\alpha^{-1}A\psi_2)_{\mathcal{D}(A^{1/2})}$$

$$= \left(\begin{pmatrix} \varphi_1 \\ \varphi_2 \end{pmatrix}, \begin{pmatrix} -c^2\alpha^{-1}A\psi_2 \\ \psi_1 - \delta\alpha^{-1}A\psi_2 \end{pmatrix} \right)_{\mathcal{D}(A^{1/2}) \times \mathcal{D}(A^{1/2})}$$

$$= \left(\begin{pmatrix} \varphi_1 \\ \varphi_2 \end{pmatrix}, \begin{pmatrix} 0 & -c^2\alpha^{-1}A \\ I & -\delta\alpha^{-1}A \end{pmatrix} \begin{pmatrix} \psi_1 \\ \psi_2 \end{pmatrix} \right)_{\mathcal{D}(A^{1/2}) \times \mathcal{D}(A^{1/2})}.$$

Therefore
$$\mathcal{A}^* = \begin{pmatrix} 0 & -c^2\alpha^{-1}A \\ I & -\delta\alpha^{-1}A \end{pmatrix},$$

with
$$\mathcal{D}(\mathcal{A}^*) = \{(u, v) \in \mathcal{D}(A^{1/2}) \times \mathcal{D}(A^{1/2}), c^2 u + \delta v \in \mathcal{D}(A^{3/2})\}.$$

We then find out the point spectrum of $\mathcal{A}^*$.
Let $\varphi = (\varphi_1, \varphi_2)^T \in \mathcal{D}(\mathcal{A}^*)$. We seek to the describe the values of $\lambda \in \mathbb{C}$ such that

$$\mathcal{A}^*\varphi = \lambda\varphi. \tag{3.44}$$

We compute:
$$\mathcal{A}^*\varphi = \begin{pmatrix} 0 & -c^2\alpha^{-1}A \\ I & -\delta\alpha^{-1}A \end{pmatrix} \begin{pmatrix} \varphi_1 \\ \varphi_2 \end{pmatrix} = \begin{pmatrix} -c^2\alpha^{-1}A\varphi_2 \\ \varphi_1 - \delta\alpha^{-1}A\varphi_2 \end{pmatrix}.$$

Therefore the Eq. (3.44) will be satisfied if and only if
$$-c^2\alpha^{-1}A\varphi_2 = \lambda\varphi_1$$

and
$$\varphi_1 - \delta\alpha^{-1}A\varphi_2 = \lambda\varphi_2.$$

Decoupling gives:
$$A\varphi_2 = \frac{-\alpha\lambda^2}{c^2 + \delta\lambda}\varphi_2$$

and the last equation means that φ_2 is an eigenvector of A and because of that must be associated with some eigenvalue μ_n through the quadratic equation

$$\alpha\lambda^2 + \delta\mu_n\lambda + c^2\mu_n = 0$$

which implies that
$$\sigma_p(\mathcal{A}^*) = \overline{\sigma_p(\mathcal{A})},$$

completing the proof. □

Part (a) then follows directly from Lemma 3.8 because we know that $\lambda \in \sigma_r(\mathcal{A})$ if and only if $\overline{\lambda} \in \sigma_p(\mathcal{A}^*)(= \overline{\sigma_p(\mathcal{A})}$, in our case). However, we know that if $\overline{\lambda} \in \sigma_p(\mathcal{A})$, so is λ. Therefore, since $\sigma_p(\mathcal{A}) \cap \sigma_r(\mathcal{A}) = \emptyset$, it follows $\sigma_r(\mathcal{A}) = \emptyset$.

Now since the parameters $\delta, \alpha, c^2 > 0$ are fixed, we can see that $\sigma_p(\mathcal{A})$ is eventually real, which means that no matter how we pick those parameters, since $\mu_n \to +\infty$ as $n \to \infty$ we will always be able to find an index N such that from that index on all the eigenvalues will be real.

It is also clear to see that the point spectrum of $\mathcal{A}$ is on the left side of the complex plane. In fact, it follows from the formula (3.43) that in case λ_n is complex we have

$$\text{Re}(\lambda_n) = -\frac{\delta\mu_n}{2\alpha} < 0.$$

For the real ones, the "$-$" case of the formula (3.43) we have nothing to check because clearly $\lambda_n < 0$. For the "$+$" case we just notice that

$$\sqrt{\delta^2\mu_n^2 - 4\alpha c^2\mu_n} < \delta\mu_n$$

and the strict inequality guarantees that $\lambda_n < 0$. Therefore, in order to describe the continuous spectrum of $\mathcal{A}$ we just analyze the limit

$$\lim_{n\to\infty} \frac{-\delta\mu_n \pm \sqrt{\delta^2\mu_n^2 - 4\alpha c^2\mu_n}}{2\alpha}.$$

Two basic limit arguments show that

$$\lim_{n\to\infty} \frac{-\delta\mu_n + \sqrt{\delta^2\mu_n^2 - 4\alpha c^2\mu_n}}{2\alpha} = -\frac{c^2}{\delta}$$

and

$$\lim_{n\to\infty} \frac{-\delta\mu_n - \sqrt{\delta^2\mu_n^2 - 4\alpha c^2\mu_n}}{2\alpha} = -\infty,$$

which implies $\sigma_c(\mathcal{A}) \supset \left\{-\frac{c^2}{\delta}\right\}$, since $\sigma(\mathcal{A})$ is closed and $\sigma_r(\mathcal{A}) = \emptyset$ with $-\delta^{-1}c^2$ not an eigenvalue of $\mathcal{A}$. In order to complete the proof of part (b) we need to show that $-\frac{c^2}{\delta}$ is *the only element* in the continuous spectrum of $\mathcal{A}$. To establish this we shall show that any $\lambda \notin \sigma_p(\mathcal{A}) \cup \left\{-\dfrac{c^2}{\delta}\right\}$ is in the resolvent set of $\mathcal{A}$. Let $(f, g)^T \in \mathcal{D}(A^{1/2}) \times \mathcal{D}(A^{1/2})$. We need to prove that there exists $(u, v)^T \in \mathcal{D}(\mathcal{A})$ such that

$$\mathcal{A}(u, v)^T - \lambda(u, v)^T = (f, g)^T.$$

After writing down explicitly the equation we obtain

$$\begin{aligned}\lambda u - v &= f \\ \alpha \lambda v + A(c^2 u + \delta v) &= \alpha g\end{aligned} \qquad (3.45)$$

which leads to solvability of

$$(c^2 + \delta \lambda) Au + \alpha \lambda^2 u = \alpha \lambda f + \alpha g + \delta A f, \ in \ [\mathcal{D}(A^{1/2})]'$$

or equivalently

$$(c^2 + \delta \lambda) u + \alpha \lambda^2 A^{-1} u = \alpha \lambda A^{-1} f + \alpha A^{-1} g + \delta f \equiv F \qquad (3.46)$$

and further because $c^2 + \delta \lambda \neq 0$,

$$u + \dfrac{\alpha \lambda^2}{c^2 + \delta \lambda} A^{-1} u = (c^2 + \delta \lambda)^{-1} F. \qquad (3.47)$$

Note that $F \in \mathcal{D}(A^{1/2})$ and we are looking for a solution $u \in \mathcal{D}(A^{1/2})$. Since A^{-1} is compact on $L(\mathcal{D}(A^{1/2}))$, unique solvability of (3.47) if and only if the operator

$$I + \dfrac{\alpha \lambda^2}{c^2 + \delta \lambda} A^{-1}$$

is injective. On the other hand, the latter takes place if and only if $-\dfrac{\alpha \lambda^2}{c^2 + \delta \lambda} \notin \sigma_p(A)$. This is also true due to the fact that $\lambda \notin \sigma_p(\mathcal{A})$. In view of the above, we obtain $u \in \mathcal{D}(A^{1/2})$ and therefore $v = \lambda u - f \in \mathcal{D}(A^{1/2})$.

To conclude we need to assert that $(u, v)^T \in \mathcal{D}(\mathcal{A})$. For the latter we just notice that $A(c^2 u + \delta v) = \alpha g - \alpha \lambda v \in \mathcal{D}(A^{1/2})$-as desired by the characterization of the domain of $\mathcal{A}$. The proof of the Proposition is thus complete.

3.6 Proof of Proposition 2.4

As in the proof of Proposition 2.3, proving Proposition 2.4 relies on the analysis of point spectrum of $\mathcal{A}^\tau$ along with the asymptotics. To begin with, the point spectrum of $\mathcal{A}^\tau$ and $(\mathcal{A}^\tau)^*$ coincide and it is given by the set

$$\sigma_p(\mathcal{A}^\tau) = \sigma_p((\mathcal{A}^\tau)^*) = \{\lambda \in \mathbb{C}; \ \tau \lambda^3 + \alpha \lambda^2 + b^\tau \mu_n \lambda + c^2 \mu_n = 0, \ n \in \mathbb{N}\},$$

which is a consequence of basic algebraic manipulation.

The exact same argument as for Part (a) in Proposition 2.3 shows Part (a) here.

Now, with empty residual spectrum we know that the points in the continuous spectrum, if any, needs to be in the approximate point spectrum. By using the exact same process as in [28] (Theorem 5.2, Part (b), (b_1) and (b_2), p. 1913 and 1914) one can show that $-\dfrac{c^2}{b^\tau}$ is an eigenvalue of $\mathcal{A}^\tau$ in case $\gamma^\tau = 0$ and a limit of eigenvalues in case $\gamma^\tau > 0$. Thus $-\dfrac{c^2}{b^\tau} \in \sigma_c(\mathcal{A}^\tau)$ in case $\gamma^\tau > 0$. We shall show now that $-\dfrac{c^2}{b^\tau}$ coincides with the point in continuous spectrum $\sigma_c(\mathcal{A}^\tau)$ in case $\gamma^\tau > 0$. This is to say $\sigma_c(\mathcal{A}^\tau) = \left\{-\dfrac{c^2}{b^\tau}\right\}$. For this, it is sufficient to show that every $\lambda \in \mathbb{C}$ different from $-\dfrac{c^2}{b^\tau}$ and outside the point spectrum of $\mathcal{A}^\tau$ belongs to the resolvent set. We need to prove that there exist solution $(u, v, w)^T \in \mathcal{D}(\mathcal{A}^\tau)$ As before, we consider the system:

$$
\begin{aligned}
v - \lambda u &= f \in \mathcal{D}(A^{1/2}) \\
w - \lambda v &= g \in \mathcal{D}(A^{1/2}) \\
-\tau^{-1}[c^2 Au + b^\tau Av + \alpha w] - \lambda w &= h \in L^2(\Omega).
\end{aligned}
\tag{3.48}
$$

Collecting the terms yields:

$$(\tau^{-1}c^2 + \lambda\tau^{-1}b^\tau)Au + (\lambda^3 + \tau^{-1}\alpha\lambda^2)u = \tau^{-1}(\alpha\lambda + b^\tau)Af + \tau^{-1}\alpha Ag + \lambda^2 f + \lambda g + h,$$

where the equation is defined on $[\mathcal{D}(A^{1/2})]'$. Since $\lambda \neq -\dfrac{c^2}{b^\tau}$, the above can be written as

$$u + d(\lambda, \tau) A^{-1} u = F(f, g, h) \in \mathcal{D}(A^{1/2}), \tag{3.49}$$

where

$$d(\lambda, \tau) = \dfrac{\lambda^3 + \tau^{-1}\alpha\lambda^2}{\tau^{-1}c^2 + \lambda\tau^{-1}b^\tau} = \dfrac{\lambda^3\tau + \alpha\lambda^2}{c^2 + \lambda b^\tau}$$

and

$$F(f, g, h) = \tau^{-1}(\alpha\lambda + b^\tau)Af + \tau^{-1}\alpha Ag + \lambda^2 f + \lambda g + h.$$

Since A^{-1} is compact in $L(\mathcal{D}(A^{1/2}))$, unique solvability [for $u \in \mathcal{D}(A^{1/2})$ of (3.49) is equivalent to the injectivity of $I + d(\lambda, \tau)A^{-1}$. The latter is equivalent to the fact that $-d(\lambda, \tau) \notin \sigma_p(A^\tau)$, which in turn is equivalent to $\mu_n + d(\lambda, \tau) \neq 0$. This last condition is guaranteed by the fact that $\lambda \notin \sigma_p(\mathcal{A}^\tau)$. Thus there exists a unique $u \in \mathcal{D}(A^{1/2})$ solving (3.49). Going back to (3.48) we obtain the improved regularity $v \in \mathcal{D}(A^{1/2})$, $w \in \mathcal{D}(A^{1/2})$ and also $c^2 u + b^\tau v \in \mathcal{D}(A)$. Hence $(u, v, w)^T \in \mathcal{D}(\mathcal{A}^\tau)$ as desired. The proof of equivalence $\sigma_c(\mathcal{A}^\tau) = \{-\dfrac{c^2}{b^\tau}\}$ is completed.

In order to complete the proof of Proposition 2.4 it suffices to prove the part (c). Here, the aim is to show that when $\tau \to 0$ the hyperbolic branch of the spectrum of $\mathcal{A}^\tau$ escapes to $-\infty$. For this we show that for n large, the equation

$$\tau\lambda^3 + \alpha\lambda^2 + (b^\tau \mu_n)\lambda + c^2\mu_n = 0. \tag{3.50}$$

has two complex roots whose imaginary parts approach $\pm\infty$.

The argument is as follows: define a number θ_n such that

$$\theta_n \approx -\frac{\gamma^\tau c^4}{\mu_n b^3} \text{ for } n \text{ large.}$$

We claim that

$$-\frac{c^2}{b^\tau} + \theta_n \approx \lambda_n^{0,\tau} \text{ for } n \text{ large,}$$

that is, for n large $-\frac{c^2}{b^\tau} + \theta_n$ is *almost* a root of (3.50).
Indeed, notice that (3.50) can be rewritten as

$$\tau\lambda^3 + \alpha\lambda^2 + (b^\tau \mu_n)\lambda + c^2 \mu_n = \left(\lambda + \frac{c^2}{b^\tau} - \theta_n\right) q_n(\lambda) + r_n(\lambda),$$

where

$$q_n(\lambda) = \tau\lambda^2 + (\gamma^\tau + \tau\theta_n)\lambda + b^\tau \mu_n - (\gamma^\tau + \tau\theta_n)\left(\frac{c^2}{b^\tau} - \theta_n\right)$$

and

$$r_n(\lambda) = (\gamma^\tau + \tau\theta_n)\left(\frac{c^2}{b^\tau} - \theta_n\right)^2 + b^\tau \mu_n \theta_n.$$

Since $\mu_n \to \infty$ as $n \to \infty$, we have $\theta_n \approx 0$ which implies

$$r_n(\lambda) \approx \gamma^\tau \frac{c^4}{b^{2\tau}} - \gamma^\tau \frac{c^4}{b^{2\tau}} = 0 \text{ for } n \text{ large,}$$

where we have used $b^\tau \mu_n \theta_n \approx -\frac{\gamma^\tau c^4}{b^{2\tau}}$. This proves the claim made above.

As a consequence, for n large we have

$$\tau\lambda^3 + \alpha\lambda^2 + (b^\tau \mu_n)\lambda + c^2 \mu_n \approx \left(\lambda + \frac{c^2}{b^\tau} - \theta_n\right)\left(\tau\lambda^2 + (\gamma^\tau + \tau\theta_n)\lambda + b^\tau \mu_n - (\gamma^\tau + \tau\theta_n)\left(\frac{c^2}{b^\tau} - \theta_n\right)\right).$$

Therefore, for n large the two other roots of the equation (which are complex) are approximately the two roots of

$$\tau\lambda^2 + (\gamma^\tau + \tau\theta_n)\lambda + b^\tau \mu_n - (\gamma^\tau + \tau\theta_n)\left(\frac{c^2}{b^\tau} - \theta_n\right).$$

Then, a basic result for quadratic equation yields

$$2\text{Re}(\lambda_n^{\tau,1}) = 2\text{Re}(\lambda_n^{\tau,2}) = -\frac{\gamma^\tau + \tau\theta_n}{\tau} \to -\frac{\gamma^\tau}{\tau} \text{ as } n \to \infty$$

and

$$|\text{Img}(\lambda_n^{\tau,1})| = |\text{Img}(\lambda_n^{\tau,1})| \to +\infty \text{ as } n \to \infty.$$

The proof of part (c) is then complete.

4 Appendix

Lemma 4.1 (The energy identity) *For all $U_0^\tau \in \mathbb{H}_0$ we have*

$$\frac{d}{dt} E_1^\tau(t) + \gamma^\tau \|u_{tt}^\tau(t)\|_2^2 = 0. \tag{4.1}$$

Proof We first consider strong solutions with initial data in $\mathcal{D}(\mathcal{A}^\tau)$. This implies $U^\tau(t) \in \mathbb{H}_0$ and $u_t(t), u_{tt}(t) \in \mathcal{D}(A^{1/2})$, $u_{ttt}(t) \in H$. For these elements the following calculus is justifiable. Notice that the expansion of $E_1^\tau(t)$ is

$$E_1^\tau(t) = \frac{\tau}{2} \|u_{tt}^\tau(t)\|_2^2 + \frac{b^\tau}{2} \|A^{1/2} u_t^\tau(t)\|_2^2 + \frac{c^4}{2b^\tau} \|A^{1/2} u^\tau(t)\|_2^2$$
$$+ c^2 (Au_t^\tau(t), u^\tau(t)) + \frac{\tau c^2}{b^\tau} (u_{tt}^\tau(t), u_t^\tau(t)) + \frac{\alpha c^2}{2 b^\tau} \|u_t^\tau(t)\|_2^2. \tag{4.2}$$

Firs, taking the L^2-inner product of (2.1) with u_{tt}^τ gives

$$\frac{d}{dt} \left(\frac{\tau}{2} \|u_{tt}^\tau\|_2^2 + c^2 (Au_t^\tau, u^\tau) + \frac{b^\tau}{2} \|A^{1/2} u_t^\tau\|_2^2 \right) + \alpha \|u_{tt}^\tau(t)\|^2 - c^2 (Au_t^\tau(t), u_t^\tau(t)) = 0. \tag{4.3}$$

Next similarly, taking the L^2-inner product of (2.1) with u_t^τ gives

$$\frac{d}{dt} \left(\tau (u_{tt}^\tau, u_t^\tau) + \frac{\alpha}{2} \|u_t^\tau\|_2^2 + \frac{c^2}{2} \|A^{1/2} u^\tau\|_2^2 \right) - \tau \|u_{tt}^\tau(t)\|_2^2 + b^\tau \|A^{1/2} u_t^\tau(t)\|_2^2 = 0. \tag{4.4}$$

Combining (4.3) and $\frac{c^2}{b^\tau} \times$ (4.4), we get

$$\frac{d}{dt} \left[\frac{\tau}{2} \|u_{tt}^\tau\|_2^2 + \frac{b^\tau}{2} \|A^{1/2} u_t^\tau\|_2^2 + \frac{c^4}{2b^\tau} \|A^{1/2} u^\tau\|_2^2 + c^2 (Au_t^\tau, u_t^\tau) + \frac{\tau c^2}{b^\tau} (u_{tt}^\tau, u_t^\tau) + \frac{\alpha c^2}{2b^\tau} \|u_t^\tau\|_2^2 \right]$$
$$+ (\alpha - \frac{c^2 \tau}{b^\tau}) \|u_{tt}^\tau(t)\|_2^2 = 0. \tag{4.5}$$

By (4.2) and the definition of $\gamma^\tau = \alpha - \frac{c^2 \tau}{b^\tau}$, we obtain the identity

$$\frac{d}{dt} E_1^\tau(t) + \gamma^\tau \|u_{tt}^\tau(t)\|_2^2 = 0.$$

$\square$

It is equivalent to say that

$$E_1^\tau(t) + \gamma^\tau \int_0^t \|u_{tt}^\tau(s)\|_2^2 ds = E_1^\tau(0), \tag{4.6}$$

and the final conclusion is obtained by evoking density of $\mathcal{D}(\mathcal{A}^\tau)$ in $\mathbb{H}_0$.

References

1. Alves, M., Caixeta, A., Silva, M.J., Rodrigues, J.: Moore-Gibson-Thompson equation with memory in a history framework: a semigroup approach. Zeitschrift für angewandte Mathematik und Physik **69**(4), 106 (2018)
2. Caixeta, A.H., Domingos Cavalcanti, V.N., Lasiecka, I.: On long time behavior of Moore-Gibson-Thompson equation with molecular relaxation. Evol. Equ. Control Theory **5**(4) (2016)
3. Cattaneo, C.: Sulla Conduzione Del Calore. Atti Sem. Mat. Fis. Univ. Modena **3**, 83–101 (1948)
4. Cattaneo, C.: A form of heat-conduction equations which eliminates the paradox of instantaneous propagation. Comptes Rendus **247**, 431 (1958)
5. Charoenphon, S.: Vanishing relaxation time dynamics of the Jordan Moore-Gibson-Thompson (JMGT) equation arising in high frequency ultrasound (HFU) (Ph.D thesis). Unoversiy of Memphis (2020)
6. Chen, G., Russell, D.L.: A mathematical model for linear elastic systems with structural damping. Q. Appl. Math. **39**(4), 433–454 (1982)
7. Chen, S.P., Triggiani, R.: Proof of extensions of two conjectures on structural damping for elastic systems. Pac. J. Math. **136**(1), 15–55 (1989)
8. Christov, C., Jordan, P.: Heat conduction paradox involving second-sound propagation in moving media. Phys. Rev. Lett. **94**(15), 154301 (2005)
9. Conejero, J.A., Lizama, C., Ródenas Escribá, F.D.A.: Chaotic behaviour of the solutions of the Moore-Gibson-Thompson equation. Appl. Math. Inf. Sci. **9**(5), 2233–2238 (2015)
10. Coulouvrat, F.: On the equations of nonlinear acoustics. J. Acoust. **5**(321–359), 52 (1992)
11. Crighton, D.G.: Model equations of nonlinear acoustics. Annu. Rev. Fluid Mech. **11**(1), 11–33 (1979)
12. Dell'Oro, F., Lasiecka, I., Pata, V.: The Moore-Gibson-Thompson equation with memory in the critical case. J. Differ. Equ. **261**(7), 4188–4222 (2016)
13. Ekoue, F., d'Halloy, A.F., Gigon, D., Plantamp, G., Zajdman, E.: Maxwell-Cattaneo regularization of heat equation. World Acad. Sci. Eng. Technol. **7**, 05–23 (2013)
14. Fattorini, H.O.: The Cauchy Problem. Addison Wesley (1983)
15. Hamilton, M.F., Blackstock, D.T., et al.: Nonlinear Acoustics. Academic Press (1997)
16. Jordan, P.M.: Nonlinear acoustic phenomena in viscous thermally relaxing fluids: Shock bifurcation and the emergence of diffusive solitons. J. Acoust. Soc. Am. **124**(4), 2491–2491 (2008)
17. Kaltenbacher, B.: Mathematics of nonlinear acoustics. Evol. Equ. Control Theory **4**(4), 447–491 (2015)
18. Kaltenbacher, B., Lasiecka, I.: Global existence and exponential decay rates for the Westervelt's equation. Discret. Contin. Dyn. Syst.-Ser. S **2**(3), 503–525 (2009)
19. Kaltenbacher, B., Lasiecka, I., Marchand, R.: Wellposedness and exponential decay rates for the Moore-Gibson-Thompson equation arising in high intensity ultrasound. Control Cybern. **40**, 971–988 (2011)
20. Kaltenbacher, B., Lasiecka, I., Pospieszalska, M.K.: Well-posedness and exponential decay of the energy in the nonlinear Jordan-Moore-Gibson-Thompson equation arising in high intensity ultrasound. Math. Models Methods Appl. Sci. **22**(11), 1250035 (2012)
21. Kato, T.: Perturbation Theory for Linear Operators. Springer, Berlin Heidelberg (1976)
22. Lai, W.M., Rubin, D.H., Krempl, E., Rubin, D.: Introduction to Continuum Mechanics. Butterworth-Heinemann (2009)
23. Lasiecka, I.: Global solvability of Moore-Gibson-Thompson equation with memory arising in nonlinear acoustics. J. Evol. Equ. **17**(1), 411–441 (2017)
24. Lasiecka, I., Triggiani, R.: Control Theory for Partial Differential Equations: Volume 1, Abstract Parabolic Systems: Continuous and Approximation Theories. Cambridge University Press (2000)
25. Lasiecka, I., Wang, X.: Moore-Gibson-Thompson equation with memory, part ii: general decay of energy. J. Differ. Equ. **259**(12), 7610–7635 (2015)
26. Lasiecka, I., Wang, X.: Moore-Gibson-Thompson equation with memory, part i: exponential decay of energy. Zeitschrift für angewandte Mathematik und Physik **67**(2), 17 (2016)

27. Lunardi, A.: Analytic Semigroups and Optimal Regularity in Parabolic Problems. Birkhäuser (1995)
28. Marchand, R., McDevitt, T., Triggiani, R.: An abstract semigroup approach to the third-order Moore-Gibson-Thompson partial differential equation arising in high-intensity ultrasound: structural decomposition, spectral analysis, exponential stability. Math. Methods Appl. Sci. **35**(15), 1896–1929 (2012)
29. Moore, F.K., Gibson, W.E.: Propagation of weak disturbances in a gas subject to relaxation effects. J. Aerosp. Sci. **27**(2), 117–127 (1960)
30. Pazy, A.: Semigroups of Linear Operators and Applications to Partial Differential Equations. Springer (1983)
31. Stokes, G.G.: An examination of the possible effect of the radiation of heat on the propagation of sound. Philos. Mag. Ser. **1**(4), 305–317 (1851)
32. Straughan, B.: Heat Waves. Springer Science & Business Media (2011)
33. Triggiani, R.: On the stabilizability problem in Banach space. J. Math. Anal. Appl. **52**(3), 383–403 (1975)
34. Xu, F., Lu, T., Seffen, K.: Biothermomechanical behavior of skin tissue. Acta Mech. Sin. **24**(1), 1–23 (2008)

Applications of the Kantorovich–Rubinstein Maximum Principle in the Theory of Boltzmann Equations

Henryk Gacki and Roksana Brodnicka

Abstract A generalized version of the Tjon–Wu equation is considered. It describes the evolution of the energy distribution in a model of gas in which simultaneous collisions of many particles are permitted. Using the technique of the Kantorovich–Rubinstein maximum principle concerning the properties of probability metrics we show that the stationary solution is asymptotically stable with respect to the Kantorovich–Wasserstein distance. Our research was stimulated by the problem of stability of solutions of the same equation which was derived by Lasota (see [1]). This stability result was based on the technique of Zolotarev norms. Lasota shows that the stationary solution is exponentially stable in the Zolotarev norm of order 2.

Keywords Boltzmann type equation · Asymptotic stability · Nonlinear operator · Kantorovich–Wasserstein metric

1 Introduction

Some problems of the mathematical physics can be written as differential equations for functions with values in the space of measures. The vector space of signed measures does not have good analytical properties. For example, this space with the Fortet–Mourier, the Kantorovich–Wasserstein or Zolotarev metric is not complete. In the first part of this article we will show that this space with Fortet–Mourier metric is not complete.

There are two methods to overcome this problem. First, we may replace the original equations by the adjoint ones on the space of continuous bounded functions. Secondly, we may restrict our equations to some complete convex subsets of the

---

H. Gacki · R. Brodnicka (✉)
University of Silesia in Katowice, Institute of Mathematics, ul. Bankowa 12, 40-007 Katowice, Poland
e-mail: rbrodnicka@o2.pl

H. Gacki
e-mail: Henryk.Gacki@us.edu.pl

© Springer Nature Switzerland AG 2020
J. Banasiak et al. (eds.), *Semigroups of Operators – Theory and Applications*,
Springer Proceedings in Mathematics & Statistics 325,
https://doi.org/10.1007/978-3-030-46079-2_10

vector space of measures. This approach seems to be quite natural and it is related to the classical results concerning differential equations on convex subsets of Banach spaces (see [2]). The convex sets method in studying the Boltzmann equation was used in a series of papers (see for example : [3–7]).

The main purpose of our study is to show some application of the Kantorovich–Rubinstein maximum principle concerning the properties of probability metrics. Then we show that the Kantorovich–Rubinstein maximum principle combined with the LaSalle invariance principle used in the theory of dynamical systems allows to find new sufficient conditions for the asymptotic stability of Markov semigroups.

To illustrate the application of this criterion we will discuss an example drawn from the kinetic theory of gases. This example was stimulated by the problem of the stability of solutions of the following version of the Boltzmann equation

$$\frac{\partial u(t,x)}{\partial t} + u(t,x) = \int_x^\infty \frac{dy}{y} \int_0^y u(t, y-z) u(t,z) dz \qquad t \geq 0, \qquad x \geq 0. \qquad (1)$$

Due to the physical interpretation Eq. (1) is considered with the additional conditions

$$\int_0^\infty u(t,x) \, dx = \int_0^\infty x u(t,x) \, dx = 1. \qquad (2)$$

Equation (1) was derived by J. A. Tjon and T. T. Wu from the Boltzmann equation (see [8]). According to Barnsley and Cornille [9] it is called the *Tjon–Wu equation*. It is easy to see that the function $u_*(t,x) := \exp(-x)$ is a (stationary) solution of (1). M. F. Barnsley and G. Turchetti (see [10], p. 369) proved that this solution is stable in the class of all initial functions $u_0 := u(0, \cdot)$ satisfying the condition

$$\int_0^\infty u_0(x) e^{\frac{x}{2}} dx < \infty. \qquad (3)$$

This condition was replaced by T. Dłotko and A. Lasota (see [11], Theorem 3) by a less restrictive

$$\int_0^\infty x^n u_0(x) \, dx < \infty \qquad \text{for} \qquad n = 2, 3, \ldots . \qquad (4)$$

In 1990 Z. Kiełek (see [12], Theorem 1.1) succeeded in proving that the stationary solution u_* is asymptotically stable if (4) is satisfied for $n = 2$.

Equation (1) has a simple interpretation. For fixed $t \geq 0$ the function $u(t, \cdot)$ denotes the density distribution function of the energy of a particle in an ideal gas. In the time interval $(t, t + \Delta t)$ a particle changes its energy with the probability $\Delta t + o(\Delta t)$ and the change is equal to $[-u(t,x) + P(u(t,x))]\Delta t + o(\Delta t)$, where the operator P is given by the formula

$$(P\,v)(x) = \int_x^\infty \frac{dy}{y} \int_0^y v(y-z)\,v(z)\,dz. \tag{5}$$

In order to understand the action of P consider three independent random variables ξ_1, ξ_2 and η, such that ξ_1, ξ_2 have the same density distribution function v and η is uniformly distributed on the interval $[0, 1]$. Then $P\,v$ is the density distribution function of the random variable

$$\eta(\xi_1 + \xi_2). \tag{6}$$

Physically this means that the energies of the particles before a collision are independent quantities and that a particle after collision takes the η part of the sum of the energies of the colliding particles.

The assumption that η has the density distribution function of the form $\mathbf{1}_{[0,1]}$ is quite restrictive. In general, if η has the density distribution h, then the random variable (6) has the density distribution function

$$(P\,v)(x) = \int_0^\infty h\!\left(\frac{x}{y}\right) \frac{dy}{y} \int_0^y v(y-z)\,v(z)\,dz. \tag{7}$$

In 1999 A. Lasota and J. Traple (see [6], Theorem 1.1) studied the asymptotic behaviour of solutions of the equation

$$u' + u = P\,u, \tag{8}$$

where $u : \mathbb{R} \to L^1(\mathbb{R})$ is an unknown function and P is the operator given by formula (7). Equation (8) was studied in the spaces $L^p(\mathbb{R}_+)$ with $p = 1, 2$ and different weights. In the proof the following conditions concerning h were used

$$\int_0^\infty h(x)\,dx = 2 \int_0^\infty x\,h(x)\,dx = 1, \quad 2 \int_0^\infty x^p\,h(x)\,dx < 1, \tag{9}$$

$$\sup_x \{x\,h(x) : x \geq 0\} < \infty, \tag{10}$$

$$h(x) > 0 \quad \text{for} \quad 0 < x < x_0, \tag{11}$$

where $p > 1$ and $x_0 > 0$.

In [13] the authors proved that for the spatially homogeneous Boltzmann equation with cutoff hard potentials that solutions remain bounded from above, uniformly in time, by a Maxwellian distribution, provided the initial data have a Maxwellian upper bound.

In [14] the authors presented a simple fast spectral method for the Boltzmann collision operator with general collision kernels. Their new method can apply to arbitrary collision kernels. Moreover, a series of numerical tests is performed to illustrate the efficiency and accuracy of the proposed method.

The work [15] illustrating the use of transport equations in population surveys is very interesting. In this paper the authors formulate an individual based model to describe the phenotypic evolution in hermaphroditic populations which includes random and assortative mating of individuals. By increasing the number of individuals to infinity they obtain a nonlinear transport equation, which describes the evolution of phenotypic distribution. The main result of the paper is a theorem on asymptotic stability of trait with respect to the Fortet–Mourier metric. This theorem is applied to models with the offspring trait distribution given by additive and multiplicative random perturbations of the parental mean trait. The second model contains multiplicative noise, and it includes, as a special case, the Tjon–Wu equation. As a by-product of their investigation they give a simple proof of the theorem of Lasota and Traple (see [6, 7]) concerning asymptotic stability of this equation.

In the article [16], which is thematically related to ours, it was considered in particular the N-particle model which includes multi-particle interactions. It is shown that certain natural assumptions formally lead to class of equations which can be considered as the most general Maxwell-type model.

The object of our paper is to study a generalized version of (8) in the space $\mathcal{M}_{sig}$ of all signed measures on the real line with more general operator P. We will assume that P is a convex combination of N operators $P_1, \ldots, P_N$, where P_k for $k \geq 2$ describes the simultaneous collision of k particles and P_1 the influence of external forces.

This research was stimulated by the problem of stability of solutions of the same equation which was derived by Lasota [1]. This stability results was based on the technique of Zolotarev metrics. Lasota shows that the stationary solution is exponentially stable in the Zolotarev norm of order 2.

The basic idea of our method is to apply the Kantorovich–Rubinstein maximum principle combined with the LaSalle invariance principle. Using this method we proved that the stationary solution is asymptotically stable with respect to the Kantorovich–Wasserstein distance. Our result intersects with the one of Lasota (see [1]).

In Chap. 4 we will introduce several notations in order to define precisely the collision operator P.

2 Metrics in the Space of Measures

The classical Kantorovich–Rubinstein principle gives necessary conditions for the maxima of a linear functional acting on the space of Lipschitzian functions. The maximum value of this functional defines the Kantorovich–Wasserstein metric on the space of probability measures (see [3, 4]). An analogous result for the Fortet–Mourier metric is proved in [3].

Let (X, ϱ) be a Polish space, i.e., a separable, complete metric space. By $\mathcal{B}_X$ we denote the σ-algebra of Borel subsets of X and by $\mathcal{M}$ the family of all finite (nonnegative) Borel regular measures on X.

By $\mathcal{M}_1$ we denote the subset of $\mathcal{M}$ such that $\mu(X) = 1$ for $\mu \in \mathcal{M}_1$. The elements of $\mathcal{M}_1$ will be called *distributions*. Further let

$$\mathcal{M}_{sig} = \{\mu_1 - \mu_2 : \mu_1, \mu_2 \in \mathcal{M}\},$$

be the space of finite signed measures.

To introduce the norm in $\mathcal{M}_{sig}$ we need to define the distance between two measures $\mu_1, \mu_2 \in \mathcal{M}$. Let $(A_1, \ldots, A_n)$ be a measurable partition of X, that is

$$X = \bigcup_{i=1}^{n} A_i, \qquad A_i \cap A_j = \emptyset \quad \text{for } i \neq j, A_i \in \mathcal{B}_X.$$

We set

$$\|\mu_1 - \mu_2\|_T = \sup \left\{ \sum_{i=1}^{n} |\mu_1(A_i) - \mu_2(A_i)| \right\}, \tag{12}$$

where the supremum is taken over all possible measurable partitions of X (with arbitrary n). Let $\mu = \mu_1 - \mu_2$. The value $\|\mu\|_T$ is called the *total variation norm* of the measure μ, and the convergence with respect to this norm is called the *strong convergence of measures*.

Remark 1 It is well known that $(\mathcal{M}_{sig}, \|\cdot\|_T)$ is a Banach space.

Remark 2 For arbitrary $\mu \in \mathcal{M}_{sig}$ by μ_+ and μ_- we denote the positive and the negative part of μ. Then we set

$$\mu_+ - \mu_- = \mu \quad \text{and} \quad \mu_+ + \mu_- = \|\mu\|_T. \tag{13}$$

Let c be a fixed element of X. For every real number $\alpha \geq 1$ we define the sets $\mathcal{M}_{1,\alpha}$ and $\mathcal{M}_{sig,\alpha}$ by setting

$$\mathcal{M}_{1,\alpha} = \{\mu \in \mathcal{M}_1 : m_\alpha(\mu) < \infty\} \quad \text{and} \quad \mathcal{M}_{sig,\alpha} = \{\mu \in \mathcal{M}_{sig} : m_\alpha(\mu) < \infty\}$$

where

$$m_\alpha(\mu) = \int_X (\varrho(x, c))^\alpha \|\mu\|_T(dx).$$

As usual, $B(X)$ denotes the space of all bounded Borel measurable functions $f : X \to \mathbb{R}$, and $C(X)$ the subspace of all bounded continuous functions. Both spaces are endowed with the supremum norm

$$\|f\| = \sup_{x \in X} |f(x)|.$$

For every $f : X \to \mathbb{R}$ and $\mu \in \mathcal{M}_{sig}$ we write

$$\langle f, \mu \rangle = \int_X f(x)\mu(dx), \tag{14}$$

whenever this integral exists.

In the space $\mathcal{M}_1$ we introduce the *Fortet–Mourier metric* (see [17], Proposition 8.2) by the formula

$$\|\mu_1 - \mu_2\|_{\mathcal{F}} = \sup\{|\langle f, \mu_1 - \mu_2 \rangle| : f \in \mathcal{F}\}, \tag{15}$$

where $\mathcal{F}$ is the set of functions $f : X \to R$ satisfying

$$\|f\| \leq 1 \quad \text{and} \quad |f(x) - f(y)| \leq \varrho(x, y) \quad \text{for } x, y \in X.$$

We say that a sequence (μ_n), $\mu_n \in \mathcal{M}_1$, *converges weakly* to a measure $\mu \in \mathcal{M}_1$ if

$$\lim_{n \to \infty} \langle f, \mu_n \rangle = \langle f, \mu \rangle \quad \text{for} \quad f \in C(X). \tag{16}$$

Since X is a Polish space, condition (16) is equivalent to

$$\lim_{n \to \infty} \|\mu_n - \mu\|_{\mathcal{F}} = 0$$

(see [17], Theorem 8.3).

The classical Kantorovich–Rubinstein principle gives necessary conditions for the maxima of a linear functional acting on the space of Lipschitzian functions. The maximum value of this functional defines the Kantorovich–Wasserstein metric on the space of probability measures (see [3, 4]).

In the space $\mathcal{M}_1$ we introduce the *Kantorovich–Wasserstein metric* (see [18]) by the formula

$$\|\mu_1 - \mu_2\|_{\mathcal{K}} = \sup\{|\langle f, \mu_1 - \mu_2 \rangle| : f \in \mathcal{K}\} \quad \text{for} \quad \mu_1, \mu_2 \in \mathcal{M}_{1,1}, \tag{17}$$

where $\mathcal{K}$ is the set of functions $f : X \to R$ which satisfy the condition

$$|f(x) - f(y)| \leq \varrho(x, y) \quad \text{for} \quad x, y \in X.$$

Fix $c \in X$. It is easy to see that

Applications of the Kantorovich–Rubinstein Maximum Principle ...

$$\|\mu_1 - \mu_2\|_\mathcal{K} = \sup\{|\langle f, \mu_1 - \mu_2\rangle| : f \in \mathcal{K}_c\} \quad \text{for} \quad \mu_1, \mu_2 \in \mathcal{M}_{1,1},$$

where $\mathcal{K}_c = \{f \in \mathcal{K} : f(c) = 0\}$. Denote by $B(x, r)$ a closed ball in X with center $x \in X$ and radius r. Let $\mu \in \mathcal{M}_1$. We define the support of μ by setting

$$\operatorname{supp} \mu = \{x \in X : \mu(B(x, \varepsilon)) > 0 \quad \text{for every} \quad \varepsilon > 0\}.$$

Remark 3 Every set $\mathcal{M}_{1,\alpha}$, for $\alpha \geq 1$ contains the subset of all measures $\mu \in \mathcal{M}_1$ with a compact support. This subset is dense in $\mathcal{M}_1$ with respect to the Fortet–Mourier norm (see [19], Theorem 4, p. 237).

Now we are going to prove the main result of this chapter i.e. we prove that the space $(\mathcal{M}_{sig}, \|.\|_F)$ is not complete. In the proof of this fact we will use the following theorem:

Theorem 1 (Open mapping theorem) *Let $(X_1, \|\cdot\|_1)$ and $(X_2, \|\cdot\|_2)$ be Banach spaces. Assume that $f : X_1 \to X_2$ is a continuous and surjective linear operator. Then for every open set $U \subset X_1$ the image $f(U)$ is an open set in X_2.*

Moreover, we need to consider the following example:

Example 1 We assume that X has at least one accumulation point. Denote by x_0 one of the accumulation points in X. For this point we may find a sequence of points $(x_n)_{n\in\mathbb{N}}, x_n \neq x_0, n = 1 \ldots$, such that $\lim_{n\to\infty} \rho(x_n, x_0) = 0$. We will consider the sequence of measures defined as follows

$$\mu_0 = \delta_{x_0}, \quad \text{and} \quad \mu_n = \delta_{x_n} \quad \text{for} \quad n = 1, \ldots,$$

where δ_{x_n} and δ_{x_0} are the Dirac measures at the points x_0 and x_n, respectively.

Now we are going to show that a sequence of measures $(\delta_{x_n})_{n\in\mathbb{N}}$ is convergent to δ_{x_0} with respect to the Fortet–Mourier metric. Since $<f, \mu_n> = f(x_n)$ and $<f, \mu_0> = f(x_n)$, an elementary calculation shows that

$$|<f, \mu_n> - <f, \mu_0>| = |f(x_n) - f(x_0)| \leq \rho(x_n, x_0),$$

for $f \in \mathcal{F}$. Since $f \in \mathcal{F}$ was arbitrary, the last inequality and (17) imply

$$\|\mu_n - \mu_0\|_\mathcal{F} \leq \rho(x_n, x_0). \tag{18}$$

This shows that $(\mu_n)_{n\in\mathbb{N}}$ converges to μ_0 with respect to the Fortet–Mourier norm.

It is not difficult to verify that this sequence of measures $(\mu_n)_{n\in\mathbb{N}}$ does not converge to μ_0 in the total variation norm. The proof consists in the construction of a measurable partition of X by setting

$$A_1 = \{x_0\}, \quad A_2 = \{x_n\}, \quad A_3 = X\setminus\{x_0 \cup x_n\}.$$

From this and definition of the Dirac measures, it follows immediately that

$$\sum_{i=1}^{3} |(\mu_n - \mu_0)(A_i)| = |(\mu_n - \mu_0)(A_1)| + |(\mu_n - \mu_0)(A_2)| + |(\mu_n - \mu_0)(A_3)|$$
$$= |-\mu_0(\{x_0\})| + |\mu_{x_n}(\{x_n\})|$$
$$= |-\delta_0(\{x_0\})| + |\delta_{x_n}(\{x_n\})| = 2$$

Since the total variation norm we define as the supremum taken over all possible measurable partitions, the last inequality implies that

$$\|\mu_n - \mu_0\|_T \geq 2$$

This shows that $(\mu_n)_{n \in \mathbb{N}}$ does not converge to μ_0 with respect to the total variation norm.

Now we are ready to state the main result of this chapter. More precisely we will prove that the space of finite signed measures with Fortet–Mourier metric is not complete.

Proposition 1 *For every $\mu \in \mathcal{M}_{sig}$ the following inequality is satisfied*

$$\|\mu\|_\mathcal{F} \leq \|\mu\|_T. \tag{19}$$

Further, if X has at least one accumulation point then the convergence with respect to the Fortet–Mourier metric is strictly weaker than the convergence with respect to metric generated by the total variation norm.

Proof An important role in the proof of (19) is played by the inequality

$$|<f, \mu>| \leq \|f\| \|\mu\|_T \quad \text{where} \quad f \in B(X) \quad \text{and} \quad \mu \in \mathcal{M}_{sig}. \tag{20}$$

We will break up the proof of this inequality into three steps. We first show that (20) is satisfied for the simple function f of the form

$$f = \sum_{k=1}^{n} \lambda_k 1_{A_k} \quad \text{where} \quad \lambda_k \in \mathbb{R}, \quad A_k \in \mathcal{A} \quad \text{and} \quad k = 1, \ldots, n.$$

Obviously

$$|<f, \mu>| = \left| \int_X f(x)\mu(dx) \right| = \left| \int_X \sum_{k=1}^{n} \lambda_k 1_{A_k}(x)\mu(dx) \right| = \left| \sum_{k=1}^{n} \int_{A_k} (x)\lambda_k \mu(dx) \right|$$
$$= \left| \sum_{k=1}^{n} \lambda_k \mu(A_k) \right| \leq \max |\lambda_k| \sum_{k=1}^{n} |\mu(A_k)| \leq \|f\| \|\mu\|_T. \tag{21}$$

Let now $f : X \to R$ be an arbitrary nonnegative, bounded and integrable function. Then there is a sequence of simple functions

$$f_n(x) = \sum_i \lambda_{i,n} 1_{A_{i,n}}(x)$$

such that

$$\lim_{n\to\infty} f_n(x) = f(x) \quad \text{a.e. and} \quad |f_n| \leq |f(x)|. \qquad (22)$$

Since f_n for every $n = 1, \ldots$ is simple function, the inequality (21) imply that

$$|<f_n, \mu>| \leq \|f_n\| \|\mu\|_T$$

Consequently, from this and the Lebesgue dominated convergence theorem, we obtain inequality (20) for f satisfying condition (22).

Finally, let $f : X \to R$ be an arbitrary bounded and integrable function. We denote by f^+ and f^- the positive and negative parts of f given by the formula

$$f^+ = \max(0, f(x)) \quad \text{and} \quad f^- = \max(0, -f(x)).$$

Observe that thanks to the properties $f = f^+ - f^-$ the function f satisfies the inequality (20). Consequently, from the inequality (20) we obtain (19).

Conclusion 1 *To prove that the space $(\mathcal{M}_{sig}, \|.\|_F)$ is not complete we consider the identity function*

$$id : (\mathcal{M}_{sig}, \|.\|_T) \to (\mathcal{M}_{sig}, \|.\|_F).$$

On the contrary, let $(\mathcal{M}_{sig}, \|.\|_F)$ be a Banach space. From the Open mapping theorem it follows immediately, that the topology generated by $(\mathcal{M}_{sig}, \|.\|_T)$ and $(\mathcal{M}_{sig}, \|.\|_F)$ are equivalent. This contradicts with the fact that there exists the sequence of measures $(\mu_n)_{n\in\mathbb{N}}$ which is convergent in the Fortet–Mourier norm but not in the total variation norm—see Example 1.

3 Modelling Collision with the Influence of External Forces

In the theory of dilute gases the Boltzmann equation

$$\frac{DF(t, x, v)}{Dt} = C(F(t, x, v))$$

is studied to obtain information about the particle distribution function F that depends on time (t), position (x), and velocity (v). DF/Dt denotes the total rate of change of F due to spatial gradients and any external forces, whereas the collision operator $C(\cdot)$ determines the way in which particle collisions affect F. In the case of a spatially homogeneous gas with no external forces the Boltzmann equation reduces to

$$\frac{\partial F(t, v)}{\partial t} = C(F(t, v)). \tag{23}$$

Tjon and Wu (see [8]) have shown that in some cases Eq. (23) may be transformed into

$$\frac{\partial u(t, x)}{\partial t} + u(t, x) = \int_x^\infty \frac{dy}{y} \int_0^y u(t, y - z) u(t, z) dz \quad t \geq 0, \quad x \geq 0. \tag{24}$$

where $x = \frac{v^2}{2}$ (note that x is not a spatial coordinate) and

$$u(t, x) = const \int_x^\infty \frac{F(t, v)}{\sqrt{v - x}} dv.$$

Equation (24), called the Tjon–Wu equation, is nonlinear because of the presence of $u(t, y - z) u(t, z)$ in the integrand on the right-hand side. However, note that e^{-x} is a solution of (24). The equation describing the homogeneous model in the dilute gas with a possibility of collisions of two particles. Equation (24) governs the evolution of the density distribution function of the energy of particles imbeded in an ideal gas in the equilibrium stage. The solution u of the problem has a simple physical interpretation. Namely $u(t, \cdot)$ for fixed $t > 0$ is a probability distribution function of the energy of particles in an ideal gas. In the time interval $(t, t + \Delta t)$ a particle changes its energy with the probability $\Delta t + o(\Delta t)$ and this change is described by the collision operator

$$(Pu)(x) = \int_x^\infty \frac{dy}{y} \int_0^y u(y - z) u(z) dz. \tag{25}$$

Hence, the change is equal to $[-u(t, x) + P(u(t, x))]\Delta t + o(\Delta t)$.

In order to understand the action of P consider three independent random variables ξ_1, ξ_2 and η, such that ξ_1, ξ_2 have the same density distribution function u and η is uniformly distributed on the interval $[0, 1]$. Here we obtain that Pu is the density distribution function of the random variable

$$\eta(\xi_1 + \xi_2). \tag{26}$$

This corresponds to the physical assumption that the energies of the particles before a collision are independent quantities and that a particle after collision takes η part of the sum of the energies of the colliding particles.

The assumption that η has the density distribution function of the form $\mathbf{1}_{[0,1]}$ is quite restrictive. Moreover there are no physical reasons which will allow us to assume that the distribution of energy of particles can be described only by density (so by the absolutely continuous measure).

Following this physical interpretation, Gacki in 2007 (see [3]) considered the evolutionary Boltzmann-type equation

$$\frac{d\psi}{dt} + \psi = P\psi \quad \text{for} \quad t \geq 0 \tag{27}$$

with the initial condition

$$\psi(0) = \psi_0, \tag{28}$$

where $\psi_0 \in \mathcal{M}_1(\mathbb{R}_+)$ and $\psi : \mathbb{R}_+ \to \mathcal{M}_{sig}(\mathbb{R}_+)$ is an unknown function. Moreover

$$P : \mathcal{M}_1(\mathbb{R}_+) \to \mathcal{M}_1(\mathbb{R}_+)$$

is analogous to (25), but in this case P is an operator acting on the space of probability measures. More precisely an operator P is acting on the subset $D \subset \mathcal{M}_1(\mathbb{R}_+)$ given by formula

$$D := \{\mu \in \mathcal{M}_1 : m_1(\mu) = 1\}, \quad \text{where} \quad m_1(\mu) = \int_0^\infty x\mu(dx). \tag{29}$$

Equation (27) was studied in the space $\mathcal{M}_1(\mathbb{R}_+)$. The operator P describes the collision of two particles in general situation. To describe the collision operator in this case we start from recalling the convolution operator of order n and the linear operator P_φ which is related to multiplication of the random variables.

For every $n \in \mathbb{N}$ we define the convolution operator of order n, $P_{*n} : \mathcal{M}_{sig} \to \mathcal{M}_{sig}$, by the formula

$$P_{*1}\mu := \mu, \quad P_{*(n+1)}\mu := \mu * P_{*n}\mu \quad \text{for} \quad \mu \in \mathcal{M}_{sig}. \tag{30}$$

Remark 4 Observe that P_{*n} is not the nth power of P_{*1} but $P_{*n}\mu$ is the nth convolution power of μ.

It is easy to verify that $P_{*n}(\mathcal{M}_1) \subset \mathcal{M}_1$ for every $n \in \mathbb{N}$. Moreover, $P_{*n}|_{\mathcal{M}_1}$ has a simple probabilistic interpretation. Namely, if $\xi_1, \ldots, \xi_n$ are independent random variables with the same distribution μ, then $P_{*n}\mu$ is the distribution of $\xi_1 + \cdots + \xi_n$.

The second class of operators we are going to study is related to multiplication of the random variables. The formal definition is as follows. With given $\mu, \nu \in \mathcal{M}_{sig}$, we define their elementary product $u \circ v$ by

$$(u \circ v)(A) := \int_{\mathbb{R}_+} \int_{\mathbb{R}_+} \mathbf{1}_A(xy)\mu(dx)v(dy) \quad \text{for} \quad A \in \mathcal{B}_{\mathbb{R}_+}. \tag{31}$$

It follows that

$$\langle f, u \circ v \rangle = \int_{\mathbb{R}_+} \int_{\mathbb{R}_+} f(xy)\mu(dx)v(dy) \tag{32}$$

for every Borel measurable $f : \mathbb{R}_+ \to \mathbb{R}$ such that $(x, y) \mapsto f(xy)$ is integrable with respect to the product of $|\mu|$ and $|v|$. For fixed $\varphi \in \mathcal{M}_1$ we define a linear operator $P_\varphi : \mathcal{M}_{sig} \to \mathcal{M}_{sig}$ by

$$P_\varphi \mu := \varphi \circ \mu \quad \text{for} \quad \mu \in \mathcal{M}_{sig}. \tag{33}$$

Again, as in the case of convolution, from this definition it follows that $P_\varphi(\mathcal{M}_1) \subset \mathcal{M}_1$. For $\mu \in \mathcal{M}_1$ the measure $P_\varphi \mu$ has an immediate probabilistic interpretation. If φ and μ are the distributions of random variables ξ and η respectively, then $P_\varphi \mu$ is the distribution of the product $\xi \eta$.

We will give a precise definition of P:

$$P := P_\varphi P_{*2}, \tag{34}$$

where $\varphi \in \mathcal{M}_1$ and $m_1(\varphi) = 1/2$. Form (34) it follows that $P(\mathcal{M}_1) \subset \mathcal{M}_1$. Further using (30) and (33) it is easy to verify that for $\mu \in D$,

$$m_1(P_{*2}\mu) = 2 \quad \text{and} \quad m_1(P_\varphi \mu) = 1/2. \tag{35}$$

In order to understand the action of P consider three independent random variables ξ_1, ξ_2 and η, such that ξ_1, ξ_2 have the same distribution μ and η has the distribution φ. Then $P\mu$ is the distribution of the random variable

$$\eta(\xi_1 + \xi_2). \tag{36}$$

Physically this means that the energies of the particles before a collision are independent quantities and that a particle after collision takes the η part of the sum of the energies of the colliding particles.

Remark 5 Evidently every fixed point of the operator P is a stationary solution of Eq. (8).

We will show that if the Eq. (27) has a stationary measure u_* such that $supp\, u_* = \mathbb{R}_+$ (that means $u_*(B(x, \varepsilon)) > 0$ for every $\varepsilon > 0$ and $x \geq 0$), then this measure is asymptotically stable.

Similar problem for the Eq. (8) was studied by Lasota and Traple (see [6], Theorem 3.3). The positivity of u_* plays an important role in the proof of the stability. Namely, it allows to apply the maximum and invariance principle in order to show that the

Kantorovich–Wasserstein metric distance between u_* and an arbitrary solution u is decreasing in time. We start with two simple lemmas concerning the supports of $P\mu$ (more details see [3]).

Lemma 1 *Assume that the measure $\varphi \in \mathcal{M}_1$ satisfies the following conditions*

$$\varphi \neq \delta_{\frac{1}{2}} \tag{37}$$

and

$$m_1(\varphi) = \frac{1}{2}. \tag{38}$$

Then there exists a number $\beta > 1$ such that the following implication holds:

If $v \in D$ and $\operatorname{supp} v \supset (a, b)$, then $\operatorname{supp} Pv \supset (\beta a, \beta b)$.

The following result may be proved in much the same way as Lemma 1.

Lemma 2 *Assume that there is $\sigma_0 > 0$ such that $(0, \sigma_0) \subset \operatorname{supp} \varphi$. Then for every $v \in \mathcal{M}$ there exists a number $\sigma > 0$ such that*

$$\operatorname{supp} Pv \supset (0, \sigma). \tag{39}$$

whenever $v \neq \delta_0$.

We are in a position to formulate the following theorem.

Theorem 2 *Let φ be a probability measure and let $m_1(\varphi) = \frac{1}{2}$. Assume that:*

(i) There is $\sigma_0 > 0$ such that

$$(0, \sigma_0) \subset \operatorname{supp} \varphi. \tag{40}$$

(ii) The operator P has a fixed point $v \in \mathcal{M}$ such that $v \neq \delta_0$.
Then

$$\operatorname{supp} v = \mathbb{R}_+. \tag{41}$$

Proof From Lemmas 1 and 2 it follows that

$$\operatorname{supp} v \supset (0, \beta^n \sigma) \quad \text{for} \quad n \in \mathbb{N}.$$

Since $\beta > 1$, this completes the proof. $\square$

Remark 6 When $\varphi \in \mathcal{M}_1$ and $m_1(\varphi) = \frac{1}{2}$, then the operator P given by (34) is nonexpansive on D with respect to the Kantorovich–Wasserstein metric i.e.

$$\|Pv - Pw\|_\mathcal{K} \leq \|v - w\|_\mathcal{K} \quad \text{for} \quad v, w \in D. \tag{42}$$

In fact, using conditions $m_1(\varphi) = \frac{1}{2}$, $m_1(v+w) = 2$ it is easy to show that the function $\tilde{f} : \mathbb{R}_+ \to \mathbb{R}$ given by the formula

$$\tilde{f}(x) = \int\limits_{\mathbb{R}_+}\int\limits_{\mathbb{R}_+} f\big((x+y)z\big)\varphi(dz)\big(w(dy) + v(dy)\big) \quad \text{for} \quad x \in \mathbb{R}_+$$

belongs to $\mathcal{K}$ for $f \in \mathcal{K}$. Furthermore, from the definition of $\tilde{f}$ it follows that

$$\langle f, Pv - Pw \rangle = \langle \tilde{f}, v - w \rangle \quad \text{for} \quad f \in \mathcal{K}, \quad v, w \in D.$$

Finally, according to the last equality we obtain

$$\|Pv - Pw\|_{\mathcal{K}} = \sup\{|\langle f, Pv - Pw\rangle| : f \in \mathcal{K}\}$$
$$\leq \sup\{|\langle g, v - w\rangle| : g \in \mathcal{K}\} = \|v - w\|_{\mathcal{K}}.$$

We will consider a generalized version of (27) in the space $\mathcal{M}_{sig}$ of all signed measures on the real line with more general operator P. We will assume that P is a convex combination of N operators $P_1, \ldots, P_N$, where P_k for $k \geq 2$ describes the simultaneous collision k particles and P_1 the influence of external forces.

More precisely we consider a differential equation with operators of the form $P_\varphi P_{*n}$ on the right-hand side. Let $(\varphi_1, \varphi_2, \ldots, \varphi_N)$ be a sequence of probability measures and $(c_1, c_2, \ldots, c_N)$ a sequence of real numbers such that

$$\sum_{n=1}^{N} c_n = 1, \quad c_n \geq 0 \quad \text{for} \quad n \in N. \tag{43}$$

and

$$\sum_{n=1}^{N} n c_n m_1(\varphi_n) = 1. \tag{44}$$

We will consider the differential equation

$$\frac{d\psi}{dt} + \psi = \sum_{n=1}^{N} c_n P_{\varphi_n} P_{*n} \psi \quad \text{for} \quad t \geq 0 \tag{45}$$

with the initial problem

$$\psi(0) = \psi_0, \tag{46}$$

$\psi(0) = \psi_0$
where $\psi : \mathbb{R}_+ \to \mathcal{M}_{sig}(\mathbb{R}_+)$ is an unknown function and $\psi_0 \in D$. It is convenient to write the operator appearing on the right-hand side of (45) in the form

$$P = \sum_{n=1}^{N} c_n P_{\varphi_n} P_{*n}, \quad \text{where} \quad P_n = P_{\varphi_n} P_{*n} \qquad (47)$$

We will show that if the Eq. (45) has a stationary measure u_* such that $supp\, u_* = \mathbb{R}_+$ (that means $u_*(B(x, \varepsilon)) > 0$ for every $\varepsilon > 0$ and $x \geq 0$), then this measure is asymptotically stable.

Remark 7 The positivity of u_* plays an important role in the proof of the stability. Namely, it allows to apply the maximum and invariance principle in order to show that the Kantorovich–Wasserstein distance between u_* and an arbitrary solution ψ is decreasing in time.

Before formulating the main result of this chapter concerning nonexpansive operator P we present some simple properties of the operators P_{φ_n} and P_{*n} where $n \in \mathbb{N}$. Using the interpretation of P_{*n} and formulas (43), (44) and (30) it is easy to verify that the operator P_{*n} has the following properties

$$m_1(P_{*n}\mu) = n m_1(\mu) \quad \text{for} \quad \mu \in \mathcal{M}_1, \qquad (48)$$

$$m_s(P_{*n}\mu) = n^s m_1(\mu) \quad \text{for} \quad \mu \in \mathcal{M}_1, s \geq 1, \qquad (49)$$

$$\|P_{*n}\mu_1 - P_{*n}\mu_2\|_{\mathcal{K}} \leq n\|\mu_1 - \mu_2\|_{\mathcal{K}}, \qquad (50)$$

for $\mu_1, \mu_2 \in \mathcal{M}_1$ such that $m_i(\mu) < \infty$ for $i = 1, 2$.

Again using this interpretation and formulas (30) and (33) it can be verified that the operator P_φ satisfies the following

$$m_s(P_\varphi \mu) = m_s(\varphi)(m_s(\mu)) \quad \text{for} \quad \mu \in \mathcal{M}_1, s \geq 1, \qquad (51)$$

$$\|P_\varphi \mu_1 - P_\varphi \mu_2\|_{\mathcal{K}} \leq m_1(\varphi)\|\mu_1 - \mu_2\|_{\mathcal{K}}, \qquad (52)$$

for $\mu_1, \mu_2 \in \mathcal{M}_1$ such that $m_i(\mu) < \infty$ for $i = 1, 2$.

Using (48), (51) and (44) we obtain

$$m_1(P\mu) = \sum_{n=1}^{N} c_n m_1(\varphi_n) n m_1(\mu) = m_1(\mu), \qquad (53)$$

for $\mu \in \mathcal{M}_1$ such that $m_1(\mu) < \infty$ Now we are ready to state the main result of this chapter. According to (44), (52) we have

$$\|P\mu_1 - P\mu_2\|_{\mathcal{K}} \leq \sum_{n=1}^{N} c_n \|P_{\varphi_n} P_{*n}\mu_1 - P_{\varphi_n} P_{*n}\mu_2\|_{\mathcal{K}}$$

$$\leq \sum_{n=1}^{N} c_n m_1(\varphi_n) n \|\mu_1 - \mu_2\|_{\mathcal{K}} = \|\mu_1 - \mu_2\|_{\mathcal{K}}$$

Corollary 1 *Thus P is a nonexpansive operator with respect to the Kantorovich–Wasserstein metric.*

4 Criterion for the Asymptotic Stability of Trajectories

We start from recalling some standard facts concerning theory of dynamical systems.

Let X be a Hausdorff topological space. Further, let T be a *nontrivial semigroup* of nonnegative real numbers. More precisely we assume that T has the following properties:
$$\{0\} \subsetneq T \subset \mathbb{R}_+$$
and
$$t_1 + t_2 \in T, \quad t_1 - t_2 \in T \quad \text{for} \quad t_1, t_2 \in T, \ t_1 \geq t_2. \tag{54}$$

A family of operators $(P^t)_{t \in T}$ is called a *semigroup* if
$$P^{t+s} = P^t P^s \quad \text{for} \quad t, s \in T$$
and $P^0 = I$ where I is the identity operator.

A semigroup $(P^t)_{t \in T}$ is called a *semidynamical system* if the transformation $\mathcal{M}_{sig} \ni \mu \to P^t \mu \in \mathcal{M}_{sig}$ is continuous for every $t \in T$.

If a semidynamical system $(P^t)_{t \in T}$ is given, then for every fixed $\mu \in \mathcal{M}_{sig}$ the function $T \ni t \to P^t \mu \in \mathcal{M}_{sig}$ will be called a *trajectory* starting from μ and denoted by $(P^t \mu)$. A point $\nu \in \mathcal{M}_{sig}$ is called a *limiting point* of a trajectory $(P^t \mu)$ if there exists a sequence (t_n), $t_n \in T$, such that $t_n \to \infty$ and
$$\lim_{n \to \infty} P^{t_n} \mu = \nu.$$

The set of all limiting points of the trajectory $(P^t \mu)$ will be denoted by $\Omega(\mu)$.

We say that a trajectory $(P^t \mu)$ is *sequentially compact* if for every sequence (t_n), $t_n \in T$, $t_n \to \infty$, there exists a subsequence (t_{k_n}) such that the sequence $(P^{t_{k_n}} \mu)$ is convergent to a point $\nu \in \mathcal{M}_{sig}$.

Remark 8 If the trajectory $(P^t \mu)$ is sequentially compact, then $\Omega(\mu)$ is a nonempty, sequentially compact set.

A point $\mu_* \in \mathcal{M}_{sig}$ is called *stationary* (or *invariant*) with respect to a semidynamical system $(P^t)_{t \in T}$ if
$$P^t \mu_* = \mu_* \quad \text{for} \quad t \in T. \tag{55}$$

A semidynamical system $(P^t)_{t \in T}$ is called *asymptotically stable* if there exists a stationary point $x_* \in X$ such that

$$\lim_{t \to \infty} P^t \mu = \mu_* \quad \text{for} \quad \mu \in \mathcal{M}_{sig}. \tag{56}$$

Remark 9 Since X is a Hausdorff space, an asymptotically stable dynamical system has exactly one stationary point.

A function $d : X \times X \to \mathbb{R}^+$ is called a *distance* if d is continuous and if

$$d(x, y) = 0 \Leftrightarrow x = y, \quad \text{for} \quad x, y \in X. \tag{57}$$

Let $(P^t)_{t \in T}$ be a semidynamical system which has at least one sequentially compact trajectory and $\mathcal{Z}$—the set of all $\mu \in \mathcal{M}_{sig}$ such that the trajectory $(P^t \mu)$ is sequentially compact. $\mathcal{Z}$ is a nonempty set, so

$$\Omega = \bigcup_{\mu \in \mathcal{Z}} \Omega(\mu) \neq \emptyset.$$

We will use some criterion for the asymptotic stability of trajectories. This result can be stated as follows.

Theorem 3 *Let $x_* \in \Omega$ be fixed. Assume that for every $x \in \Omega$, $x \neq x_*$ there is $t(x) \in T$ such that*

$$d(S^{t(x)}x, S^{t(x)}x_*) < d(x, x_*). \tag{58}$$

Further assume that the semidynamical system $(S^t)_{t \in T}$ is nonexpansive with respect to distance d, i.e.,

$$d(S^t x, S^t y) \leq d(x, y) \quad \text{for} \quad x, y \in \mathcal{M}_{sig} \quad \text{and} \quad t \in T. \tag{59}$$

Then x_* is a stationary point of $(S^t)_{t \in T}$ and

$$\lim_{t \to \infty} d(S^t z, x_*) = 0 \quad \text{for} \quad z \in Z. \tag{60}$$

By Z we denote the set of all $z \in \mathcal{M}_{sig}$ such that the trajectory $(S^t z)$ is compact.

To illustrate the application of this results in the next section we will discuss an example drawn from the kinetic theory of gases related to the Boltzmann-type equation.

5 Asymptotic Stability of the Nonlinear Boltzmann-Type Equation

In this section we show that the Eq. (45) may by considered in a convex closed subset of a vector space of signed measures. This approach seems to be quite natural and it is related to the classical results concerning the semigroups and differential equations on convex subsets of Banach spaces (see [2, 7]).

We start with recalling some known results related with ordinary differential equations in Banach spaces. For details see [2].

Let $(E, \|\cdot\|)$ be a Banach space and let $\tilde{D}$ be a closed, convex, nonempty subset of E. In the space E we consider *an evolutionary differential equation*

$$\frac{du}{dt} = -u + \tilde{P}u \quad \text{for} \quad t \in \mathbb{R}_+ \tag{61}$$

with the initial condition

$$u(0) = u_0, \quad u_0 \in \tilde{D}, \tag{62}$$

where $\tilde{P} : \tilde{D} \to \tilde{D}$ is a given operator.

A function $u : \mathbb{R}_+ \to E$ is called *a solution of problem* (61), (62) if it is strongly differentiable on $\mathbb{R}_+$, $u(t) \in \tilde{D}$ for all $t \in \mathbb{R}_+$ and u satisfies relations (61), (62).

We start from the following theorem which is usually stated in the case $E = \tilde{D}$.

Theorem 4 *Assume that the operator $\tilde{P} : \tilde{D} \to \tilde{D}$ satisfies the Lipschitz condition*

$$\|\tilde{P}v - \tilde{P}w\| \leq l\|v - w\| \quad \text{for} \quad u, w \in \tilde{D}, \tag{63}$$

where l is a nonnegative constant. Then for every $u_0 \in \tilde{D}$ there exists a unique solution u of problem (61), (62).

The standard proof of the Theorem 4 is based on the fact, that a function $u : \mathbb{R}_+ \to \tilde{D}$ is the solution of (61), (62) if and only if it is continuous and satisfies the integral equation

$$u(t) = e^{-t}u_0 + \int_0^t e^{-(t-s)}\tilde{P}u(s)\,ds \quad \text{for} \quad t \in \mathbb{R}_+. \tag{64}$$

Due to completeness of $\tilde{D}$ the integral on the right hand side is well defined and Eq. (64) may be solved by the method of successive approximations.

Observe that, thanks to the properties of $\tilde{D}$, for every $u_0 \in \tilde{D}$ and every continuous function $u : \mathbb{R}_+ \to \tilde{D}$ the right hand side of (64) is also a function with values in $\tilde{D}$.

The solutions of (64) generate a semigroup of operators $(\tilde{P}^t)_{t \geq 0}$ on $\tilde{D}$ given by the formula

$$\tilde{P}^t u_0 = u(t) \quad \text{for} \quad t \in \mathbb{R}_+, \quad u_0 \in \tilde{D}. \tag{65}$$

In the proof of the main result (Theorem 6) we will use the following.

Theorem 5 *Let P be operator given by (47). Moreover, let $(\varphi_1, \varphi_2, \ldots, \varphi_N)$ be a sequence of probability measures and $(c_1, c_2, \ldots, c_N)$ a sequence of real numbers. Assume that the conditions (43) and (44) are satisfied. Let 0 be an accumulation point of supp $\sum_{n=1}^{N} c_n \varphi_n$. Further let $v, w \in D$ be such that $v \neq w$ and*

$$supp(v + w) = \mathbb{R}_+ \qquad (66)$$

Then

$$\|Pv - Pw\|_{\mathcal{K}} < \|v - w\|_{\mathcal{K}} \qquad (67)$$

The proof of this theorem will be omitted. Nevertheless, it is worth mentioning that in the proof of this theorem, the crucial role is played by the Kantorovich–Wesserstein maximum principle. For details see [3, 4] or [6].

We finish this section with the main result which is a sufficient condition for the asymptotic stability of solutions of the Eq. (45) with respect to the Kantorovich–Wasserstein metric.

Equation (45) together with the initial condition (46) may be considered in a convex subset D of the vector space of signed measures.

Corollary 2 *If $\varphi_n \in \mathcal{M}_1, n \in \{1, 2, \ldots, N\}$ and the conditions (44) and (43) are satisfied then for every $\psi_0 \in D$ there exists a unique solution ψ of problem (45), (46).*

The solutions of (45), (46) generate a semigroup of Markov operators $(P^t)_{t \geq 0}$ on D given by

$$P^t u_0 = u(t) \quad \text{for} \quad t \in \mathbb{R}_+, \quad u_0 \in D. \qquad (68)$$

Now using Theorem 5 we can easily derive the following main result of this thesis.

Theorem 6 *Let P be operator given by (47). Moreover, let $(\varphi_1, \varphi_2, \ldots, \varphi_N)$ be a sequence of probability measures and $(c_1, c_2, \ldots, c_N)$ a sequence of real numbers. Moreover, let the conditions (43) and (44) be satisfied and let 0 be an accumulation point of supp $\sum_{n=1}^{N} c_n \varphi_n$. If P has a fixed point $\psi_* \in D$ such that*

$$supp\ \psi_* = \mathbb{R}_+ \qquad (69)$$

then

$$\lim_{t \to \infty} \|\psi(t) - \psi_*\|_{\mathcal{K}} = 0 \qquad (70)$$

for every compact solution ψ of (45), (46).

Proof First we have to prove that $(P^t)_{t\geq 0}$ is nonexpansive on D with respect to the Kantorovich–Wasserstein metric. For this purpose let $v_0, \vartheta_0 \in D$ be given. For $t \in \mathbb{R}_+$ define
$$v(t) = P^t \eta_0 - P^t \vartheta_0.$$

Using (64), Corollary 2 and (68) it is easy to see that

$$v(t) = e^{-t} v_0 + \int_0^t e^{-(t-s)} (P(P^s \eta_0) - P(P^s \vartheta_0)) ds \quad \text{for } t \in \mathbb{R}_+. \tag{71}$$

From Corollary 1 it follows immediately that

$$\|v(t)\|_\mathcal{K} \leq e^{-t} \|v(0)\|_\mathcal{K} + \int_0^t e^{-(t-s)} \|v(s)\|_\mathcal{K} ds \quad \text{for } t \in \mathbb{R}_+. \tag{72}$$

Consequently

$$f(t) \leq e^{-t} \|v(0)\|_\mathcal{K} + \int_0^t f(s) ds \quad \text{for } t \in \mathbb{R}_+, \tag{73}$$

where $f(t) = e^t \|v(t)\|_\mathcal{K}$. From the Gronwall inequality it follows that

$$f(t) \leq e^{-t} \|v(0)\|_\mathcal{K} \tag{74}$$

This is equivalent to the fact that $(P^t)_{\geq t}$ is nonexpansive on D with respect to the Kantorovich–Wasserstein metric. To end the proof it is sufficient to verify condition (58) of Theorem 3.

From (64) and Corollary 1 it follows immediately that

$$\|P^t \psi_0 - \psi_*\|_T \leq e^{-t} \|\psi_0 - \psi_*\|_T$$
$$+ \int_0^t e^{-(t-s)} \|P^s \psi_0 - \psi_*\|_T \, ds \quad \text{for } \psi_0 \in D \text{ and } t > 0.$$

From nonexpansiveness of $(P^t)_{\geq t}$ above inequality may be rewritten in the form

$$\|P^t \psi_0 - \psi_*\|_T \leq e^{-t} \|\psi_0 - \psi_*\|_T + (1 - e^{-t}) \|\psi_0 - \psi_*\|_T \tag{75}$$
$$= \|\psi_0 - \psi_*\|_T \quad \text{for } \psi_0 \in D \text{ and } t > 0.$$

From (69) and Theorem 5 for P^t, we will get that Markov operator P^t is strongly contracting. Consequently, in (75) we have a strict inequality, because $P^t(\psi_*) = \psi_*$. An application of Theorem 3 completes the proof. □

Remark 10 It is worth noting that:

- Every solution of the equation $P\mu = \mu$ is a stationary solution of Eq. (45).
- We have many possibilities to apply the criterion written in Theorem 6. For example, if we consider the Eq. (45) with the following assumption:

$$2m_r(\varphi_n) < 1, \quad \text{where } r > 1 \text{ and } n = 1, 2, \ldots, N,$$

then for every $\psi_0 \in D$ the solution of (45) and (46) is compact (see [3]).

Remark 11 • We showed that if Eq. (45) has a stationary solution ψ_* such that $\operatorname{supp} \psi_* = \mathbb{R}_+$, then this measure is asymptotically stable.

The positivity of ψ_* plays an important role in the proof of the stability. Namely, it allows us to apply the maximum principle in order to show that the Kantorovich–Wasserstein distance between ψ_* and an arbitrary solution u is decreasing in time.
- If we assume that $c_1 = 0$, $c_2 = 1$ and $c_k = 0$ for $3 \le k \le N$, then we obtain the Eq. (27) as a special type of the Eq. (45).
- Let $c_1 = 0$, $c_2 = 1$ and $c_k = 0$ for $3 \le k \le N$.

1. Observe that in the case of the classical linear Tjon–Wu type Eq. (1) the measure φ_2 is absolutely continuous with density $\mathbf{1}_{[0,1]}$. Moreover, $u_*(t, x) := \exp(-x)$ is the density function of the stationary solution of (1). This is a simple illustration of the situation described by (69).
2. In this case the condition (69) is not particularly restrictive because in Lasota's and Traple's paper (see [20]) it has been proved that the stationary solution ψ_* has the following property: Either ψ_* is supported at one point or $supp\, \psi_* = \mathbb{R}_+$. The first case holds if and only if $\varphi = \delta_{\frac{1}{2}}$. But this case is forgettable as a physical model of particle collisions because it is more restrictive than the model described by the classical Tjon–Wu equation.

References

1. Lasota, A.: Asymptotic stability of some nonlinear Boltzmann-type equations. J. Math. Anal. Appl. **268**, 291–309 (2002)
2. Crandall, M.G.: Differential equations on convex sets. J. Math. Soc. Jpn. **22**, 443–455 (1970)
3. Gacki, H.: Applications of the Kantorovich–Rubinstein maximum principle in the theory of Markov semigroups. Dissertationes Math. **448**, 1–59 (2007)
4. Gacki, H., Lasota, A.: A nonlinear version of the Kantorovich–Rubinstein maximum principle. Nonlinear Anal. **52**, 117–125 (2003)
5. Lasota, A.: Invariant principle for discrete time dynamical systems. Univ. Iagell. Acta Math. **31**, 111–127 (1994)
6. Lasota, A., Traple, J.: An application of the Kantorovich–Rubinstein maximum principle in the theory of the Tjon–Wu equation. J. Differ. Equ. **159**, 578–596 (1999)
7. Lasota, A., Traple, J.: Asymptotic stability of differential equations on convex sets. J. Dyn. Differ. Equ. **15**, 335–355 (2003)

8. Tjon, J.A., Wu, T.T.: Numerical aspects of the approach to a Maxwellian equation. Phys. Rev. A **19**, 883–888 (1979)
9. Barnsley, M.F., Cornille, H.: General solution of a Boltzmann equation and the formation of Maxwellian tails. Proc. Roy. Lond. A **374**, 371–400 (1981)
10. Barnsley, M.F., Turchetti, G.: New results on the nonlinear Boltzmann equation. In: Bardos, C., Bessis, D. (eds.) Bifurcation Phenomena in Mathematical Physics and Related Topics, pp. 351–370. Reidel, Boston (1980)
11. Dłotko, T., Lasota, A.: On the Tjon–Wu representation of the Boltzmann equation. Ann. Polon. Math. **42**, 73–82 (1983)
12. Kiełek, Z.: Asymptotic behaviour of the Tjon–Wu equation. Ann. Polon. Math. **52**, 109–118 (1990)
13. Gamba, I.M., Panferov, V., Villani, C.: Upper Maxwellians bounds for the spatially homogeneous Boltzmann equation. Arch. Rat. Mech. Anal **194**, 253–282 (2009)
14. Gamba, I.M., Haack, J.R., Hauck, C.D., Hu, J.: A fast spectral method for the Boltzmann collision operator with general collision kernels. SIAM J. Sci. Comput. **39**(4), B658–B674 (2017). ArXiv:1610.00397 [math.NA]
15. Rudnicki, R., Zwoleński, P.: Model of phenotypic evolution in hermaphroditic populations. J. Math. Biol. **70**, 1295–1321 (2015)
16. Bobylev, A.V., Cercignani, C., Gamba, I.M.: On the self-similar asymptotics for generalized non-linear kinetic Maxwell models. Commun. Math. Phys. **291**, 599–644 (2009). (original version at arXiv:math-ph/0608035)
17. Dudley, R.M.: Probabilities and Metrics. Aarhaus Universitet, Aarhaus (1976)
18. Rachev, S.T.: Probability Metrics and the Stability of Stochastic Models. Wiley, New York (1991)
19. Billingsley, H.P.: Convergence of Probability Measures. Willey, New York (1968)
20. Lasota, A., Traple, J.: Properties of stationary solutions of a generalized Tjon–Wu equation. J. Math. Anal. Appl. **335**, 669–682 (2007)

Propagators of the Sobolev Equations

Evgeniy V. Bychkov

Abstract In this paper a initial-boundary value problem for the Sobolev equation is investigated. This problem is a part of more general mathematical model of wave propagation in uniform incompressible rotating with constant angular velocity Ω fluid. The studied problem may be obtained from it if we direct Oz axis collinear to Ω. In addition an initial-boundary value problem for the Sobolev equation represents a model describing oscillations in stratified fluid. The solution of this problem is called an inertial (gyroscopic) wave because it arises by virtue of the Archimedes law and under influence of inertial forces. The paper shows that the relative spectrum of the pencil of operators entering the Sobolev equation is bounded. Then, based on the theory of relatively polynomially bounded pencils of operators and the theory of Sobolev type equations of higher order, the propagators of the Sobolev equation, given in a cylinder and in a parallelepiped are constructed.

Keywords The relatively polynomially bounded pencils of operators · The Sobolev equation · Propagators

1 Introduction

In internal waves the maximum vertical displacement of particles takes place not on the surface of the liquid, but inside it. For example, the internal waves arise at the interface of two liquids with different densities. In the ocean this example is observed at the location of desalinated water over heavier water with greater salinity, such a phenomenon has gained name "dead water". In this place a part of power of the ship engine is consumed, on excitation of internal waves, resulting in a decrease of sheep's speed. In the simplest case the two-layer fluid model of internal waves is quite similar to the surface waves. They also concentrate near the interface. Assuming that on both sides of the border the fluid fills entirely each half space, the dispersion relation for

E. V. Bychkov (✉)
South Ural State University, Lenina, 76, Chelyabinsk, Russia
e-mail: bychkovev@susu.ru

internal waves will be identical to the wave ratio $\omega^2 = gk$ [1] for gravitational waves but with a different effective value gravity acceleration.

The mathematical model of waves in homogeneous incompressible rotating with a constant angular velocity Ω liquid is described by a linear system hydrodynamic equations (the Sobolev system of equations [2])

$$\begin{cases} v_t + \frac{1}{\rho_0}\nabla p + 2[\Omega \times v] = 0, \\ \rho_t = 0, \\ \nabla v = 0, \end{cases} \quad (1)$$

where $v = \{u, v, w\}$ is a vector of velocity, $\rho_0 = $ const is an equilibrium density, buoyancy frequency is equal to zero. By directing the Oz axis collinear to the Ω vector, we can obtain the equation for the vertical velocity component of the fluid particles (the Sobolev equation [3])

$$\Delta w_{tt} + F^2 w_{zz} = 0, \quad (2)$$

where $2[\Omega \times v] = \{-Fv, Fu, 0\}$, $F = \Omega$ is the Coriolis parameter. The wave solutions that satisfy (2) are called inertial (or gyroscopic) waves propagating on the surface of a rotating fluid. In [2] there was obtained a solution of equation (2) in an unbounded domain using the Green function method. The behavior of solutions of two-dimensional Hamiltonian systems arising in the theory of small oscillations of rotating ideal fluid was described, and a mathematical model of the incipience of a vortex structure was constructed in [4].

In this paper, we study Eq. (2) in a bounded domain $D \subset \mathbf{R}^3$ with a smooth boundary ∂D. On the boundary of D we set the Dirichlet condition

$$w(x, t) = 0, \quad (x, t) \in \partial D \times \mathbf{R} \quad (3)$$

and at the initial moment of time we set the Cauchy condition

$$w(x, 0) = w_0, \quad w_t(x, 0) = w_1. \quad (4)$$

We solve problem (2)–(4) in the framework of the theory of relatively polynomially bounded operator pencils developed by Zamyshlyaeva [5, 6]. The work also relies on the theory of operator (semi)groups [7–9]. Sobolev type equations and equations unsolved with respect to the highest time derivative were studied in detail in [3, 9]. In [3] various classes of Sobolev type equations are introduced and, in particular, Eq. (2) is assigned to simple Sobolev type equations.

The article in addition to the introduction (the first section) consists of three sections. In the second section we introduce the basic definitions and concepts of the theory of relatively polynomially bounded operator pencils. In the third section, the general solution for an abstract operator-differential equation is given. In the fourth section, the mathematical model is reduced to the Cauchy problem for an abstract operator-differential equation and the propagators for the Sobolev equation (2) are constructed when D is a parallelepiped and a cylinder.

2 (A, p)-Bounded Operator Pencils

Let $\mathscr{U}, \mathscr{F}$ be Banach spaces and operators $A, B_0, B_1, \ldots, B_{n-1} \in \mathscr{L}(\mathscr{U}; \mathscr{F})$ (linear and bounded). By $\vec{B}$ denote a pencil formed by operators $B_{n-1}, \ldots, B_1, B_0$. The sets $\rho^A(\vec{B}) = \{\mu \in \mathbf{C} : (\mu^n A - \mu^{n-1} B_{n-1} - \cdots - \mu B_1 - B_0)^{-1} \in \mathscr{L}(\mathscr{U}; \mathscr{F})\}$ and $\sigma^A(\vec{B}) = \overline{\mathbf{C}} \setminus \rho^A(\vec{B})$ are called an *A-resolvent set* and an *A-spectrum* of the pencil $\vec{B}$ respectively. The operator-function of a complex variable $R_\mu^A(\vec{B}) = (\mu^n A - \mu^{n-1} B_{n-1} - \cdots - \mu B_1 - B_0)^{-1}$ with the domain $\rho^A(\vec{B})$ is called an *A-resolvent* of the pencil $\vec{B}$.

Definition 1 The operator pencil $\vec{B}$ is called *polynomially bounded with respect to an operator A* (or *polynomially A-bounded*) if $\exists a \in \mathbf{R}_+ \; \forall \mu \in \mathbf{C} \; (|\mu| > a) \Rightarrow (R_\mu^A(\vec{B}) \in \mathscr{L}(\mathscr{F}; \mathscr{U}))$.

Remark 1 If there exists an operator $A^{-1} \in \mathscr{L}(\mathscr{F}; \mathscr{U})$ then the pencil $\vec{B}$ is *A-bounded*.

Following the necessary condition of projectors existences has been found out in [5].

$$\int_\gamma \mu^k R_\mu^A(\vec{B}) d\mu \equiv O, \quad k = 0, 1, \ldots, n-2, \tag{5}$$

where the circuit $\gamma = \{\mu \in \mathbf{C} : |\mu| = r > a\}$.

Lemma 1 ([5]) *Let the operator pencil $\vec{B}$ be polynomially A-bounded and condition (5) be fulfilled. Then the operators*

$$P = \frac{1}{2\pi i} \int_\gamma R_\mu^A(\vec{B}) \mu^{n-1} A d\mu, \quad Q = \frac{1}{2\pi i} \int_\gamma \mu^{n-1} A R_\mu^A(\vec{B}) d\mu$$

are projectors in spaces $\mathscr{U}$ and $\mathscr{F}$ respectively.

Denote $\mathscr{U}^0 = \ker P$, $\mathscr{F}^0 = \ker Q$, $\mathscr{U}^1 = \operatorname{im} P$, $\mathscr{F}^1 = \operatorname{im} Q$. According to Lemma 1 $\mathscr{U} = \mathscr{U}^0 \oplus \mathscr{U}^1$, $\mathscr{F} = \mathscr{F}^0 \oplus \mathscr{F}^1$. By A^k (B_l^k) denote the restriction of operator A (B_l) on $\mathscr{U}^k$, $k = 0, 1$; $l = 0, 1, \ldots, n-1$.

Theorem 1 ([5]) *Let the operator pencil $\vec{B}$ be polynomially A-bounded and condition (5) be fulfilled. Then*

(i) $A^k \in \mathscr{L}(\mathscr{U}^k; \mathscr{F}^k)$, $k = 0, 1$;
(ii) $B_l^k \in \mathscr{L}(\mathscr{U}^k; \mathscr{F}^k)$, $k = 0, 1$, $l = 0, 1, \ldots, n-1$;
(iii) *operator* $(A^1)^{-1} \in \mathscr{L}(\mathscr{F}^1; \mathscr{U}^1)$ *exists*;
(iv) *operator* $(B_0^0)^{-1} \in \mathscr{L}(\mathscr{F}^0; \mathscr{U}^0)$ *exists*.

Using Theorem 1 construct operators $H_0 = (B_0^0)^{-1} A^0 \in \mathscr{L}(\mathscr{U}^0)$, $H_1 = (B_0^0)^{-1} B_1^0 \in \mathscr{L}(\mathscr{U}^0)$,…, $H_{n-1} = (B_0^0)^{-1} B_{n-1}^0 \in \mathscr{L}(\mathscr{U}^0)$ and $S_0 = (A^1)^{-1} B_0^1 \in \mathscr{L}(\mathscr{U}^1)$, $S_1 = (A^1)^{-1} B_1^1 \in \mathscr{L}(\mathscr{U}^1)$,…, $S_{n-1} = (A^1)^{-1} B_{n-1}^1 \in \mathscr{L}(\mathscr{U}^1)$.

Definition 2 Define the *family of operators* $\{K_q^1, K_q^2, \ldots, K_q^n\}$ as follows:

$$K_0^s = O, \; s \neq n, \; K_0^n = I,$$
$$K_1^1 = H_0, \; K_1^2 = -H_1, \ldots, K_1^s = -H_{s-1}, \ldots, K_1^n = H_{n-1},$$
$$K_q^1 = K_{q-1}^n H_0, \; K_q^2 = K_{q-1}^1 - K_{q-1}^n H_1, \ldots, K_q^s = K_{q-1}^{s-1} - K_{q-1}^n H_{s-1}, \ldots,$$
$$K_q^s = K_{q-1}^{n-1} - K_{q-1}^n H_{n-1}, \; q = 1, 2, \ldots.$$

The A-resolvent can be represented by a Laurent series [5]

$$(\mu^n A - \mu^{n-1} B_{n-1} - \cdots - \mu B_1 - B_0)^{-1} = -\sum_{q=0}^{\infty} \mu^q K_q^n (B_0^0)^{-1} (I - Q) +$$

$$\sum_{q=1}^{\infty} \mu^{-q} (\mu^{n-1} S_{n-1} + \cdots + \mu S_1 + S_0)^q L_1^{-1} Q.$$

Using this representation we classify the character of the point at infinity of the A-resolvent of the operator pencil $\vec{B}$.

Definition 3 The point ∞ is called

- a *removable singularity of an A-resolvent of the pencil* $\vec{B}$, if $K_1^s \equiv O$, $s = 1, 2, \ldots, n$;
- a *pole of order* $p \in \mathbb{N}$ *of an A-resolvent of the pencil* $\vec{B}$, if $\exists p$ such that $K_p^s \not\equiv O$, $s = 1, 2, \ldots, n$, but $K_{p+1}^s \equiv O$, $s = 1, 2, \ldots, n$;
- an *essential singularity of an A-resolvent of the pencil* $\vec{B}$, if $K_q^n \not\equiv O$ for all $q \in \mathbb{N}$.

Further a removable singularity of an A-resolvent of the pencil $\vec{B}$ will be called a pole of order 0 for brevity. If the operator pencil $\vec{B}$ is polynomially A-bounded and the point ∞ is a pole of order $p \in \{0\} \cup \mathbb{N}$ of an A-resolvent of the pencil $\vec{B}$ then the operator pencil $\vec{B}$ is called *polynomially (A, p)-bounded*.

Theorem 2 ([10]) *Let A, B_{n-1}, ..., B_1, $B_0 \in \mathscr{L}(\mathscr{U}; \mathscr{F})$ and A be a Fredholm operator. Then the following statements are equivalent:*
(i) The lengths of all chains of the $\vec{B}$-adjoined vectors of the operator A are bounded by number $(p + n - 1) \in \{0\} \cup \mathbb{N}$ and the chain of length $(p + n - 1)$ exists.
(ii) The operator pencil $\vec{B}$ is polynomially (A, p)-bounded.

3 The Cauchy Problem for the Sobolev Type Equation of the Second Order

Let $\mathscr{U}$, $\mathscr{F}$ be Banach spaces and A, B_1, $B_0 : \mathscr{U} \to \mathscr{F}$ be linear and bounded operators. Consider the Cauchy problem

$$u(0) = u_0, \quad \dot{u}(0) = u_1 \qquad (6)$$

for the Sobolev type equation of the second order

$$A\ddot{u} = B_1 \dot{u} + B_0 u. \qquad (7)$$

A solution of (6), (7) in the frame of theory of degenerate groups was obtained in [5] under condition of relative polynomially boundedness of operator pencil $\vec{B}$.

Theorem 3 ([10]) *Let the pencil $\vec{B}$ be (A, p)-bounded, condition (5) be satisfied. Then there exists a unique solution of (6), (7) given by*

$$u(t) = U_1^t u_1 + U_0^t u_0,$$

where U_k^t, $(k = 0, 1)$ are the propagators of the form:

$$\begin{aligned} U_0^t &= \tfrac{1}{2\pi i} \int_\gamma R_\mu^A(\vec{B})(\mu A - B_1) e^{\mu t} d\mu, \\ U_1^t &= \tfrac{1}{2\pi i} \int_\gamma R_\mu^A(\vec{B}) A e^{\mu t} d\mu. \end{aligned} \qquad (8)$$

Here circuit $\gamma \in \mathbb{C}$ is a boundary of the domain containing the A-spectrum of the operator pencil $\vec{B}$, and initial data is such that $u_k \in \operatorname{im} U_k^{\cdot}$, $k = 0, 1$.

4 Propagators of the Sobolev Equation

Domains with an analytic boundary for which the operator $\Delta^{-1}\dfrac{\partial^2}{\partial z^2}$ has continuous spectrum were constructed in [11]. Therefore, it is necessary to select the domain carefully. We consider two cases. The first one when the domain D is a parallelepiped and the second one when the domain D is a cylinder. Take such space over domain D that operator Δ^{-1} is a compact operator and the operator $\dfrac{\partial^2}{\partial z^2}$ is bounded, and therefore their composition is a compact operator. Thus, the spectrum of $\Delta^{-1}\dfrac{\partial^2}{\partial z^2}$ is bounded. Later we will show that the A-spectrum of the operator pencil $\vec{B}$ coincides with the spectrum of the operator $\Delta^{-1}\dfrac{\partial^2}{\partial z^2}$.

Theorem 4 *Let the domain D be a parallelepiped $[0, a] \times [0, b] \times [0, c]$ and initial data $w_1, w_0 \in W_2^{l+2}(D)$. Then there exists a unique solution $w \in \mathscr{C}^2(0, T; W_2^{l+2}(D))$ of the problem (2)–(4).*

Introduce spaces $\mathscr{U} = \{u \in W_2^{l+2}(D) : u(x, y, z, t) = 0, (x, y, z) \in \partial D\}$, $\mathscr{F} = W_2^l(D)$ and define operators

$$A = \Delta, \quad B_1 = O, \quad B_0 = -F^2 \frac{\partial^2}{\partial z^2}.$$

For any $l \in \{0\} \cup \mathbb{N}$ operators $A, B_1, B_0 \in \mathscr{L}(\mathscr{U}, \mathscr{F})$. Therefore, we have reduced the mathematical model (2)–(4) to the Cauchy problem (6) for the abstract Eq. (7).

Denote by $-\lambda_{k,m,n}^2 = -\left(\frac{\pi k}{a}\right)^2 - \left(\frac{\pi m}{b}\right)^2 - \left(\frac{\pi n}{c}\right)^2$ eigenvalues of the Dirichlet problem for the Laplace operator Δ. Denote by $\varphi_{k,m,n} = \sin\left(\frac{\pi kx}{a}\right) \sin\left(\frac{\pi my}{b}\right) \sin\left(\frac{\pi nz}{c}\right)$ the corresponding to $-\lambda_{k,m,n}^2$ eigenfunctions.

Since $\{\varphi_{k,m,n}\} \subset C^\infty(D)$ then

$$\mu^2 A - \mu B_1 - B_0 = \sum_{k,m,n=1}^{\infty} [-\lambda_{k,m,n}^2 \mu^2 - F^2 \left(\frac{\pi n}{c}\right)^2)] < \varphi_{k,m,n}, \cdot > \varphi_{k,m,n}$$

where $< \cdot, \cdot >$ is an inner product in $L^2(D)$. Equation

$$\lambda_{k,m,n}^2 \mu^2 + F^2 \left(\frac{\pi n}{c}\right)^2 = 0,$$

determines the A-spectrum of the operator pencil $\vec{B}$:

$$\mu_{k,m,n}^{\pm} = \pm \frac{F\pi n}{c\sqrt{\lambda_{k,m,n}^2}} i.$$

Therefore $\sigma^A(\vec{B}) = \{\mu_{k,m,n}^{\pm}\}$ is the A-spectrum of operator pencil $\vec{B}$ and it is bounded. Since the operator A is continuously invertible condition (5) is satisfied and therefore the conditions of Lemma 1 hold.

Construct the propagators according to (8). Due to the discreteness of the relative spectrum of the operator pencil of $\vec{B}$, we obtain

$$U_0^t w_0 = \sum_{k,m,n=1}^{\infty} \cos\left(\frac{F\pi n}{c\sqrt{\lambda_{k,m,n}^2}}t\right) < \varphi_{k,m,n}, w_0 > \varphi_{k,m,n},$$
$$U_1^t w_1 = \sum_{k,m,n=1}^{\infty} \frac{F\pi n}{c\sqrt{\lambda_{k,m,n}^2}} \sin\left(\frac{F\pi n}{c\sqrt{\lambda_{k,m,n}^2}}t\right) < \varphi_{k,m,n}, w_1 > \varphi_{k,m,n}.$$
(9)

Therefore solution of problem (2)–(4) has form

$$w(x,t) = U_0^t w_0 + U_1^t w_1.$$

Theorem 5 *Let the domain D be a cylinder of height h and radius R_0 and initial data $w_1, w_0 \in W_2^{l+2}(D)$. Then there exists a unique solution $w \in \mathscr{C}^2(0, T; W_2^{l+2}(D))$ of the problem (2)–(4).*

The Laplace operator in cylindrical coordinates (r, φ, z) has the form $\Delta = \frac{\partial^2}{\partial^2 r} + \frac{1}{r}\frac{\partial}{\partial r} + \frac{1}{r^2}\frac{\partial^2}{\partial^2 \varphi} + \frac{\partial^2}{\partial^2 z}$.

Mathematical model (2)–(4) can be reduced to the Cauchy problem (6) for Eq. (7) in the same way. Introduce spaces $\mathscr{U} = \{u \in W_2^{l+2}(D) : u(r, \varphi, z, t) = 0, (r, \varphi, z) \in \partial D\}$, $\mathscr{F} = W_2^l(D)$ and define operators

$$A = \Delta, \quad B_1 = O, \quad B_0 = -F^2 \frac{\partial^2}{\partial z^2}.$$

For any $l \in \{0\} \cup \mathbb{N}$ operators $A, B_1, B_0 \in \mathscr{L}(\mathscr{U}, \mathscr{F})$. In this case

$$-\lambda_{k,m,n}^2 = -\left(\frac{v_{0m}^k}{R_0}\right)^2 - \left(\frac{\pi n}{h}\right)^2$$

are the eigenvalues of the Dirichlet problem for the Laplace operator. Denote by

$$\varphi_{k,m,n} = J_m\left(\frac{v_{0m}^k r}{R_0}\right)(\cos(m\varphi) + \sin(m\varphi))\sin\left(\frac{\pi n z}{h}\right)$$

the corresponding to $\{-\lambda_{k,m,n}^2\}$ orthonormal in $L^2(D)$ eigenfunctions. Here v_{0m}^k is the kth of the Bessel function of the first kind J_m.

Since $\{\varphi_{k,m,n}\} \subset C^{\infty}(D)$ then

$$\mu^2 A - \mu B_1 - B_0 = \sum_{k,m,n=1}^{\infty} [-\lambda_{k,m,n}^2 \mu^2 - F^2\left(\frac{\pi n}{h}\right)^2] < \varphi_{k,m,n}, \cdot > \varphi_{k,m,n}$$

where $< \cdot, \cdot >$ is an inner product in $L^2(D)$. Equation

$$\lambda_{k,m,n}^2 \mu^2 + F^2 \left(\frac{\pi n}{h}\right)^2 = 0$$

determine the A-spectrum of the operator pencil $\vec{B}$:

$$\mu_{k,m,n}^{\pm} = \pm \sqrt{\frac{F^2}{\left(\frac{v_0^k}{R_0}\right)^2 \left(\frac{h}{\pi n}\right)^2 + 1}} \cdot i.$$

The A-spectrum $\sigma^A(\vec{B}) = \{\mu_{k,m,n}^{\pm}\}$ is bounded, since $|\mu_{k,m,n}^{1,2}| \leq F$.

Since the operator A is continuously invertible condition (5) is satisfied and therefore the conditions of Lemma 1 hold. Construct the propagators according to (8). Due to discreteness of the relative spectrum of the operator pencil of $\vec{B}$, we obtain

$$U_0^t w_0 = \sum_{k,m,n=1}^{\infty} \cos\left(\frac{F n \pi}{h \sqrt{\lambda_{k,m,n}^2}} t\right) < \varphi_{k,m,n}, w_0 > \varphi_{k,m,n},$$

$$U_1^t w_1 = \sum_{k,m,n=1}^{\infty} \frac{F n \pi}{h \sqrt{\lambda_{k,m,n}^2}} \sin\left(\frac{F n \pi}{h \sqrt{\lambda_{k,m,n}^2}} t\right) < \varphi_{k,m,n}, w_1 > \varphi_{k,m,n}.$$

(10)

And the solution of (2)–(4) in the cylinder is given by

$$w(x,t) = U_0^t w_0 + U_1^t w_1.$$

Acknowledgements The work was supported by Act 211 Government of the Russian Federation, contract no 02.A03.21.0011.

References

1. Brekhovskikh, L.M., Goncharov, V.V.: Introduction to the Mechanics of Continuous Media (as Applied to the Theory of Waves). M.: Nauka (1982)
2. Sobolev, S.L.: On a new problem of mathematical physics. Izv. Akad. Nauk SSSR, Ser. Mat. **18**, 3–50 (1954)
3. Demidenko, G.V., Uspenskii, S.V.: Partial Differential Equations and Systems Not Solvable with Respect to the Highest Order Derivative. Basel, Hong Kong, Marcel Dekker, Inc, N.Y. (2003)
4. Fokin, M.V.: Hamiltonian systems in the theory of small oscillations of a rotating ideal fluid I. Math. Works **4**(2), 155–206 (2001)
5. Zamyshlyaeva, A.A.: Phase spaces of a class of linear equations of Sobolev type of the second order. Comput. Technol. **8**(4), 45–54 (2003)
6. Sviridyuk, G.A., Zamyshlyaeva, A.A.: Phase spaces of a class linear Sobolev type equations of higher order. Differ. Equ. **42**(2), 252–260 (2006)
7. Vasilyev, V.V., Keirn, S.G., Piskarev, S.I.: Operator semigroups, cosine operator functions and linear differential equations. Itogi Nauki i Tekhn. Ser. Mat. anal., VINITI. **28**, 87–202 (1990)

8. Melnikova, I.V., Filinkov, A.I.: Integrated semigroups and C-semigroups. Well-posedness and regularization of differential-operator problems. Russian Math. Surv. **49**(6), 111–150 (1994)
9. Sviridyuk, G.A., Fedorov, V.E.: Linear Sobolev Type Equations and Degenerate Semigroups of Operators. VSP, Utrecht; Boston; Köln; Tokyo (2003)
10. Zamyshlyaeva, A.A.: Linear Sobolev Type Equations of High Order. Publ., Center of SUSU, Chelyabinsk (2012)
11. Fokin, M.V.: On the spectrum of one operator. Differ. Equ. **7**(1), 135–141 (1971)

The Fourth Order Wentzell Heat Equation

Gisèle Ruiz Goldstein, Jerome A. Goldstein, Davide Guidetti, and Silvia Romanelli

Dedicated to Jan Kisynski on his 85th birthday

Abstract We consider the Wentzell Laplacian and its square in general domains. Applications to the Cauchy problems associated with Wentzell heat, wave and plate equations are presented.

Keywords Wentzell Laplacian · Wentzell boundary condition · Fourth order equation · Plate equation · Telegraph equation · Asymptotic parabolicity

1 The Wentzell Laplacian in General Domains

Let Ω be a uniformly regular domain of $\mathbf{R}^n$ of class $C^{2+\varepsilon}$ (see [1] Definition 1) having a nonempty boundary $\partial \Omega$ which is a uniformly regular domain of class $C^{2+\varepsilon}$, where

G. R. Goldstein (✉) · J. A. Goldstein
Department of Mathematical Sciences, University of Memphis,
373 Dunn Hall, Memphis, TN 38152-3240, USA
e-mail: ggoldste@memphis.edu

J. A. Goldstein
e-mail: jgoldste@memphis.edu

D. Guidetti
Dipartimento di Matematica, Università di Bologna, Piazza di Porta S. Donato 5,
40126 Bologna, Italy
e-mail: davide.guidetti@unibo.it

S. Romanelli
Dipartimento di Matematica, Università degli Studi di Bari Aldo Moro,
via E. Orabona 4, 70125 Bari, Italy
e-mail: silvia.romanelli@uniba.it

Ω could be bounded or unbounded. Simple examples of unbounded domains are exterior domains (in which case $\partial\Omega$ is bounded) and half spaces (in which case $\partial\Omega$ is unbounded). If Ω is unbounded, assume $\int_\Omega dx = \infty$.

Let $\mathscr{A}(x)$ be an $n \times n$ real, bounded, symmetric, positive definite matrix for $x \in \overline{\Omega}$, of class $C^{2+\varepsilon}$; in symbols, $\mathscr{A} \in C^{2+\varepsilon}(\overline{\Omega}, \mathscr{M}_n)$, and

$$\alpha_0 |\xi|^2 \leq \xi \cdot \mathscr{A}(x)\xi \leq \alpha_1 |\xi|^2$$

holds for all $x \in \overline{\Omega}$, all $\xi \in \mathbf{R}^n$ and constants

$$0 < \alpha_0 \leq \alpha_1 < \infty.$$

Similarly, let $\mathscr{B}(x)$ be an $(n-1) \times (n-1)$ matrix having the same properties (with the same α_0, α_1) for $x \in \partial\Omega$, $\mathscr{B} \in C^{2+\varepsilon}(\partial\Omega, \mathscr{M}_{n-1})$.

Define the distributional partial differential operators

$$L := \nabla \cdot \mathscr{A}(x)\nabla, \quad L_\partial := \nabla_\tau \cdot \mathscr{B}(x)\nabla_\tau$$

where ∇ (resp. ∇_τ) is the gradient on Ω (resp. the tangential gradient on $\partial\Omega$). Note that $L = \Delta$ (the Laplacian) if $\mathscr{A} = I$ and $L_\partial = \Delta_{LB}$ (the Laplace - Beltrami operator) if $\mathscr{B} = I$.

The Wentzell heat equation is

$$\frac{\partial u}{\partial t} = Lu \quad in \quad \Omega \times [0, \infty)$$

$$\frac{\partial(\gamma u)}{\partial t} = -\beta(x)\partial_\nu^{\mathscr{A}} u - \gamma_0(x)(\gamma u) + q\beta(x)L_\partial(\gamma u) \quad on \quad \partial\Omega, \tag{1}$$

where $\gamma = trace$,

$$u(0, x) = u_0(x) \quad in \quad \overline{\Omega}$$

where $u_0 \in C(\overline{\Omega}) \cap L^\infty(\Omega)$. Here $\beta, \gamma_0 \in C^{2+\varepsilon}(\partial\Omega, \mathbf{R})$, $\partial^\alpha \beta, \partial^\alpha \gamma_0 \in L^\infty(\partial\Omega)$, for $|\alpha| \leq 2$ and $q \in [0, \infty)$. Also

$$\partial_\nu^{\mathscr{A}} u(x) = (\mathscr{A}(x)\nabla u) \cdot \nu,$$

with ν being the outer unit normal vector to $\partial\Omega$ at x. But our main interest is in $q > 0$.

Take $\beta > 0$ with $\eta \leq \beta(x) \leq \frac{1}{\eta}$ for some $\eta > 0$ and all $x \in \partial\Omega$. Let

$$X_p := L^p(\overline{\Omega}, d\mu), \quad 1 \leq p < \infty,$$

where $d\mu := dx|_\Omega \times \frac{dS}{\beta}|_{\partial\Omega}$. Write $U = (u_1, u_2)$ for $U \in X_p$ where

$$u_1 \in L^p(\Omega, dx), \quad u_2 \in L^p\left(\partial\Omega, \frac{dS}{\beta}\right).$$

Let
$$X_\infty = C(\overline{\Omega}),$$

where $C(\overline{\Omega})$ is the space of all continuous functions on $\overline{\Omega}$ if Ω is bounded, and

$$X_\infty = C_0(\overline{\Omega}),$$

where $C_0(\overline{\Omega})$ is the space of all continuous functions on $\overline{\Omega}$ vanishing at infinity, if Ω is unbounded. We identify $u \in X_\infty$ with $U = (u, \gamma u) \in C_0(\overline{\Omega}) \times C(\partial\Omega) = X_\infty$. Note that X_∞ is contained in and is dense in X_p for all $p \geq 1$, if Ω is bounded. But in the unbounded domain case, we do not have $X_p \subset X_r$ for $p \neq r$. The norms are given by

$$\|U\|_{X_p} = \left[\int_\Omega |u_1(x)|^p\, dx + \int_{\partial\Omega} |u_2(x)|^p \frac{dS}{\beta(x)}\right]^{\frac{1}{p}}, \quad 1 \leq p < \infty$$

and
$$\|U\|_{X_\infty} = \|u_1\|_{C_0(\overline{\Omega}) \cap L^\infty(\Omega)}$$

for
$$U = \begin{pmatrix} u_1 \\ u_2 \end{pmatrix} = \begin{pmatrix} u_1 \\ \gamma u_1 \end{pmatrix}$$

in the latter case.

Under all the previous assumptions we have the following result.

Theorem 1.1 *Let*

$$A_p = \begin{pmatrix} L & 0 \\ -\beta \partial_\nu^{\mathscr{A}} & -\gamma_0 + q\beta L_\partial \end{pmatrix} \tag{2}$$

have a suitable domain (as indicated below) in X_p, $1 \leq p \leq \infty$. Then A_p is densely defined and quasi dissipative on X_p for all p, $1 \leq p \leq \infty$. Its closure G_p is quasi m-dissipative for all p, $1 \leq p \leq \infty$. In fact, $G_p - \omega I$ is m-dissipative on X_p where $\omega = \|(\gamma_0)_-\|_\infty$, the sup norm of the negative part of γ_0. Thus $\omega = 0$ if $\gamma_0 \geq 0$. G_p is the X_p-closure of the operator matrix in (2) on

$$\mathscr{D}_{0,p} = \{u \in C_c^2(\overline{\Omega}) : (u|_\Omega, u|_{\partial\Omega}) \in X_p,\ (1)\ \text{holds on}\ \partial\Omega\}.$$

Here $C_c^2(\overline{\Omega})$ denotes the space of all C^2 functions on $\overline{\Omega}$ having compact support. Furthermore, $G_p^* = G_q$ for $\frac{1}{p} + \frac{1}{q} = 1$ and $1 < p < \infty$; and $G_2 = G_2^* \leq \omega I$, where as above $\omega = \|(\gamma_0)_-\|_\infty$.

The theorem was proved in several steps. It was proved in bounded domains assuming $q = 0$ in [7]. But with minor modifications the proof works for Ω unbounded with uniformly regular boundary. We proved the extension to $q > 0$ in 2005 after hearing a lecture by Alain Miranville. The extension to unbounded domains was mentioned in [3, 11] for $p = 2$. The extension to unbounded domains for $1 \leq p \leq \infty$ involves only minor inessential changes in the existing proofs, see [6].

These semigroups have analytic extensions. The simplest case is worked out in [5] for Ω bounded. $\{T_p(t) = e^{tA_p} : t \geq 0\}$ has an analytic extension with sector of analyticity

$$\Sigma(\theta_p) = \{z \in \mathbf{C} : Re\, z > 0, |arg(z)| < \theta_p\}$$

and $\theta_p \geq \frac{\pi}{p}$ for $2 \leq p < \infty$; $\theta_p = \theta_{p'}$ for $1 < p \leq 2$ with $\frac{1}{p} + \frac{1}{p'} = 1$. This is also valid in unbounded domains. The proof is based on the Stein interpolation theorem.

In case Ω is bounded and $\partial\Omega$ and all the coefficients are C^∞, then the value of θ_p exceeds $\frac{\pi}{p}$ if $p > 2$. We conjecture it is $\frac{\pi}{2}$ for all $p \in [1, \infty]$, but the proof is not yet finished. In any case, due to the applications in Sect. 3 below, our interest here largely lies in unbounded domains. For related results concerning Wentzell Laplacians see also [4, 12, 13, 16, 17].

2 The Square of the Wentzell Laplacian

Let A generate a (C_0) semigroup on a complex Banach space X, analytic in the sector $\Sigma(\alpha)$ where $\frac{\pi}{4} < \alpha \leq \frac{\pi}{2}$. Then J. A. Goldstein in [14] proved that $-A^2$ generates a (C_0) semigroup on X, analytic in the sector $\Sigma(\alpha_1)$ where $\alpha_1 = 2\alpha - \frac{\pi}{2}$. In particular, for $\alpha = \frac{\pi}{2}$, then $\alpha_1 = \frac{\pi}{2}$. For $\alpha = \frac{\pi}{p}$ which is the case of G_p on X_p in the previous section for $p \geq 2$, then $\alpha > \frac{\pi}{4}$ if and only if $p < 4$, so that $|\frac{1}{p} - \frac{1}{2}| < \frac{1}{4}$ or $\frac{4}{3} < p < 4$. Let G_p be the closure of the Wentzell Laplacian discussed in Sect. 1 on X_p for $\frac{4}{3} < p < 4$. The Cauchy problem for $-A_p^2$ is

$$\frac{\partial u}{\partial t} = -(\nabla \cdot \mathscr{A}(x)\nabla)^2 u, \quad u(x, 0) = u_0(x),$$

and

$$(BC1) \qquad \frac{\partial(\gamma u)}{\partial t} = \beta \partial_\nu^{\mathscr{A}} u + \gamma_0(x)(\gamma u) - q\beta L_\partial(\gamma u),$$

$$(BC2) \qquad \frac{\partial(\gamma Lu)}{\partial t} = \beta \partial_\nu^{\mathscr{A}} Lu + \gamma_0(x)(\gamma Lu) - q\beta L_\partial(\gamma Lu).$$

This can be written as

$$\frac{\partial u}{\partial t} = -A_p^2 U \quad in \quad X_p, \quad \frac{4}{3} < p < 4$$

where G_p, the closure of A_p, generates a semigroup analytic in X_p for any $q \geq 0$ and $G_2 = G_2^*$. But no quasi dissipativity results hold for G_p for $p \neq 2$.

Remark 2.1 The inhomogeneous problem

$$\frac{\partial u}{\partial t} = Lu + h_1(x,t), \quad u(x,0) = u_0(x)$$

$$\frac{\partial(\gamma u)}{\partial t} = \beta \partial_\nu^{\mathscr{A}} u + \gamma_0(x)(\gamma u) - q\beta L_\partial(\gamma u) + h_2(x,t),$$

can be written as

$$\frac{\partial U}{\partial t} = A_p U + H(t), \quad U(0) = U_0$$

with $H = (h_1, h_2)$, and the unique mild solution in X_p associated with G_p is

$$U(t) = e^{tG_p} U_0 + \int_0^t e^{(t-s)G_p} H(s)\, ds$$

for all $p \in [1, \infty]$.

Remark 2.2 Replace $L = \nabla \cdot \mathscr{A}(x)\nabla$ and $L_\partial = \nabla_\tau \cdot \mathscr{B}(x)\nabla_\tau$ by

$$\widetilde{L} = L + \Sigma_{j=1}^n a_j(x)\partial_{x_j} + c_0(x) = L + a \cdot \nabla + c_0$$

$$\widetilde{L}_\partial = L_\partial + b\nabla_\tau + b_0,$$

with each $a_j, b_j \in C^{2+\varepsilon}(\overline{\Omega})$ and $\partial^\alpha a_j, \partial^\alpha b_j \in L^\infty(\partial\Omega)$ for each α with $|\alpha| \leq 2$. Then $\widetilde{L}, \widetilde{L}_\partial$ are Kato perturbations of L, L_∂ respectively, so that the resulting problem $\frac{\partial U}{\partial t} = \widetilde{G}_p U$ is governed by a (C_0) semigroup analytic in $\Sigma(\theta_p)$ for all $p \in (1, \infty)$. It follows that $-\widetilde{G}_p^{\,2}$ generates a (C_0) semigroup analytic in $\Sigma(\widetilde{\theta}_p)$ for $\frac{4}{3} < p < 4$ where

$$\widetilde{\theta}_p = \frac{\pi}{2}\left(\frac{4}{p} - 1\right)$$

for $2 \leq p < 4$ and $\widetilde{\theta}_p = \widetilde{\theta}_{p'}$. Then

$$\frac{\partial U}{\partial t} = -\widetilde{G}_p^{\,2} U$$

is a fourth order equation in which $\widetilde{G}_p^2$ is not symmetric, and $-\widetilde{G}_p^2$ can be added to a Kato perturbation P making $-\widetilde{G}_p^2 + P$ a (C_0) and analytic semigroup generator which is not a perfect square.

3 The Wentzell Wave and Plate Equations

Let G be the closure of the Wentzell Laplacian on X_2. Then, according to Theorem 1.1, one has $G = G^* \leq 0$ on X_2 when $\gamma_0 \geq 0$, which we assume. The wave, plate and telegraph equations are, respectively,

$$\partial_t^2 = Gu, \tag{3}$$

$$\partial_t^2 = -G^2 u, \tag{4}$$

$$\partial_t^2 u + 2a\partial_t u = -G^2 u. \tag{5}$$

Let u satisfy the Schrödinger equation

$$\partial_t u = iGu, \quad u(0) = f.$$

Then for $f \in D(G^2)$, u satisfies the plate equation

$$\begin{cases} \partial_t^2 u = -G^2 u, \\ u(0) = f, \\ u'(0) = iGf. \end{cases}$$

This is (4) with special initial conditions. For (4) we can, more generally, require

$$\begin{cases} u(0) = f, \\ u'(0) = g. \end{cases}$$

We will also consider the strongly damped Wentzell plate equation

$$\partial_t^2 u + 2a(-G)^\alpha \partial_t u = -G^2 u \tag{6}$$

for $a > 0$ and $0 \leq \alpha < 1$, with $\alpha > 0$ being the strongly damped case and $\alpha = 0$ being the damped (or telegraph) case.

As noted previously, G is the closure of the operator

$$\begin{pmatrix} L & 0 \\ -\beta\partial_\nu^{\mathscr{A}} & -\gamma_0 + q\beta L_\partial \end{pmatrix}$$

and G^2 is the closure of

$$\begin{pmatrix} L^2 & 0 \\ \beta\partial_\nu^{\mathscr{A}} L + (-\gamma_0 + q\beta L_\partial)(\beta\partial_\nu^{\mathscr{A}}) & (-\gamma_0 + q\beta L_\partial)^2 \end{pmatrix}.$$

We work in X_2 and recall $G = G^* \leq 0$ because $\gamma_0 \geq 0$. The Cauchy problems for (3)–(6) are all wellposed.

Our interest here is to prove that, in certain unbounded domains, the solutions of (5) are asymptotically equal to the solutions of

$$2a\partial_t u = -G^2 u$$

with analogous results holding for (6). The idea of this asymptotically parabolic nature of damped wave (or plate) equations goes back to Taylor [18], although many attribute it to Cattaneo [2].

Theorem 3.1 *Let $T = T^* \geq 0$ on X_2 with $0 \notin \sigma_p(T)$ and $0 = \min \sigma(T)$. In other words, 0 is in the spectrum of T but is not an eigenvalue. Consider the wellposed Cauchy problem*

$$\begin{cases} u'' + 2\mathscr{B}u' + T^2 u = 0, \\ u(0) = f, \\ u'(0) = g \end{cases}$$

where $(f, g) \neq (0, 0)$, $\mathscr{B} = aT^\alpha$, $0 \leq \alpha < 1$ and $a > 0$. Let $c = a^{\frac{1}{1-\alpha}}$. Using the spectral theorem for T, define

$$h = \frac{1}{2}\chi_{(0,c)}(T)\{(a^2 T^{2\alpha} - T^2)^{1/2}(aT^\alpha f + g) + f\}. \tag{7}$$

This h in (7) is well defined, assuming $f \in D(T^2)$, $g \in D(T)$ and $c \notin \sigma_p(T)$. If v is the unique solution of

$$\begin{cases} 2aT^\alpha v' = -T^2 v, \\ v(0) = h, \end{cases}$$

then $u(t) \approx v(t)$, i.e., u is asymptotically equivalent to v. Since both $u(t)$ and $v(t)$ tend to 0 as $t \to \infty$, $u(t) \approx v(t)$ as $t \to \infty$ means

$$\lim_{t\to\infty} \frac{\|u(t) - v(t)\|}{\|v(t)\|} = 0.$$

Our convention for square roots of selfadjoint operators T is

$$T = T_+ - T_- = T\chi_{[0,\infty)}(T) + T\chi_{(-\infty,0]}(T)$$

with $T_\pm = T_\pm^* \geq 0$, $T_+ T_- = 0$. Then $T_\pm^{1/2}$ is the unique nonnegative selfadjoint square root of $T_\pm$ and we define

$$T^{1/2} := T_+^{1/2} + iT_-^{1/2}.$$

Proof This was proved in [3] for $\alpha = 0$ and in [11] for $0 < \alpha < 1$. Moreover, there are error estimates of the form

$$\frac{\|u(t) - v(t)\|}{\|v(t)\|} \leq C_n e^{-\epsilon_n t}$$

for $f, g \in E_n$ where $E_n \subset E_{n+1}$, $\cup_{n=1}^\infty E_n$ is dense in X_2, and $E_n, C_n, \epsilon_n > 0$ can be constructed explicitly. We want to apply this to $T = -G$ so that $T^2 = G^2$. What we must check is that $0 \in \sigma(T) \setminus \sigma_p(T)$. The strategy for this comes from [3]. By the operational calculus associated with the spectral theorem, $0 \in \sigma(T) \setminus \sigma_p(T)$ holds if and only if $0 \in \sigma(T^2) \setminus \sigma_p(T^2)$.

Our additional assumption on Ω is that it contains arbitrarily large balls, that is, for each $R > 0$, there exists $x_R \in \Omega$ such that

$$\overline{B}(x_R, R) = \{x \in \mathbf{R}^n : |x - x_R| \leq R\} \subset \Omega.$$

It is easy to see $0 \notin \sigma_p(G)$. Because if $GU = 0$, then

$$0 = <GU, U> = -\int_\Omega |\sqrt{\mathscr{A}}\nabla u|^2 dx - q\int_{\partial\Omega} |\nabla_\tau u|^2 dS - \int_{\partial\Omega} \gamma_0 |u(x)|^2 \frac{dS}{\beta(x)}$$

which implies $\nabla u \equiv 0$ and so $u = k$, a constant, in $\overline{\Omega}$ since the semigroup is analytic. If Ω has infinite volume, then k must be 0 (since we only consider the problem in X_2). If Ω has finite volume, $\int_\Omega dx < \infty$, then again $u = 0$ provided $\gamma_0 > 0$ at a point (and therefore on a set of positive measure on $\partial\Omega$), which we assume in this case.

To see that $\inf \sigma(G) = 0$ we recall a construction from [3]. Let $R > 0$ be given and choose $B(x_R, R)$ having closure contained in Ω. For definiteness take $R > 10$; later we will let R go to ∞. Let e be a unit vector in $\mathbf{R}^n$ and consider the line $D_L := \{x_R + xe : x \in \mathbf{R}\}$. It intersects $B(x_R, R)$ in the diameter $D := [x_l = x_R - 10Re, x_r = x_R + 10Re]$. Define g_1 on D by

$$g_1 \equiv 0 \quad \text{on } [x_l, x_l + e] \cup [x_r - e, x_r],$$

$$g_1 \equiv 1 \quad \text{on} \quad [x_l + 2e, x_r - 2e],$$

$$0 \leq g_1 \leq 1, \qquad g_1 \in C^\infty(D).$$

For $y \in e^\perp \subset \mathbf{R}^n$ let

$$g_2(y) = 1 \text{ for } |y| < 1,$$

$$g_2(y) = 0 \text{ for } |y| > 2,$$

and $g_2 \in C_c^\infty(e^\perp)$ with $0 \leq g_2(y) \leq 1$ for all y. Finally define $g : B(x_R, R) \to \mathbf{R}$, by

$$g(x_R + se, y) = g_1(x_R + se)g_2(y).$$

Then $g \in C_c^\infty(\Omega)$, $g(x_R + se, y) > 0$ if and only if $1 < s < R - 1$ and $|y| < 2$; $g(x + se, y) = 0$ otherwise. Then

$$0 \leq \inf \sigma(G) \leq \inf_{R \geq 10} \frac{<Gg, g>_{X_2}}{\|g\|_{X_2}^2}.$$

But

$$<Gg, g>_{X_2} = <Lg, g>_{L^2(\Omega)}$$

$$= \int_{B(0,2)} \int_1^{R-1} L(g_1(x_R + se)g_2(y))ds dy$$

$$= \int_{B(0,2)} (\int_1^2 + \int_{R-2}^{R-1}) L\phi(g_1(x_R + se)g_2(r))r^{n-2} ds\, i\, dr\omega_n$$

$$= C_1,$$

where C_1 is a positive constant, since g_2 is a radial function and ω_n is the surface area of the unit sphere in $\mathbf{R}^n$. Note that the boundary conditions trivially hold since $g \in C_c^2(B(x_R, R)) \subset C_c^2(\Omega)$.

Next,

$$\|g\|_{X_2}^2 = \|g\|_{L^2(\Omega)}^2 \geq \int_1^{R-1} \int_{B(0,1) \subset \mathbf{R}^{n-1}} 1 \, ds\, dy$$

$$= (R - 2)\omega_{n-1},$$

whence

$$\frac{<Gg, g>}{\|g\|_{X_2}^2} \leq \frac{C_1}{(R - 2)\omega_{n-1}} \to 0$$

as $R \to \infty$.

The above proof works in dimension $n \geq 2$. The case $n = 1$ is easier. The only proper domain with an arbitrary big ball is a semiinfinite interval Ω of the form $(-\infty, b)$ or (b, ∞) with $b \in \Omega$. Then $\partial \Omega = \{b\}$. In the above proof, $e^\perp$ is $\{0\}$ and the proof simplifies greatly. The details are omitted.

This completes the proof that the solution of

$$\begin{cases} u_{tt} + 2au_t + G^2 u = 0, \\ u(0) = f, \\ u'(0) = g \end{cases}$$

is asymptotically equal to the solution of

$$\begin{cases} 2av_t + G^2 u = 0, \\ v(0) = h. \end{cases}$$

This also proves that the solution of

$$\begin{cases} u_{tt} + 2a(-G)^\alpha u_t + G^2 u = 0, \\ u(0) = f, \\ u'(0) = g \end{cases}$$

is asymptotically equal to the solution of

$$\begin{cases} 2a(-G)^\alpha v_t + G^2 v = 0, \\ v(0) = h \end{cases}$$

or

$$\begin{cases} 2aw_t + (-G)^{2-\alpha} w = 0, \\ w(0) = (-G)^\alpha h. \end{cases}$$

This is a fractional heat equation. □

Remark 3.1 We proved that the bottom of the spectrum of the selfadjoint operator G is 0 by using the numerical range criterion,

$$\min \sigma(G) = \inf\{< Gu, u >: \|u\| = 1\}.$$

But we really know very little else about the spectrum $\sigma(G)$. It may contain many eigenvalues or no eigenvalues, in the general unbounded domain case.

Acknowledgements Part of this paper was done during the period when G. R. Goldstein and J. A. Goldstein were Visiting Professors at the University of Bari Aldo Moro. The results obtained are part of the research plan of D. Guidetti and S. Romanelli as members of G.N.A.M.P.A. (Istituto Nazionale di Alta Matematica).

References

1. Browder, F.E.: On the spectral theory of elliptic differential operators I. Math. Ann. **142**, 22–130 (1961)
2. Cattaneo, C.: Sulla conduzione del calore. Atti Sem. Mat. Fis. Univ. Modena **3**, 83–101 (1948–1949)
3. Clarke, T., Goldstein, G.R., Goldstein, J.A., Romanelli, S.: The Wentzell telegraph equation: asymptotics and continuous dependence on the boundary conditions. Comm. Appl. Anal. **15**, 313–324 (2011)
4. Coclite, G.M., Favini, A., Gal, C.G., Goldstein, G.R., Goldstein, J.A., Obrecht, E., Romanelli, S.: The role of Wentzell boundary conditions in linear and nonlinear analysis. In: Sivasundaran, S. (ed.) Advances in Nonlinear Analysis: Theory, Methods and Applications, vol. 3, pp. 279–292 (2009)
5. Favini, A., Goldstein, G.R., Goldstein, J.A., Obrecht, E., Romanelli, S.: Elliptic operators with Wentzell boundary conditions, analytic semigroups and the angle concavity theorem. Math. Nachr. **283**, 504–521 (2010)
6. Favini, A., Goldstein, G.R., Goldstein, J.A., Obrecht, E., Romanelli, S.: Nonsymmetric elliptic operators with Wentzell boundary conditions in general domains. Commun. Pure Appl. Anal. **15**, 2475–2487 (2016)
7. Favini, A., Goldstein, G.R., Goldstein, J.A., Romanelli, S.: The heat equation with generalized Wentzell boundary conditions. J. Evol. Equ. **2**, 1–19 (2002)
8. Favini, A., Goldstein, G.R., Goldstein, J. A., Romanelli, S.: Fourth order ordinary differential operators with general Wentzell boundary conditions. In: Favini, A., Lorenzi, A. (eds.) Differential Equations: Direct and Inverse Problems, vol. 2, pp. 61–74. Chapman and Hall/CRC, Boca Raton (2006)
9. Favini, A., Goldstein, G.R., Goldstein, J.A., Romanelli, S.: Classification of general Wentzell boundary conditions for fourth order operators in one space dimension. J. Math. Anal. Appl. **333**, 219–235 (2007)
10. Favini, A., Goldstein, G.R., Goldstein, J.A., Romanelli, S.: Fourth order operators with general Wentzell boundary conditions. Rocky Mount. J. Math. **38**, 445–460 (2008)
11. Fragnelli, G., Goldstein, G.R., Goldstein, J.A., Romanelli, S.: Asymptotic parabolicity for strongly damped wave equations. In: Spectral Analysis, Differential Equations, and Mathematical Physics, A Festschrift in Honor of Fritz Gesztesy's 60th Birthday, Proceedings of Symposia in Pure Mathematics, vol. 87, pp. 119–131. American Mathematical Society (2013)
12. Goldstein, G.R.: Derivation and physical interpretation of general boundary conditions. Adv. Diff. Eq. **11**, 457–480 (2006)
13. Goldstein, G.R., Goldstein, J.A., Guidetti, D., Romanelli, S.: Maximal regularity, analytic semigroups, and dynamic and general Wentzell boundary conditions with boundary diffusion. Annali Mat. Pura e Appl. 199, 127–146 (2020). https://doi.org/10.1007/s10231-019-00868-3
14. Goldstein, J.A.: Some remarks on infinitesimal generators of analytic semigroups. Proc. Am. Math. Soc. **22**, 91–93 (1969)
15. Goldstein, J.A.: Semigroups of Linear Operators and Applications, 2nd edn. Dover Publications, Mineola, New York (2017)

16. Guidetti, D.: Abstract elliptic problems depending on a parameter and parabolic problems with dynamic boundary conditions. In: New Prospects in Direct, Inverse and Control Problems for Evolution Equations, Springer INdAM Series, vol. 10, pp. 161–202. Springer (2014)
17. Guidetti, D.: Parabolic problems with general Wentzell boundary conditions and diffusion on the boundary. Commun. Pure Appl. Anal. **15**, 1401–1417 (2016)
18. Taylor, G.I.: Diffusion by continuous movements. Proc. Lond. Math. Soc. **S2–20**, 196–212 (1922)

Nonlinear Semigroups and Their Perturbations in Hydrodynamics. Three Examples

Piotr Kalita, Grzegorz Łukaszewicz, and Jakub Siemianowski

Abstract In this paper we present some mutual relations between semigroup theory in the context of the theory of infinite dimensional dynamical systems and the mathematical theory of hydrodynamics. These mutual relations prove to be very fruitful, enrich both fields and help to understand behaviour of solutions of both infinite dimensional dynamical systems and hydrodynamical equations. We confine ourselves to present these connections on some recent developments in the important problem of heat transport in incompressible fluids which features all main aspects of chaotic dynamics. To be specific, we consider the Rayleigh–Bénard problem for the two and three-dimensional Boussinesq systems for the Navier–Stokes and micropolar fluids and two-dimensional thermomicropolar fluid. Each of the three examples is remarkably distinct from the other two in the context of the semigroup theory.

Keywords Semigroup · Eventual semigroup · Attractor · Rayleigh–Bénard problem · Boussinesq system · Micropolar fluid

2010 Mathematics Subject Classification 76F35 · 76E15 · 37L30 · 35Q30 · 35Q79

P. Kalita
Faculty of Mathematics and Computer Science, Jagiellonian University,
ul. Łojasiewicza 6, 30-348 Kraków, Poland
e-mail: piotr.kalita@ii.uj.edu.pl

G. Łukaszewicz (✉)
Faculty of Mathematics, Informatics, and Mechanics, University of Warsaw,
ul. Banacha 2, 02-097 Warszawa, Poland
e-mail: glukasz@mimuw.edu.pl

J. Siemianowski
Faculty of Mathematics and Computer Sciences, Nicolaus Copernicus University,
Chopina 12/18, 87-100 Toruń, Poland
e-mail: jsiem@mat.umk.pl

1 Introduction

Evolutionary systems of hydrodynamical equations can be formulated as the Cauchy problem

$$\frac{du}{dt} + N(u, \mu) = f(t), \quad t > \tau \tag{1.1}$$

$$u(\tau) = u_0 \in H \tag{1.2}$$

where, for example in the case of the celebrated Navier–Stokes equation, u is the velocity field of the fluid, f is the acting external force, and μ is a positive parameter describing the fluid viscosity. The evolutionary equation is then just the Newtonian equation of balance of linear momentum ($m\dot{v} = F$), and u_0 is the velocity field distribution at the initial time τ, an element of an infinite dimensional phase space H. The aim is to prove the existence and properties of the velocity field u for times $t > \tau$ in some functional framing.

In this way, we reduce the considerations of a hydrodynamical system to that of a chaotic (dissipative) dynamical system in some infinite dimensional phase space. Here enters the classical nonlinear semigroup theory in several contexts (assuming that the solution exists for all times $t > \tau$). First, we distinguish autonomous and nonautonomous cases, where the external force does not depend or depend on time, respectively, second, we distinguish the cases where the solution u is uniquely determined or not. We shall comment on these distinctions and related theories in Sect. 6.

Below we consider the autonomous case where we can associate to system (1.1) a single or multivalued semigroup. Recall that for the Navier–Stokes system there is a huge difference in our understanding of its solutions in the cases of two and three-dimensional space domains, respectively.

Besides of the fundamental problems of existence of global in time solutions, their uniqueness and continuous dependence on the data (initial conditions, external forces), we are interested in their bifurcations with respect to system parameters (e.g., viscosity), time asymptotics (behaviour of solutions for large time), energy dissipation rate, and other features, allowing us, in particular, to better understand the physical phenomenon of the turbulence of fluid motions. Such a research is also a certain test of the chosen theoretical fluid model. A deep and complex interplay between theoretical and practical aspects of fluid dynamics is presented in [5].

Due to the complexity of hydrodynamical equations and our limited knowledge of the behaviour of fluid motion, one of the fundamental objects to study in the frame of the related theory of dynamical systems is the global attractor, a subset of (usually) the phase space of initial conditions, compact, invariant, and attracting bounded subsets of the phase space. Its existence and properties (e.g., structure, finiteness of its fractal dimension) allow us to understand many features of the chaotic dynamical system and thus the turbulent fluid flow. We refer the interested reader to the literature devoted to this topic, e.g., [1, 7, 20, 24, 36, 44, 49].

In this paper we present some of these problems confining ourselves to our recent results on the systems of equations describing the heat transfer in fluids, the problem that possesses all the main features of chaotic dynamical systems, cf. [25]. We consider the Rayleigh–Bénard heat convection problem for the model of micropolar fluid as well as that for the classical Navier–Stokes model with the aim to show their chaotic properties and to compare them, as the former model can be regarded as a perturbation of the latter. We consider these systems in the frame of the theory of infinite dimensional dynamical systems and related single and multi-valued semigroups, associated with two and three dimensional cases, respectively. Then, we present some results for the thermomicropolar model in two dimensions.

We aim at simplicity of the exposition, in most cases omitting the long proofs and pedantic details. For these we refer the reader to related literature.

In this place we would like to present in a few words the considered micropolar model. It is both a simple and significant generalization of the Navier–Stokes model of classical hydrodynamics, which has much more applications than the classical model due to the fact that the latter cannot describe (by definition) fluids with microstructure. In general, individual particles of such complex fluids (e.g., polymeric suspensions, blood, liquid crystals) may be of different shape, may shrink and expand, or change their shape, and moreover, they may rotate, independently of the rotation and movement of the fluid. To describe accurately the behavior of such fluids one needs a theory that takes into account geometry, deformation, and intrinsic motion of individual material particles. In the framework of continuum mechanics several such theories have appeared, e.g., theories of simple microfluids, simple deformable directed fluids, micropolar fluids, dipolar fluids, to name some of them.

One of the best-established theories of fluids with microstructure is the theory of micropolar fluids developed in [21], and studied from mathematical point of view in [34]. Physically, micropolar models represent fluids consisting of rigid, randomly oriented (or spherical) particles suspended in a viscous medium, where the deformation of the particles is ignored. This constitutes a substantial generalization of the Navier–Stokes model and opens a new field of potential applications including a large number of complex fluids.

The system of equations of the three-dimensional Boussinesq system for the micropolar model, in dimensionless form, reads [27],

$$\frac{1}{\Pr}(u_t + (u \cdot \nabla)u) - (1 + K)\Delta u + \frac{1}{\Pr}\nabla p = 2K \operatorname{rot} \gamma + e_3 \operatorname{Ra} T, \quad (1.3)$$

$$\operatorname{div} u = 0, \quad (1.4)$$

$$\frac{M}{\Pr}(\gamma_t + u \cdot \nabla \gamma) - L\Delta \gamma + G\nabla \operatorname{div} \gamma + 4K\gamma = 2K \operatorname{rot} u, \quad (1.5)$$

$$T_t + u \cdot \nabla T - \Delta T = 0. \quad (1.6)$$

The meaning of all unknowns and parameters in the above problem will be explained in Sect. 2, where also the associated boundary conditions will be provided. We will compare solutions of the above system with that for the Newtonian fluid, given by

$$\frac{1}{\text{Pr}}(u_t + (u \cdot \nabla)u) - \Delta u + \frac{1}{\text{Pr}}\nabla p = e_3 \text{Ra} T, \qquad (1.7)$$

$$\text{div}\, u = 0, \qquad (1.8)$$

$$T_t + u \cdot \nabla T - \Delta T = 0. \qquad (1.9)$$

Observe that if (u, γ, T) solves (1.3)–(1.6) with $K = 0$ then (u, T) solves (1.7)–(1.9) so that it is natural to ask about properties of solutions to (1.3)–(1.6) when $K \to 0$. We can formulate the problem in the language of related semigroups and ask about relations between semigroups $S_K(t)$ and $S_0(t)$ and the corresponding global attractors $\mathcal{A}_K$ and $\mathcal{A}_0$ for small K.

As there is essential difference between two and three-dimensional problems, we treat the two cases separately. In Sect. 2 we consider the two-dimensional case, and in Sects. 3 and 4 the three-dimensional one. Section 5 is devoted to the two-dimensional thermomicropolar model. In the final Sect. 6 we provide additional comments and set the problems in a larger context.

2 Two-Dimensional Problem

Problem formulation. Let $\Omega = (0, A) \times (0, 1) \subset \mathbb{R}^2$. The boundary of Ω is divided into three parts, $\partial \Omega = \overline{\Gamma_B} \cup \overline{\Gamma_T} \cup \overline{\Gamma_L}$, where $\Gamma_B = (0, A) \times \{0\}$ is the bottom, $\Gamma_T = (0, A) \times \{1\}$ is the top, and $\Gamma_L = \{0, A\} \times (0, 1)$ is the lateral boundary. We use the notation

$$\text{rot}\, \gamma = \left(\frac{\partial \gamma}{\partial x_2}, -\frac{\partial \gamma}{\partial x_1} \right) \quad \text{and} \quad \text{rot}\, u = \frac{\partial u_2}{\partial x_1} - \frac{\partial u_1}{\partial x_2},$$

for a scalar function γ defined on Ω, and a vector function u defined on Ω with values in $\mathbb{R}^2$, respectively. We consider the following Rayleigh–Bénard problem for micropolar fluid, where $(x, t) \in \Omega \times (0, \infty)$.

In the two-dimensional setting system (1.3)–(1.6) reduces to

$$\frac{1}{\text{Pr}}(u_t + (u \cdot \nabla)u) - (1 + K)\Delta u + \frac{1}{\text{Pr}}\nabla p = 2K \text{rot}\, \gamma + e_2 \text{Ra} T, \qquad (2.1)$$

$$\text{div}\, u = 0, \qquad (2.2)$$

$$\frac{M}{\text{Pr}}(\gamma_t + u \cdot \nabla \gamma) - L\Delta \gamma + 4K\gamma = 2K \text{rot}\, u, \qquad (2.3)$$

$$T_t + u \cdot \nabla T - \Delta T = 0. \qquad (2.4)$$

The unknowns in the above equations are the velocity field $u : \Omega \times (0, \infty) \to \mathbb{R}^2$, $u = (u_1, u_2)$, the pressure $p : \Omega \times (0, \infty) \to \mathbb{R}$, the temperature $T : \Omega \times (0, \infty) \to \mathbb{R}$, and the microrotation field $\gamma : \Omega \times (0, \infty) \to \mathbb{R}$. We assume the boundary conditions $u = 0$ and $\gamma = 0$ on $\Gamma_B \cup \Gamma_T$, $T = 0$ on Γ_T, and $T = 1$ on Γ_B. On Γ_L we impose the periodic conditions on the functions u, T, γ and their normal derivatives as well as the pressure p such that all boundary integrals on Γ_L in the weak formulation cancel. All physical constants $K, L, M, \text{Pr}, \text{Ra}$ are assumed to be positive numbers, save for K, where we allow for the case $K = 0$, in which the micropolar model reduces to the Boussinesq one, $e_2 = (0, 1)$. The relevant constant is the Rayleigh number Ra which is proportional to the difference between temperatures at the bottom and at the top parts of the boundary and thus is the driving force for the system. Observe that if $\text{Ra} = 0$ then $(u, \gamma, T) = (0, 0, 1 - x_2)$ is a stationary solution of the system.

Finally we impose the initial conditions

$$u(x, 0) = u_0(x), \quad T(x, 0) = T_0(x), \quad \gamma(x, 0) = \gamma_0(x) \quad \text{for} \quad x \in \Omega.$$

Introducing the new variable θ by

$$T(x_1, x_2, t) = \theta(x_1, x_2, t) + 1 - x_2,$$

we obtain the system

$$\frac{1}{\text{Pr}}(u_t + (u \cdot \nabla)u) - (1 + K)\Delta u + \frac{1}{\text{Pr}}\nabla p = 2K \operatorname{rot} \gamma + e_2 \text{Ra}\theta, \tag{2.5}$$

$$\operatorname{div} u = 0, \tag{2.6}$$

$$\frac{M}{\text{Pr}}(\gamma_t + u \cdot \nabla \gamma) - L\Delta \gamma + 4K\gamma = 2K \operatorname{rot} u, \tag{2.7}$$

$$\theta_t + u \cdot \nabla \theta - \Delta \theta = u_2, \tag{2.8}$$

with homogeneous boundary condition for θ at both the bottom and the top parts of the boundary.

Existence and uniqueness of the weak solution, existence of attractors. We have to introduce some notations. Let $(\cdot, \cdot)$ denote the scalar product in $L^2(\Omega)$ or, depending on the context, in $L^2(\Omega)^2$. Let $\widetilde{V}$ be the space of divergence-free (with respect to the x variable) functions which are restrictions to $\overline{\Omega}$ of functions from $C^\infty(\mathbb{R} \times [0, 1])^2$ which are equal to zero on $\mathbb{R} \times \{0, 1\}$, and, together with all their derivatives, A-periodic with respect to the first variable. Similarly, by $\widetilde{W}$ we define the space of all functions which are restrictions to $\overline{\Omega}$ of functions from $C^\infty(\mathbb{R} \times [0, 1])$ which are A-periodic with all derivatives with respect to the first variable and satisfying the homogeneous Dirichlet boundary conditions on bottom and top boundaries $\mathbb{R} \times \{0, 1\}$. Define the spaces

$$V = \{\text{ closure of } \widetilde{V} \text{ in } H^1(\Omega)^2\} \quad \text{and} \quad H = \{\text{ closure of } \widetilde{V} \text{ in } L^2(\Omega)^2\},$$

$$W = \{\text{ closure of } \widetilde{W} \text{ in } H^1(\Omega)\} \quad \text{and} \quad E = \{\text{ closure of } \widetilde{W} \text{ in } L^2(\Omega)\}.$$

Moreover, let $P : L^2(\Omega)^2 \to H$ be the Leray–Helmholtz projection. The duality pairings between V and its dual V' as well as between W and its dual W' will be denoted by $\langle \cdot, \cdot \rangle$.

Definition of the weak solution. The weak form of (2.5)–(2.8) is obtained in a standard way. We assume that

$$u_0 \in H, \quad \gamma_0 \in E, \quad \text{and} \quad \theta_0 \in E,$$

and look for $u \in L^2_{loc}([0, \infty); V)$, $\gamma \in L^2_{loc}([0, \infty); W)$, and $\theta \in L^2_{loc}([0, \infty); W)$ with $u_t \in L^2_{loc}([0, \infty); V')$, $\gamma_t \in L^2_{loc}([0, \infty); W')$, and $\theta_t \in L^2_{loc}([0, \infty); W')$ such that $u(0) = u_0, \gamma(0) = \gamma_0, \theta(0) = \theta_0$, and for all test functions $v \in V, \xi \in W, \eta \in W$ and a.e. $t > 0$ there holds

$$\frac{1}{\Pr}(\langle u_t(t), v \rangle + ((u(t) \cdot \nabla)u(t), v)) + (1 + K)(\nabla u(t), \nabla v) = 2K(\operatorname{rot} \gamma(t), v) + \operatorname{Ra}(\theta(t), v_2), \tag{2.9}$$

$$\frac{M}{\Pr}(\langle \gamma_t(t), \xi \rangle + (u(t) \cdot \nabla \gamma(t), \xi)) + L(\nabla \gamma(t), \nabla \xi) + 4K(\gamma(t), \xi) = 2K(\operatorname{rot} u(t), \xi), \tag{2.10}$$

$$\langle \theta_t(t), \eta \rangle + (u(t) \cdot \nabla \theta(t), \eta) + (\nabla \theta(t), \nabla \eta) = (u_2(t), \eta). \tag{2.11}$$

Existence, uniqueness, and regularity of weak solutions for the above problem has been established in [47], also see [48], in the following theorem

Theorem 2.1 *The problem of heat convection in micropolar fluid has a unique weak solution. This solution has the regularity*

$$u \in L^2_{loc}([\eta, \infty); H^2(\Omega)^2), \quad \gamma \in L^2_{loc}([\eta, \infty); H^2(\Omega)), \quad \text{and} \quad \theta \in L^2_{loc}([\eta, \infty); H^2(\Omega)).$$

Moreover the mapping $S_K(t)(u_0, \gamma_0, \theta_0) = (u(t), \gamma(t), \theta(t))$ is continuous as a map from $H \times E \times E$ into itself and the family of mappings $\{S_K(t)\}_{t \geq 0}$ is a semigroup.

In [47, Sects. 4 and 5] the following result has been proved,

Theorem 2.2 *The solution semigroup $\{S_K(t)\}_{t \geq 0}$ for a heat convection problem in a micropolar fluid has a global attractor $\mathcal{A}_K$ which is compact in $H \times E \times E$ and bounded in $V \times W \times W$ and has finite fractal (Hausdorff and upper box counting) dimension.*

We recall that a global attractor for a semigroup of mappings $\{S(t)\}_{t \geq 0}$ leading from X to itself is a set $\mathcal{A} \subset X$ which is compact in X, invariant, i.e. $S(t)\mathcal{A} = \mathcal{A}$ for every $t \geq 0$, and attracting, i.e. $\lim_{t \to \infty} \operatorname{dist}_X(S(t)B, \mathcal{A}) = 0$ for every bounded $B \subset X$, where $\operatorname{dist}_X(A, B) = \sup_{a \in A} \inf_{b \in B} \|a - b\|_X$ is the Hausdorff semidistance between two sets A and B in a Banach space X.

Using the standard bootstrapping argument it is possible to prove that the weak solutions for the considered problem become smooth instantaneously, and the attractor contains in fact the smooth functions, see [29] for the argument in the case of the more general model of thermomicropolar fluids.

We note that the projection of $\mathcal{A}_0$ on the variables (u, θ) is the extensively studied Rayleigh–Bénard attractor [13–15, 20, 37].

The following result has been proved in [26].

Theorem 2.3 *The following bound holds for the fractal dimension of the attractor for the Rayleigh-Bénard convection in micropolar fluid,*

$$d_F(\mathcal{A}_K) \leq C(|\Omega|)\left(\sqrt{\mathrm{Gr}}\left(\mathrm{Pr} + \frac{M}{L} + H(K,L)\right)\right.$$
$$\left. + \frac{\mathrm{Gr}}{\sqrt[4]{\mathrm{Ra}}}\left(\mathrm{Pr} + \frac{M}{L} + \sqrt{H(K,L)}\right)\sqrt{\mathrm{Pr} + \frac{M}{L} + H(K,L)\sqrt[4]{H(K,L)}}\right),$$

where $\mathrm{Gr} = \mathrm{Ra}/\mathrm{Pr}$ *is the Grashof number and*

$$H(K, L) = 1 - \frac{KL\pi^2}{4K + (K+1)L\pi^2}.$$

The number π^2 *which appears here is the first eigenvalue of the Stokes operator with the considered boundary conditions. The function H decreases in both variables, with*

$$H(0, L) = H(K, 0) = 1 \quad \text{and} \quad \lim_{K \to \infty, L \to \infty} H(K, L) = 0.$$

The estimate obtained in the above Theorem is not optimal. In fact, if damping parameters K and L are sufficiently large with respect to the driving parameter Ra then the critical Rayleigh number (which depends on K and L representing the fluid properties) is also arbitrarily large. In fact for any given Rayleigh number we can find K and L large enough such that the critical Rayleigh number is higher and then $\mathcal{A} = \{0\}$ and hence the dimension of the attractor also equals zero, cf. [26, Theorem 3.3]. In this case the null stationary solution is globally stable. However, the upper estimate of the global attractor $\mathcal{A}_K$ confirms our expectations based on our understanding of the physics of the considered problem.

If $K = 0$ and $M = 0$, i.e. there are no micropolar effects, we obtain the bound

$$d_F(\mathcal{A}) \leq C(A)\left(\sqrt{\mathrm{Gr}}(1 + \mathrm{Pr}) + \frac{\mathrm{Gr}}{\sqrt[4]{\mathrm{Ra}}}(1 + \mathrm{Pr})^{\frac{3}{2}}\right),$$

an improvement over the estimate obtained by Foias, Manley, and Temam [23].

The global attractors $\mathcal{A}_K$ converge upper-semicontinuously to the global attractor of the limit problem when $K = 0$, cf. [26],

Theorem 2.4 *Let* Ra, Pr, M, L *be fixed positive constants. Then*

$$\lim_{K \to 0^+} \mathrm{dist}_{H \times E \times E}(\mathcal{A}_K, \mathcal{A}_0) = 0.$$

The limit problem, for $K = 0$, reads

$$\frac{1}{\Pr}(u_t + (u \cdot \nabla)u) - \Delta u + \frac{1}{\Pr}\nabla p = e_2 \mathrm{Ra}\theta, \qquad (2.12)$$

$$\mathrm{div}\, u = 0, \qquad (2.13)$$

$$\frac{M}{\Pr}(\gamma_t + u \cdot \nabla \gamma) - L\Delta\gamma = 0 \qquad (2.14)$$

$$\theta_t + u \cdot \nabla\theta - \Delta\theta = u_2. \qquad (2.15)$$

It follows that the projection of the global attractor $\mathcal{A}_0$ on the variables u, θ is simply the Rayleigh–Bénard attractor.

In general, the results on the upper-semicontinous convergence of the global attractors are standard and the techniques to prove them are well established, see for instance [24, 43, 49]. For micropolar fluids without heat convection the corresponding result was already obtained in [35]. We prove the upper-semicontinuous convergence result to make our comparison between the heat convection in micropolar and Newtonian fluid more complete. In [26] we have used the following theorem, [24, Theorem 2.5.2],

Theorem 2.5 *Let $\{S_K(t)\}_{t\geq 0}$ be a family of semigroups on the Banach space X parameterized by $K \geq 0$. Assume that the following three assertions hold.*

(i) *For every $K \geq 0$ the semigroup $\{S_K(t)\}_{t\geq 0}$ has a global attractor $\mathcal{A}_K$.*
(ii) *There exists a bounded set $B \subset X$ such that $\bigcup_{K \in [0,1]} \mathcal{A}_K \subset B$.*
(iii) *For any $\epsilon > 0$ and $t \geq 0$ there exists $K_0(\epsilon, t) < 1$ such that for every $K \in [0, K_0]$ and for every $x \in \mathcal{A}_K$ there holds $\|S_K(t)x - S_0(t)x\|_X \leq \epsilon$.*

Then

$$\lim_{K \to 0^+} \mathrm{dist}_X(\mathcal{A}_K, \mathcal{A}_0) = 0.$$

We have proved the following result

Lemma 2.6 *Let $K \in [0, 1]$ and let $(u_0, \gamma_0, \theta_0) \in \mathcal{A}_K$. Let $(u^K, \gamma^K, \theta^K)$ be the trajectory (in $\mathcal{A}_K$) of the problem (2.5)–(2.8) with the initial data $(u_0, \gamma_0, \theta_0)$ and let $(u^0, \gamma^0, \theta^0)$ be the trajectory of (2.12)–(2.15) with the same data. Then there exists a constant C dependent on A, $\Pr$, Ra, M, L, but not on K, such that for every $t > 0$*

$$\|u^K(t) - u^0(t)\|_{L^2}^2 + \|\gamma^K(t) - \gamma^0(t)\|_{L^2}^2 + \|\theta^K(t) - \theta^0(t)\|_{L^2}^2 \leq KC(1+t)e^{C(1+t)}.$$

Thus, the trajectories of both systems are arbitrarily close to each other on a given finite time interval $[0, t]$ if only the parameter K is sufficiently small. In particular the assertion (iii) of Theorem 2.5 holds, and assertion (ii) is easily proved.

The result of Lemma 2.6 can be also elegantly expressed in the language of semigroups and their perturbations.

3 Three-Dimensional Problem

Multivalued eventual semiflows. As it has been mentioned in the introduction, the three-dimensional problem essentially differs from its two-dimensional counterpart. Namely, now we do not know if its solutions are uniquely determined by initial conditions but become unique after some time of evolution. We have to work with some generalization of single valued semiflows. To this end in [27] we introduced the notion of a *multivalued eventual semiflow*. It would also prove useful in many similar situations when we do not know whether a given multivalued semiflow satisfies the translation property, e.g., as in the case of the three-dimensional Navier–Stokes system, but which satisfies the translation property for large times, uniformly for bounded sets of initial data.

Below we state some properties of such semiflows and their attractors.

For a complete metric space (X, ϱ_X), we define by $\mathcal{P}(X)$ the family of its nonempty subsets, and by $\mathcal{B}(X)$ the family of nonempty and bounded subsets.

The following definition is the slight relaxation of the multivalued semiflow definition due to Melnik and Valero [40, 41].

Definition 3.1 Let (X, ϱ_X) be a complete metric space. The family of mappings $\{S(t)\}_{t \geq 0}$ such that $S(t) : X \to \mathcal{P}(X)$ is a multivalued *eventual* semiflow if

(i) $S(0)v = \{v\}$ for every $v \in X$.
(ii) For every $B \in \mathcal{B}(X)$ bounded there exists time $t_1(B)$ such that for every $t \geq t_1(B), s \geq 0$ there holds $S(s+t)B \subset S(s)S(t)B$.

The difference between the above definition and that of Melnik and Valero [40, 41] is that property (ii) is assumed to hold for every $t \geq 0$ in [40, 41] and not for just $t \geq t_1(B)$. As it turns out, it suffices to relax the definition to $t \geq t_1(B)$, and the result of [40, 41] on the global attractor existence remains valid.

We pass to the definition of a global attractor for multivalued eventual semiflow (cf. [27]). In this definition, firstly, following [8], we do not impose invariance of the global attractor. We only assume that it is a minimal compact attracting set. Secondly, we follow [1] and assume that our attractor is compact in a "smaller space" Y while it attracts bounded sets from a "bigger" space X. Specifically we make the standing assumptions, that (X, ϱ_X) and (Y, ϱ_Y) are complete metric spaces such that $Y \subset X$ and the identity $i : Y \to X$ is continuous.

Definition 3.2 Let $\{S(t)\}_{t \geq 0}$ be a multivalued eventual semiflow in X such that $S(t)B \subset Y$ for every $B \in \mathcal{B}(X)$ and $t \geq t_0(B)$. The set $\mathcal{A} \subset Y$ is called a (X, Y)-global attractor for $\{S(t)\}_{t \geq 0}$ if

(i) $\mathcal{A} \in \mathcal{B}(Y)$ is compact in Y,
(ii) there holds $\lim_{t \to \infty} \text{dist}_Y(S(t)B, \mathcal{A}) = 0$ for every $B \in \mathcal{B}(X)$,
(iii) if, for a closed in Y set $\overline{\mathcal{A}}$ there holds $\lim_{t \to \infty} \text{dist}_Y(S(t)B, \overline{\mathcal{A}}) = 0$ for every $B \in \mathcal{B}(X)$, then $\mathcal{A} \subset \overline{\mathcal{A}}$.

Let $B \in \mathcal{B}(X)$. Assuming that there exists $t_0(B)$ such that for every $t \geq t_0$ there holds $S(t)B \subset Y$ we define the ω-limit set in Y by

$$\omega_Y(B) = \bigcap_{s \geq t_0(B)} \overline{\bigcup_{t \geq s} S(t)B}^Y.$$

We have the following theorem.

Theorem 3.3 [27] *Let $\{S(t)\}_{t \geq 0}$ be multivalued eventual semiflow in X. Assume that the following two conditions hold.*

(i) *The family $\{S(t)\}_{t \geq 0}$ is (X, Y)- dissipative, i.e., there exists a set $B_0 \in \mathcal{B}(Y)$ such that for every $B \in \mathcal{B}(X)$ there exists $t_0(B)$ such that*

$$\bigcup_{t \geq t_0} S(t)B \subset B_0.$$

(ii) *The family $\{S(t)\}_{t \geq 0}$ is Y- asymptotically compact on B_0, i.e., if $t_n \to \infty$, then every sequence $v_n \in S(t_n)B_0$ is relatively compact in Y.*

Then the family $\{S(t)\}_{t \geq 0}$ has a (X, Y)-global attractor $\mathcal{A} \subset \overline{B_0}^Y$ which is given by $\mathcal{A} = \omega_Y(B_0)$.

We pass to the global attractor invariance. Namely, the following theorem holds.

Theorem 3.4 [27] *Assume, in addition to the assumptions of Theorem 3.3, that for every $t \geq 0$ the restriction $S(t)|_Y$ is a single valued semigroup of (Y, Y)-continuous maps. Then, the (X, Y)-global attractor $\mathcal{A}$ is an invariant set, that is $S(t)\mathcal{A} = \mathcal{A}$ for every $t \geq 0$.*

Proof Since the assumptions of Theorem 3.3 imply that the single valued semigroup $\{S(t)|_Y\}_{t \geq 0}$ has a bounded absorbing set in Y and is asymptotically compact on Y, the result follows from a very well known abstract theorem on the global attractor existence, cf., e.g., [44, 49]. □

We conclude this section with results on the relation between global attractors and complete (eternal) bounded trajectories.

Definition 3.5 The function $u : \mathbb{R} \to X$ is an X-bounded (respectively, Y-bounded) eternal trajectory if for every $t \in \mathbb{R}$ and every $s > t$ there holds $u(s) = S(s-t)u(t)$ and the set $\{u(t)\}_{t \in \mathbb{R}}$ is bounded in X (respectively, Y).

Since the family $\{S(t)|_Y\}_{t \geq 0}$ is a semigroup on Y with the global attractor $\mathcal{A}$, the following theorem follows from the well known result for the semigroups, see, e.g., [7, Lemma 1.4 and Theorem 1.7].

Theorem 3.6 *Under the assumptions of Theorem 3.4, for every $v \in \mathcal{A}$ and for every $t \in \mathbb{R}$ there exists an eternal Y-bounded (and hence also X-bounded) trajectory $u : \mathbb{R} \to Y$ such that $u(t) = v$ and $u(s) \in \mathcal{A}$ for every $s \in \mathbb{R}$.*

On the other hand, if the function $u : \mathbb{R} \to X$ is a complete X-bounded trajectory, then the next theorem shows that u has values in Y and $u(t) \in \mathcal{A}$ for every $t \in \mathbb{R}$. The proof mostly follows the lines of [7, Theorem 1.7], with the modification that we consider only X-bounded and not Y-bounded trajectories.

Theorem 3.7 [27] *Under the assumptions of Theorem 3.4, for every complete X-bounded trajectory $u : \mathbb{R} \to X$ there holds $u(t) \in \mathcal{A}$ for every $t \in \mathbb{R}$.*

Proof Denote $B = \bigcup_{t \in \mathbb{R}} \{u(t)\}$. This is a bounded set in X. We show that this set is actually bounded in Y. Indeed, take s such that $t - s \geq t_0(B)$. It follows that

$$u(t) = S(t-s)u(s) \in \bigcup_{r \geq t_0} S(r)B \subset B_0.$$

This means that u is Y-bounded. Now, for $s \in \mathbb{R}$ and $t \geq s + t_0(B)$ there holds

$$\operatorname{dist}_Y(u(t), \mathcal{A}) \leq \operatorname{dist}_Y(S(t-s)u(s), \mathcal{A}) \leq \operatorname{dist}_Y(S(t-s)B, \mathcal{A}).$$

Passing with s to $-\infty$ we deduce that

$$\operatorname{dist}_Y(u(t), \mathcal{A}) = 0,$$

whence $u(t) \in Y$ as $\mathcal{A}$ is Y-closed. □

Now we can come back to our hydrodynamical problem.

Weak and strong solutions. After the homogenization of the boundary data for the temperature, the problem (1.3)–(1.6) reads

$$\frac{1}{\Pr}(u_t + (u \cdot \nabla)u) - (1+K)\Delta u + \nabla p = 2K \operatorname{rot} \gamma + e_3 \operatorname{Ra} \theta, \tag{3.1}$$

$$\operatorname{div} u = 0, \tag{3.2}$$

$$\frac{M}{\Pr}(\gamma_t + (u \cdot \nabla)\gamma) - L\Delta\gamma - G\nabla \operatorname{div} \gamma + 4K\gamma = 2K \operatorname{rot} u, \tag{3.3}$$

$$\theta_t + u \cdot \nabla \theta - \Delta \theta = u_3. \tag{3.4}$$

Setting $K = 0$ we obtain the system

$$\frac{1}{\Pr}(u_t + (u \cdot \nabla)u) - \Delta u + \nabla p = e_3 \operatorname{Ra} \theta, \tag{3.5}$$

$$\operatorname{div} u = 0, \tag{3.6}$$

$$\frac{M}{\Pr}(\gamma_t + (u \cdot \nabla)\gamma) - L\Delta\gamma - G\nabla \operatorname{div} \gamma = 0, \tag{3.7}$$

$$\theta_t + u \cdot \nabla \theta - \Delta \theta = u_3, \tag{3.8}$$

formally corresponding to the Newtonian fluid, that is $K = 0$. Then, (3.5), (3.6), and (3.8) constitute the well known Boussinesq system (1.7)–(1.9), while the equation

(3.7) can be independently solved for γ once the solution of the system of the remaining three equations is known. All definitions and results stated below are valid also for the case $K = 0$.

We define the weak solution to the above problem (3.1)–(3.4) as the approximative limit of subsequences of the (u, γ)-Galerkin problems.

The function spaces V, H, W_k, and E_k, $k = 1, 3$, are defined in quite the same way as those defined in Sect. 2, with the difference that now: first, the domain Ω is three-dimensional and not two-dimensional, and second, the variable γ is a three-dimensional vector while the variable θ stays a scalar, and so we distinguish these cases by adding indexes 3 or 1 instead of writing W and E as in the previous section. In this section the symbol P denotes the three-dimensional Leray–Helmholtz projection.

Definition 3.8 [27] Let

$$u_0 \in H, \quad \gamma_0 \in E_3, \quad \text{and} \quad \theta_0 \in E_1.$$

The triple of functions (u, γ, θ) such that

$u \in L^2_{loc}([0, \infty); V) \cap C_w([0, \infty); H)$, with $u_t \in L^{4/3}_{loc}([0, \infty); V')$ and $u(0) = u_0$,

$\gamma \in L^2_{loc}([0, \infty); W_3) \cap C_w([0, \infty); E_3)$, with $\gamma_t \in L^{4/3}_{loc}([0, \infty); W'_3)$ and $\gamma(0) = \gamma_0$,

$\theta \in L^2_{loc}([0, \infty); W_1) \cap C_w([0, \infty); E_1)$, with $\theta_t \in L^{4/3}_{loc}([0, \infty); W'_1)$, and $\theta(0) = \theta_0$,

is called a weak solution of the problem (3.1)–(3.4) if for all test functions $v \in V$, $\xi \in W_3$, $\eta \in W_1$ and a.e. $t > 0$ there holds

$$\frac{1}{\Pr}(\langle u_t(t), v\rangle + ((u(t) \cdot \nabla)u(t), v)) + (1 + K)(\nabla u(t), \nabla v) = 2K(\operatorname{rot}\gamma(t), v) + \operatorname{Ra}(\theta(t), v_3), \tag{3.9}$$

$$\frac{M}{\Pr}(\langle\gamma_t(t), \xi\rangle + (u(t) \cdot \nabla\gamma(t), \xi)) + L(\nabla\gamma(t), \nabla\xi) + G(\operatorname{div}\gamma, \operatorname{div}\xi) + 4K(\gamma(t), \xi)$$
$$= 2K(\operatorname{rot} u(t), \xi), \tag{3.10}$$

$$\langle\theta_t(t), \eta\rangle + (u(t) \cdot \nabla\theta(t), \eta) + (\nabla\theta(t), \nabla\eta) = (u_3(t), \eta) \tag{3.11}$$

and if (u, γ, θ) is the limit of approximative problems in the sense that there exist sequences of initial data

$$H^n \ni u_0^n \to u_0 \in H \quad \text{strongly in } H,$$
$$E_3^n \ni \gamma_0^n \to \gamma_0 \in E_3 \quad \text{strongly in } E_3,$$
$$E_1^n \ni \theta_0^n \to \theta_0 \in E_1 \quad \text{strongly in } E_1$$

such that if $(u^n, \gamma^n, \theta^n)$ are corresponding solutions to (u, γ)-Galerkin problems with initial data $(u_0^n, \gamma_0^n, \theta_0^n)$, then, for a subsequence of indices denoted by n_k there holds

$$u^{n_k} \to u \quad \text{weakly in } L^2_{loc}([0, \infty); V) \text{ and weakly} - * \text{ in } L^\infty_{loc}([0, \infty); H), \tag{3.12}$$

$$\gamma^{n_k} \to \gamma \quad \text{weakly in } L^2_{loc}([0, \infty); W_3) \text{ and weakly} - * \text{ in } L^\infty_{loc}([0, \infty); E_3), \tag{3.13}$$

$$\theta^{n_k} \to \theta \quad \text{weakly in } L^2_{loc}([0, \infty); W_1) \text{ and weakly} - * \text{ in } L^\infty_{loc}([0, \infty); E_1). \tag{3.14}$$

The following result on the existence of the weak solution given in Definition 3.8 is standard, cf. [22].

Lemma 3.9 *For every initial data $u_0 \in H$, $\gamma_0 \in E_3$, and $\theta_0 \in E_1$ there exists a weak solution to problem (3.1)–(3.4) given by Definition 3.8.*

We can now to pass to the definition of the strong solution of the problem. For the details we refer the reader to [27].

Definition 3.10 Let

$$u_0 \in V, \quad \gamma_0 \in W_3, \quad \text{and} \quad \theta_0 \in W_1.$$

The triple of functions (u, γ, θ) such that

$u \in L^2_{loc}([0, \infty); D(-P\Delta)) \cap C([0, \infty); V)$, with $u_t \in L^2_{loc}([0, \infty); H)$ and $u(0) = u_0$,
$\gamma \in L^2_{loc}([0, \infty); D_3(-\Delta)) \cap C([0, \infty); W_3)$, with $\gamma_t \in L^2_{loc}([0, \infty); E_3)$ and $\gamma(0) = \gamma_0$,
$\theta \in L^2_{loc}([0, \infty); D_1(-\Delta)) \cap C([0, \infty); W_1)$, with $\theta_t \in L^2_{loc}([0, \infty); E_1)$, and $\theta(0) = \theta_0$,

is called a strong solution of the problem (3.1)–(3.4) if for all test functions $v \in H$, $\xi \in E_3$, $\eta \in E_1$ and a.e. $t > 0$ there holds

$$\frac{1}{\Pr}((u_t(t), v) + ((u(t) \cdot \nabla)u(t), v)) + (1 + K)(-P\Delta u(t), v) = 2K(\text{rot } \gamma(t), v) + \text{Ra}(\theta(t), v_3), \tag{3.15}$$

$$\frac{M}{\Pr}((\gamma_t(t), \xi) + (u(t) \cdot \nabla \gamma(t), \xi)) + L(-\Delta \gamma(t), \xi) + G(-\nabla \text{div } \gamma(t), \xi) + 4K(\gamma(t), \xi)$$
$$= 2K(\text{rot } u(t), \xi), \tag{3.16}$$

$$(\theta_t(t), \eta) + (u(t) \cdot \nabla \theta(t), \eta) + (-\Delta \theta(t), \eta) = (u_3(t), \eta). \tag{3.17}$$

Remark 3.11 We stress that, to our knowledge, nothing is known on the existence of the strong solution *for every* initial data $u_0 \in V$, $\gamma_0 \in W_3$, and $\theta_0 \in W_1$. If, however, the constants of the problem satisfy some restriction which will be given later, and the initial data is sufficiently small, such strong solution always exists. We will prove that it always exists on the global attractor for the weak solutions.

The next result establishes the weak-strong uniqueness property of strong and weak solutions.

Lemma 3.12 [27] *If (u, γ, θ) is a strong solution then it is also a weak solution and it is moreover unique in the class of the weak solutions.*

4 Global Attractors, Their Existence and Invariance in 3D

In this section we deal with the existence of the global attractor and its properties for the three-dimensional problem.

Existence of a global attractor. From now on we assume that the following assumption (H) holds,

(H) $\quad L \geq \dfrac{16}{3\pi^2} K \quad$ and $\quad \Pr \geq 2c_1 \operatorname{Ra} D^{3/2} \sqrt{A}, \quad$ where $\quad D = \max\{2, \dfrac{M}{L}\}, A = |\Omega|,$

where c_1 is a constant independent of the problem data. To prove the existence of the global attractor for the three-dimensional problem we use Theorem 3.3 with $X = H \times E_3 \times E_1$ and Y being a complete metric space, a subset of $Z = V \times W_3 \times W_1$ equipped with the norm topology. More precisely,

$$Y = \overline{\bigcup_{t \geq t_1(B_1)} S(t) B_1}^{Z},$$

where B_1 is a bounded set in Z absorbing bounded sets in X for the multivalued map defined below. We can prove that $Y \subset B_1$. Note that as B_1 is Z-bounded, then Y, considered as a metric space, is also bounded in the metric given by the norm of Z. We will use Theorem 3.3 with the whole space Y as B_0.

To this end, we first need to show that the multivalued map

$$S(t)(u_0, \gamma_0, \theta_0) = \{(u(t), \gamma(t), \theta(t)) : u, \gamma, \theta \text{ is a weak solution}$$
$$\text{given by Definition 3.8 with initial data } (u_0, \gamma_0, \theta_0) \in X\}, \tag{4.1}$$

is a multivalued eventual semiflow, and that it satisfies the dissipativity and asymptotic compactness properties of Theorem 3.3. There holds the following lemma.

Lemma 4.1 [27] *The family of multivalued mappings $\{S(t)\}_{t \geq 0}$ defined by (4.1) is a multivalued eventual semiflow.*

The next lemma states that the restriction of any weak solution to the interval $[t, \infty)$, where $t \geq t_1(B)$ is in fact strong.

Lemma 4.2 [27] *Let (u, γ, θ) be the weak solution with the initial data $(u_0, \gamma_0, \theta_0) \in B \in \mathcal{B}(X)$. For every $t \geq t_1(B)$ the translated restrictions*

$$(u(\cdot + t), \gamma(\cdot + t), \theta(\cdot + t))|_{[0,\infty)}$$

are strong solutions.

The next lemma ensures that Y is absorbing, and hence it can be used as B_0 in Theorem 3.3.

Lemma 4.3 [27] *For every $B \subset \mathcal{B}(X)$ there exists $t_0 = t_0(B)$ such that*

$$\bigcup_{t \geq t_0} S(t) B \subset Y.$$

The proof of the asymptotic compactness is based on the energy equation method, see for instance [4]. Note that in Theorem 3.3 it is only sufficient to obtain the asymptotic compactness for the initial data in the absorbing set B_0. The following lemma asserts the asymptotic compactness for the initial data in any set which is bounded in X.

Lemma 4.4 *Assume that $B \in \mathcal{B}(X)$ and $(w_n, \xi_n, \eta_n) \in S(t_n) B$ with a sequence $t_n \to \infty$. Then the sequence (w_n, ξ_n, η_n) is relatively compact in Y, i.e. w_n is relatively compact in V, ξ_n is relatively compact in W_3, and η_n is relatively compact in W_1.*

Thus, Theorem 3.3, together with Lemmas 4.1, 4.3, and 4.4 implies the existence of the (X, Y)-global attractor $\mathcal{A}$.

Theorem 4.5 *The multivalued eventual semiflow $\{S(t)\}_{t \geq 0}$ defined by (4.1) has a (X, Y)-global attractor $\mathcal{A}$.*

Invariance of the global attractor $\mathcal{A}$. We start from the observation which follows from Lemma 4.2 and the weak-strong uniqueness property obtained in Lemma 3.12.

Lemma 4.6 *Let (u, γ, θ) be a weak solution given by Definition 3.8. There exists t_0, which can be chosen uniformly with respect to bounded sets in X of initial data, such that for every $t \geq t_0$, this solution restricted to $[t, \infty)$ is in fact strong. Moreover, for any $t \geq t_0$ and any $s \geq 0$ the set $S(s)(u(t), \gamma(t), \theta(t))$ is a singleton contained in Y.*

Note that in the above result the time $t_0(B)$ can be made exactly the same as in Lemma 4.3. The following lemma ensures the continuous dependence of the strong solutions on the initial data.

Lemma 4.7 *Let $(u^1, \gamma^1, \theta^1)$ and $(u^2, \gamma^2, \theta^2)$ be two strong solutions such that*

$$\|(u^1(t), \gamma^1(t), \theta^1(t))\|_Z \leq C \quad \text{for} \quad t \in [0, T], \tag{4.2}$$

$$\|(u^2(t), \gamma^2(t), \theta^2(t))\|_Z \leq C \quad \text{for} \quad t \in [0, T]. \tag{4.3}$$

Then there exists a constant R dependent only on C and T such that

$$\|(u^1(t), \gamma^1(t), \theta^1(t)) - (u^2(t), \gamma^2(t), \theta^2(t))\|_Z \leq R\|(u^1(0), \gamma^1(0), \theta^1(0)) - (u^2(0), \gamma^2(0), \theta^2(0))\|_Z,$$

for every $t \in [0, T]$.

The multivalued eventual semiflow given by weak solutions is in fact a single valued semiflow in Y when we restrict it to Y. We have

Lemma 4.8 *Let $(u_0, \gamma_0, \theta_0) \in Y$. There exists a strong solution $(u(t), \gamma(t), \theta(t))$ with the initial data $(u_0, \gamma_0, \theta_0)$ which is also a unique weak solution, and for every $t \geq 0$ there holds $(u(t), \gamma(t), \theta(t)) \in Y$. In consequence, $S(t)$ restricted to Y is a single valued semiflow and has values in Y.*

Having the above results we can prove the invariance of the attractor.

Lemma 4.9 *Let $\mathcal{A}$ be the (X, Y)-global attractor obtained in Theorem 4.5. For every $t \geq 0$ there holds $S(t)\mathcal{A} = \mathcal{A}$, and $S(t)$ is a single valued semigroup on $\mathcal{A}$.*

Proof As $\mathcal{A}$ is a compact set contained in Y, and by Lemma 4.8, $S(t)$ is single-valued on Y, it also must be that $S(t)$ is single valued on $\mathcal{A}$. The fact that on Y the multivalued maps $S(t)$ are actually governed by the strong solutions and have valued in Y implies that $S(t)|_Y$ is a semigroup. Since Y is a bounded set in Z, Lemma 4.7 implies that $S(t)|_Y$ are (Y, Y) continuous maps. Hence, Theorem 3.4 implies that $\mathcal{A}$ is an invariant set. The proof is complete. □

Remark 4.10 We have only proved that the attractor $\mathcal{A}$ is a compact set in the topology of $Z = V \times W_3 \times W_1$. We note that it is possible to continue the bootstrapping argument, which would lead us to further regularity of the attractor $\mathcal{A}$ in higher order Sobolev spaces. It is also possible, proceeding in a now well established way to get its finite dimension. This would give an example of a problem without known solution uniqueness, which has the attractor of finite dimension. The solutions are unique on the attractor, but, in contrast to examples of [30], the semigroup does not instantaneously enter the regime where the solution has to stay unique. Rather than that, such regime is entered after some time uniformly with respect to bounded sets of initial data.

Upper semicontinuous convergence of attractors in 3D. Our main aim now is the proof of the following convergence

$$\lim_{K \to 0_+} \text{dist}_X(\mathcal{A}_K, \mathcal{A}_0) = 0, \quad (4.4)$$

under the assumptions that the parameters Pr, Ra, M, L, G, A are fixed and satisfy the above assumption (H), and the parameter K varies in the interval

$$K \in \left[0, L\frac{3\pi^2}{16}\right],$$

so that (H) always holds. Let C denote a generic constant independent of K, but possibly dependent on parameters Pr, Ra, M, L, G, A. Moreover let $L\frac{3\pi^2}{16} = K_{max}$. For every $K \in [0, K_{max}]$ let $\mathcal{A}_K$ denote the corresponding global attractor, which exists by Theorem 4.5 and is invariant by Lemma 4.9. The corresponding semigroup is $\{S_K(t)\}_{t \geq 0}$. Note that while the Banach spaces $X = H \times E_3 \times E_1$ and $Z = V \times W_3 \times W_1$ are independent of K, the metric space Y is dependent on K and hence it will be denoted by Y^K.

To prove the Hausdorff upper-semicontinuous convergence (4.4) it suffices to prove that the global attractors $\mathcal{A}_K$ converge to $\mathcal{A}_0$ upper semicontinuously in Kuratowski sense and then use the relation between these convergences for a family of the uniformly bounded nonempty compact sets, given in Proposition 4.13 below.

Let us recall the two notions of convergence and their mutual relations.

Definition 4.11 Let (X, ϱ) be a metric space and let $\{A_K\}_{K \in [0, K_{max}]}$ be sets in X. We say that the family $\{A_K\}_{K \in (0, K_{max}]}$ converges to A_0 upper-semicontinuously in Hausdorff sense if
$$\lim_{K \to 0^+} \text{dist}_X(A_K, A_0) = 0.$$

Definition 4.12 Let (X, ϱ) be a metric space and let $\{A_K\}_{K \in [0, K_{max}]}$ be sets in X. We say that the family $\{A_K\}_{K \in (0, K_{max}]}$ converges to A_0 upper-semicontinuously in Kuratowski sense if
$$X - \limsup_{K \to 0^+} A_K \subset A_0,$$

where $X - \limsup_{K \to 0^+} A_K$ is the Kuratowski upper limit defined by

$$X - \limsup_{K \to 0^+} A_K = \{x \in X : \lim_{n \to \infty} \rho(x_n, x) = 0, x_n \in A_{K_n}, K_n \to 0 \text{ as } n \to \infty\}.$$

The following useful result holds, cf., [19, Proposition 4.7.16], whose short proof we provide for clarity.

Proposition 4.13 *Assume that the sets* $\{A_K\}_{K \in [0, K_{max}]}$ *are nonempty and compact and the set* $\bigcup_{K \in (0, K_{max}]} A_K$ *is relatively compact. If the family* $\{A_K\}_{K \in (0, K_{max}]}$ *converges to* A_0 *upper-semicontinuously in Kuratowski sense then* $\{A_K\}_{K \in (0, K_{max}]}$ *converges to* A_0 *upper-semicontinuously in Hausdorff sense.*

Proof Since all A_K are compact sets then for every K there exist $x_K \in A_K$ such that $\text{dist}_X(A_K, A_0) = \text{dist}_X(x_K, A_0)$. Let $K_n \to 0$ be a sequence with a subsequence K_τ. As $\bigcup_{K \in (0, K_{max}]} A_K$ is relatively compact, there exists $x \in X$ such that $x_{K_\tau} \to x$, for another subsequence, where by the Kuratowski upper semicontinuous convergence there must hold $x \in A_0$. So,

$$\text{dist}_X(A_{K_\tau}, A_0) = \text{dist}_X(x_{K_\tau}, A_0) \leq \rho(x_{K_\tau}, x) \to 0.$$

Hence $\text{dist}_X(A_{K_n}, A_0) \to 0$ for the whole sequence K_n and the assertion is proved. □

Returning to our hydrodynamical problem, in [27] we have established the following theorem.

Theorem 4.14 *The global attractors $\mathcal{A}_K$ converge to $\mathcal{A}_0$ upper semicontinuously in Kuratowski sense.*

In consequence (4.4) holds.

Relation between $\mathcal{A}_0$ and the attractor for the Newtonian fluid. It is natural to ask about the relation between the global attractor for the Rayleigh–Bénard problem for the Newtonian fluid and the $\Pi_{(u,\theta)}$ projection of the global attractor $\mathcal{A}_0$ on variables (u, θ). In [27] we have proved the following property of the global attractor $\mathcal{A}_0$.

Theorem 4.15 *If $(u^0, \gamma^0, \theta^0)$ belong to $\mathcal{A}_0$ then $\gamma^0 = 0$.*

Moreover,

Theorem 4.16 *Let $\Pr \geq 4\sqrt{2}c_1 \mathrm{Ra}\sqrt{A}$ and let M, L be any nonnegative numbers such that $M \leq 2L$. The projection $\Pi_{(u,\theta)}\mathcal{A}_0$ is the $(H \times E_1, \Pi_{(u,\theta)}Y^0)$-global attractor for the three-dimensional Rayleigh–Bénard problem for the Newtonian fluid, which is furthermore the invariant set.*

We remark that for Newtonian fluids governed by equations (3.5), (3.6), and (3.8) we defined the weak solutions as the limit of u-Galerkin approximative problems analogously to Definition 3.8, and the strong solutions analogously to Definition 3.10 with the equation for γ removed from the system. This allowed us to define the appropriate $(H \times E_1, \Pi_{(u,\theta)}Y^0)$-global attractor for multivalued eventual semiflow governed by the weak solutions of the problem for the Newtonian fluid, where the semiflow, denoted by $\{S_{NEWT}(t)\}_{t \geq 0}$, $S_{NEWT}(t) : H \times E_1 \to \mathcal{P}(H \times E_1)$, associates to the initial data the value of the weak solution at time t.

5 The Thermomicropolar Model

In this section we consider an extension of the model (1.3)–(1.6) introduced in Sect. 1. We name it *thermomicropolar* and confine ourselves here to its two-dimensional case. It differs only slightly from system (2.1)–(2.4), however this difference makes the system mathematically much more involved. It reads [30],

$$\frac{1}{\Pr}(u_t + (u \cdot \nabla)u) - (1+K)\Delta u + \frac{1}{\Pr}\nabla p = 2K \operatorname{rot} \gamma + e_2 \mathrm{Ra} T, \quad (5.1)$$

$$\operatorname{div} u = 0, \quad (5.2)$$

$$\frac{M}{\Pr}(\gamma_t + u \cdot \nabla \gamma) - L\Delta \gamma + 4K\gamma = 2K \operatorname{rot} u, \quad (5.3)$$

$$T_t + u \cdot \nabla T - \Delta T = D \operatorname{rot} \gamma \cdot \nabla T. \quad (5.4)$$

The new term which produces problems is the one on the right-hand side of (5.4). Trying to define a weak solution as in Sect. 1 we see that it is impossible as the new term does not belong to L^2 over the spatial domain and so we cannot test equation (5.4) by a function from W, and also in consequence obtain the uniqueness result by using the energy method. In fact, the situation for the two-dimensional thermomicropolar model is similar to that for the three-dimensional micropolar one in that we do not know if the weak solutions are unique but know this for the strong solutions, that the weak solution become strong, and that the global attractors for the associated multi and single-valued semiflows coincide. The setting below comes from [30].

We recall some useful definitions and results from the theory of *m-semiflows*, cf. [16, 17, 28, 40]. In the following, X is a Banach space. Let $\mathcal{B}(X)$ denote the family of nonempty and bounded subsets of X.

Definition 5.1 A family $\{S(t)\}_{t\geq 0}$ of multivalued maps $S(t) : X \to 2^X \setminus \{\emptyset\}$ is an m-semiflow if

(i) For any $x \in X$ we have $S(0)x = \{x\}$.
(ii) For any $s, t \geq 0$ and $x \in X$ we have $S(t+s)x \subset S(t)S(s)x$.

A set $B_0 \in \mathcal{B}(X)$ is *absorbing* if for every bounded set $B \subset X$ there exists $t_B \geq 0$ such that
$$\bigcup_{t \geq t_B} S(t)B \subset B_0.$$

As customary, the main object of study is the so-called global attractor whose attraction property is defined in terms of the Hausdorff semidistance in X.

Definition 5.2 The set $\mathcal{A} \subset X$ is a global attractor for an m-semiflow $\{S(t)\}_{t\geq 0}$ if

(i) $\mathcal{A}$ is a compact set in X.
(ii) $\mathcal{A}$ uniformly attracts all bounded sets in X, i.e.,
$$\lim_{t\to\infty} \text{dist}_X(S(t)B, \mathcal{A}) = 0 \quad \text{for every} \quad B \in \mathcal{B}(X).$$

(iii) $\mathcal{A}$ is the smallest (in the sense of inclusion) closed set which has the property (ii).

Note that we do not impose *any continuity or closed graph type condition* on $S(t)$. This way, the attractor will be the minimal closed attracting set, but it does not have to be invariant (neither positively nor negatively semi-invariant), see [8, 17]. In our case, the following sufficient condition for the existence of the global attractor is enough. The theorem has been proved in [17, Proposition 4.2], where more general formalism of pullback attractors is considered.

Theorem 5.3 *If the m-semiflow $\{S(t)\}_{t\geq 0}$ possesses a compact absorbing set B_0 then it has a global attractor $\mathcal{A}$.*

Clearly we have $\mathcal{A} \subset B_0$ as B_0 is the compact absorbing (and hence also attracting) set.

We also briefly recall some notions concerning singe-valued semiflows and their attractors. Note that, in contrast to the multivalued case, we include the continuity in the definition of the semiflow.

Definition 5.4 A family $\{S(t)\}_{t \geq 0}$ of maps $S(t) : X \to X$ is a semiflow if

(i) For any $x \in X$ we have $S(0)x = x$.
(ii) For any $s, t \geq 0$ and $x \in X$ we have $S(t+s)x = S(t)S(s)x$.
(iii) For any $t \geq 0$ the mapping $S(t)$ is continuous.

Definition 5.5 The set $\mathcal{A} \subset X$ is a global attractor for a semiflow $\{S(t)\}_{t \geq 0}$ if

(i) $\mathcal{A}$ is a compact set in X.
(ii) $\mathcal{A}$ uniformly attracts all bounded sets in X, i.e.,

$$\lim_{t \to \infty} \text{dist}_X(S(t)B, \mathcal{A}) = 0 \quad \text{for every} \quad B \in \mathcal{B}(X).$$

(iii) $\mathcal{A}$ is invariant, i.e., $S(t)\mathcal{A} = \mathcal{A}$ for all $t \geq 0$.

The following result on the existence of a global attractor is classical, cf., [36, 44, 49].

Theorem 5.6 *If the semiflow* $\{S(t)\}_{t \geq 0}$ *possesses a compact absorbing set* B_0*, then it has a global attractor* $\mathcal{A}$*.*

It is well known that $\mathcal{A}$ is the smallest closed attracting set, so we must have $\mathcal{A} \subset B_0$, as B_0 is closed and attracting.

In [30] we have proved the following results.

Theorem 5.7 *Let* $u_0 \in V$, $\omega_0 \in W$, $\theta_0 \in W$ *and* $T > 0$. *There exists a unique strong solution* (u, ω, θ). *Moreover, the strong solution depends continuously on the initial data, namely, for each* $t \geq 0$, *the following map is continuous*

$$\mathcal{V} = V \times W \times W \ni (u_0, \omega_0, \theta_0) \mapsto (u(t), \omega(t), \theta(t)) \in \mathcal{V}$$

The uniqueness of the strong solution holds also in the class of weak solutions. Moreover if $(u_0, \omega_0, \theta_0) \in H \times E \times E = \mathcal{H}$ *and* (u, ω, θ) *is the weak solution with the initial data* $(u_0, \omega_0, \theta_0)$, *then for any* $\epsilon > 0$ *this solution restricted to* $[\epsilon, T]$ *is strong.*

Theorem 5.8 *Weak solutions generate an m-semiflow* $\{S_\mathcal{H}(t)\}_{t \geq 0}$ *and strong solutions generate a semiflow* $\{S_\mathcal{V}(t)\}_{t \geq 0}$. *We have* $S_\mathcal{V}(t) = S_\mathcal{H}(t)|_\mathcal{V}$ *for all* $t \geq 0$. *There exists an invariant set* $\mathcal{A}$ *compact in* $\mathcal{V}$ *and bounded in* $D(-P\Delta) \times D(-\Delta) \times D(-\Delta)$ *which is the global attractor both for* $\{S_\mathcal{H}(t)\}_{t \geq 0}$ *and* $\{S_\mathcal{V}(t)\}_{t \geq 0}$.

For the details we refer the reader to [30].

6 Comments and Bibliographical Notes

In this place we would like to recapitulate the results presented in previous sections and make some remarks on related problems.

In Sects. 2 and 5 we consider two-dimensional problems, and in Sects. 3 and 4 the related three-dimensional problem.

While for two-dimensional problems the weak and strong solutions exist globally with respect to time for all initial conditions from the considered functional spaces, and the strong solutions are unique, in the three-dimensional problem we only know that the strong unique solution exists for some smaller set of initial conditions, including the global attractor. This situation touches the famous millennium problem for the Navier-Stokes equations. Note that for the micropolar model the weak solution is uniquely determined by initial conditions whereas for the thermomicropolar model the uniqueness issue is open.

As concerns the time asymptotic of solutions we have proved the existence of corresponding global attractors, which is a consequence of the dissipativity of the considered dynamical systems. Moreover, the global attractors for weak and strong solutions coincide. This would suggest that the global attractor, which is a compact and invariant subset of the phase space, is in some sense the core of the system, revealing its behaviour after a large time of evolution. While evidently no one would oppose it, a known shadowing theorem and simple examples, cf. [44], show that for individual trajectories the situation is more complicated. In particular it is not true that for a given trajectory converging to the global attractor there exists one lying on the attractor to which it eventually converges. The mentioned theorem assures that such situation holds but only locally, the closer is a trajectory to the attractor, the longer is the time interval on which it has its "shadow" on the attractor.

We can prove also that the corresponding semigroups or multivalued semigroups become invertible on the global attractor $\mathcal{A}$, with $S(-t) = S(t)^{-1}$ for $t > 0$ (for the backward uniqueness property cf. [49]). From the continuity of the operators $S(t)$ for $t > 0$ and compactness of the global attractor it follows then that $\{S(t) : -\infty < t < \infty\}$ is the group of homeomorphisms on the compact metric space $\mathcal{A}$, and so we are in the framework of the theory of topological dynamics, cf. [42].

As to our results about structural stability of the considered systems, we proved that the trajectories of the perturbed and original systems starting from the same initial condition stay close to each other of distance less than ϵ on any bounded interval $[0, t]$ if only the perturbation parameter $K = K(t, \epsilon)$ is small enough. Moreover, the corresponding global attractors $\mathcal{A}_K$ converge upper-semicontinuously to the attractor $\mathcal{A}_0$ when $K \to 0$.

At the end we would like to make some remarks on infinite dimensional dynamical systems without uniqueness and also on nonautonomous ones.

There are several mathematical formalisms useful to study the global attractors for autonomous evolution problems without uniqueness of solutions. Among them there are generalized semiflows introduced and studied by Ball [4], multivalued semiflows (m-semiflows) anticipated by Babin and Vishik [2] and introduced and studied by

Melnik and Valero in [40] and [41], trajectory attractors introduced independently by Chepyzhov and Vishik [9], Málek and Nečas [38] and Sell [46], evolutionary systems introduced by Cheskidov and Foiaş and [12] and later extended by Cheskidov in [11].

As concerns the attractors for the non-autonomous dynamical systems the first attempts to extend the notion of global attractor to the non-autonomous case led to the concept of the so-called uniform attractor (see [10]). The conditions ensuring the existence of the uniform attractor parallel those for autonomous systems. In this theory non-autonomous systems are lifted in [50] to autonomous ones by expanding the phase space. Then, the existence of uniform attractors relies on some compactness property of the solution operator associated to the resulting system. One disadvantage of uniform attractors is that they do not need to be invariant unlike the global attractor for autonomous systems. Moreover, they demand rather strong conditions on the time-dependent data.

At the same time, the theory of pullback (or cocycle) attractors has been developed for both the non-autonomous and random dynamical systems (see [18, 32, 33, 45]), and has shown to be very useful in the understanding of the dynamics of non-autonomous dynamical systems.

More information on non-autonomous infinite dimensional dynamical systems can be found in monographs [7, 31] and the review article [3]. Two-dimensional problems with delay have been studied in [6, 39]. In all the above mentioned theories hydrodynamic played a principal role being not only one of the fundamental examples from physics but, in view of its mathematical intricacies, the important challenge to develop them.

Acknowledgments We thank anonymous reviewers for their careful reading and valuable comments. This work was supported by National Science Center (NCN) of Poland under projects No. DEC-2017/25/B/ST1/00302 and UMO-2016/22/A/ST1/00077.

References

1. Babin, A.V., Vishik, M.I.: Attractors of Evolution Equations. North Holland, Amsterdam, London, New York, Tokyo (1992)
2. Babin, A.V., Vishik, M.I.: Maximal attractors of semigroups corresponding to evolution differential equations. Math. USSR Sb. **126**, 397–419 (1985)
3. Balibrea, F., Caraballo, T., Kloeden, P.E., Valero, J.: Recent developments in dynamical systems: three perspectives. Int. J. Bifurcat. Chaos **20**, 2591–2636 (2010)
4. Ball, J.M.: Continuity properties and global attractors of generalized semiflows and the Navier-Stokes equations. J. Nonlinear Sci. **7**, 475–502 (1997)
5. Birkhoff, G.: Hydrodynamics. A Study in Fact Logic and Similitude. Princeton University Press (1960)
6. Caraballo, T., Real, J.: Asymptotic behaviour of two-dimensional Navier-Stokes equations with delays. P. Roy. Soc. Lond. A Math. **459**, 3181–3194 (2003)
7. Carvalho, A.N., Langa, J.A., Robinson, J.C.: Attractors for Infinite-dimensional Non-autonomous Dynamical Systems. Springer (2013)
8. Chepyzhov, V.V., Conti, M., Pata, V.: A minimal approach to the theory of global attractors. Discret. Contin. Dyn. Syst. **32**, 2079–2088 (2012)

9. Chepyzhov, V.V., Vishik, M.I.: Trajectory attractors for evolution equations. CR. Acad. Sci. I-Math. **321**, 1309–1314 (1995)
10. Chepyzhov, V.V., Vishik, M.I.: Attractors for Equations of Mathematical Physics. American Mathematical Society, Providence, RI (2002)
11. Cheskidov, A.: Global attractors of evolution systems. J. Dyn. Differ. Equ. **21**, 249–268 (2009)
12. Cheskidov, A., Foiaş, C.: On global attractors of the 3D Navier-Stokes equations. J. Differ. Equ. **231**, 714–754 (2006)
13. Constantin, P., Doering, C.R.: Heat transfer in convective turbulence. Nonlinearity **9**, 1049–1060 (1996)
14. Constantin, P., Doering, C.R.: Variational bounds on energy dissipation in incompressible flows. III. Convection. Phys. Rev. E **53**, 5957–5981 (1996)
15. Constantin, P., Foias, C., Temam, R.: On the dimension of the attractors in two-dimensional turbulence. Physica D: Nonlinear Phenom. **30**, 284–296 (1988)
16. Coti Zelati, M.: On the theory of global attractors and Lyapunov functionals. Set-Valued Var. Anal. **21**, 127–149 (2013)
17. Coti Zelati, M., Kalita, P.: Minimality properties of set-valued processes and their pullback attractors. SIAM J. Math. Anal. **47**, 1530–1561 (2015)
18. Crauel, H., Debussche, A., Flandoli, F.: Random attractors. J. Dyn. Differ. Equ. **9**(2), 307–341 (1997)
19. Denkowski, Z., Migórski, S., Papageorgiou, N.S.: An Introduction to Nonlinear Analysis: Theory. Kluwer Academic Publishers (2003)
20. Doering, C.R., Gibbon, J.D.: Applied Analysis of the Navier-Stokes equations. Cambridge Texts in Applied Mathematics (1995)
21. Eringen, A.C.: Theory of micropolar fluids. J. Math. Mech. **16**, 1–16 (1966)
22. Foias, C., Manley, O., Rosa, R., Temam, R.: Navier–Stokes Equations and Turbulence. Cambridge University Press (2001)
23. Foias, C., Manley, O., Temam, R.: Attractors for the Bénard problem: existence and physical bounds on their fractal dimensions. Nonlinear Anal. Theory Methods Appl. **11**, 939–967 (1987)
24. Hale, J.: Asymptotic Behavior of Dissipative Systems. American Mathematical Society (1988)
25. Kadanoff, L.P.: Turbulent heat flow: structures and scaling. Phys. Today **54**, 34–39 (2001)
26. Kalita, P., Langa, J., Łukaszewicz, G.: Micropolar meets Newtonian. The Rayleigh-Bénard problem. Physica D **392**, 57–80 (2019)
27. Kalita, P., Łukaszewicz, G.: Micropolar meets Newtonian in three dimensions. The Raylcigh Bénard problem for large Prandtl numbers. arXiv:1902.07765
28. Kalita, P., Łukaszewicz, G.: Global attractors for multivalued semiflows with weak continuity properties. Nonlinear Anal. **101**, 124–143 (2014)
29. Kalita, P., Łukaszewicz, G., Siemianowski, J.: On Rayleigh-Bénard problem for thermomicropolar fluid. Topol. Methods Nonlinear Anal. **52**, 477–514 (2018)
30. Kalita, P., Łukaszewicz, G., Siemianowski, J.: On relation between attractors for single and multivalued semiflows for a certain class of PDEs. Discret. Contin. Dyn. Syst. B **24**, 1199–1227 (2019)
31. Kloeden, P.E., Rasmussen, M.: Nonautonomous Dynamical Systems, Mathematical Surveys and Monographs, vol. 176. American Mathematical Society (1987)
32. Kloeden, P.E., Schmalfuß, B.: Asymptotic behaviour of nonautonomous difference inclusions. Syst. Control Lett. **33**(4), 275–280 (1998)
33. Langa, J.A., Schmalfuß, B.: Finite dimensionality of attractors for non-autonomous dynamical systems given by partial differential equations. Stoch. Dynam. **4**(3), 385–404 (2004)
34. Łukaszewicz, G.: Micropolar Fluids - Theory and Applications. Birkhäuser Basel (1999)
35. Łukaszewicz, G.: Long time behavior of 2D micropolar fluid flows. Math. Comput. Model. **34**, 487–509 (2001)
36. Łukaszewicz, G., Kalita, P.: Navier–Stokes Equations. An Introduction with Applications. Springer (2016)
37. Ma, T., Wang, S.: Dynamic bifurcation and stability in the Rayleigh-Bénard convection. Commun. Math. Sci. **2**, 159–193 (2004)

38. Málek, J., Nečas, J.: A finite-dimensional attractor for three-dimensional flow of incompressible fluids. J. Differ. Equ. **127**, 498–518 (1996)
39. Marín-Rubio, P., Real, J., Valero, J.: Pullback attractors for a two-dimensional Navier–Stokes model in an infinite delay case. Nonlinear Anal.-Theor. **74**, 2012–2030 (2011)
40. Melnik, V.S., Valero, J.: On attractors of multivalued semiflows and differential inclusions. Set-Valued Anal. **6**, 83–111 (1998)
41. Melnik, V.S., Valero, J.: Addendum to "On Attractors of Multivalued Semiflows and Differential Inclusions" [Set-Valued Anal. **6**, 83–111]. Set-Valued Anal. **16**(2008), 507–509 (1998)
42. Nemytskii, V.V., Stepanow, V.V.: Qualitative Theory of Differential Equations. Princeton U.P. (1960)
43. Raugel, G.: Global attractors in partial differential equations. Handbook of Dynamical Systems, vol. 2, pp. 887–982. North-Holland (2002)
44. Robinson, J.C.: Infinite-Dimensional Dynamical Systems. Cambridge University Press, Cambridge, UK (2001)
45. Schmalfuß, B.: Attractors for non-autonomous dynamical systems. In: Fiedler, B., Gröger, K., Sprekels, J., (eds.) Proceedings of Equadiff, vol. 99, pp. 684–689. World Scientific, Berlin (2000)
46. Sell, G.R.: Global attractors for the three dimensional Navier-Stokes equations. J. Dyn. Differ. Equ. **8**, 1–33 (1996)
47. Tarasińska, A.: Global attractor for heat convection problem in a micropolar fluid. Math. Meth. Appl. Sci. **29**, 1215–1236 (2006)
48. Tarasińska, A.: Deterministic and statistical solutions of micropolar fluid equations, PhD Thesis, University of Warsaw (2010)
49. Temam, R.: Infinite Dimensional Dynamical Systems in Mechanics and Physics. Springer (1997)
50. Vishik, M.I.: Asymptotic Behaviour of Solutions of Evolutionary Equations. Cambridge University Press, Cambridge (1992)

Method of Lines for a Kinetic Equation of Swarm Formation

Adrian Karpowicz and Henryk Leszczyński

Abstract Kinetic equations with drift and without drift are approximated by the method of lines. Its stability is proved in $L^1 \cap L^\infty$. Because integrals are calculated over unbounded domains, we apply Gauss-Hermite quadratures.

Keywords Kinetic equation · Method of lines · Gauss-Hermite quadrature · Stability

2010 Mathematics Classication 47G20, 65M12 (65M20)

1 Introduction

Let $f = f(t, x, v)$ be a probability density of individuals at $t \geq 0$ at position $x \in \mathbb{R}^d$ and with velocity of individual $v \in \mathbb{R}^d$. The evolution of populations at the mesoscopic scale is defined by the nonlinear integro-differential Boltzmann–like equation, see Parisot, Lachowicz (2015)

$$\frac{\partial f}{\partial t}(t, x, v) + v \cdot \nabla_x f(t, x, v) = \frac{1}{\epsilon} Q[f](t, x, v) \qquad (1)$$
$$= \frac{1}{\epsilon} \int_{\mathbb{R}^d} (T[f(t, x, \cdot)](w, v) f(t, x, w) - T[f(t, x, \cdot)](v, w) f(t, x, v)) \, dw$$

with the initial data $f(0, x, v) = f_0(x, v)$. Many numerical methods for kinetic equations are applied in the literature and practice. Let us mention [1, 5]. Our numerical approach is different from typical numerics, because we use a semidiscretisation in

---

A. Karpowicz · H. Leszczyński (✉)
Institute of Mathematics, University of Gdańsk, Wita Stwosza 57, 80-308 Gdańsk, Poland
e-mail: hleszcz@mat.ug.edu.pl

A. Karpowicz
e-mail: Adrian.Karpowicz@mat.ug.edu.pl

© Springer Nature Switzerland AG 2020
J. Banasiak et al. (eds.), *Semigroups of Operators – Theory and Applications*,
Springer Proceedings in Mathematics & Statistics 325,
https://doi.org/10.1007/978-3-030-46079-2_14

knots being zeros of orthogonal polynomials. The numerical method of lines (MOL) is a technique for solving partial differential equations by discretising in all but one dimension. In this paper we study MOL for some special case of Eq. (1). We use roots of Hermite's polynomials to construct of the mesh and Gauss-Hermite quadrature to approximate of integrals. The Gauss-Hermite quadrature is based on the formula

$$\int_{-\infty}^{+\infty} e^{-v^2} f(v) dv \approx \sum_{i=1}^{N} A_i f(v_i),$$

where v_i for $i = 1, \ldots, N$ are the roots of the Hermite polynomial $H_N(v)$. It is known that $A_i > 0$ and $\sum_{i=1}^{N} A_i = \sqrt{\pi}$. Hermite's polynomials are orthogonal with respect to the weight function e^{-v^2}, i.e. we have

$$\int_{\mathbb{R}} e^{-v^2} H_n(v) H_m(v) \, dv = 0 \quad \text{for} \quad n \neq m.$$

2 The Space Homogeneous Case

In this section we focus on the space homogeneous case and the positive gregarious case, i.e. all functions and parameters are assumed to be independent of x, $\sigma = 1$, $\gamma > 1$. The main Eq. (1) reads

$$\partial_t f(t, v) = (\beta * f)(t, v) f^\gamma(t, v) - (\beta * f^\gamma)(t, v) f(t, v), \qquad (2)$$
$$f(0, v) = f_0(v). \qquad (3)$$

for $t \in [0, T)$ and $v \in \mathbb{R}$. By $(\beta * f)$ we denote the convolution-like product in $\mathbb{R}$, i.e.

$$(\beta * f)(t, v) = \int_{\mathbb{R}} \beta(v, w) f(t, w) \, dw.$$

We say that $f \in L^\infty([0, T] \times \mathbb{R}) \cap L^1([0, T] \times \mathbb{R}) \cap C([0, T] \times \mathbb{R})$ is a solution of problem (2), (3) if it satisfies the integral equation

$$f(t, v) = f_0(v) + \int_0^t \left[(\beta * f)(s, v) f^\gamma(s, v) - (\beta * f^\gamma)(s, v) f(s, v) \right] ds.$$

Method of Lines for a Kinetic Equation of Swarm Formation

Assumption 1 Suppose that $\beta \in L^\infty(\mathbb{R}^2) \cap C(\mathbb{R}^2)$, $\beta(v, w) = \beta(w, v)$ and $0 \leq \beta(v, w) \leq \|\beta\|_{L^\infty}$, where $\|\cdot\|_{L^\infty}$ is the supremum norm on $\mathbb{R}^2$.

In Ref. [2] (Theorem 2.5) a local in time existence and uniqueness result was proved and the existence time was estimated from below. It is proved there that if β satisfies Assumption 1 and the probability density f_0 is in $L^\infty(\mathbb{R})$, there exist $T > 0$ and a unique solution of problem (2), (3) in $C^1([0, T]; L^\infty(\mathbb{R}))$. The solution is a probability density.

Consider problem (2), (3). The first step is a discretisation of the variable v. For this purpose, one constructs a mesh $\{v_i\}_{i=1}^N$, where v_i are the roots of the Hermite polynomial $H_N(v)$. The second step is an approximation of integrals by the Gauss-Hermite quadrature. Denote $f_i(t) = f(t, v_i)$ and $\beta_{ij} = \beta(v_i, v_j)$. Then the method of lines (2), (3) can be written as follows

$$\frac{d}{dt} f_i(t) = f_i^\gamma(t) \sum_{j=1}^N A_j e^{v_j^2} \beta_{ij} f_j(t) - f_i(t) \sum_{j=1}^N A_j e^{v_j^2} \beta_{ij} f_j^\gamma(t), \quad (4)$$

$$f_i(0) = f_0(v_i), \quad (5)$$

for $i = 1, 2, \ldots, N$.

Proposition 1 Let Assumption 1 be satisfied and $f_0 \in L^\infty(\mathbb{R}) \cap L^1(\mathbb{R}) \cap C(\mathbb{R})$ is a probability density. Then there exists a unique solution of problem (4), (5) on $[0, T^*)$, where $T^* \in (0, T]$ can be chosen independent of N. Moreover, the solution is non-negative.

Proof It is easy to see that there is a unique solution of problem (4), (5) on some interval $[0, T_W)$, where $T_W \in (0, T]$. It is an elementary check that this solution preserves non-negativity of the initial datum. We prove that T_W can be chosen independent of N. The problem (4) and (5) is equivalent to the integral equations

$$f_i(t) = f_i(0) + \int_0^t \left[f_i^\gamma(s) \sum_{j=1}^N A_j e^{v_j^2} \beta_{ij} f_j(s) - f_i(s) \sum_{j=1}^N A_j e^{v_j^2} \beta_{ij} f_j^\gamma(s) \right] ds \quad (6)$$

for $i = 1, 2, \ldots, N$. Define a norm

$$\|f(t)\| = \max \left\{ \max_{i=1,\ldots,N} |f_i(t)|, \sum_{i=1}^N A_i e^{v_i^2} |f_i(t)| \right\}.$$

Then

$$|f_i(t)| \leq |f_i(0)| + \int_0^t \left| f_i^\gamma(s) \sum_{j=1}^N A_j e^{v_j^2} \beta_{ij} f_j(s) \right| ds \qquad (7)$$

$$\leq \|f(0)\| + \|\beta\|_{L^\infty} \int_0^t \|f^\gamma(s)\| \cdot \|f(s)\|_1 ds$$

$$\leq \|f_0\| + \|\beta\|_{L^\infty} \int_0^t \|f(s)\|^{\gamma+1} ds.$$

Multiplying both sides of inequality (7) by $A_i e^{v_i^2}$ and adding, we get

$$\sum_{i=1}^N A_i e^{v_i^2} |f_i(t)| \leq \sum_{i=1}^N A_i e^{v_i^2} |f_i(0)| + \sum_{i=1}^N A_i e^{v_i^2} \int_0^t \left| f_i^\gamma(s) \sum_{j=1}^N A_j e^{v_j^2} \beta(v_i, v_j) f_j(s) \right| ds$$

$$\leq \|f(0)\| + \|\beta\|_{L^\infty} \int_0^t \|f(s)\|^{\gamma-1} \cdot \|f(s)\|^2 ds$$

$$\leq \|f_0\| + \|\beta\|_{L^\infty} \int_0^t \|f(s)\|^{\gamma+1} ds.$$

Thus, we arrive at the inequality

$$\|f(t)\| \leq \|f(0)\| + \|\beta\|_{L^\infty} \int_0^t \|f(s)\|^{\gamma+1} ds. \qquad (8)$$

The norm $\|f(0)\|$ can be estimated by a constant independent of N, because

$$\sum_{i=1}^N A_i e^{v_i^2} |f_i(0)| \to \int_{-\infty}^{+\infty} f_0(v) dv = 1 \text{ as } N \to \infty.$$

Therefore $\|f(t)\|$ can be also estimated by another constant independent of N on some interval.

Now, we are ready to formulate a stability criterion for MOL. Consider the method of lines with the perturbed right-hand side

$$\frac{d}{dt} f_i(t) = f_i^\gamma(t) \sum_{j=1}^N A_j e^{v_j^2} \beta_{ij} f_j(t) - f_i(t) \sum_{j=1}^N A_j e^{v_j^2} \beta_{ij} f_j^\gamma(t) + \xi_i(t), \quad (9)$$

for $i = 1, \ldots, N$. Here $\xi = (\xi_i)_{i=1}^N \in L^\infty([0, T], \mathbb{R}^N)$ is a continuous function.

Theorem 1 *Suppose that Assumption 1 is satisfied and $f_0 \in L^\infty(\mathbb{R}) \cap L^1(\mathbb{R}) \cap C(\mathbb{R})$ is a probability density. Let f, $\bar{f}$ be solutions of (4) and (9), respectively. If*

$$\|\bar{f}(0) - f(0)\| \leqslant \varepsilon \quad \text{and} \quad \|\xi(t)\| \leqslant \varepsilon \quad (10)$$

then

$$\|\bar{f}(t) - f(t)\| \to 0 \quad \text{as} \quad N \to \infty \quad \text{and} \quad \varepsilon \to 0.$$

Proof We see from (9) and (10) that

$$\|\bar{f}(t)\| \leq \|f_0\| + \varepsilon + \|\beta\|_{L^\infty} \int_0^t \left[\|\bar{f}(s)\|^{\gamma+1} + \varepsilon\right] ds. \quad (11)$$

Let ρ be a solution of the equation

$$\rho(t) = \rho(0) + \|\beta\|_{L^\infty} \int_0^t \rho^{\gamma+1}(s) ds, \quad (12)$$

where $\rho(0) > \max\{\|f(0)\| + \varepsilon(BT+1), 1\}$. Applying the comparison theorem for ODE's and (8), (11), (12) we obtain $\|f(t)\| \leq \rho(t)$ and $\|\bar{f}(t)\| \leq \rho(t)$. From (4) and (9), we get

$$\bar{f}_i(t) - f_i(t) = \bar{f}_i(0) - f_i(0) \quad (13)$$

$$+ \int_0^t \left[\left(\bar{f}_i^\gamma(s) \sum_{j=1}^N A_j e^{v_j^2} \beta_{ij} \bar{f}_j(s) - f_i^\gamma(s) \sum_{j=1}^N A_j e^{v_j^2} \beta_{ij} f_j(s) \right) \right.$$

$$\left. - \left(\bar{f}_i(s) \sum_{j=1}^N A_j e^{v_j^2} \beta_{ij} \bar{f}_j^\gamma(s) - f_i(s) \sum_{j=1}^N A_j e^{v_j^2} \beta_{ij} f_j^\gamma(s) \right) + \xi_i(s) \right] ds$$

Applying the mean value theorem, we get

$$|\bar{f}_i^\gamma(s)\sum_{j=1}^N A_j e^{v_j^2}\beta_{ij}\bar{f}_j(s) - f_i^\gamma(s)\sum_{j=1}^N A_j e^{v_j^2}\beta_{ij} f_j(s)| \qquad (14)$$

$$= |(\bar{f}_i^\gamma(s) - f_i^\gamma(s))\sum_{j=1}^N A_j e^{v_j^2}\beta_{ij}\bar{f}_j(s) + f_i^\gamma(s)\sum_{j=1}^N A_j e^{v_j^2}\beta_{ij}(\bar{f}_j(s) - f_j(s))|$$

$$\leqslant \gamma\rho^{\gamma-1}(s)\|\bar{f}(s) - f(s)\|\|\beta\|_{L^\infty}\rho(s) + \rho^\gamma(s)\|\beta\|_{L^\infty}\|\bar{f}(s) - f(s)\|$$

Analogously we have

$$|\bar{f}_i(s)\sum_{j=1}^N A_j e^{v_j^2}\beta_{ij}\bar{f}_j^\gamma(s) - f_i(s)\sum_{j=1}^N A_j e^{v_j^2}\beta_{ij} f_j^\gamma(s)| \qquad (15)$$

$$\leqslant \gamma\rho^{\gamma-1}(s)\|\bar{f}(s) - f(s)\|\|\beta\|_{L^\infty}\rho(s) + \rho^\gamma(s)\|\beta\|_{L^\infty}\|\bar{f}(s) - f(s)\|$$

We substitute (14) and (15) in (13), we obtain

$$\left|\bar{f}_i(t) - f_i(t)\right| \leqslant \left|\bar{f}_i(0) - f_i(0)\right| + \int_0^t \left(2\|\beta\|_{L^\infty}(1+\gamma)n^\gamma(s)\|\bar{f}(s) - f(s)\| + |\xi_i(s)|\right)ds \qquad (16)$$

Multiplying both sides of inequality (13) by $A_i e^{v_i^2}$, adding them and noting estimaties (14) and (15), we get

$$\sum_{i=1}^N A_i e^{v_i^2}\left|\bar{f}_i(t) - f_i(t)\right| \leq \sum_{i=1}^N A_i e^{v_i^2}\left|\bar{f}_i(0) - f_i(0)\right| \qquad (17)$$

$$+ \sum_{i=1}^N A_i e^{v_i^2} \int_0^t \Big[\|\beta\|_{L^\infty}(\gamma+1)\rho^\gamma(s)|\bar{f}_i(s) - f_i(s)|$$

$$+ \|\beta\|_{L^\infty}(1+\gamma)\rho^{\gamma-1}(t)f_i(s)\|\bar{f}(s) - f(s)\| + |\xi_i(s)|\Big]ds$$

$$\leq \|\bar{f}(0) - f(0)\| + \int_0^t \Big[2\|\beta\|_{L^\infty}(1+\gamma)\rho^\gamma(s)\|\bar{f}(s) - f(s)\| + \|\xi(s)\|\Big]ds$$

From (16) and (17) we obtain

$$\|\bar{f}(t) - f(t)\| \leq \|\bar{f}(0) - f(0)\| + \int_0^t \Big[2\|\beta\|_{L^\infty}(\gamma+1)n^\gamma(s)\|\bar{f}(s) - f(s)\| + \|\xi(s)\|\Big]ds.$$

Consider a comparison equation

$$\omega(t) = \omega(0) + \int_0^t \left[2\|\beta\|_{L^\infty}(\gamma+1)\rho^\gamma(s)\omega(s) + \|\xi(s)\|\right] ds, \qquad (18)$$

where $\omega(0) = \varepsilon$ and ρ to satisfy (12), that is $\|\beta\|_{L^\infty}\rho^{\gamma+1}(t) = \rho'(t)$. By a comparison theorem for ODE's, we obtain

$$\|\bar{f}(t) - f(t)\| \leq \left(\omega(0) + \int_0^t \|\xi(\tau)\| d\tau\right) \rho^{2(\gamma+1)}(t),$$

that is $\|\bar{f}(t) - f(t)\| \to 0$ as $\varepsilon \to 0$.

3 The Space Non-homogeneous Case

We can apply the technique developed in previous section to the following version of Eq. (1)

$$\frac{\partial f}{\partial t}(t,x,v) + v \cdot \nabla f(t,x,v) \qquad (19)$$
$$= f^\gamma(t,x,v) \int_{\mathbb{R}^2} \beta(v,w) f(t,y,w) dy dw - f(t,x,v) \int_{\mathbb{R}^2} \beta(v,w) f^\gamma(t,y,w) dy dw$$

$$f(0,x,v) = f_0(x,v), \qquad (20)$$

where $t \in [0,T]$, $x \in \mathbb{R}$, $v \in \mathbb{R}$. Here f_0 is a probability densite on $\mathbb{R}^2$. We consider this equation along characteristics and change vaiables in integrals transforms the problem (19) and (20) into

$$\frac{df}{dt}(t, x+tv, v) = f^\gamma(t, x+tv, v) \int_{\mathbb{R}^2} \beta(v,w) f(t, y+tw, w) dy dw \qquad (21)$$

$$- f(t, x+tv, v) \int_{\mathbb{R}^2} \beta(v,w) f^\gamma(t, y+tw, w) dy dw,$$

$$f(0,x,v) = f_0(x,v). \qquad (22)$$

Consider the method of lines. The first step is a discretisation of v and x. For this purpose, fix $N > 1$ and construct meshes $\{v_i\}_{i=1}^N$ and $\{x_i\}_{i=1}^N$, where v_i and x_i are the roots of the Hermite polynomial $H_N(v)$. The second step is approximation of integrals by Gauss-Hermite quadrature. To simplify notations, we denote $f_{ik}^\#(t) =$

$f(t, x_k + tv_i, v_i)$ and $\beta_{ij} = \beta(v_i, v_j)$. Then the method of lines (21), (22) can be written as the ODE problem

$$\frac{d}{dt} f_{ik}^{\#}(t) = \left(f_{ik}^{\#}(t)\right)^{\gamma} \sum_{j,l=1}^{N} A_j A_l e^{v_j^2 + v_l^2} \beta_{ij} f_{jl}^{\#}(t) \tag{23}$$

$$- f_{ik}^{\#}(t) \sum_{j,l=1}^{N} A_j A_l e^{v_j^2 + v_l^2} \beta_{ij} \left(f_{jl}^{\#}(t)\right)^{\gamma},$$

$$f_{ik}^{\#}(0) = f_0(0, x_k, v_i). \tag{24}$$

We define norms

$$\|f^{\#}(t)\|_{\infty} = \max_{i,k} |f_{ik}^{\#}(t)|, \quad \|f^{\#}(t)\|_1 = \sum_{j,l=1}^{N} A_j A_l e^{v_j^2 + v_l^2} |f_{jl}^{\#}(t)|$$

and $\|f^{\#}(t)\| = \max\{\|f^{\#}(t)\|_{\infty}, \|f^{\#}(t)\|_1\}$. The solution of the problem (23), (31) preserves non-negativity of the initial datum. Because of this property, we get

$$f_{ik}^{\#}(t) \leq f_{ik}^{\#}(0) + \|\beta\|_{L^\infty} \int_0^t \left(f_{ik}^{\#}(s)\right)^{\gamma} \sum_{j,l=1}^{N} A_j A_l e^{v_j^2 + v_l^2} f_{jl}^{\#}(s) ds. \tag{25}$$

Hence

$$\|f^{\#}(t)\|_{\infty} \leq \|f^{\#}(0)\|_{\infty} + \|\beta\|_{L^\infty} \int_0^t \|f^{\#}(s)\|_{\infty}^{\gamma} \cdot \|f^{\#}(s)\|_1 ds.$$

Multiplying both sides of inequality (25) by $A_i A_k e^{v_i^2 + v_k^2}$ and adding them, we get

$$\|f^{\#}(t)\|_1 \leq \|f^{\#}(0)\|_1 + \|\beta\|_{L^\infty} \int_0^t \|f^{\#}(s)\|_{\infty}^{\gamma-1} \cdot \|f^{\#}(s)\|_1^2 ds.$$

By the above inequalities, we obtain

$$\|f^{\#}(t)\| \leq \|f^{\#}(0)\| + \|\beta\|_{L^\infty} \int_0^t \|f^{\#}(s)\|^{\gamma+1} ds.$$

Now, we can apply the technique developed in the previous section. We consider the following version of Eq. (1)

$$\frac{\partial f}{\partial t}(t,x,v) + v \cdot \nabla f(t,x,v) \tag{26}$$
$$= f^\gamma(t,x,v) \int_\mathbb{R} \beta(v,w) f(t,x,w) dw - f(t,x,v) \int_\mathbb{R} \beta(v,w) f^\gamma(t,x,w) dw,$$
$$f(0,x,v) = f_0(x,v), \tag{27}$$

where $t \in [0,T]$, $x \in \mathbb{R}$, $v \in \mathbb{R}$. Considering this equation along characteristics transforms problem (26) and (27) into

$$\frac{df}{dt}(t, x+tv, v) = f^\gamma(t, x+tv, v) \int_\mathbb{R} \beta(v,w) f(t, x+tv, w) dw \tag{28}$$
$$- f(t, x+tv, v) \int_\mathbb{R} \beta(v,w) f^\gamma(t, x+tv, w) dw,$$
$$f(0, x, v) = f_0(x, v). \tag{29}$$

Consider again the method of lines. To simplify notations, we put $f_i^\#(t,x) = f(t, x+tv_i, v_i)$ and $\beta_{ij} = \beta(v_i, v_j)$. Then the method of lines (28), (29) can be written as follows

$$\frac{d}{dt} f_i^\#(t,x) = \left(f_i^\#(t,x)\right)^\gamma \sum_{j=1}^N A_j e^{v_j^2} \beta_{ij} f_j^\#(t, x+t(v_i-v_j)) \tag{30}$$
$$- f_i^\#(t,x) \sum_{j=1}^N A_j e^{v_j^2} \beta_{ij} \left(f_j^\#(t, x+t(v_i-v_j))\right)^\gamma,$$
$$f_i^\#(0,x) = f_0(0, x, v_i), \tag{31}$$

This method of lines is much more involving than previous cases. We put off its analysis to a further research.

4 Numerical Experiment

We apply the results presented in Sect. 2 to a problem (2), (3), where $\beta = 1$ and f_0 is a density of probability. In this case, we can solve this problem. Indeed, we can rewrite problem (2), (3) in the following form

$$f(t,v) = \frac{f_0(v) e^{-z(t)}}{\left(1 - (\gamma-1) f_0^{\gamma-1}(v) u(t)\right)^{\frac{1}{\gamma-1}}}, \tag{32}$$

where

$$z(t) = \int_0^t \int_\mathbb{R} f^\gamma(s,v) dv ds \quad \text{and} \quad u(t) = \int_0^t e^{-(\gamma-1)z(s)} ds$$

is a solution of the problem

$$\frac{du}{dt} = \left(\int_{\mathbb{R}} \frac{f_0(v)dv}{\left(1-(\gamma-1)f_0^{\gamma-1}(v)u\right)^{\frac{1}{\gamma-1}}} \right)^{1-\gamma}, \quad u(0) = 0. \tag{33}$$

For more details we refer the reader to [3, 4].

Consider the problem with $\beta = 1$

$$\partial_t f(t,v) = f^2(t,v) - f(t,v) \int_{\mathbb{R}} f^2(t,w)\,dw, \tag{34}$$

$$f_0(v) = \begin{cases} \frac{v+2}{3}, & v \in [-2,-1), \\ \frac{1}{3}, & v \in [-1,1], \\ \frac{-v+2}{3}, & v \in (1,2], \\ 0, & v \notin [-2,2]. \end{cases} \tag{35}$$

Solving the ODE problem (33) we get the following algebraic equation

$$\frac{2}{3-u} - \frac{2}{u} - \frac{6}{u^2}\ln\left(1-\frac{u}{3}\right) = t.$$

We use the Newton method to solve this equation (for fixed t). We regard this solution as exact. By (32), we get the solution of (34), (35).

Now, we take a mesh $\{v_i\}_{i=1}^{20}$, where v_i are the roots of Hermite's polynomial $H_{20}(v)$. We use a 4th order Runge Kutta method to solve the problem (4), (5). In Table we give experimental values of the maximal relative error for $h = 0.001$.

t_i	0.1	0.2	0.3	0.4	0.5
Max relative error	0.000479	0.000967	0.001464	0.001971	0.002489
L_∞ error	0.000160	0.000325	0.000494	0.000667	0.000845
L_1 error	0.000476	0.000958	0.001446	0.001942	0.002445

t_i	0.6	0.7	0.8	0.9	1.0
Max relative error	0.003017	0.003556	0.004108	0.004672	0.005249
L_∞ error	0.001029	0.001218	0.001412	0.001612	0.001818
L_1 error	0.002955	0.003475	0.004002	0.004539	0.005086

References

1. Dimarco, G., Pareschi, L.: Numerical methods for kinetic equation. Acta Numer. **23**, 369–520 (2014)
2. Lachowicz, M., Parisot, M.: A kinetic model for the formation of Swarms with nonlinear interactions. Kinet. Relat. Models **9**, 131–164 (2016)

3. Lachowicz, M., Leszczyński, H., Parisot, M.: A simple kinetic equation of swarm formation: blow-up and global existence. Appl. Math. Lett. **57**, 104–107 (2016)
4. Lachowicz, M., Leszczyński, H., Parisot, M.: Blow-up and global existence for a kinetic equation of swarm formation. Models Methods Appl. Sci. **27**, 1153–1175 (2017)
5. Lo Schiavo, M.: Population kinetic models for social dynamics: dependence on structural parameters. Comp. Math. Appl. **44**, 1129–1146 (2002)

Degenerate Matrix Groups and Degenerate Matrix Flows in Solving the Optimal Control Problem for Dynamic Balance Models of the Economy

Alevtina V. Keller and Minzilia A. Sagadeeva

Abstract Dynamic interindustry balance models are described by differential equations of the first order, and the matrix at the derivative, which is a matrix of specific capital expenditures, can be degenerated. The stationary case, including optimal control problems, is well studied for such models. We consider the non-stationary case, where one of the matrices is multiplied by the scalar function that depends on time. In the stationary case, the results of the study of optimal control problems for degenerate balance interindustry models are presented by methods of the theory of degenerate matrix groups. Highlight that there is the importance of considering this problem for Leontief type models. Only for this particular case there is a convergence of numerical solutions to the optimal control problem to the exact one. We introduce the concept of flow of degenerate matrices and use them to construct the solutions of the non-stationary Leontief type system. We use these solutions in order to investigate an optimal control problem for non-stationary Leontief type systems of the specified type and we have proved the existence of a unique solution to the problem. The obtained results are illustrated by the example of solving an optimal control problem for the classical Leontief model in the non-stationary case.

1 Introduction

W.W. Leontief's balance interindustry models, both static and dynamic, are used to model economic systems of different levels: countries, regions, enterprises [1, 2]. These models are based on the balance method, i.e. the method that compares

---

A. V. Keller
Department of Applied Mathematics and Mechanics, Voronezh State Technical University, 14, Moscow ave, Voronezh 394026, Russian Federation
e-mail: alevtinak@inbox.ru

M. A. Sagadeeva (✉)
Department of Mathematical and Computational Modelling, South Ural State University (National Research University), 76, Lenina ave, Chelyabinsk 454080, Russian Federation
e-mail: sagadeevama@susu.ru

© Springer Nature Switzerland AG 2020
J. Banasiak et al. (eds.), *Semigroups of Operators – Theory and Applications*,
Springer Proceedings in Mathematics & Statistics 325,
https://doi.org/10.1007/978-3-030-46079-2_15

available material, labor and financial resources and needs for them. If we describe the economic system as a whole, then by the balance model we mean a system of equations, each of which expresses the requirement of a balance between the amount of resources produced by individual economic entities and the total need for this resource. Static models can be developed only for individual periods, and within the framework of these models there is no relationship between previous or subsequent periods, i.e. the use of such models introduces certain simplifications and reduces the possibilities of analysis.

In contrast to static models, dynamic ones are constructed in order to reflect not the state, but the process of economic development, as well as to establish a direct relationship between previous and subsequent stages of development, and, therefore, to bring analysis based on an economic-mathematical model to the real conditions of economic system development. In dynamic models, production capital investments are distinguished from the composition of the final product, and their structure and impact on the growth of production are investigated. The model given by a dynamic system of equations is based on a mathematical relationship between the value of capital investments (savings) and the increase in production.

In $\mathbb{R}^n$ consider the basic equation of the dynamic balance model

$$x(t) = c(t)Ax(t) + L\dot{x}(t) + b(t),$$

where $x(t)$ is a vector-function of gross output, A is a matrix of unit direct costs, L is a specific capital expenditure matrix, $b(t)$ is a vector-function of final demand. Convert this equation to

$$L\dot{x}(t) = a(t)Mx(t) + f(t) + Bu(t), \tag{1}$$

where L, M and B are square matrices of order n, such as $a(t)M = E_n - c(t)A$, and $\det L = 0$. Here and below denote E_n is an identity matrix of order n. The function $a : [0, \tau] \to \mathbb{R}_+$ is a scalar function describing the time variation of the parameters that correspond to the relationship between the states of the system under study, and the matrix M is (L, p)-regular (i.e. there exists $\mu \in \mathbb{C}$ such that $\det(\mu L - M) \neq 0$, and the infinity point is the pole of $(\mu L - M)^{-1}$ of order p ($p = \overline{0, n-1}$)). The vector function $f : [0, \tau] \to \mathbb{R}^n$ describes external influences on the system, and the vector function $u : [0, \tau] \to \mathbb{R}^n$ describes the control action on the system. Note that the degeneracy condition of the system $\det L = 0$ is one of the distinguishing features of balance models of the economy, since resources of a certain type cannot be stored [1]. In addition, note that balance models often have a non-stationary form, i.e. the matrices included in system (1) depend on time (see, for example, [3]). However, in this case, in order to obtain a constructive solution, we need for the special conditions on these matrices [4, 5]. In this paper, we consider a system of form (1) and assume that the non-stationary part of the matrices can be averaged and reduced to some general time dependence as the function $a(t)$. Moreover, it is clear that a system of form (1) is a generalization of the stationary case, since we obtain the stationary case, if we take $c(t) \equiv 1$.

Stationary systems of form (1) under the condition $\det L = 0$ do not have a single generally accepted name. The systems are called algebraic-differential systems [4], differential-algebraic systems [5], degenerate systems of ordinary differential equations [6]. The paper [7] was the first to call a system of form (1) by a *Leontief type system*, since the prototype of such a system is the famous balance model "input –output" proposed by Leontief [1]. Note that Leontief type system is a particular case of linear Sobolev type equation [8–10]. Despite this, such systems are a separate topic of research. Using the methods of the Sobolev type theory, it was possible to show the convergence of numerical solutions of the optimal control problem to exact ones for Leontief type systems [11]. Besides these systems simulate not only economic systems [6, 7], but also technical ones [12, 13].

In order to construct a numerical solution to problems for stationary Leontief type systems, a number of studies [6, 7, 14] uses the methods of the theory of degenerate (semi) groups proposed by G.A. Sviridyuk [10] and developed by his followers [15–20]. At the same time, in order to investigate numerically the problems for Leontief type systems, the initial Showalter–Sidorov condition [21]

$$[(\nu L - M)^{-1} L]^{p+1} (x(0) - x_0) = 0 \quad \text{with} \quad \nu \in \mathbb{C}: \ \det(\nu L - M) \neq 0 \quad (2)$$

is often used, since this condition allows to remove the restriction of the initial data matching, for example, in the case of the initial Cauchy condition. In addition, in modern studies on Sobolev type equations, the initial Showalter–Sidorov condition is considered as more natural one in order to investigate various applied problems [21].

In the case of balance models, the problem on optimal control is important. In order to formulate the problem on optimal control of the solutions to Showalter–Sidorov problem (2) for Leontief type system (1), we introduce the penalty functional

$$J(u) = \alpha \sum_{q=0}^{1} \int_0^\tau \|Cx^{(q)}(u,t) - z_0^{(q)}(t)\|_3^2 dt + (1-\alpha) \sum_{q=0}^{\theta} \int_0^\tau \langle N_q u^{(q)}(t), u^{(q)}(t) \rangle_{\mathfrak{U}} dt,$$

where $\alpha \in (0, 1]$, $\tau \in \mathbb{R}_+$, $\mathfrak{U}$ and $\mathfrak{Z}$ are Hilbert spaces, C and N_q are square matrices of order n, and the parameter $\theta \in \{0, 1, \ldots, p+1\}$ is chosen according to the applied sense of the penalty functional. Here $z_0(t)$ describes the planned dynamics of states, to which the system is brought by the control $u(t)$. At the same time, the magnitude of this control action is also estimated. It is necessary to find the optimal control $\hat{u} \in \mathfrak{U}_{ad}$ satisfying the condition

$$J(\hat{u}) = \min_{u \in \mathfrak{U}_{ad}} J(u)$$

for the above functional $J(u)$ and such that $x(\hat{u})$ satisfies problem (1), (2) almost everywhere on $(0, \tau)$. Here the set $\mathfrak{U}_{ad}$ is some convex and compact subset of admissible controls in the control space $\mathfrak{U}$. The optimal control problem for the stationary linear Sobolev type equation with the Cauchy condition was first considered in [22]. Later, the optimal control problem for Sobolev type equations was studied in many papers (see, for example, [17, 23–25]). For Leontief type systems, the optimal control problem was investigated, for example, in the papers [11, 16, 26]. The main goal of this article is to construct a solution to the problem of optimal control of solutions to problem (2) for non-stationary Leontief type systems of form (1) by methods of degenerate flows of solving matrices, which are analogs of the degenerate flows of solving operators [24, 27].

Besides the introduction, conclusion and bibliography, the article contains five sections. Section 1 describes relatively p-regular matrices, which we use in order to consider degenerate groups and flows of matrices in Sect. 2. In Sect. 3, we obtain solutions to the initial problems for non-stationary Leontief type systems. Based on the constructed solutions, we investigate an optimal control problem for non-stationary Leontief type systems in Sect. 4. Finally, in Sect. 5, we investigate a problem on optimal control of solutions to the classical Leontief model [1] in the non-stationary case.

2 Relatively p-Regular Matrices

Let $\mathbb{M}_{n \times m}$ be the set of matrices of size $n \times m$, and $L, M \in \mathbb{M}_{n \times n}$ be square matrices of order n. Following [10, 13], by L-*resolvent set* and L-*spectrum* of matrix M we mean the sets $\rho^L(M) = \{\mu \in \mathbb{C} : \det(\mu L - M) \neq 0\}$ and $\sigma^L(M) = \mathbb{C} \setminus \rho^L(M)$, respectively. It is easy to see that either $\rho^L(M) = \emptyset$, or the L-spectrum of M consists of a finite set of points ([10, 13]). In addition, note that the sets $\rho^L(M)$ and $\sigma^L(M)$ do not depend on basis. Hereinafter, we assume that $\rho^L(M) \neq \emptyset$.

For the complex variable $\mu \in \mathbb{C}$, by an L-*resolvent*, *right L-resolvent* and *left L-resolvent* of matrix M we mean the matrix-valued functions $(\mu L - M)^{-1}$, $R^L_\mu(M) = (\mu L - M)^{-1} L$, $L^L_\mu(M) = L(\mu L - M)^{-1}$ with the domain $\rho^L(M)$, respectively. Also, by virtue of the results of the papers [10, 13], the right and left L-resolvents of matrix M are holomorphic in $\rho^L(M)$.

Definition 1 A matrix M is called L-*regular*, if $\rho^L(M) \neq \emptyset$, and is called (L, p)-*regular*, if $p \in \mathbb{N}$ equals to the pole order in ∞ for the function $(\mu L - M)^{-1}$.

Remark 1 If infinity is a removable singular point of the L-resolvent of the matrix M, then $p = 0$. For square matrices, the parameter p cannot exceed the dimension of the space n.

Remark 2 The term "L-regular matrix M" is equivalent to the term "regular pencil of matrices $\mu L - M$" proposed by Weierstrass [28]. In addition, relatively p-regular matrices ($p = \overline{0, n-1}$) are a special case of relatively p-bounded operators in the theory of Sobolev type equations [10, 13].

In the complex plane $\mathbb{C}$, we consider a closed contour that bounds the domain containing the relative spectrum $\sigma^L(M)$ of the matrix M. Then the following integrals make sense as integrals of holomorphic functions on a closed contour:

$$P = \frac{1}{2\pi i} \int_\gamma R_\mu^L(M) d\mu, \quad Q = \frac{1}{2\pi i} \int_\gamma L_\mu^L(M) d\mu.$$

If the matrix M is (L, p)-regular $(p = \overline{0, n-1})$, then the matrices P and Q are projectors [10, 13].

Since matrices are a finite-dimensional analogue of bounded operators, then the following statement is true by virtue of [27].

Lemma 1 *Let the matrix M be (L, p)-regular $(p = \overline{0, n-1})$. Then*

$$P = \lim_{k \to \infty} \left(k R_k^L(M)\right)^{p+1}, \quad Q = \lim_{k \to \infty} \left(k L_k^L(M)\right)^{p+1}.$$

Example 1 To illustrate these concepts, we consider a degenerate system of ordinary differential equations

$$\begin{cases} \dot{x}_1 = x_1, \\ \dot{x}_3 = x_2, \\ 0 = x_3. \end{cases}$$

Let $x(t) = \text{col}(x_1(t), x_2(t), x_3(t))$. Then we can rewrite this system as $L\dot{x} = Mx$ with matrices

$$L = \begin{pmatrix} 1 & 0 & 0 \\ 0 & 0 & 1 \\ 0 & 0 & 0 \end{pmatrix} \quad \text{and} \quad M = E_3 = \begin{pmatrix} 1 & 0 & 0 \\ 0 & 1 & 0 \\ 0 & 0 & 1 \end{pmatrix}.$$

For such matrices L and M, the L-resolvent, the right L-resolvent and the left L-resolvent of the matrix M have the form

$$(\mu L - M)^{-1} = \begin{pmatrix} \frac{1}{\mu-1} & 0 & 0 \\ 0 & -1 & -\mu \\ 0 & 0 & -1 \end{pmatrix} \quad \text{and} \quad R_\mu^L(M) = L_\mu^L(M) = \begin{pmatrix} \frac{1}{\mu-1} & 0 & 0 \\ 0 & 0 & -1 \\ 0 & 0 & 0 \end{pmatrix}.$$

So we can conclude that $p = 1$. Since $\det(\mu L - M) = \frac{1}{\mu-1}$, then the L-spectrum of matrix M consists of one point $\sigma^L(M) = \{1\}$. Thus, the matrix M is $(L, 1)$-regular. The projection P from Lemma 1 has the form

$$P = \begin{pmatrix} 1 & 0 & 0 \\ 0 & 0 & 0 \\ 0 & 0 & 0 \end{pmatrix} = \lim_{k \to \infty} (k R_k^L(M))^2 = \lim_{k \to \infty} \begin{pmatrix} \frac{k^2}{(k-1)^2} & 0 & 0 \\ 0 & 0 & 0 \\ 0 & 0 & 0 \end{pmatrix}.$$

Note that $\ker L = \text{span}\left\{\begin{pmatrix}0\\1\\0\end{pmatrix}\right\}$ and $\ker P = \text{span}\left\{\begin{pmatrix}0\\1\\0\end{pmatrix}, \begin{pmatrix}0\\0\\1\end{pmatrix}\right\}$.

Denote by L_k and M_k ($k = 0, 1$) the restriction of matrices L and M to $\ker P$ and $\text{im } P$, respectively.

Lemma 2 ([13]) *Let the matrix M be (L, p)-regular ($p = \overline{0, n-1}$). Then there exist matrices L_1^{-1} and M_0^{-1}.*

3 Degenerate Groups and Flows Of Matrices

Definition 2 An one-parameter family $X^\bullet : \mathbb{R} \to \mathbb{M}_{n \times n}$ is called a *degenerate group of matrices*, if the following conditions are satisfied:
(i) $X^0 = P$;
(ii) $X^t X^s = X^{t+s}$ for all $t, s \in \mathbb{R}$.

A degenerate group of matrices is called *analytic*, if the group admits analytic extension to the whole complex plane $\mathbb{C}$ with preserving properties (i) and (ii) of Definition 2.

Theorem 1 ([10, 13]) *Let the matrix M be (L, p)-regular ($p = \overline{0, n-1}$). Then there exists analytic groups $\{X^t \in \mathbb{M}_{n \times n} : t \in \mathbb{R}\}$ and $\{Y^t \in \mathbb{M}_{n \times n} : t \in \mathbb{R}\}$, and matrices of these groups are given by integrals of the Dunford–Taylor type:*

$$X^t = \frac{1}{2\pi i} \int_\gamma R_\mu^L(M) e^{\mu t} d\mu, \quad Y^t = \frac{1}{2\pi i} \int_\gamma L_\mu^L(M) e^{\mu t} d\mu, \tag{3}$$

where the closed contour γ bounds the domain containing the relative spectrum $\sigma^L(M)$ of the matrix M.

Based on the results of Theorem 1, we obtain the following theorem on the solvability of the stationary inhomogeneous equation of the form

$$L\dot{x}(t) = Mx(t) + f(t). \tag{4}$$

Theorem 2 ([10, 13]) *Let the matrix M be (L, p)-regular, $p = \overline{0, n-1}$. Then for any $x_0 \in \mathbb{R}^n$, $Qf \in C(\mathbb{R}^n)$ and $(E_n - Q)f \in C^{p+1}(\mathbb{R}^n)$ there exists a unique solution $x \in C^1(\mathbb{R}^n)$ to Showalter–Sidorov problem (2), (4), and the solution has the form*

$$x(t) = X^t P x_0 + \int_0^t X^{t-s} L_1^{-1} Qf(s) ds - \sum_{k=0}^p (M_0^{-1} L_0)^k M_0^{-1} ((E_n - Q)f)^{(k)}(t),$$

where the matrices L_1^{-1} and M_0^{-1} are the same as in Lemma 2.

By virtue of the results of the paper [27], the following statement is true.

Corollary 1 *Let the matrix M be (L, p)-regular ($p = \overline{0, n-1}$). Then the matrices of the groups $\{X^t \in \mathbb{M}_{n \times n} : t \in \mathbb{R}\}$ and $\{Y^t \in \mathbb{M}_{n \times n} : t \in \mathbb{R}\}$ are given by the Hille–Widder–Post approximations:*

$$X^t = \lim_{k \to \infty} \left(\frac{k}{t} R^L_{\frac{k}{t}}(M)\right)^k, \quad Y^t = \lim_{k \to \infty} \left(\frac{k}{t} L^L_{\frac{k}{t}}(M)\right)^k.$$

Remark 3 Highlight that the use of such constructions allows us to conduct numerical studies of Leontief type systems. This method, despite the difficulty of theoretical constructions, is simpler in technical execution, since it is essentially based on multiplying matrices. That is, it doesn't use a special type of transformation of the source system, as is done, for example, in the papers [4, 5].

Definition 3 A two-parameter family $X(\cdot, \cdot) : \mathbb{R} \times \mathbb{R} \to \mathbb{M}_{n \times n}$ is called a *degenerate flow of matrices*, if the following conditions are satisfied:
(i) $X(t, t) = P$;
(ii) $X(t, s)X(s, \tau) = X(t, \tau)$.

A degenerate flow of matrices is called *analytic*, if the operators of the flow allow the analytic extension in the whole complex plane $\mathbb{C}$ with preserving properties (i) and (ii) of Definition 3.

Let the matrix M be (L, p)-regular ($p = \overline{0, n-1}$) and the function $a \in C(\mathbb{R}; \mathbb{R})$. By analogy with (3), consider

$$X(t, s) = \frac{1}{2\pi i} \int_\gamma R^L_\mu(M) \exp\left(\mu \int_s^t a(\zeta)d\zeta\right) d\mu, \quad s < t, \qquad (5)$$

where $s, t \in \mathbb{R}$ and the closed contour γ is the same as in Theorem 1.

Since matrices are a finite-dimensional analogue of bounded operators, then the following statement is true by virtue of [27].

Theorem 3 *Let the matrix M be (L, p)-regular ($p = \overline{0, n-1}$) and the function $a \in C(\mathbb{R}; \mathbb{R})$, then the family $\{X(t, s) \in \mathbb{M}_{n \times n} : t, s \in \mathbb{R}\}$ given by formula (5) is an analytic degenerate flow of matrices.*

Remark 4 By analogy with (5), we can set the matrix flow by the following formula

$$Y(t, s) = \frac{1}{2\pi i} \int_\gamma L^L_\mu(M) \exp\left(\mu \int_s^t a(\zeta)d\zeta\right) d\mu, \quad s < t.$$

The proof of the properties of this family is analogous to the proof of Theorem 3.

Similar to Corollary 1, the following corollary is true.

Corollary 2 *Let the matrix M be (L, p)-regular ($p = \overline{0, n-1}$) and the function $a \in C(\mathbb{R}; \mathbb{R}_+)$. Then the matrices of flows $\{X(t, s) \in \mathbb{M}_{n \times n} : t, s \in \mathbb{R}\}$ and $\{Y(t, s) \in \mathbb{M}_{n \times n} : t, s \in \mathbb{R}\}$ are given by the Hille–Widder–Post approximations:*

$$X(t, s) = \lim_{k \to \infty} \left(\left(L - \frac{1}{k} M \int_s^t a(\zeta) d\zeta \right)^{-1} L \right)^k, \quad Y(t, s) = \lim_{k \to \infty} \left(L \left(L - \frac{1}{k} M \int_s^t a(\zeta) d\zeta \right)^{-1} \right)^k.$$

4 Solvability of Initial Problems for Non-stationary Leontief Type Systems

Let $L, M \in \mathbb{M}_{n \times n}$ be square matrices of order n. On the interval $\mathfrak{J} \subset \mathbb{R}$, consider the Cauchy problem

$$x(t_0) = x_0 \quad (t_0 \in \mathfrak{J}) \tag{6}$$

for the homogeneous non-stationary equation

$$L\dot{x}(t) = a(t) M x(t), \tag{7}$$

where the function $a : \mathfrak{J} \to \mathbb{R}_+$ should be determined.

Definition 4 By a solution to Eq. (7) we mean a vector function $x \in C^1(\mathfrak{J}; \mathbb{R}^n)$ that satisfies this equation on $\mathfrak{J}$. A solution to Eq. (7) is called a *solution to Cauchy problem* (6), (7), if the solution additionally satisfies condition (6).

A closed set $\mathfrak{P} \subset \mathbb{R}^n$ is called the *phase space* of Eq. (7), if
(i) any solution $x(t)$ to Eq. (7) belongs to $\mathfrak{P}$ (pointwise);
(ii) for any $x_0 \in \mathfrak{P}$, there exists a unique solution to Cauchy problem (6) for Eq. (7).

Along with Eq. (7), we consider the equation

$$L(\nu L - M)^{-1} \dot{y}(t) = a(t) M (\nu L - M)^{-1} y(t), \tag{8}$$

which is equivalent to Eq. (7) for $\nu \in \rho^L(M)$.

By virtue of the results of the paper [27], the following statement is true.

Theorem 4 *Let the matrix M be (L, p)-regular ($p = \overline{0, n-1}$) and the function $a \in C(\mathbb{R}, \mathbb{R}_+)$. Then the phase space of Eq. (7) is the set $\mathrm{im}\ P$, and the phase space of Eq. (8) is the set $\mathrm{im}\ Q$.*

A flow of operators $X(\cdot, \cdot) : \mathbb{R} \times \mathbb{R} \to \mathbb{M}_{n \times n}$ is called a *flow of solving matrices* of Eq. (7), if the vector function $x(t) = X(t, t_0) x_0$ is a solution to Eq. (7) (in the sense of Definition 4) for any $x_0 \in \mathbb{R}^n$.

Consider the Showalter–Sidorov problem

$$P(x(t_0) - x_0) = 0 \qquad (9)$$

for the inhomogeneous equation

$$L\dot{x}(t) = a(t)Mx(t) + f(t) \qquad (10)$$

with the function $f : \mathfrak{J} \to \mathbb{R}^n$. Denote $(E_n - Q)f(t) = f^0(t)$.

Definition 5 A solution to Eq. (10) is called a *solution to Showalter–Sidorov problem* (9), (10), if the solution additionally satisfies condition (9).

Theorem 5 *Suppose that $[t_0, \tau] \subset \mathfrak{J}$, the matrix M is (L, p)-regular ($p = \overline{0, n-1}$), and the function $a \in C^{p+1}([t_0, \tau]; \mathbb{R}_+)$. Then for any function $f : [t_0, \tau] \to \mathbb{R}^n$ such that $Qf \in C^1([t_0, \tau]; \operatorname{im} Q)$ and $f^0 \in C^{p+1}([t_0, \tau]; \ker Q)$, and for any initial value $x_0 \in \mathbb{R}^n$ there exists a unique solution $x \in C^1([t_0, \tau]; \mathbb{R}^n)$ to Showalter–Sidorov problem (9) for Eq. (10), and the solution has the form*

$$x(t) = X(t, t_0)Px_0 + \int_{t_0}^{t} X(t, s) L_1^{-1} Qf(s)\,ds -$$
$$- \sum_{k=0}^{p} (M_0^{-1} L_0)^k M_0^{-1} \left(\frac{1}{a(t)} \frac{d}{dt}\right)^k \frac{(E_n - Q)f(t)}{a(t)}, \qquad (11)$$

where the expression $\left(\dfrac{1}{a(t)} \dfrac{d}{dt}\right)^q$ in the last term means the sequential application of this operator q times.

If, in addition, the matching condition

$$(E_n - Q)x_0 = -\sum_{q=0}^{p} (M_0^{-1} L_0)^q M_0^{-1} \left(\frac{1}{a(t)} \frac{d}{dt}\right)^q \frac{(E_n - Q)f(t)}{a(t)} \bigg|_{t=t_0}$$

is fulfilled, then function (11) is a unique solution to Cauchy problem (6) for Eq. (10).

By virtue of Corollary 2, the following statement holds.

Corollary 3 *Let the matrix M be such that $\det M \neq 0$, then under the conditions of Theorem 5, the solution $x \in C^1([t_0, \tau]; \mathbb{R}^n)$ to problem (9), (10) can also be found by the formula*

$$x(u,t) = \lim_{k\to\infty} x_k(u,t) = \lim_{k\to\infty} \left[\left(\left(L - \frac{1}{k}M \int_{t_0}^{t} a(\zeta)d\zeta \right)^{-1} L \right)^{k} x_0 \right.$$

$$+ \int_{t_0}^{t} \left(\left(L - \frac{1}{k}M \int_{s}^{t} a(\zeta)d\zeta \right)^{-1} L \right)^{k} \left(L - \frac{1}{k}M \right)^{-1} \left(kL_k^L(M) \right)^{p} f(s)ds$$

$$\left. - \sum_{q=0}^{p} (M^{-1}\tilde{Q}L)^q M^{-1}\tilde{Q} \left(\frac{1}{a(t)} \frac{d}{dt} \right)^q \frac{f(t)}{a(t)} \right], \tag{12}$$

where $\tilde{Q} = \left(E_n - \left(kL_k^L(M) \right)^{p+1} \right)$.

Remark 5 The condition $\det M \neq 0$ does not reduce the generality of the result. Indeed, let us make the change of variables $x(t) = e^{\eta t} y(t)$ in the system $L\dot{x}(t) = Mx(t) + f(t)$. Then, in the right-hand side, we obtain the matrix $\tilde{M} = M - \eta L$ such that $\det \tilde{M} \neq 0$.

5 Optimal Control Problem

Suppose that $\mathfrak{X} = \mathfrak{Y} = \mathfrak{U} = \mathbb{R}^n$. On the interval $[0, \tau) \subset \mathbb{R}_+$ ($\tau < +\infty$), consider the Showalter–Sidorov problem

$$P(x(0) - x_0) = 0 \tag{13}$$

for the equation

$$L\dot{x}(t) = a(t)Mx(t) + f(t) + Bu(t), \tag{14}$$

where the matrices $L, M, B \in \mathbb{M}_{n\times n}$, the scalar function $a : [0, \tau) \to \mathbb{R}_+$, the vector functions $u : [0, \tau) \to \mathbb{R}^n$, $f : [0, \tau) \to \mathbb{R}^n$ should be determined.

Definition 6 A vector function $x \in H^1(\mathfrak{X}) = \{x \in L_2(0, \tau; \mathfrak{X}) : \dot{x} \in L_2(0, \tau; \mathfrak{X})\}$ is called a *strong solution to Eq.* (14), if the function turns the equation into an identity almost everywhere on $(0, \tau)$. A strong solution $x = x(t)$ to system (14) is called a *strong solution to Showalter–Sidorov problem* (13), (14), if the solution satisfies (13).

Construct the space

$$H^{p+1}(\mathfrak{Y}) = \{\xi \in L_2(0, \tau; \mathfrak{Y}) : \xi^{(p+1)} \in L_2(0, \tau; \mathfrak{Y}), \ p = \overline{0, n-1}\},$$

which is Hilbert (since $\mathfrak{Y}$ is Hilbert) with the scalar product

$$[\xi, \eta] = \sum_{q=0}^{p+1} \int_0^\tau \langle \xi^{(q)}, \eta^{(q)} \rangle_{\mathfrak{Y}} \, dt.$$

By virtue of Theorem 5, the following statement is true.

Theorem 6 *Let the matrix M be (L, p)-regular ($p = \overline{0, n-1}$) and the function $a \in C^{p+1}([0, \tau); \mathbb{R}_+)$. Then for any $x_0 \in \mathfrak{X}$, $f \in H^{p+1}(\mathfrak{Y})$ and for any $u \in H^{p+1}(\mathfrak{U})$, there exists a unique solution $x \in H^1(\mathfrak{X})$ to Showalter–Sidorov problem (13) for system (14), and the solution has form (12), where $t_0 = 0$ and the function $f(t)$ is substituted by the function $f(t) + Bu(t)$.*

Let $\mathfrak{Z}$ be a Hilbert space, and the matrix $C \in \mathbb{M}_{n \times n}$. Construct a penalty functional

$$J(u) = \alpha \sum_{q=0}^1 \int_0^\tau \|Cx^{(q)} - z_0^{(q)}\|_{\mathfrak{Z}}^2 dt + (1-\alpha) \sum_{q=0}^\theta \int_0^\tau \langle N_q u^{(q)}, u^{(q)} \rangle_{\mathfrak{U}} dt, \quad (15)$$

where $\alpha \in (0, 1]$, $0 \le \theta \le p+1$, $N_q \in \mathbb{M}_{n \times n}$ ($q = 0, 1, \ldots, p+1$) are positive defined matrices, $z_0(t)$ is the planned dynamics of the states of the system, which is a function that belongs to a certain Hilbert observation space $\mathfrak{Z}$. Note that $z \in H^1(\mathfrak{Z})$, if $x \in H^1(\mathfrak{X})$. If the parameter $\alpha = 1$ in functional (15) then the second term vanishes, and we have the problem of the *hard optimal control*.

Similarly to $H^{p+1}(\mathfrak{Y})$, define the space $H^{p+1}(\mathfrak{U})$, which is Hilbert, since $\mathfrak{U}$ is Hilbert. In the space $H^{p+1}(\mathfrak{U})$, consider a closed and convex subset $H^{p+1}_\partial(\mathfrak{U}) = \mathfrak{U}_{ad}$, which is a *set of admissible controls*.

Definition 7 A vector function $\hat{u} \in \mathfrak{U}_{ad}$ is called an *optimal control* of solutions to problem (13), (14), if

$$J(\hat{u}) = \min_{u \in \mathfrak{U}_{ad}} J(u), \quad (16)$$

where the functions $x(u) \in \mathfrak{X}$ and $u \in \mathfrak{U}_{ad}$ are such that $x(u) \in \mathfrak{X}$ is a solution to problem (13), (14).

Finally, we give the main result of the section.

Theorem 7 *Let the matrix M be (L, p)-regular ($p = \overline{0, n-1}$), and the function $a \in C^{p+1}([0, \tau); \mathbb{R}_+)$. Then for any $x_0 \in \mathfrak{X}$, $z_0 \in \mathfrak{Z}$, $f \in H^{p+1}(\mathfrak{Y})$, there exists a unique optimal control $\hat{u} \in \mathfrak{U}_{ad}$ of solutions to problem (13), (14), (16) with the functional (15).*

Proof of this theorem is carried out similarly to the general case (see for example [24]) and is based on a linear dependence of the solution $x(u, t)$ of the system on the control function $u(t)$ and the properties of the penalty functional (15).

6 Optimal Control Problem for Non-stationary Leontief Model

In order to illustrate the obtained theoretical results, we consider the optimal control problem for the non-stationary Leontief system. As a Leontief system, we take a dynamic balance model that describes the relationship between the three sectors of the economy: agriculture, industry, and household. In this model [1], the matrices L and M have the form

$$L = \begin{pmatrix} \dfrac{7}{20} & \dfrac{1}{20} & \dfrac{21}{20} \\ \dfrac{1}{100} & \dfrac{103}{200} & \dfrac{8}{25} \\ 0 & 0 & 0 \end{pmatrix}, \quad M = \begin{pmatrix} \dfrac{3}{4} & \dfrac{-1}{5} & \dfrac{-11}{20} \\ \dfrac{-7}{25} & \dfrac{22}{35} & \dfrac{-3}{5} \\ \dfrac{-4}{15} & \dfrac{-2}{15} & \dfrac{13}{15} \end{pmatrix}.$$

Note that the third row of the matrix L is zero, since labor cannot be stored.

In order to apply the results obtained above, first of all, we construct the relative resolvent of the matrices. Construct the determinant

$$\det(\mu L - M) = \frac{-43\,295\mu^2 + 99\,225\mu - 21\,744}{140\,000}.$$

The L-spectrum of the matrix M consists of two points:

$$\sigma^L(M) = \{0.245419;\ 2.046416\}.$$

And then the L-resolvents of the matrix M has the form

$$(\mu L - M)^{-1} = \frac{5}{3(-43\,295\mu^2 + 99\,225\mu - 21\,744)}$$

$$\cdot \begin{pmatrix} 4(9\,760 - 10\,269\mu) & 280(55\mu + 74) & 3(-14\,693\mu^2 + 13\,181\mu + 13\,040) \\ 56(141\mu + 604) & 280(151 - 175\mu) & 42(-203\mu^2 + 659\mu + 1\,208) \\ 32(538 - 357\mu) & 560(23 - 5\mu) & 3(5\,033\mu^2 - 17\,423\mu + 11\,632). \end{pmatrix}$$

Therefore, by Definition 1, we obtain that $p = 0$, and the matrix M is $(L, 0)$-regular. In system (14), for the matrices L and M given above, we take

$$a(t) = \frac{1}{\sqrt{t+1}}, \quad f(t) = \begin{pmatrix} 0 \\ 0 \\ 0 \end{pmatrix}, \quad B = \begin{pmatrix} 1 & 0 & 0 \\ 0 & 1 & 0 \\ 0 & 0 & 0 \end{pmatrix}$$

and obtain the system

Fig. 1 Graph of the component of the planned dynamics $z_{01}(t)$ (black line) and graph of the state function $x_1(t)$ (grey line)

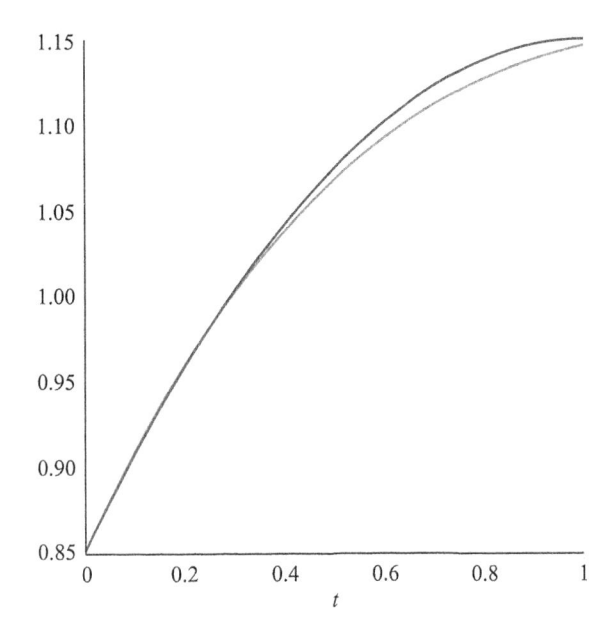

$$\begin{cases} \dfrac{7}{20}\dot{x}_1(t) + \dfrac{1}{20}\dot{x}_2(t) + \dfrac{21}{20}\dot{x}_3(t) = \dfrac{1}{\sqrt{t+1}}\left(\dfrac{3}{4}x_1(t) - \dfrac{1}{5}x_2(t) - \dfrac{11}{20}x_3(t)\right) + u_1(t), \\ \dfrac{1}{100}\dot{x}_1(t) + \dfrac{103}{200}\dot{x}_2(t) + \dfrac{8}{25}\dot{x}_3(t) = \dfrac{1}{\sqrt{t+1}}\left(-\dfrac{7}{25}x_1(t) + \dfrac{22}{35}x_2(t) - \dfrac{3}{5}x_3(t)\right) + u_2(t), \\ 0 = \dfrac{1}{\sqrt{t+1}}\left(-\dfrac{4}{15}x_1(t) - \dfrac{2}{15}x_2(t) + \dfrac{13}{15}x_3(t)\right). \end{cases} \quad (17)$$

Consider the problem on optimal control of the solutions to this system for the initial values $x_0 = \mathrm{col}(0.85; 0; 0.2)$. As control functions, we take third-degree polynomials with indefinite coefficients. For functional (15), we set the parameters

$$\alpha = 0.9, \quad \tau = 1, \quad C = E_3, \quad z_0(t) = \begin{pmatrix} 0.85 + 0.6t - 0.3t^2 \\ 0.2t - 0.1t^2 \\ 0 \end{pmatrix}, \quad \theta = 0, \quad N_0 = E_3.$$

For these values, all the conditions of Theorems 5, 6, and 7 are fulfilled, and there exists a unique solution to the optimal control problem. For the chosen values of the parameters, we obtain the following solution to problem (16) with penalty functional (15) for solutions to system (17):

$$\hat{u}_1(t) = 0.849207 + 0.633840t - 0.446622t^2 + 0.110294t^3,$$
$$\hat{u}_2(t) = -0.000245 + 0.105757t + 0.170562t^2 + 0.000703t^3,$$

and the value of the functional $J(\hat{u}) = 0.020483$. Figure 1 shows the graphs of the first component of the planned dynamics $z_0(t)$ and the state function $x_1(t)$, the difference between which is minimized when solving the optimal control problem for the chosen values of the parameters.

7 Conclusion

The constructed families of matrices allow to study numerically the balance models of the economy. This fact is important in order to consider various problems, including optimal control problems. In the general case of matrices with time-dependent coefficients, in order to obtain a numerical solution, either a theoretical study is carried out or additional conditions about matrices are imposed. In our case, there is no need to impose these specific requirements on the type of matrices.

References

1. Leontief, W.W.: Input-Output Economics. Oxford University Press, Oxford (1986)
2. McConnell, C.R., Brue, S.L., Flynn, S.M.: Economics: Principles, Problems, and Policies. McGraw-Hill, New York (2009)
3. Pospelov, I.G.: Intensive quantities in an economy and conjugate variables. Math. Notes **94**(1), 146–156 (2013). https://doi.org/10.1134/S0001434613070134
4. Boyarintsev, YuE, Chistyakov, V.F.: Algebra-Differential Systems: The Methods of Solution and Research. Nauka, Novosibirsk (1998). (in Russian)
5. Bulatov, M.V., Chistyakov, V.F.: A numerical method for solving differential-algebraic equations. Comput. Math. Math. Phys. **42**(4), 439–449 (2002)
6. Burlachko, I.V., Sviridyuk, G.A.: An algorithm for solving the cauchy problem for degenerate linear systems of ordinary differential equations. Comput. Math. Math. Phys. **43**(11), 1613–1619 (2003)
7. Sviridyuk, G.A., Brychev, S.V.: Numerical solution of systems of equations of Leontief type. Russian Math. **47**(8), 44–50 (2003)
8. Al'shin, A.B., Korpusov, M.O., Sveshnikov, A.G.: Blow-up in Nonlinear Sobolev Type Equations. De Gruyter, Berlin (2011)
9. Demidenko, G.V., Uspenskii, S.V.: Partial Differential Equations and Systems not Solvable with Respect to the Highest-Order Derivative. Marcel Dekker Inc, New York, Basel, Hong Kong (2003)
10. Sviridyuk, G.A., Fedorov, V.E.: Linear Sobolev Type Equations and Degenerate Semigroups of Operators. VSP, Utrech, Boston, Koln (2003)
11. Sviridyuk, G.A., Keller, A.V.: On the numerical solution convergence of optimal control problems for Leontief type system. J. Samara State Tech. Univ. Ser.: Phys. Math. Sci. **14**(2), 24–33 (2011). https://doi.org/10.14498/vsgtu951 (in Russian)
12. Shestakov, A.L., Sviridyuk, G.A., Keller, A.V.: Theory of optimal measurements. J. Comput. Eng. Math. **1**(1), 3–15 (2014)
13. Shestakov, A.L., Sviridyuk, G.A., Khudyakov, YuV: Dynamical measurements in the view of the group operators theory. Semigroups Oper.-Springer Proc. Math. Stat. **113**, 273–286 (2015). https://doi.org/10.1007/978-3-319-12145-1_17

14. Keller, A.V.: The Leontief type systems: classes of problems with the Showalter–Sidorov intial condition and numerical solving. Bull. Irkutsk State Univ. Ser.: Math. **3**(2), 30–43 (2010) (in Russian)
15. Bychkov, E.V.: On a Semilinear Sobolev-type mathematical model. Bull. South Ural. State Univ. Ser.: Math. Model. Program. Comput. Softw. **7**(2), 111–117 (2014). https://doi.org/10.14529/mmp140210 (in Russian)
16. Keller, A.V.: On the computational efficiency of the algorithm of the numerical solution of optimal control problems for models of Leontieff type. J. Comput. Eng. Math. **2**(2), 39–59 (2015). https://doi.org/10.14529/jcem150205
17. Manakova, N.A.: Mathematical models and optimal control of the filtration and deformation processes. Bull. South Ural State Univ. Ser.: Math. Model. Program. Comput. Softw. **8**(3), 5–24 (2015). https://doi.org/10.14529/mmp150301 (in Russian)
18. Sukacheva, T.G., Kondyukov, A.O.: On a class of Sobolev-type equations. Bull. South Ural State Univ. Ser.: Math. Model. Program. Comput. Softw. **7**(4), 5–21 (2014). https://doi.org/10.14529/mmp140401
19. Zagrebina, S.A.: A multipoint initial-final value problem for a linear model of plane-parallel thermal convection in viscoelastic incompressible fluid. Bull. South Ural State Univ. Ser.: Math. Model. Program. Comput. Softw. **7**(3), 5–22 (2014). https://doi.org/10.14529/mmp140301
20. Zamyshlyaeva, A.A.: The higher-order Sobolev-type models. Bull. South Ural State Univ. Ser.: Math. Model. Program. Comput. Softw. **7**(2), 5–28 (2014). https://doi.org/10.14529/mmp140201 (in Russian)
21. Sviridyuk, G.A., Zagrebina, S.A.: The Showalter–Sidorov problem as a phenomena of the Sobolev-type equations. Bull. Irkutsk State Univ. Ser.: Math. **3**(1), 104–125 (2010) (in Russian)
22. Sviridyuk, G.A., Efremov, A.A.: Optimal control of Sobolev-type linear equations with relatively p-sectorial operators. Differ. Equ. **31**(11), 1882–1890 (1995)
23. Manakova, N.A.: Method of decomposition in the optimal control problem for Semilinear Sobolev type models. Bull. South Ural State Univ. Ser.: Math. Model. Program. Comput. Softw. **8**(2), 133–137 (2015). https://doi.org/10.14529/mmp150212 (in Russian)
24. Sagadeeva, M.A., Zagrebina, S.A., Manakova, N.A.: Optimal control of solutions of a multipoint initial-final problem for non-autonomous evolutionary Sobolev type equation. Evol. Equ. Control. Theory **8**(3), 473–488 (2019). https://doi.org/10.3934/eect.2019023
25. Tsyplenkova, O.N.: Optimal control in higher-order Sobolev-type mathematical models with (A, p)-bounded operators. Bull. South Ural State Univ. Ser.: Math. Model. Program. Comput. Softw. **7**(2), 129–135 (2014). https://doi.org/10.14529/mmp140213 (in Russian)
26. Keller, A.V.: Numerical solution of the optimal control problem for degenerate linear system of equations with Showalter–Sidorov initial conditions. Bull. South Ural. State Univ. Ser.: Math. Model. Program. Comput. Softw. **1**(2), 50–56 (2008) (in Russian)
27. Sagadeeva, M.A.: Degenerate flows of solving operators for nonstationary Sobolev type equations. Bull. South Ural State Univ. Ser.: Math. Mech. Phys. **9**(1), 22–30 (2017). https://doi.org/10.14529/mmph170103 (in Russian)
28. Gantmacher, F.R.: The Theory of Matrices. Reprinted by American Mathematical Society, AMS Chelsea Publishing (2000)

Degenerate Holomorphic Semigroups of Operators in Spaces of K-"Noises" on Riemannian manifolds

Olga G. Kitaeva, Dmitriy E. Shafranov, and Georgy A. Sviridyuk

Abstract We investigate the degenerate holomorphic resolving semigroups for the linear stochastic Sobolev type models with relatively p-sectorial operator in the spaces of smooth differential forms defined on a smooth compact oriented Riemannian manifold without boundary. To this end, in the space of differential forms, we use the pseudo-differential Laplace–Beltrami operator instead of the usual Laplace operator. The Cauchy condition and the Showalter–Sidorov condition are used as the initial conditions for an abstract Sobolev type model. Since "white noise" of the model is non-differentiable in the usual sense, we use the derivative of stochastic process in the Nelson–Gliklikh sense. In order to investigate the stability of solutions, we establish that there exist the exponential dichotomies splitting the space of solutions into stable and unstable invariant subspaces. As an example, we use a stochastic version of the Dzektser equation in the space of differential forms defined on the 2-dimensional torus, which is a smooth compact oriented Riemannian surface without boundary.

Keywords Sobolev type equations · Additive "white noise" · Nelson–Gliklikh derivative · Riemannian manifold · Differential form

1 Introduction

Degenerate holomorphic semigroups of operators arise as resolving semigroups of Sobolev type equations of the form $L\overset{\circ}{u} = Mu$, where the operator $M \in \mathcal{C}l(\mathcal{U}; \mathcal{F})$ (i.e. M is linear, closed, and densely defined) is strongly p-sectorial, $p \in \{0\} \bigcup \mathbb{N}$,

---

O. G. Kitaeva · D. E. Shafranov (✉) · G. A. Sviridyuk
South Ural State University, 76 Lenina av., Chelyabinsk, Russia
e-mail: shafranovde@susu.ru

O. G. Kitaeva
e-mail: kitaevaog@susu.ru

G. A. Sviridyuk
e-mail: sviridiukga@susu.ru

© Springer Nature Switzerland AG 2020
J. Banasiak et al. (eds.), *Semigroups of Operators – Theory and Applications*,
Springer Proceedings in Mathematics & Statistics 325,
https://doi.org/10.1007/978-3-030-46079-2_16

relative to the operator $L \in \mathscr{L}(\mathscr{U}; \mathscr{F})$ (i.e. L is linear and continuous), $\mathscr{U}$ and $\mathscr{F}$ are the Banach spaces [1]. The paper [2] was the first to consider degenerate holomorphic semigroups of operators as resolving semigroups of stochastic Sobolev type equations

$$L \overset{o}{\eta} = M\eta, \qquad (1)$$

where the operator $M \in \mathscr{C}l(\mathbf{U_K L}_2; \mathbf{F_K L}_2)$ is strongly p-sectorial, $p \in \{0\} \cup N$, relative to the operator $L \in \mathscr{L}(\mathbf{U_K L}_2; \mathbf{F_K L}_2)$, $\eta = \eta(t)$ is the desired stochastic process, $\overset{o}{\eta} = \overset{o}{\eta}(t)$ is *the Nelson–Gliklikh derivative* of η, $\mathbf{U_K L}_2(\mathbf{F_K L}_2)$ are the spaces of $\mathscr{U}(\mathscr{F})$-*valued random* $\mathbf{K}$-*variables*, $\mathscr{U}(\mathscr{F})$ is the real separable Hilbert space, $\mathbf{K} = \{\lambda_k\} \subset R_+$ is the monotone sequence such that $\sum_{k=1}^{\infty} \lambda_k^2 < +\infty$.

Along with the weakened (in the sense of S.G. Krein) Cauchy problem

$$\lim_{t \to 0+} (\eta(t) - \eta_0) = 0 \qquad (2)$$

for Eq. (1), the paper [2] consider the weakened Showalter–Sidorov problem

$$\lim_{t \to 0+} P(\eta(t) - \eta_0) = 0 \qquad (3)$$

for the inhomogeneous stochastic Sobolev type equation

$$L \overset{o}{\eta} = M\eta + N \overset{o}{W}_k. \qquad (4)$$

Here $P \in \mathscr{L}(\mathbf{U_K L}_2)$ is the relatively sectorial projector constructed using the operators L and M, and $\overset{o}{W}_k = \overset{o}{W}_k(t)$ is the Nelson–Gliklikh [3] derivative of the Wiener $\mathbf{K}$-process $W_k = W_k(t), t \in \{0\} \cup R_+$. In the paper [4], the stochastic $\mathbf{K}$-process $\overset{o}{W}_k$ is called white $\mathbf{K}$-"noise".

The Cauchy problem and the Showalter–Sidorov problem for stochastic Sobolev type equations were studied in various aspects [5–7]. However, concrete interpretations of abstract schemes are always presented by non-classical equations of mathematical physics given on bounded areas of the space R^d. We are the first, at least for relatively strong p-sectorial operators, to consider the Dzektser linear equation as a concrete interpretation. The equation simulates the processes of filtration on a connected compact oriented Riemannian manifold without boundary [8]. In addition, we are the first to consider the dichotomies of solutions to equation (1) occurring in its phase space [9]. In general, we continue the series of articles [10, 11], where the delineated above range of questions is addressed to the Cauchy problem and the Showalter–Sidorov problem for Eqs. (1) and (4) in the case of (L, p)-bounded of the operator $M \in \mathscr{L}(\mathbf{U_K L}_2; \mathbf{F_K L}_2)$, $p \in \{0\} \cup N$ [1].

The paper, in addition to references, contains five sections. The introduction is the first section. In the second section of this paper, we will describe differential forms with coefficients from a specially selected spaces of $\mathbf{K}$-"noises" obtain by using

Nelson–Gliklikh derivative. In Sect. 3, we will research the exponential dichotomy of linear Sobolev type equations with p-sectorial operators. In the 4th section, we will find the solution of Showalter–Sidorov problem and present the example for Dzektser equation. The conclusion presents several directions for further research is in the 5th section. Note also that the list of references does not pretend to be complete, but reflects only the personal preferences of the authors.

2 Spaces of Random K-Variables and K-"Noises" on Riemannian Manifolds

Let $\Omega \equiv (\Omega, \mathscr{A}, \mathbf{P})$ be a complete probability space with a probability measure $\mathbf{P}$ associated with a σ-algebra $\mathscr{A}$ of subsets of the set Ω. Let $\mathbf{R}$ be the set of real numbers endowed with the σ-algebra. Then the mapping $\xi : \mathscr{A} \to \mathbf{R}$ is called a *random variable*. Consider the set of random variables $\{\xi\}$ having zero expectation ($\mathbf{E}\xi = 0$) and finite variance ($\mathbf{D}\xi < +\infty$). The set forms Hilbert space $\mathbf{L}_2$ with scalar product $(\xi_1, \xi_2) = \mathbf{E}\xi_1\xi_2$ and norm $\|\xi\|_{\mathbf{L}_2}$. Let $\mathscr{A}_0$ be a σ-subalgebra of the σ-algebra $\mathscr{A}$. Construct a subspace $\mathbf{L}_2^0 \subset \mathbf{L}_2$ of random variables measurable with respect to $\mathscr{A}_0$. Denote the orthoprojector by $\Pi : \mathbf{L}_2 \to \mathbf{L}_2^0$. Let $\xi \in \mathbf{L}_2$ be a random variable, then $\Pi\xi$ is called *conditional expectation* and is denoted by $\mathbf{E}(\xi|\mathscr{A}_0)$.

Further, a measurable mapping $\eta : \mathscr{I} \times \mathscr{A} \to \mathbf{R}$, where $\mathscr{I} \subset \mathbf{R}$ is an interval, we call a *stochastic process*, a random variable $\eta(\cdot, \omega)$ we call a *cut* of the stochastic process, and a function $\eta(t, \cdot)$, $t \in \mathscr{I}$ we call a *trajectory* of the stochastic process. A stochastic process $\eta = \eta(t, \cdot)$ is called *continuous,* if almost sure (i.e. for almost all $\omega \in \mathscr{A}$) the trajectories $\eta(t, \omega)$ are continuous functions. The set $\{\eta = \eta(t, \omega)\}$ of continuous stochastic processes forms Banach space $\mathbf{CL}_2$ with norm

$$\|\eta\|_{\mathbf{CL}_2} = \sup_{t \in \mathscr{I}} (\mathbf{D}\eta(t, \omega))^{1/2}. \tag{5}$$

An example of a continuous stochastic process is the Wiener process

$$\beta(t, \omega) = \sum_{k=0}^{\infty} \xi_k(\omega) \sin \frac{\pi}{2}(2k + 1)t, \ t \in \{0\} \cup R_+, \tag{6}$$

describing Brownian motion in the Einstein–Smoluchowski model [12]. Here the coefficients $\{\xi_k = \xi_k(\omega)\} \subset \mathbf{L}_2$ are pairwise uncorrelated random variables such that $\mathbf{D}\xi_k^2 = [\frac{\pi}{2}(2k + 1)]^{-2}, k \in \{0\} \bigcup N$. For the space $\mathbf{L}_2$, note that two random variables ξ_1 and ξ_2 are uncorrelated if they are orthogonal. Indeed, $0 = cov(\xi_1, \xi_2) = \mathbf{E}\xi_1\xi_2 = (\xi_1, \xi_2) = 0$.

Consider a stochastic process $\eta \in \mathbf{CL}_2$. The *Nelson–Gliklikh derivative* of the stochastic process η at a point $t \in \mathscr{I}$ is a random variable

$$\overset{o}{\eta}(t,\cdot) = \frac{1}{2}\left(\lim_{\Delta t \to 0+} \mathbf{E}_t^\eta\left(\frac{\eta(t+\Delta t,\cdot) - \eta(t,\cdot)}{\Delta t}\right) + \lim_{\Delta t \to 0+} \mathbf{E}_t^\eta\left(\frac{\eta(t,\cdot) - \eta(t-\Delta t,\cdot)}{\Delta t}\right)\right), \quad (7)$$

if the limit exists in the sense of a uniform metric on R. Here $\mathbf{E}_t^\eta = \mathbf{E}(\cdot|\mathcal{N}_t^\eta)$, and $\mathcal{N}_t^\eta \subset \mathcal{A}$ is the σ-algebra generated by the random variable $\eta(t,\omega)$.

We say that the Nelson–Gliklikh derivative $\overset{o}{\eta}(\cdot,\omega)$ of the stochastic process $\eta(\cdot,\omega)$ exists (or almost sure exists) on the interval $\mathcal{I}$, if there exist the Nelson–Gliklikh derivatives $\overset{o}{\eta}(\cdot,\omega)$ at all (or almost all) points of $\mathcal{I}$. The set of continuous stochastic processes that have continuous Nelson–Gliklikh derivatives $\overset{o}{\eta} \in \mathbf{CL}_2(\mathcal{I})$ forms Banach space $\mathbf{C}^1\mathbf{L}_2(\mathcal{I})$ with norm

$$\|\eta\|_{\mathbf{C}^1\mathbf{L}_2} = \sup_{t \in \mathcal{I}}(\mathbf{D}\eta(t,\omega) + \mathbf{D}\overset{o}{\eta}(t,\omega))^{1/2}. \quad (8)$$

Hence, by induction, we define Banach spaces $\mathbf{C}^l\mathbf{L}_2(\mathcal{I})$, $l \in N$, of stochastic processes that have continuous Nelson–Gliklikh derivatives on $\mathcal{I}$ up to the order $l \in N$ inclusive. The norms of the spaces are given by the formulas

$$\|\eta\|_{\mathbf{C}^l\mathbf{L}_2} = \sup_{t \in \mathcal{I}}\left(\sum_{k=1}^l \mathbf{D}\overset{o}{\eta^l}(t,\omega)\right)^{1/2}, \quad (9)$$

where $\overset{o}{\eta^0} \equiv \eta$. Since "white noise" $\overset{o}{\beta}(t,\omega) = \dfrac{\beta(t,\omega)}{2t}$ belongs to each of the spaces $\mathbf{C}^l\mathbf{L}_2(R_+)$, $l \in \{0\} \bigcup N$ (see [1–5]), then, for brevity, all these spaces are called *spaces of "noises"*.

Therefore, we constructed space of random variables $\mathbf{L}_2$ and spaces of "noises" $\mathbf{C}^l\mathbf{L}_2(\mathcal{I})$, $l \in \{0\} \bigcup N$. Let us construct the space of *random* $\mathbf{K}$*-variables*. Let $\mathcal{H}$ be a separable Hilbert space with an orthonormal basis $\{\varphi_k\}$, $\mathbf{K} = \{\lambda_k\} \subset R_+$ be a monotone sequence such that $\sum_{k=1}^\infty \lambda_k^2 < +\infty$, $\{\xi_k = \xi_k(\omega)\} \subset \mathbf{L}_2$ be a sequence of random variables such that $\|\xi_k\|_{\mathbf{L}_2} \leq C$ for all $C \in R_+$ and $k \in N$. Construct a $\mathcal{H}$-valued random $\mathbf{K}$-variable

$$\xi(\omega) = \sum_{k=1}^\infty \lambda_k \xi_k(\omega)\varphi_k.$$

Completion of the linear span of the set $\{\lambda_k \xi_k \varphi_k\}$ in the norm

$$\|\eta\|_{\mathbf{H_K L}_2} = \left(\sum_{k=1}^\infty \lambda_k^2 \mathbf{D}\xi_k\right)^{1/2} \quad (10)$$

is called the *space of ($\mathcal{H}$-valued) random* $\mathbf{K}$*-variables* and is denoted by $\mathbf{H_K L}_2$. It is easy to see that the space $\mathbf{H_K L}_2$ is Hilbert, and the constructed above random

K-variable $\xi = \xi(\omega) \in \mathbf{H_K L_2}$. Similarly, we define the Banach space of ($\mathcal{H}$-valued) **K**-"noises" as completion of the linear span of the set $\{\lambda_k \xi_k \varphi_k\}$ in the norm

$$\|\eta\|_{\mathbf{C}^l \mathbf{H_K L_2}} = \sup_{t \in \mathscr{I}} \left(\sum_{k=1}^{\infty} \lambda_k^2 \sum_{m=1}^{l} \overset{o}{\mathbf{D}} \eta_k^m \right)^{1/2}, \tag{11}$$

where the sequence $\{\eta_k\} \subset \mathbf{C}^l \mathbf{L}_2$ $l \in \{0\} \bigcup N$. It is easy to see that the vector

$$\eta(t, \omega) = \sum_{k=1}^{\infty} \lambda_k \eta_k(t, \omega) \varphi_k \tag{12}$$

belongs to the space $\mathbf{C}^l \mathbf{H_K L_2}(\mathscr{I})$, if the sequence $\{\eta_k\} \subset \mathbf{C}^l \mathbf{L}_2$ of vectors and all their Nelson–Gliklikh derivatives up to the order $l \in \{0\} \bigcup N$ inclusively is uniformly bounded in the norm $\| \cdot \|_{\mathbf{C}^l \mathbf{L}_2}$. An example of a vector that belongs to all spaces is the ($\mathcal{H}$-valued) Wiener **K**-process

$$W_\mathbf{K}(t, \omega) = \sum_{k=1}^{\infty} \lambda_k \beta_k(t, \omega) \varphi_k, \tag{13}$$

where $\{\beta_k\} \subset \mathbf{C}^l \mathbf{L}_2$ is a sequence of Brownian motions.

Finally, consider the space of random **K**-variables and spaces of **K**-"noises" defined on Riemannian manifolds. We choose the magnificent monograph [13] for references. Let M be a connected oriented compact Riemannian manifold without boundary having class C^∞ and dimension d. On the manifold M, consider the vector space $E^q(M)$ of q-forms

$$a = \sum_{i_1 < i_2 < \cdots, < i_q} a_{i_1, i_2, \ldots, i_q} dx_{i_1} \wedge dx_{i_2} \wedge \cdots \wedge dx_{i_q}, \tag{14}$$

where $a_{i_1, i_2, \ldots, i_q} \in C^\infty, q = \{0, 1, \ldots, d\}$. On the space E^q, define the scalar product

$$<a, b> = \int_M a \wedge *b, \tag{15}$$

where $*$ is the Hodge operator, which each q-form on M associates with $(p-q)$-form, moreover, $** = (-1)^{q(p-q)}$. Further, define the Laplace–Beltrami operator on the space $E^q(M)$ by the formula $\Delta = d\delta + \delta d$, where d is an exterior derivative operator, $\delta = (-1)^{n(k+1)+1} * d*$. This operator is self-adjoint in the space $E^q(M)$, $q \in \{0, 1, \ldots, p\}$, i.e. $<\Delta a, b> = <a, \Delta b>$. Define scalar products $<a, b>_0 = <a, b>$, $<a, b>_1 = <a, b>_0 + <\Delta a, b>_0$, $<a, b>_2 = <a, b>_1 + <\Delta a, \Delta b>_0$. Denote by H_k^q the (real) Hilbert space, which is completion of E^q in the norm $\| \cdot \|_k$, induced by the scalar product $<\cdot, \cdot>_k$, $k \in \{0, 1, 2\}$,

$q \in \{0, 1, \ldots, p\}$. H_k^q is the separable Hilbert space with basis formed by the eigenfunctions of the Laplace–Beltrami operator, which are orthonormal by $<\cdot, \cdot>_k$, $k \in \{0, 1, 2\}$.

Similarly to the reasoning above, consider *spaces of random* **K**-*variables* defined on the manifold M. Namely, let $\mathbf{H}_{0\mathbf{K}}^q \mathbf{L}_2$ and $\mathbf{H}_{2\mathbf{K}}^q \mathbf{L}_2$ be the spaces of differential q-forms with coefficients of $\mathcal{H}$-valued random **K**-variables, where $\mathbf{K} = \{\lambda_k\}$ is a monotone sequence such that $\sum_{k=1}^{\infty} \lambda_k^2 < +\infty$. The elements of these spaces are vectors $\alpha = \sum_{k=1}^{\infty} \lambda_k \xi_k \varphi_k$ and $\beta = \sum_{k=1}^{\infty} \lambda_k \xi \psi_k$, respectively, where $\{\varphi_k\}$ and $\{\psi_k\}$ are the operator eigenvectors, which are orthonormal with respect to $<\cdot, \cdot>_0$ and $<\cdot, \cdot>_2$. Finally, consider the spaces of **K**-"noises" $\mathbf{C}^l \mathbf{H}_{0\mathbf{K}}^q \mathbf{L}_2$ and $\mathbf{C}^l \mathbf{H}_{0\mathbf{K}}^q \mathbf{L}_2$, where $l \in \{0\} \bigcup N$, $q \in \{0, 1, \ldots, d\}$, and $\mathscr{I} \subset R$ is an interval.

3 Degenerate Holomorphic Groups and Semigroups of Operator in Spaces of Random K-Variables

Let $\mathscr{U}$ and $\mathscr{F}$ be real separable Hilbert spaces. Denote by $\mathscr{L}(\mathscr{U}; \mathscr{F})$ the space of linear continuous operators, and denote by $\mathscr{C}l(\mathscr{U}; \mathscr{F})$ the space of linear closed and densely defined operators. Construct Hilbert spaces $\mathbf{U}_\mathbf{K} \mathbf{L}_2$ and $\mathbf{F}_\mathbf{K} \mathbf{L}_2$, where $\mathbf{K} = \{\lambda_k\} \subset R$ is a monotone sequence such that $\sum_{k=1}^{\infty} \lambda_k^2 < +\infty$. The following lemma is correct.

Lemma 1 *(i) The operator* $A \in \mathscr{L}(\mathscr{U}; \mathscr{F})$ *iff* $A \in \mathscr{L}(\mathbf{U}_\mathbf{K} \mathbf{L}_2; \mathbf{F}_\mathbf{K} \mathbf{L}_2)$.
(ii) The operator $A \in \mathscr{C}l(\mathscr{U}; \mathscr{F})$ *iff* $A \in \mathscr{C}l(\mathbf{U}_\mathbf{K} \mathbf{L}_2; \mathbf{F}_\mathbf{K} \mathbf{L}_2)$.

Here the abbreviation *"iff"* means "if and only if".

Definition 1 The operator $M \in \mathscr{C}l(\mathbf{U}_\mathbf{K} \mathbf{L}_2; \mathbf{F}_\mathbf{K} \mathbf{L}_2)$ is called *p-sectorial with respect to the operator* $L \in \mathscr{L}(\mathbf{U}_\mathbf{K} \mathbf{L}_2; \mathbf{F}_\mathbf{K} \mathbf{L}_2)$ (for shortness, (L, p)-*sectorial*), $p \in \{0\} \cup N$, if

(i) there exist the constants $a \in R$ and $\Theta \in (\frac{\pi}{2}, \pi)$ such that the sector

$$S_{a,\Theta}^L(M) = \{\mu \in C : |arg(\mu - a)| < \Theta, \mu \neq a\} \subset \rho^L(M), \quad (16)$$

(ii) there exists the constant $K \in R_+$ such that

$$max\{\|R_{(\mu,p)}^L(M)\|_{\mathbf{U}_\mathbf{K} \mathbf{L}_2}, \|L_{(\mu,p)}^L(M)\|_{\mathbf{F}_\mathbf{K} \mathbf{L}_2}\} \leq \frac{K}{\prod_{q=0}^{p} |\mu_q - a|} \quad (17)$$

for any $\mu_0, \mu_1, \ldots, \mu_p \in S_{a,\Theta}^L(M)$.

Here $\rho^L(M) = \{\mu \in C : (\mu L - M)^{-1} \in \mathscr{L}(\mathbf{F_K L_2}; \mathbf{U_K L_2})\}$ is the L-resolvent set, and $\sigma^L(M) = C \setminus \rho^L(M)$ is the L-spectrum of the operator M. Suppose that $\mu_q \in \rho^L(M)$, $q = 0, 1, \ldots, p$. Operator functions $R_\mu^L(M) = (\mu L - M)^{-1} L$ and $L_\mu^L(M) = L(\mu L - M)^{-1}$ are called the *right L-resolvent* and the *left L-resolvent* of the operator M, respectively, and $R_{(\mu, p)}^L(M) = \prod_{q=0}^{p} (\mu_q L - M)^{-1} L$ and $L_{(\mu, p)}^L(M) = \prod_{q=0}^{p} L(\mu_q L - M)^{-1}$ are called the *right (L, p)-resolvent* and the *left (L, p)-resolvent* of the operator M, respectively.

Definition 2 A mapping $V^\bullet \in C(R_+; \mathscr{L}(\mathbf{H_K L_2}))$ $(V^\bullet \in C(R; \mathscr{L}(\mathbf{H_K L_2})))$ is called a *semigroup (group)* in the Hilbert space $\mathbf{H_K L_2}$, if

$$V^s V^t = V^{s+t} \quad \forall s, t \in R_+ \ (\forall s, t \in R). \tag{18}$$

Let us identify the semigroup (group) with its graph $\{V^t : t \in R_+\}$ ($\{V^t : t \in R\}$). A semigroup $\{V^t : t \in R_+\}$ is called *holomorphic*, if the semigroup is analytic in some sector containing the ray R_+, and condition (18) holds. A group is called *holomorphic*, if the group is analytic in the entire complex plane C, and condition (18) holds. A semigroup (group) is called *uniformly bounded*, if $\|V^t\|_{\mathbf{H_K L_2}} \leq const$ on any compact in R_+ (R).

Theorem 1 *Suppose that the operator M is (L, p)-sectorial, then there exists a holomorphic and uniformly bounded semigroup of operators*

$$U^t = \frac{1}{2\pi i} \int_\Gamma R_\mu^L(M) e^{\mu t} d\mu \ \left(F^t = \frac{1}{2\pi i} \int_\Gamma L_\mu^L(M) e^{\mu t} d\mu \right), \tag{19}$$

where the contour $\Gamma \subset \rho^L(M)$ is such that $|\arg \mu| \to \Theta$ for $\mu \to \infty$, $\mu \in \Gamma$, $t \in R_+$.

The set $\ker V^\bullet = \{v \in \mathbf{H_K L_2} : V^t v = 0 \ \exists t \in R_+\}$ is called the *kernel*, and the set $\operatorname{im} V^\bullet = \{v \in \mathbf{H_K L_2} : \lim_{t \to 0_+} V^t v = v\}$ is called the *image* of the holomorphic semigroup $\{V^t : t \in R_+\}$. Denote by $\mathbf{U_K^1 L_2}$ ($\mathbf{F_K^1 L_2}$) the subspace that is the closure of $\operatorname{im} R_{(\mu, p)}^L(M)$ ($\operatorname{im} L_{(\mu, p)}^L(M)$) in the norm of the space $\mathbf{U_K L_2}$ ($\mathbf{U_K L_2}$).

Theorem 2 *Suppose that the operator M is (L, p)-sectorial, then $\operatorname{im} U^\bullet = \mathbf{U_K^1 L_2}$ ($\operatorname{im} F^\bullet = \mathbf{F_K^1 L_2}$).*

Let the operator M be (L, p)-sectorial. Suppose that $\ker U^\bullet = \mathbf{U_K^0 L_2}$ and $\ker F^\bullet = \mathbf{F_K^0 L_2}$. Denote by L^k (M^k) the restriction of the operator L (M) on $\mathbf{U_K^k L_2}$ ($\operatorname{dom} M \cap \mathbf{U_K^k L_2}$), $k = 0, 1$.

Theorem 3 *Suppose that the operator M is (L, p)-sectorial, then*

(i) the operator $L_0 \in \mathscr{L}(\mathbf{U}_\mathbf{K}^0 \mathbf{L}_2; \mathbf{F}_\mathbf{K}^0 \mathbf{L}_2)$, and the operator $M_0 : \mathrm{dom}\, M \cap \mathbf{U}_\mathbf{K}^0 \mathbf{L}_2 \to \mathbf{F}_\mathbf{K}^0 \mathbf{L}_2$;

(ii) there exists the operator $M_0^{-1} \in \mathscr{L}(\mathbf{F}_\mathbf{K}^0 \mathbf{L}_2; \mathbf{U}_\mathbf{K}^0 \mathbf{L}_2)$;

(iii) the operator $H = M_0^{-1} L_0 \in \mathscr{L}(\mathbf{U}_\mathbf{K}^0 \mathbf{L}_2)$ is nilpotent, and the degree of its nilpotency does not exceed p.

The proof of Theorems 1, 2, and 3 follows from the corresponding results [1] and Lemma 1.

Suppose that the conditions

$$\mathbf{U}_\mathbf{K} \mathbf{L}_2 = \mathbf{U}_\mathbf{K}^0 \mathbf{L}_2 \oplus \mathbf{U}_\mathbf{K}^1 \mathbf{L}_2, \quad \mathbf{F}_\mathbf{K} \mathbf{L}_2 = \mathbf{F}_\mathbf{K}^0 \mathbf{L}_2 \oplus \mathbf{F}_\mathbf{K}^0 \mathbf{L}_2 \tag{20}$$

are fulfilled, and there exists the operator

$$L_1^{-1} \in \mathscr{L}(\mathbf{F}_\mathbf{K}^1 \mathbf{L}_2; \mathbf{U}_\mathbf{K}^1 \mathbf{L}_2). \tag{21}$$

Remark 1 Conditions (20) are fulfilled, for example, if the operator M is strongly (L, p)-sectorial on the right (left) [1]. Condition (21) is fulfilled, if either the operator M is strongly (L, p)-sectorial, or the operator M is (L, p)-sectorial, conditions (20) are fulfilled, and $\mathrm{im}\, L = \mathbf{U}_\mathbf{K}^1 \mathbf{L}_2$. Conditions (20) are equivalent to the existence of the projectors $P : \mathbf{U}_\mathbf{K} \mathbf{L}_2 \to \mathbf{U}_\mathbf{K}^0 \mathbf{L}_2$ and $Q : \mathbf{F}_\mathbf{K} \mathbf{L}_2 \to \mathbf{F}_\mathbf{K}^0 \mathbf{L}_2$.

Suppose that the operator M is (L, p)-sectorial, and conditions (20) and (21) are fulfilled. Equation (1) can be reduced to a pair of equations

$$R_\alpha^L \overset{o}{\eta}_u = (\alpha L - M)^{-1} M \eta_u, \tag{22}$$

$$L_\alpha^L \overset{o}{\eta}_f = M(\alpha L - M)^{-1} \eta_f, \tag{23}$$

which is equivalent to (1). Equations (22) and (23) can be considered as specific interpretations of the equation

$$A \overset{o}{\zeta} = B\zeta, \tag{24}$$

where the operators $A, B \in \mathscr{L}(\mathbf{H}_\mathbf{K} \mathbf{L}_2)$. A stochastic K-process $\zeta \in \mathbf{C}^1(R_+; \mathbf{H}_\mathbf{K} \mathbf{L}_2)$ is called a *solution to equation* (24), if substitution of ζ in (24) gives an identity. Consider the Cauchy problem

$$\zeta(0) = \zeta_0 \tag{25}$$

for Eq. (24), where $\zeta_0 = \sum_{k=0}^\infty \lambda_k \zeta_{0k} \varphi_k$ is a random K-variable. A solution to equation (24) is called a *solution to problem* (24), (25), if equality (25) holds for some random K-variable $\zeta_0 \in \mathbf{H}_\mathbf{K} \mathbf{L}_2$.

Remark 2 In the definition of a solution, we use equality of stochastic processes. We say that the stochastic processes $\zeta_1 = \zeta_1(t, \omega)$ and $\zeta_2 = \zeta_2(t, \omega)$ are *equivalent*, if almost sure (i.e. for almost all $\omega \in \mathscr{A}$) the trajectories $\zeta_1 = \zeta_1(t, \cdot)$ and $\zeta_2 = \zeta_2(t, \cdot)$ coincide.

Definition 3 A set $\mathscr{P} \subset \mathbf{H_K L_2}$ is called the *phase space* of equation (24), if
(i) almost sure each trajectory of the solution $\zeta = \zeta(t)$ to Eq. (24) belongs to $\mathscr{P}$, i.e. $\zeta(t) \in \mathscr{P}, t \in R_+$, for almost all trajectories;
(ii) for $\zeta_0 \in \mathscr{P}$, there exists a solution to problem (24), (25).

Theorem 4 *Suppose that the operator M is (L, p)-sectorial. Then the phase space $\mathscr{P}$ of Eq. (24) is the image $\operatorname{im} V^\bullet$ of the semigroup of the form (19).*

Remark 3 If the image of the semigroup $\{U^t : t \in R_+\}$ coincides with the phase space of Eq. (24), then (see [1]) the semigroup is called the *resolving semigroup of Eq. (24)*.

Remark 4 Under conditions of Theorem 4, the solution to problem (24), (25) has the form $\zeta(t) = U^t \zeta_0 = \sum_{k=0}^{\infty} \lambda_k \zeta_{0k} V^t \varphi_k$ for any $\zeta_0 \in \mathscr{P}$.

Theorem 5 *Suppose that the operator M is (L, p)-sectorial, and conditions (20), (21) are fulfilled, then*
(i) the projectors P, Q can be represented as

$$P = U^0 = s - \lim_{t \to 0_+} U^t, \quad Q = F^0 = s - \lim_{t \to 0_+} F^t;$$

(ii) the operators $M_k \in \mathscr{C}l(\operatorname{dom} M \cap \mathbf{U_K^k L_2}; \mathbf{F_K^k L_2})$, $k = 1, 2$;
(iii) the operator $S = L_1^{-1} M_1$ is sectorial.

Suppose that L-spectrum of the operator M is such that

$$\left.\begin{array}{c} \sigma^L(M) = \sigma_1^L(M) \cup \sigma_2^L(M), \quad \sigma_1^L(M) \neq \varnothing, \\ \text{there exists the closed loop } \gamma \in C \\ \text{that bounds a region containing } \sigma_1^L(M), \text{ and} \\ \sigma_2^L(M) \text{ does not belong to this region,} \\ \gamma \cap \sigma_2^L(M) = \varnothing. \end{array}\right\} \quad (26)$$

Theorem 6 *Suppose that condition (26) is fulfilled, then there exists a holomorphic group*

$$U_1^t = \frac{1}{2\pi i} \int_\gamma R_\mu^L(M) e^{\mu t} d\mu \left(F_1^t = \frac{1}{2\pi i} \int_\gamma L_\mu^L(M) e^{\mu t} d\mu \right), \quad t \in R_+, \quad (27)$$

where the contour $\gamma \subset \rho^L(M)$ is taken from (26).

Definition 4 A subspace $\mathscr{D} \subset \mathbf{U_K L_2}$ is called the *invariant space* of equation (1), if there exist the solution η to Eq. (1) with the Cauchy condition $\eta(0) = \eta_0$ is such that $\eta \in \mathbf{C}^1(R_+; \mathscr{D})$ for any $\eta_0 \in \mathscr{D}$.

Note that if Eq. (1) has the phase space $\mathscr{P}$, then any invariant space $\mathscr{I}$ of this equation is such that $\mathscr{I} \subset \mathscr{P}$.

Definition 5 *Solutions* $\eta = \eta(t)$ *to Eq. (1) have exponential dichotomy, if*
(i) the phase space $\mathscr{P}$ of Eq. (1) splits into the direct sum of two invariant spaces (i.e. $\mathscr{P} = \mathscr{I}^+ \oplus \mathscr{I}^-$), and
(ii) there exist the constants $N_k \in R_+, v_k \in R_+, k = 1, 2$, such that

$$\|\eta^1(t)\|_{\mathbf{U_K L_2}} \le N_1 e^{-v_1(s-t)} \|\eta^1(s)\|_{\mathbf{U_K L_2}} \quad \text{for} \quad s \ge t,$$
$$\|\eta^2(t)\|_{\mathbf{U_K L_2}} \le N_2 e^{-v_2(t-s)} \|\eta^2(s)\|_{\mathbf{U_K L_2}} \quad \text{for} \quad t \ge s,$$

where $\eta_1 = \eta_1(t) \in \mathscr{I}^+$ and $\eta_2 = \eta_2(t) \in \mathscr{I}^-$ for all $t \in R$. The space $\mathscr{I}^+$ ($\mathscr{I}^-$) is called the *stable (unstable) invariant space* of Eq. (1).

Theorem 7 *Suppose that the operator M is (L, p)-sectorial, $p \in \{0\} \bigcup N$, and $\sigma^L(M) \bigcap \{iR\} = \emptyset$. Then solutions $\eta = \eta(t)$ to Eq. (1) have exponential dichotomy.*

By analogy with Sect. 2, we transfer all reasonings on the spaces $\mathbf{H}_{0\mathbf{K}}^q \mathbf{L}_2$ and $\mathbf{H}_{2\mathbf{K}}^q \mathbf{L}_2$ of random $\mathbf{K}$-variables defined on the manifold $\mathscr{M}$. There exist continuous and dense embeddings $\mathbf{H}_{2\mathbf{K}}^q \mathbf{L}_2 \subset \mathbf{H}_{1\mathbf{K}}^q \mathbf{L}_2 \subset \mathbf{H}_{0\mathbf{K}}^q \mathbf{L}_2$, and for any $q = 0, 1, \ldots, d$ the spaces can be splitted as follows:

$$\mathbf{H}_{l\mathbf{K}}^q \mathbf{L}_2 = \mathbf{H}_{l\mathbf{K}_\Delta}^{q\perp} \mathbf{L}_2 \oplus \mathbf{H}_{l\mathbf{K}_\Delta}^q \mathbf{L}_2, \tag{28}$$

where $\mathbf{H}_{l\mathbf{K}_\Delta}^{q\perp} \mathbf{L}_2 = (I - \mathbf{P}_\Delta)[\mathbf{H}_{l\mathbf{K}}^q \mathbf{L}_2]$, $l = 0, 1, 2$, and $\mathbf{P}_\Delta$ is a projector on the subspace of harmonic forms. Suppose that the operators $L \in \mathscr{L}(\mathbf{H}_{l\mathbf{K}_\Delta}^{2\perp} \mathbf{L}_2; \mathbf{H}_{l\mathbf{K}_\Delta}^{0\perp} \mathbf{L}_2)$, $M \in \mathscr{Cl}(\mathbf{H}_{l\mathbf{K}_\Delta}^{2\perp} \mathbf{L}_2; \mathbf{H}_{l\mathbf{K}_\Delta}^{0\perp} \mathbf{L}_2)$.

Theorem 8 *Suppose that the operator M is (L, p)-sectorial. Then the phase space of Eq. (1) coincides with the image of the semigroup of the form (19).*

Since the set $\{\mu \in \sigma^L(M) : Re\mu = 0\}$ does not belong to the space $\mathbf{H}_{l\mathbf{K}_\Delta}^{2\perp} \mathbf{L}_2$, the theorem below follows from Theorem 7.

Theorem 9 *Suppose that the operator M is (L, p)-sectorial. Then there exist the stable $\mathscr{I}^+$ and unstable $\mathscr{I}^-$ invariant spaces of Eq. (1), and solutions $\eta = \eta(t)$ to Eq. (1) have exponential dichotomy.*

4 Degenerate Holomorphic Semigroups in the Spaces of K-"Noises"

Consider the holomorphic semigroups constructed in Sect. 3 in the case of inhomogeneous stochastic Sobolev type equation. Suppose that the operator M is (L, p)-sectorial, $p \in \{0\} \cup N$, conditions (20) and (21) are fulfilled, and $\Theta \in \mathbf{C}^l(\mathscr{I}; \mathbf{F_K L_2})$ is a stochastic K-process, where $\mathscr{I} \subset R$. Consider the inhomogeneous equation

$$L \overset{o}{\eta} = M\eta + \Theta, \qquad (29)$$

and use Theorem 3 in order to reduce Eq. (29) to the equivalent form:

$$H \overset{o}{\eta}{}^0 = \eta^0 + M_0^{-1}(I - Q)\Theta^0, \qquad (30)$$

$$\overset{o}{\eta}{}^1 = S\eta^1 + L_1^{-1} Q\Theta^1. \qquad (31)$$

Remark 5 Since the operator H is nilpotent, the Cauchy problem $\eta^0(0) = \eta_0^0$ for Eq. (30) is unsolvable, if

$$\eta_0^0 \neq -\sum_{q=0}^{p} H^p M_0^{-1} \frac{d^q \Theta^0}{dt^q}(0). \qquad (32)$$

Therefore, additional conditions on η_0 are necessary for solvability of the Cauchy problem $\eta(0) = \eta_0$ for Eq. (29). The conditions depend on the right-hand side of the equation. These difficulties can be avoided, if the initial conditions are considered to be the Showalter-Sidorov conditions. Suppose that $\lim_{t \to 0+} (R_\alpha^L(M))^{p+1} (\eta(L) - \eta_0) = 0$ for Eq. (2). According to Theorem 3 and conditions (20) and (21), Showalter-Sidorov condition (33) is equivalent to the condition

$$\lim_{t \to 0+} P(\eta(0) - \eta_0) = 0. \qquad (33)$$

A solution $u = u(t)$ to Eq. (29) is called a *solution to problem* (29), (33), if met condition (33).

Theorem 10 *Let the operator M be strongly (L, p)-sectorial, $p \in \{0\} \cup N$. Then, for any stochastic K-process $\Theta \in \mathbf{C}^l(\mathscr{I}; \mathbf{F_K L_2})$ and any $\eta_0 \in \mathbf{U_K L_2}$, there exists almost sure the unique solution $\eta \in C^1(\mathscr{I}; \mathbf{U_K L_2})$ to Showalter-Sidorov problem (33) for Eq. (29) of the form*

$$u(t) = U^t u_0 - \sum_{q=0}^{p} H^p M_0^{-1} \frac{d^q \Theta^0}{dt^q}(t) + \int_0^t U^{t-s} \Theta ds. \qquad (34)$$

Remark 6 In real life, we can study only trajectories of random processes. Therefore, we understand almost sure uniqueness of the solution to problem (29), (33) as follows. If $\eta_1 = \eta_1(t, \omega)$ and $\eta_2 = \eta_2(t, \omega)$ are the solutions to problem (29), (33), then almost sure the trajectories $\eta_1 = \eta_1(t, \cdot)$ and $\eta_2 = \eta_2(t, \cdot)$ coincide.

Example 1 Consider an application of the obtained results to the equation simulated the change in free surface of filtered liquid in the spaces of **K**-"noises" defined on smooth compact oriented Riemannian manifold without boundary, see Sect. 2. We take a two-dimensional torus $T^2 = [0, 2\pi] \times [0, \pi]$ (with 0 and 2π and 0 and π identified) as such a manifold, and consider the stochastic variant of the Dzektser equation

$$(\lambda - \Delta)u_t = \alpha \Delta u - \beta \Delta^2 u. \tag{35}$$

Define the operators L and M by the formulas

$$L = (\lambda - d\delta - \delta d), \quad M = \alpha(d\delta + \delta d) - \beta(d\delta + \delta d)^2. \tag{36}$$

See [8] for the equation in the deterministic case on Riemannian manifold without boundary.

Lemma 2 *For any $\alpha, \beta, \lambda \in R \setminus \{0\}$ and $\lambda \neq \frac{\alpha}{\beta}$, operator M be $(L, 0)$-sectorial.*

Taking into account the general representation of the Laplace–Beltrami operator on the manifold Ω_d with Riemannian metric g as $\Delta_{\Omega d} = \frac{1}{\sqrt{|g|}} \partial_i(\sqrt{|g|} g^{ij} \partial_j)$, we represent the Laplace–Beltrami operator on the two-dimensional torus with coordinates x, y as follows:

$$\Delta_{T^2} = (8\pi^2 \partial_x^2 - 4\pi \partial_x \partial_y + \partial_y^2). \tag{37}$$

The eigenvalues v_l are calculated by the Courant–Fischer formulas

$$v_l(a, b) = \min_{E \subset Z^2, |E|=l+1} \max_{(c_1, c_2) \subset E} 4\pi^2 \left[c_1^2 \left(1 + \frac{a^2}{b^2}\right) - 2c_1 c_2 \frac{a}{b^2} + \frac{c^2}{b^2} \right] \tag{38}$$

for $a = 2\pi, b = \pi$ and we can calculated, for example, for 0-form eigenvalues

$$v_{n,m} = \left(\frac{n}{2}\right)^2 + (m)^2, \quad n, m \in N, \ x \in [0, 2\pi), \ y \in [0, \pi) \tag{39}$$

and eigenfunctions

$$u_{n,m}(x, y) = \sin \frac{n}{2} x \cdot \sin my. \tag{40}$$

The obtained sequence of eigenvalues $\{v_l\}$ is non-negative and nondecreasing, has finite multiplicity, and converges only to ∞, while the sequence of corresponding eigenfunctions $\{v_l\}$ forms the necessary orthonormal basis in $\mathscr{U}^1 = \mathbf{H}_{0\mathbf{K}\Delta}^{q\perp} \mathbf{L}_2$, see Sect. 2.

Since dimension of the manifold is $d = 2$, we obtain the solutions of two types: for 0-forms (and for 2-forms that are isomorphic to them), and for 1-forms. The results of the paper [2] are correct for 0-forms (and for 2-forms that are isomorphic to them), and there exists a solution for any η_0 from the phase space $\mathscr{P}$. In the case of 1-forms, it is necessary to establish equalities in all terms of the differential forms thats way eigenfunctions

$$u_{n,m}(x, y) = \sin \frac{n}{2}x \cdot \sin my dx, u_{n,m}(x, y) = \cos \frac{n}{2}x \cdot \sin my dx, \quad (41)$$

$$u_{n,m}(x, y) = \sin \frac{n}{2}x \cdot \sin my dy, u_{n,m}(x, y) = \cos \frac{n}{2}x \cdot \sin my dy. \quad (42)$$

Theorem 11 *For any $\alpha, \beta, \lambda \in R \setminus \{0\}$ and $\lambda \neq \frac{\alpha}{\beta}$, there exists a solution $\eta = \eta(t)$ to the Cauchy problem $\eta(0) = \eta_0 \in \mathscr{P}$ for Eq. (35) of the form*

$$\eta(t) = \sum_{n=1}^{\infty}{}' \left(\sum_{m=1}^{\infty} exp\left(\frac{\alpha v_{n,m}}{\lambda - v_{n,m}}t\right) \lambda_m \xi_m <u_{m,n}, u_{m,n}>_0 u_{m,n} \right). \quad (43)$$

5 Conclusions

Note various papers on both the applications of stochastic equations [14–16] and the theoretical investigations [17–19]. As a whole, the results of the paper are in general trend of research on Sobolev type stochastic equations that is outlined during last years. The next steps in the study of stochastic equations of Sobolev type should be made in the direction of the case of p-radial operator [6].

Acknowledgements The work was supported by Act 211 Government of the Russian Federation, contract no. 02.A03.21.0011.

References

1. Sviridyuk, G.A., Fedorov, V.E.: Linear Sobolev Type Equations and Degenerate Semigroups of Operators. VSP, Utrecht-Boston-Koln-Tokyo (2003)
2. Favini, A., Sviridyuk, G.A., Manakova, N.A.: Linear Sobolev type equations with relatively p-sectorial operators in space of "Noises". Abstract Appl. Anal. Article ID **69741**, 8p (2015). https://doi.org/10.1155/2015/697410
3. Gliklikh, YuE, Mohammed, S.E.A.: Stochastic delay equations and inclusions with mean derivatives on Riemannian manifolds. Glob. Stoch. Anal. 1(1), 49–56 (2014)
4. Sviridyuk, G.A., Manakova, N.A.: The dynamical models of Sobolev type with Showalter–Sidorov condition and additive "noise". Bulle. South Ural State Univ. Ser. Math. Model. Program. Comput. Softw. Chelyabinsk. 1(7) 90–103 (2014). https://doi.org/10.14529/mmp140108
5. Favini, A., Sviridyuk, G.A., Zamyshlyaeva, A.A.: One class of sobolev type equations of higher order with additive "white noise". Commun. Pure Appli. Anal. Springer. 1(15), 185–196 (2016)

6. Favini, A., Sviridyuk, G.A., Sagadeeva, M.A.: Linear Sobolev type equations with relatively p-radial operators in space of "Noises". Mediter. J. Math. **6**(13), 4607–4621 (2016)
7. Favini, A., Zagrebina, S.A., Sviridyuk, G.A.: Multipoint initial-final value problems for dynamical Sobolev-type equations in the space of noises. Electron. J. Differ. Equ. **128**, 1–10 (2018)
8. Shafranov, D.E.: On the Cauchy problem for the equation of free surface of filtred fluid on the manifolds. Bull. South Ural State Univ. Ser. Math. Model. Program. Comput. Softw. Chelyabinsk. **27**(127), 117–120 (2008). (in Russian)
9. Sviridyuk, G.A., Keller, A.V.: Invariant spaces and dichotomies of solutions of a class of linear equations of Sobolev type. Izv. Vyssh. Uchebn. Zaved. Mat. **5**, 60–68 (1997)
10. Shafranov, D.E., Kitaeva, O.G.: The Barenblatt-Zheltov-Kochina model with the Showalter-Sidorov condition and additive "white noise" in spaces of differential forms on Riemannian manifolds without boundary. Glob. Stoch. Anal. **2**(5), 145–159 (2018)
11. Kitaeva, O.G., Shafranov, D.E., Sviridiuk, G.A.: Exponential Dichotomies in the Barenblatt–Zheltov–Kochina model in spaces of differential forms with "Noise". Bull. South Ural State Univ. Ser. Math. Modell. Program. Comput. Softw. (Bulletin SUSU MMCS). **2**(12), 47–57 (2019). https://doi.org/10.14529/mmp190204
12. Sagadeeva, M.A.: Reconstruction of observation from distorted data for the optimal dinamic measurement problem. Bull. South Ural State Univ. Ser. Math. Model. Program. Comput. Softw. (Bulletin SUSU MMCS). **2**(12), 82–96 (2019). https://doi.org/10.14529/mmp190207
13. Warner, F.W.: Foundations of Differentiable Manifolds and Lie Groups. Springer, New York (1983)
14. Banasiak, J., Lachowicz, M., Moszynski, M.: Chaotic behavior of semigroups related to the process of gene amplification-deamplification with cell proliferation. Math. Biosci. **2**(206), 200–2015 (2007)
15. Banasiak, J., Lamb, W.: Analytic fragmentation semigroups and continuous coagulation-fragmentation equations with unbounded rates. J. Math. Anal. Appl. **1**(391), 312–322 (2012)
16. Banasiak, J., Falkiewicz, A., Namayanja, P.: Asymptotic state lumping in transport and diffusion problems on networks with applications to population problems. Math. Models Methods Appl. Sci. **2**(26), 215–247 (2016)
17. Melnikova, I.V.: Abstract stochastic equations II Solutions spaces of abstract stochastic distributions. J. Math. Sci. **5**(116), 3620–3656 (2003)
18. Zakirova, G.A., Manakova, N.A., Sviridyuk, G.A.: The asymptotics of eigenvalues of a differential operator in the stochastic models with "white noise". Appl. Math. Sci. **8**(173–176), 8747–8754 (2014)
19. Zamyshlyaeva, A.A., Keller, A.V., Syropiatov, M.B.: Stochastic model of optimal dynamic measurements. Bull. South Ural State Univ. Ser. Math. Model. Program. Comput. Softw. (Bulletin SUSU MMCS). **2**(11), 147–153 (2018). https://doi.org/10.14529/mmp180212

Optimal Energy Decay in a One-Dimensional Wave-Heat-Wave System

Abraham C. S. Ng

Abstract Harnessing the abstract power of the celebrated result due to Borichev and Tomilov (Math. Ann. 347:455–478, 2010, no. 2), we study the energy decay in a one-dimensional coupled wave-heat-wave system. We obtain a sharp estimate for the rate of energy decay of classical solutions by first proving a growth bound for the resolvent of the semigroup generator and then applying the asymptotic theory of C_0-semigroups. The present article can be naturally thought of as an extension of a recent paper by Batty, Paunonen, and Seifert (J. Evol. Equ. 16:649–664, 2016) which studied a similar wave-heat system via the same theoretical framework.

Keywords Coupled wave-heat equation · Energy · Rates of decay · C_0-semigroups · Resolvent estimates

2010 Mathematics Subject Classifications 35M33 · 35B40 · 47D06 (34K30)

1 Introduction

In this article, we apply the theorem of Borichev–Tomilov [6, Theorem 4.1] to a one-dimensional system with coupled wave and heat parts. This application is modelled upon the 2016 paper of Batty, Paunonen, and Seifert [4] where the 'optimal energy decay in a one-dimensional coupled wave-heat system' with finite Neumann wave and Dirichlet heat parts was studied by analysing the following system:

A. C. S. Ng (✉)
St Edmund Hall, University of Oxford, Queen's Lane, Oxford OX1 4AR, UK
e-mail: abraham.ng@maths.ox.ac.uk

$$\begin{cases} u_{tt}(\xi,t) = u_{\xi\xi}(\xi,t), & \xi \in (-1,0),\ t > 0, \\ w_t(\xi,t) = w_{\xi\xi}(\xi,t), & \xi \in (0,1),\ t > 0, \\ u_t(0,t) = w(0,t), \quad u_\xi(0,t) = w_\xi(0,t), & t > 0, \\ u_\xi(-1,t) = 0, \quad w(1,t) = 0, & t > 0, \\ u(\xi,0) = u(\xi), \quad u_t(\xi,0) = v(\xi) & \xi \in (-1,0), \\ w(\xi,0) = w(\xi), & \xi \in (0,1), \end{cases} \qquad (1)$$

where the initial data u, v, and w lived in $H^1(-1,0)$, $L^2(-1,0)$ and $L^2(0,1)$ respectively. The energy was then defined, given a vector of initial data $x = (u,v,w)$, as

$$E_x(t) = \frac{1}{2}\int_{-1}^{1} |u_\xi(\xi,t)|^2 + |u_t(\xi,t)|^2 + |w(\xi,t)|^2\, d\xi, \quad t \geq 0,$$

with all the functions being understood to have been extended by zero in ξ to the interval $(-1,1)$. If the solution is sufficiently regular, a routine calculation via integration by parts shows that

$$E'_x(t) = -\int_0^1 |w_\xi(\xi,t)|^2\, d\xi, \quad t \geq 0,$$

and, in particular, that the energy of any such solution is non-increasing with respect to time. The main goal of analysing such a model is to quantitatively estimate the rate of energy decay of a given solution.

The system (1) was first studied (with Dirichlet boundary at $\xi = -1$ and a slightly different coupling condition) in [13], yielding the sharp decay rate $E_x(t) = O(t^{-4})$, $t \to \infty$ (see below for the meaning of 'big O' notation). The approach in [13] relied on a rather complicated spectral analysis used in conjunction with the theory of Riesz spectral operators. In contrast to [13], however, the approach in [4] was based on the semigroup methods of non-uniform stability pioneered by Batty and Duyckaerts in [3], widely popularised by Borichev and Tomilov in [6], and largely completed by Rozendaal, Seifert, and Stahn in [10], greatly simplifying the analysis necessary to obtain the rate of decay.

The motivation of studying models like this and, in particular, the one in this article presented below, stems mainly from the study of fluid-structure models where, often in higher-dimensional settings, the Navier–Stokes equations (the fluid half) are coupled with the nonlinear elasticity equation (the structure half). We refer to [4, Sect. 1] and [2] for surveys of similar problems (see also [5] where the same approach with suitable adjustments is applied to study a wave-heat system on a rectangular domain and [9] where a similar method is used to study a one-dimensional wave-heat system with infinite heat part).

In this article, we add an extra wave component to the system (1) and take Dirichlet boundary conditions on both ends, analysing the following wave-heat-wave system:

Optimal Energy Decay in a One-Dimensional Wave-Heat-Wave System

$$\begin{cases} u_{tt}(\xi,t) = u_{\xi\xi}(\xi,t), & \xi \in (0,1), t > 0, \\ w_t(\xi,t) = w_{\xi\xi}(\xi,t), & \xi \in (1,2), t > 0, \\ \tilde{u}_{tt}(\xi,t) = \tilde{u}_{\xi\xi}(\xi,t), & \xi \in (2,3), t > 0, \\ u(0,t) = \tilde{u}(3,t) = 0, & t > 0, \\ u_t(1,t) = w(1,t), \quad u_\xi(1,t) = w_\xi(1,t), & t > 0, \\ \tilde{u}_t(2,t) = w(2,t), \quad \tilde{u}_\xi(2,t) = w_\xi(2,t), & t > 0, \\ u(\xi,0) = u(\xi), \quad u_t(\xi,0) = v(\xi), & \xi \in (0,1), \\ w(\xi,0) = w(\xi), & \xi \in (1,2), \\ \tilde{u}(\xi,0) = \tilde{u}(\xi), \quad \tilde{u}_t(\xi,0) = \tilde{v}(\xi), & \xi \in (2,3). \end{cases} \quad (2)$$

The initial data is required to satisfy $u = u(\xi, 0) \in H^1(0,1)$, $v = u_t(\xi, 0) \in L^2(0,1)$, $w=w(\xi,0) \in L^2(1,2)$, $\tilde{u}=\tilde{u}(\xi,0) \in H^1(2,3)$, and $\tilde{v} = \tilde{u}_t(\xi, 0) \in L^2(2,3)$.

As in [4], the aim here is to find a quantitative estimate for the rate of energy decay of a given solution. Given a vector of initial data $x = (u, v, w, \tilde{u}, \tilde{v})$ satisfying the conditions above, we similarly define the energy of the corresponding solution as

$$E_x(t) = \frac{1}{2} \int_0^3 |u_\xi(\xi,t)|^2 + |u_t(\xi,t)|^2 + |w(\xi,t)|^2 + |\tilde{u}_\xi(\xi,t)|^2 + |\tilde{u}_t(\xi,t)|^2 \, d\xi, \quad t \geq 0.$$

Again, all functions have been extended by zero in ξ to the interval $(0,3)$. Provided we have sufficient regularity of the solution, a simple calculation via integration by parts shows that

$$E'_x(t) = \operatorname{Re}\left\{ \tilde{u}_\xi(3,t)\overline{\tilde{u}_t(3,t)} - u_\xi(0,t)\overline{u_t(0,t)} \right\} - \int_1^2 |w_\xi(\xi,t)|^2 \, d\xi, \quad t \geq 0.$$

Since $u_t(0,t) = \frac{\partial}{\partial t} u(0,t) = \tilde{u}_t(3,t) = \frac{\partial}{\partial t} \tilde{u}(3,t) = 0$ for $t > 0$, the energy of any such solution is non-increasing with respect to time. The remaining sections are devoted to obtaining a sharp quantitative estimate for the rate of this decay for classical solutions of (2), but first, we detail below, the mostly standard notation used in this article.

Closely following the notation of [4], the domain, kernel, range, spectrum, and range of a closed operator A acting on a Hilbert space (always complex by assumption) will be denoted by $D(A)$, $\operatorname{Ker} A$, $\operatorname{Ran} A$, $\sigma(A)$ and $\rho(A)$ respectively. For $\lambda \in \rho(A)$, we write $R(\lambda, A)$ to signify the resolvent operator $(\lambda - A)^{-1}$. For $\lambda \in \mathbb{C}$, we define the square root $\sqrt{\lambda}$ by taking the branch cut along the negative real axis, that is, for $\lambda = re^{i\theta}$ where $r \geq 0$ and $\theta \in (-\pi, \pi]$, we let $\sqrt{\lambda} = r^{1/2} e^{i\theta/2}$. We also denote the closed complex left half-plane by $\mathbb{C}_- := \{z \in \mathbb{C} : \operatorname{Re} z < 0\}$. Finally, given two functions $f, g : (0, \infty) \to [0, \infty]$ and $a \in [0, \infty]$ fixed, we write $f(t) = O(g(t))$, $t \to \infty$, to indicate that there exists some constant $C > 0$ such that $f(t) \leq Cg(t)$ for all t sufficiently large, the so-called 'big O notation'. If g is strictly positive for all sufficiently large $t > 0$, we write $f(t) = o(g(t))$, $t \to \infty$, to mean that $f(t)/g(t) \to 0$ as $t \to \infty$, the so-called 'little o notation'. If p and q are non-negative real-valued quantities, the notation $p \lesssim q$ denotes that $p \leq Cq$ for some constant $C > 0$ that is independent of any varying parameters in a given context.

2 Well-Posedness—The Semigroup and Its Generator

In this section, we first prove that (2) is well posed and has solution given by the orbits of a C_0-semigroup of contractions $(T(t))_{t \geq 0}$, before turning to analyse the spectrum of the generator A of $(T(t))_{t \geq 0}$.

2.1 Existence of the Semigroup

We start by recasting (2) into an abstract Cauchy problem in order to later apply the methods of non-uniform stability. Consider the Hilbert space

$$X_0 = H^1(0, 1) \times L^2(0, 1) \times L^2(1, 2) \times H^1(2, 3) \times L^2(2, 3)$$

and define

$$X = \{(u, v, w, \tilde{u}, \tilde{v}) \in X_0 : u(0) = \tilde{u}(3) = 0\}$$

endowed with the norm (and corresponding inner product) given by

$$\|(u, v, w, \tilde{u}, \tilde{v})\|_X^2 = \|u'\|_{L^2}^2 + \|v\|_{L^2}^2 + \|w\|_{L^2}^2 + \|\tilde{u}'\|_{L^2}^2 + \|\tilde{v}\|_{L^2}^2$$

which is non-degenerate because the fundamental theorem of calculus applied in conjunction with the boundary conditions $u(0) = \tilde{u}(3) = 0$ implies that $\|u\|_{L^2} \lesssim \|u'\|_{L^2}$ and $\|\tilde{u}\|_{L^2} \lesssim \|\tilde{u}'\|_{L^2}$. Here and in the rest of the article, the intervals for function spaces appearing as subscripts will often be omitted if they are clear from the context. Let

$$X_1 = X \cap [H^2(0, 1) \times H^1(0, 1) \times H^2(1, 2) \times H^2(2, 3) \times H^1(2, 3)]$$

and define the operator A on X by $Ax = (v, u'', w'', \tilde{v}, \tilde{u}'')$ for $x = (u, v, w, \tilde{u}, \tilde{v})$ in the domain

$$D(A) = \{(u, v, w, \tilde{u}, \tilde{v}) \in X_1 : v(0) = \tilde{v}(3) = 0, u'(1) = w'(1),$$
$$v(1) = w(1), \tilde{u}'(2) = w'(2), \tilde{v}(2) = w(2)\}.$$

Lemma 1 *The following hold:*

1. A is closed;
2. A is densely defined;
3. A is dissipative;
4. $1 - A$ is surjective.

Proof (i) Let $x_n = (u_n, v_n, w_n, \tilde{u}_n, \tilde{v}_n) \in D(A)$ be such that

$$x_n \to x = (u, v, w, \tilde{u}, \tilde{v}), \quad Ax_n = (v_n, u_n'', w_n'', \tilde{v}_n, \tilde{u}_n'') \to y = (f, g, h, \tilde{f}, \tilde{g})$$

in X. Then u_n converges to u in $H^1(0, 1)$ and u_n'' converges to g in $L^2(0, 1)$. Hence

$$\int u\varphi'' = \lim_{n\to\infty} \int u_n \varphi'' = \lim_{n\to\infty} \int u_n'' \varphi = \int g\varphi, \quad \varphi \in C_c^\infty(0, 1), \tag{3}$$

where the integral is taken over $((0, 1), d\xi)$ so that $u \in H^2(0, 1)$ and $u'' = g$. As v_n converges to both v and f in $L^2(0, 1)$, $v = f$. In particular, $v \in H^1(0, 1)$. The same argument shows that $\tilde{u} \in H^2(2, 3)$ with $\tilde{u}'' = \tilde{g}$ and $\tilde{v} = \tilde{f} \in H^1(2, 3)$.

Next, w_n converges to w and w_n'' to h in $L^2(1, 2)$. Standard Sobolev theory (see for example [7, Page 217]) ensures the existence of a constant C such that

$$\|\psi'\|_{L^2(1,2)} \leq \|\psi''\|_{L^2(1,2)} + C\|\psi\|_{L^2(1,2)}, \quad \psi \in H^2(1, 2).$$

Hence, the sequence w_n' is Cauchy and converges to some H in $L^2(1, 2)$. Using similar reasoning to that in (3), we see that $w \in H^2(1, 2)$ with $w' = H$ and $w'' = h$.

To check that the coupling conditions for x to be in the domain $D(A)$ are satisfied, it is enough to pass to a subsequence x_{n_k} that converges pointwise a.e. and note the continuity of $u', v, w', w, \tilde{u}', \tilde{v}$. It follows that $Ax = y$.

(ii) Consider the subspace X_1 equipped with the X norm, which is dense in X. The linear functional $\phi_1 : x = (u, v, w, \tilde{u}, \tilde{v}) \mapsto v(0)$ is unbounded on X_1, and hence

$$X_2 = \text{Ker } \phi_1 = \{(u, v, w, \tilde{u}, \tilde{v}) \in X_1 : v(0) = 0\}$$

is dense in X_1. Similarly,

$$X_3 = \text{Ker } \phi_2 = \{(u, v, w, \tilde{u}, \tilde{v}) \in X_2 : v(1) = w(1)\}$$

is dense in X_2 where ϕ_2 is the unbounded linear functional on X_2 defined by $x \mapsto v(1) - w(1)$. Again, by considering the unbounded linear functional $\phi_3 : x \mapsto u'(1) - w'(1)$ on X_3, we see that

$$X_4 = \text{Ker } \phi_3 = \{(u, v, w, \tilde{u}, \tilde{v}) \in X_3 : u'(1) = w'(1)\}$$

is dense in X_3. The same argument can be repeated for the coupling and boundary conditions for $w, \tilde{u}$, and $\tilde{v}$ to produce a decreasing finite chain of subspaces

$$X \supset X_1 \supset X_2 \supset \ldots \supset D(A),$$

where each subspace is dense in the preceding one under the X norm. Hence A is densely defined.

(iii) Let $x \in D(A)$. Assuming the appropriate intervals over which to take the L^2 inner products, we have, through integration by parts and the coupling and boundary conditions,

$$\langle Ax, x \rangle = \langle v', u' \rangle_{L^2} + \langle u'', v \rangle_{L^2} + \langle w'', w \rangle_{L^2} + \langle \tilde{v}', \tilde{u}' \rangle_{L^2} + \langle \tilde{u}'', \tilde{v} \rangle_{L^2}$$
$$= -\overline{\langle u'', v \rangle}_{L^2} + \langle u'', v \rangle_{L^2} - \langle w', w' \rangle_{L^2} - \overline{\langle \tilde{u}'', \tilde{v} \rangle}_{L^2} + \langle \tilde{u}'', \tilde{v} \rangle_{L^2}.$$

Hence

$$\operatorname{Re} \langle Ax, x \rangle = -\|w'\|_{L^2}^2 \leq 0,$$

showing that A is dissipative.

(iv) Though in the setting of this lemma, we only need to work with $1 - A$, we perform a procedure here with $\lambda - A$ for general $\lambda \neq 0$ in order to avoid repetition that otherwise would be inevitable in later sections. Note that we are closely following the proof of [4, Theorem 3.1].

Let $x = (u, v, w, \tilde{u}, \tilde{v})$ and $y = (f, g, h, \tilde{f}, \tilde{g})$ be in X. Then the equation $(\lambda - A)x = y$ can be rewritten as the following system of boundary value problems:

$$u'' = \lambda^2 u - \lambda f - g, \quad \xi \in (0, 1), \tag{4a}$$
$$v = \lambda u - f, \quad \xi \in (0, 1), \tag{4b}$$
$$w'' = \lambda w - h, \quad \xi \in (1, 2), \tag{4c}$$
$$\tilde{u}'' = \lambda^2 \tilde{u} - \lambda \tilde{f} - \tilde{g}, \quad \xi \in (2, 3), \tag{4d}$$
$$\tilde{v} = \lambda \tilde{u} - \tilde{f}, \quad \xi \in (2, 3), \tag{4e}$$
$$u(0) = v(0) = 0, \quad v(1) = w(1), \quad u'(1) = w'(1), \tag{4f}$$
$$\tilde{u}(3) = \tilde{v}(3) = 0, \quad \tilde{v}(2) = w(2), \quad \tilde{u}'(2) = w'(2). \tag{4g}$$

Let

$$U_\lambda(\xi) = \frac{1}{\lambda} \int_0^\xi \sinh(\lambda(\xi - r))(\lambda f(r) + g(r)) \, dr, \quad \xi \in [0, 1],$$

which has derivative

$$U_\lambda'(\xi) = \int_0^\xi \cosh(\lambda(\xi - r))(\lambda f(r) + g(r)) \, dr, \quad \xi \in [0, 1].$$

The differential equation (4a) with the boundary condition $u(0) = 0$ has the general solution

$$u(\xi) = a(\lambda) \sinh(\lambda \xi) - U_\lambda(\xi), \quad \xi \in [0, 1], \tag{5}$$

where $a(\lambda) \in \mathbb{C}$ is a parameter free to be varied. In particular,

$$u'(\xi) = \lambda a(\lambda) \cosh(\lambda \xi) - U'_\lambda(\xi), \quad \xi \in [0, 1]. \tag{6}$$

Clearly $u \in H^2(0, 1)$ and hence $v \in H^1(0, 1)$ with $v(0) = \lambda u(0) - f(0) = 0$.

Similarly, the general solution of (4d) with boundary condition $\tilde{u}(3) = 0$ can be written as

$$\tilde{u}(\xi) = \tilde{a}(\lambda) \sinh(\lambda(3 - \xi)) + \tilde{U}_\lambda(\xi), \quad \xi \in [2, 3], \tag{7}$$

where $\tilde{a}(\lambda) \in \mathbb{C}$ can be varied freely and

$$\tilde{U}_\lambda(\xi) = \frac{1}{\lambda} \int_\xi^3 \sinh(\lambda(r - \xi))(\lambda \tilde{f}(r) + \tilde{g}(r)) \, dr, \quad \xi \in [2, 3].$$

Thus

$$\tilde{u}'(\xi) = -\lambda \tilde{a}(\lambda) \cosh(\lambda(3 - \xi)) + \tilde{U}'_\lambda(\xi), \quad \xi \in [2, 3], \tag{8}$$

where

$$\tilde{U}'_\lambda(\xi) = - \int_\xi^3 \cosh(\lambda(r - \xi))(\lambda \tilde{f}(r) + \tilde{g}(r)) \, dr, \quad \xi \in [2, 3].$$

Again, it follows that $\tilde{u} \in H^2(2, 3)$ and $\tilde{v} \in H^1(2, 3)$ with $\tilde{v}(3) = 0$.

In the same spirit, let

$$W_\lambda(\xi) = \frac{1}{\sqrt{\lambda}} \int_1^\xi \sinh(\sqrt{\lambda}(\xi - r)) h(r) \, dr, \quad \xi \in [1, 2],$$

which has derivative

$$W'_\lambda(\xi) = \int_1^\xi \cosh(\sqrt{\lambda}(\xi - r)) h(r) \, dr, \quad \xi \in [1, 2].$$

The general solution of (4c) can then be written as

$$w(\xi) = b(\lambda) \cosh(\sqrt{\lambda}(\xi - 1)) + c(\lambda) \sinh(\sqrt{\lambda}(\xi - 1)) - W_\lambda(\xi), \quad \xi \in [1, 2], \tag{9}$$

where $b(\lambda), c(\lambda) \in \mathbb{C}$ are free parameters and in particular,

$$w'(\xi) = \sqrt{\lambda} b(\lambda) \sinh(\sqrt{\lambda}(\xi - 1)) + \sqrt{\lambda} c(\lambda) \cosh(\sqrt{\lambda}(\xi - 1)) - W'_\lambda(\xi), \quad \xi \in [1, 2]. \tag{10}$$

It remains to choose specific constants $a(\lambda), b(\lambda), c(\lambda)$ and $\tilde{a}(\lambda)$ in order to satisfy the coupling conditions. Using (5) and (9), the requirement $\lambda u(1) - f(1) = v(1) = w(1)$ holds if and only if

$$\lambda a(\lambda) \sinh(\lambda) - b(\lambda) = \lambda U_\lambda(1) + f(1).$$

Likewise, the conditions $u'(1) = w'(1)$, $\lambda\tilde{u}(2) - \tilde{f}(2) = w(2)$, and $\tilde{u}'(2) = w'(2)$ are equivalent to
$$\lambda a(\lambda)\cosh(\lambda) - \sqrt{\lambda}c(\lambda) = U'_\lambda(1),$$

$$\lambda\tilde{a}(\lambda)\sinh(\lambda) - b(\lambda)\cosh(\sqrt{\lambda}) - c(\lambda)\sinh(\sqrt{\lambda}) = -\lambda\tilde{U}_\lambda(2) + \tilde{f}(2) - W_\lambda(2),$$

and
$$-\lambda\tilde{a}(\lambda)\cosh(\lambda) - \sqrt{\lambda}b(\lambda)\sinh(\sqrt{\lambda}) - \sqrt{\lambda}c(\lambda)\cosh(\sqrt{\lambda}) = -\tilde{U}'_\lambda(2) - W'_\lambda(2)$$

respectively. These four equations can be written in matrix form as

$$M_\lambda \cdot \begin{pmatrix} a(\lambda) \\ b(\lambda) \\ c(\lambda) \\ \tilde{a}(\lambda) \end{pmatrix} = \mathbf{b}, \tag{11}$$

where

$$M_\lambda = \begin{pmatrix} \lambda\sinh(\lambda) & -1 & 0 & 0 \\ \lambda\cosh(\lambda) & 0 & -\sqrt{\lambda} & 0 \\ 0 & -\cosh(\sqrt{\lambda}) & -\sinh(\sqrt{\lambda}) & \lambda\sinh(\lambda) \\ 0 & \sqrt{\lambda}\sinh(\sqrt{\lambda}) & \sqrt{\lambda}\cosh(\sqrt{\lambda}) & \lambda\cosh(\lambda) \end{pmatrix} \tag{12}$$

and

$$\mathbf{b} = \begin{pmatrix} \lambda U_\lambda(1) + f(1) \\ U'_\lambda(1) \\ -\lambda\tilde{U}_\lambda(2) + \tilde{f}(2) - W_\lambda(2) \\ \tilde{U}'_\lambda(2) + W'_\lambda(2) \end{pmatrix}. \tag{13}$$

Thus, (11) has a solution for any given $y = (f, g, h, \tilde{f}, \tilde{g})$ in X if and only if

$$\det M_\lambda = -\lambda^2[2\sqrt{\lambda}\cosh(\sqrt{\lambda})\cosh(\lambda)\sinh(\lambda) + \sinh(\sqrt{\lambda})(\lambda\sinh^2(\lambda) + \cosh^2(\lambda))]$$

is non-zero. For $\lambda = 1$,

$$\det M_1 = -\sinh(1)[4\cosh^2(1) - 1] \neq 0,$$

proving (4).

All the dirty work has now been done (ahead of time). The following theorem follows immediately from Lemma 1 and the Lumer–Phillips theorem.

Theorem 1 *A generates a contractive C_0-semigroup $(T(t))_{t\geq 0}$ on X.*

2.2 Spectrum of the Generator

From Theorem 1 and the Hille–Yosida theorem, we know that $\sigma(A)$ is contained in the closed left half-plane. However, we can say more about the spectrum.

Theorem 2 *The spectrum of A consists of isolated eigenvalues and is given by*

$$\sigma(A) = \{\lambda \in \mathbb{C}_- : \det M_\lambda = 0\}.$$

In particular, $\sigma(A) \cap i\mathbb{R} = \emptyset$.

We will need the following lemma in order to prove the theorem above.

Lemma 2 *If $\lambda \in \rho(A)$, then $R(\lambda, A)$ is a compact operator.*

Proof Let $\lambda \in \rho(A)$. Then $\lambda - A$ is a bijective bounded (and in particular, closed) linear map from $D(A)$ endowed with the graph norm onto X. Hence the inverse map $R(\lambda, A)$ maps X isomorphically onto $(D(A), \|\cdot\|_{D(A)})$. Since

$$\begin{aligned}\|(u, v, w, \tilde{u}, \tilde{v})\|_{D(A)} &= \|(u, v, w, \tilde{u}, \tilde{v})\|_X + \|(v, u'', w'', \tilde{v}, \tilde{u}'')\|_X \\ &\lesssim \|u'\|_{L^2} + \|v\|_{L^2} + \|w\|_{L^2} + \|\tilde{u}'\|_{L^2} + \|\tilde{v}\|_{L^2} \\ &\quad + \|v'\|_{L^2} + \|u''\|_{L^2} + \|w''\|_{L^2} + \|\tilde{v}'\|_{L^2} + \|\tilde{u}''\|_{L^2},\end{aligned}$$

it follows that $(D(A), \|\cdot\|_{D(A)})$ embeds continuously into

$$H^2(0, 1) \times H^1(0, 1) \times H^2(1, 2) \times H^2(2, 3) \times H^1(2, 3)$$

endowed with its natural norm (see [7, Page 217]). This space in turn embeds compactly into X by the Rellich–Kondrachov theorem of Sobolev theory. Stringing together these embeddings, $R(\lambda, A)$ is a compact operator on X. □

Proof of Theorem 2 We first show that not only is $\lambda - A$ surjective as shown in Lemma 1 whenever $\det M_\lambda \neq 0$, it is also injective. Indeed, suppose $(\lambda - A)x = 0$. Then, x is obtained in the same way as in the proof of Lemma 1(4) with $\mathbf{b} = 0$ in (11). As $\det M_\lambda \neq 0$, we get that $x = 0$. Hence $\lambda - A$ is closed and bijective, so has bounded inverse by the closed graph theorem. In particular, $1 \in \rho(A)$ and so the resolvent is non-empty.

The spectral theorem for compact operators used in conjunction with Lemma 2 implies that the spectrum of $R(1, A)$ consists only of eigenvalues of finite multiplicity with the only possible accumulation point being the origin. By the spectral mapping theorem for the resolvent,

$$\sigma(A) = \{1 - \nu^{-1} : \nu \in \sigma(R(1, A)) \setminus \{0\}\}$$

and furthermore, a simple calculation shows that if ν is an eigenvalue of $R(1, A)$, then $1 - \nu^{-1}$ is an eigenvalue of A. Hence $\sigma(A)$ consists only of eigenvalues of finite multiplicity with the only possible accumulation point being at infinity. Thus, $\lambda \in \sigma(A)$ if and only if $\det M_\lambda = 0$.

To show the final statement, suppose that $s \in \mathbb{R}$ with $s \neq 0$ and that $x = (u, v, w, \tilde{u}, \tilde{v}) \in \mathrm{Ker}(is - A)$. From the proof of Lemma 1(3), we have

$$0 = \mathrm{Re}\,\langle (is - A)x, x\rangle = -\mathrm{Re}\,\langle Ax, x\rangle = \|w'\|_{L^2}. \tag{14}$$

Thus, from (4c), $w = (is)^{-1}(w')' = 0$. As in the proof for Lemma 1(4), we have

$$u(\xi) = a(is)\sinh(is\xi), \quad v(\xi) = is\,u(\xi), \quad \xi \in [0, 1].$$

The coupling conditions imply that $u'(1) = v(1) = 0$. Thus,

$$is\,a(is)\cosh(is) = is\,a(is)\sinh(is) = 0,$$

implying that $a(is) = 0$. Similarly, $\tilde{a}(is) = 0$ so that $x = 0$.

Consider now the case $s = 0$. Rewriting $Ax = 0$ into component differential equations, we get that $u'' = 0$ and $v = 0$ as well as $w' = 0$ as in (14). As $u'(1) = w'(1)$ and u' is constant, $u' = 0$ and hence $u(0) = 0$ implies that $u = 0$. Similarly, $v(1) = 0$ implies that $w = 0$. The same is true for $\tilde{u}$ and $\tilde{v}$. It follows that $\sigma(A) \cap i\mathbb{R} = \emptyset$. □

3 Resolvent Estimates

We turn now to obtaining an upper bound on the growth of $\|R(is, A)\|$ as $|s| \to \infty$ which will allow us to deduce a quantitative estimate on the rate of energy decay in the next section.

Theorem 3 *We have* $\|R(is, A)\| = O(|s|^{1/2})$ *as* $|s| \to \infty$.

To prove this theorem, we will need explicit forms for the $a(\lambda), b(\lambda), c(\lambda), \tilde{a}(\lambda)$ found in the proof of Lemma 1(4) for the case where $\lambda = is$ and to this end, we invert M_λ to get that

$$(\det M_\lambda)^{-1} C^T \mathbf{b} = \begin{pmatrix} a(\lambda) \\ b(\lambda) \\ c(\lambda) \\ \tilde{a}(\lambda) \end{pmatrix}, \tag{15}$$

where C is the cofactor matrix of M_λ. First, we rewrite $\det M_\lambda$ and define two terms which are ubiquitous in this section:

$$\det M_\lambda = -\lambda^2[2\sqrt{\lambda}\cosh(\sqrt{\lambda})\cosh(\lambda)\sinh(\lambda) + \sinh(\sqrt{\lambda})(\lambda\sinh^2(\lambda) + \cosh^2(\lambda)]$$
$$= \frac{\lambda^2}{2}[-e^{\sqrt{\lambda}}(\lambda\sinh^2(\lambda) + 2\sqrt{\lambda}\cosh(\lambda)\sinh(\lambda) + \cosh^2(\lambda))$$
$$+ e^{-\sqrt{\lambda}}(\lambda\sinh^2(\lambda) - 2\sqrt{\lambda}\cosh(\lambda)\sinh(\lambda) + \cosh^2(\lambda))]$$
$$= \lambda^2[-e^{\sqrt{\lambda}}T_+^2(\lambda) + e^{-\sqrt{\lambda}}T_-^2(\lambda)],$$

where

$$T_+(\lambda) = \frac{1}{2}[\cosh(\lambda) + \sqrt{\lambda}\sinh(\lambda)], \quad T_-(\lambda) = \frac{1}{2}[\cosh(\lambda) - \sqrt{\lambda}\sinh(\lambda)].$$

The functions T_+ and T_- are useful because they obey convenient lower bounds on the one hand, and appear many times in the entries of $C = \{c_{ij}\}_{i,j}$ on the other hand. As an example of this, c_{11} is explicitly computed and stated here:

$$c_{11} = -\lambda^{3/2}[\cosh(\sqrt{\lambda})\cosh(\lambda) + \sqrt{\lambda}\sinh(\sqrt{\lambda})\sinh(\lambda)]$$
$$= -\lambda^{3/2}[e^{\sqrt{\lambda}}T_+(\lambda) + e^{-\sqrt{\lambda}}T_-(\lambda)].$$

The expressions for the other entries can be found in the appendix.

We will also need the following two lemmas, the first of which is proved in [4, Lemma 3.3] (over the interval $[-1, 0]$ rather than $[0, 1]$ or $[2, 3]$ as we have here).

Lemma 3 *There exists a constant $C \geq 0$ such that, for all $f \in H^1(0, 1), g \in L^2(0, 1), \tilde{f} \in H^1(2, 3), \tilde{g} \in L^2(2, 3),$ and $\lambda \in i\mathbb{R}$,*

$$\left|\int_0^\xi \sinh(\lambda(\xi - r))(\lambda f(r) + g(r))dr\right| \leq C\|f\|_{H^1} + \|g\|_{L^2}, \quad \xi \in [0, 1],$$

$$\left|\int_0^\xi \cosh(\lambda(\xi - r))(\lambda f(r) + g(r))dr\right| \leq C\|f\|_{H^1} + \|g\|_{L^2}, \quad \xi \in [0, 1],$$

$$\left|\int_\xi^3 \sinh(\lambda(r - \xi))(\lambda\tilde{f}(r) + \tilde{g}(r))dr\right| \leq C\|\tilde{f}\|_{H^1} + \|\tilde{g}\|_{L^2}, \quad \xi \in [2, 3],$$

$$\left|\int_\xi^3 \cosh(\lambda(r - \xi))(\lambda\tilde{f}(r) + \tilde{g}(r))dr\right| \leq C\|\tilde{f}\|_{H^1} + \|\tilde{g}\|_{L^2}, \quad \xi \in [2, 3].$$

Lemma 4 *For $\lambda \in i\mathbb{R}$ with $|\lambda| \geq \left(\frac{1}{\sqrt{2}} + 1\right)^2$, we have*

$$|T_+(\lambda)|, |T_-(\lambda)| \geq 1/4.$$

Proof We prove this for $T_+(\lambda)$ where $\lambda = is$ with $s \in \mathbb{R}$ and note that $2T_+(is) = \cos(s) + i\sqrt{is}\sin(s)$. Explicit calculation yields

$$4|T_+(\lambda)|^2 = \left|\frac{1}{\sqrt{is}}(\sqrt{is}\cos(s) - s\sin(s))\right|^2$$
$$= \frac{1}{|s|}\left(\frac{|s|}{2}\cos^2(s) + \left(\sqrt{\frac{|s|}{2}}\cos(s) - s\sin(s)\right)^2\right),$$

as $\operatorname{Re}\sqrt{\lambda} \geq 0$ for all $\lambda \in \mathbb{C}$ since we have taken the branch cut of the square root along the negative real axis. In the case where $\cos^2(s) \geq 1/2$, it follows that $4|T_+(\lambda)|^2 \geq 1/4$. However, if $\cos^2(s) < 1/2$, then $|\sin(s)|^2 \geq 1/2$, so that

$$2|T_+(\lambda)| \geq |i\sqrt{is}\sin(s)| - |\cos(s)| \geq \sqrt{\frac{|s|}{2}} - \frac{1}{\sqrt{2}} \geq \frac{1}{2}$$

whenever $|\lambda| = |s| \geq \left(\frac{1}{\sqrt{2}} + 1\right)^2$. The case for $T_-(\lambda)$ is similar. □

Proof of Theorem 3 Let $\lambda = is$ for $s \in \mathbb{R}$ and let $y = (f, g, h, \tilde{f}, \tilde{g}) \in Z$, further defining $x = (u, v, w, \tilde{u}, \tilde{v}) \in D(A)$ by $x = R(\lambda, A)y$. As $v = \lambda u - f$ and $\tilde{v} = \lambda\tilde{u} - \tilde{f}$, we have that

$$\|x\| \lesssim \|\lambda u\|_{L^2} + \|f\|_{L^2} + \|u'\|_{L^2} + \|w\|_{L^2} + \|\lambda\tilde{u}\|_{L^2} + \|\tilde{f}\|_{L^2} + \|\tilde{u}'\|_{L^2}.$$

Thus the result will follow once we have established that each of the summands in the above equation are bounded by $C\sqrt{|\lambda|}\|y\|$ for $|s| \geq N$, where $C, N > 0$ are constants independent of y.

Consider u given by (5). By Lemma 3, it is enough to consider $|\lambda a(\lambda)|$ in order to estimate $\|\lambda u\|_{L^2}$ and $\|u'\|_{L^2}$. Now

$$\lambda a(\lambda) = \frac{\lambda}{\det M_\lambda}(c_{11}b_1 + c_{21}b_2 + c_{31}b_3 + c_{41}b_4) \quad (16)$$

where b_i are the components of the vector $\mathbf{b}$ in (11). We consider each of these terms. Note that by Lemma 3, the only terms in the components of $\mathbf{b}$ that are not automatically bounded by some constant multiple of $\|y\|$ are $W_\lambda(2)$ and $W'_\lambda(2)$. Looking at the first term in (16),

$$\left|\frac{\lambda}{\det M_\lambda}c_{11}\right| = \sqrt{|\lambda|}\left|\frac{e^{\sqrt{\lambda}}T_+(\lambda) + e^{-\sqrt{\lambda}}T_-(\lambda)}{-e^{\sqrt{\lambda}}T_+(\lambda)^2 + e^{-\sqrt{\lambda}}T_-(\lambda)^2}\right| \lesssim \sqrt{|\lambda|}|T_+(\lambda)|^{-1} \lesssim \sqrt{|\lambda|},$$

since $\operatorname{Re}\sqrt{\lambda} > 0$ for $\lambda \in i\mathbb{R} \setminus \{0\}$ as before, so that $e^{\sqrt{\lambda}}$ dominates $e^{-\sqrt{\lambda}}$. Thus,

$$\left|\frac{\lambda}{\det M_\lambda}c_{11}b_1\right| \lesssim \sqrt{|\lambda|}\|y\|,$$

where the implicit constant is independent of λ and y.

Likewise,

$$\left|\frac{\lambda}{\det M_\lambda} c_{21}\right| = \left|\frac{e^{\sqrt{\lambda}}T_+(\lambda) - e^{-\sqrt{\lambda}}T_-(\lambda)}{-e^{\sqrt{\lambda}}T_+(\lambda)^2 + e^{-\sqrt{\lambda}}T_-(\lambda)^2}\right| \lesssim 1,$$

so that

$$\left|\frac{\lambda}{\det M_\lambda} c_{21} b_2\right| \lesssim \|y\|.$$

Noting that $|\cosh(\lambda)|, |\sinh(\lambda)| \leq 1$, a similar argument shows that the remaining terms in (16) that do not include $W_\lambda(2)$ and $W_\lambda(2)'$ are bounded by a constant times $\sqrt{|\lambda|}\|y\|$. Consider now

$$\left|\frac{\lambda}{\det M_\lambda} c_{31} W_\lambda(2)\right| \leq \int_1^2 \left|\frac{\sinh(\sqrt{\lambda}(2-r))h(r)}{-e^{\sqrt{\lambda}}T_+(\lambda)^2 + e^{-\sqrt{\lambda}}T_-(\lambda)^2}\right| dr$$

$$\leq \frac{1}{2}\int_1^2 \left|\frac{e^{\sqrt{\lambda}(2-r)} - e^{-\sqrt{\lambda}(2-r)}}{-e^{\sqrt{\lambda}}T_+(\lambda)^2 + e^{-\sqrt{\lambda}}T_-(\lambda)^2}\right| |h(r)|\, dr$$

$$\lesssim |T_+(\lambda)|^{-2}\|h\|_{L^2} \lesssim \|h\|_{L^2}$$

where the inequality in the final line is justified as before noting that $2 - r \in [0, 1]$.

Similarly,

$$\left|\frac{\lambda}{\det M_\lambda} c_{41} W'_\lambda(2)\right| \lesssim \sqrt{|\lambda|}\|h\|_{L^2}.$$

These inequalities combined with (16) imply that

$$|\lambda a(\lambda)| \lesssim \sqrt{|\lambda|}\|y\|, \quad |\lambda| \geq N,$$

for some constant $N > 0$ independent of y and in particular,

$$\|\lambda u\|_{L^2}, \|u'\|_{L^2} \lesssim \sqrt{|\lambda|}\|y\|, \quad |\lambda| \geq N.$$

The same arguments show that this also holds for $\|\lambda \tilde{u}\|_{L^2}$ and $\|\tilde{u}'\|_{L^2}$.

We must now estimate w given by (9), noting that

$$b(\lambda) = \frac{1}{\det M_\lambda}(c_{12}b_1 + c_{22}b_2 + c_{32}b_2 + c_{42}b_4)$$

and

$$c(\lambda) = \frac{1}{\det M_\lambda}(c_{13}b_1 + c_{23}b_2 + c_{33}b_2 + c_{43}b_4).$$

The trick to estimating w is to group the terms together in a specific way. As before, the only terms in the components of **b** which are not bounded by $\|y\|$ are $W_\lambda(2)$ and $W'_\lambda(2)$. Hence it is enough to estimate the moduli of

$$w_1(\xi) = \frac{1}{\det M_\lambda}[c_{12}\cosh(\sqrt{\lambda}(\xi-1)) + c_{13}\sinh(\sqrt{\lambda}(\xi-1))],$$

$$w_2(\xi) = \frac{1}{\det M_\lambda}[c_{22}\cosh(\sqrt{\lambda}(\xi-1)) + c_{23}\sinh(\sqrt{\lambda}(\xi-1))],$$

$$w_3(\xi) = \frac{1}{\det M_\lambda}[c_{32}\cosh(\sqrt{\lambda}(\xi-1)) + c_{33}\sinh(\sqrt{\lambda}(\xi-1))],$$

$$\omega_4(\xi) = \frac{1}{\det M_\lambda}[c_{42}\cosh(\sqrt{\lambda}(\xi-1)) + c_{43}\sinh(\sqrt{\lambda}(\xi-1))],$$

and

$$w_5(\xi) = \frac{1}{\det M_\lambda}\Big[\big(-c_{32}W_\lambda(2) + c_{42}W'_\lambda(2)\big)\cosh(\sqrt{\lambda}(\xi-1))$$
$$+ \big(-c_{33}W_\lambda(2) + c_{43}W'_\lambda(2)\big)\sinh(\sqrt{\lambda}(\xi-1)) - \det M_\lambda W_\lambda(\xi)\Big],$$

where $\xi \in [1, 2]$, since the sum of the w_i is equal to the w after removing the terms of **b** that do not include $W_\lambda(2)$ and $W'_\lambda(2)$. These removed terms can be shown to obey the desired estimates using the previous method.

Plugging in the appropriate values gives

$$w_1 = \frac{\lambda^2 \cosh(\lambda)}{\det M_\lambda}\Big[e^{\sqrt{\lambda}}T_+(\lambda)\big(\cosh(\sqrt{\lambda}(\xi-1)) - \sinh(\sqrt{\lambda}(\xi-1))\big)$$
$$- e^{-\sqrt{\lambda}}T_-(\lambda)\big(\cosh(\sqrt{\lambda}(\xi-1)) + \sinh(\sqrt{\lambda}(\xi-1))\big)\Big]$$
$$= \frac{\lambda^2 \cosh(\lambda)}{\det M_\lambda}\Big(e^{\sqrt{\lambda}}T_+(\lambda)e^{-\sqrt{\lambda}(\xi-1)} - e^{-\sqrt{\lambda}}T_-(\lambda)e^{\sqrt{\lambda}(\xi-1)}\Big)$$
$$= \cosh(\lambda)\frac{e^{\sqrt{\lambda}(2-\xi)}T_+(\lambda) - e^{-\sqrt{\lambda}(2-\xi)}T_-(\lambda)}{-e^{\sqrt{\lambda}}T_+(\lambda)^2 + e^{-\sqrt{\lambda}}T_-(\lambda)^2}.$$

Since $2 - \xi \in [0, 1]$, as in the case for u,

$$|w_1(\xi)| \lesssim |T_\pm(\lambda)|^{-1} \lesssim 1,$$

where the sign of $\pm$ is determined by that of s and the growth bound is independent of ξ. Likewise, $|w_2(\xi)| \lesssim 1$ with the bound independent of ξ. Next we have that

$$w_3 = \cosh \lambda \frac{\sqrt{\lambda} \sinh(\lambda) \cosh(\sqrt{\lambda}(\xi - 1)) + \cosh(\lambda) \sinh(\sqrt{\lambda}(\xi - 1))}{-e^{\sqrt{\lambda}} T_+(\lambda)^2 + e^{-\sqrt{\lambda}} T_-(\lambda)^2}$$

$$= \cosh(\lambda) \frac{e^{\sqrt{\lambda}(\xi-1)} T_+(\lambda) - e^{-\sqrt{\lambda}(\xi-1)} T_-(\lambda)}{-e^{\sqrt{\lambda}} T_+(\lambda)^2 + e^{-\sqrt{\lambda}} T_-(\lambda)^2}.$$

Since $\xi - 1 \in [0, 1]$, the previous argument again shows that $|w_3(\xi)| \lesssim 1$ with the bound independent of ξ. The same holds for w_4. Thus, it remains to estimate w_5 and after some simple manipulation, we can rewrite this as

$$w_5 = \frac{\lambda^2}{\det M_\lambda} \left[\frac{\cosh^2(\lambda)}{\sqrt{\lambda}} \Omega_1(\xi) + \sqrt{\lambda} \sinh^2(\lambda) \Omega_2(\xi) + \cosh(\lambda) \sinh(\lambda) \Omega_3(\xi) \right], \quad (17)$$

where

$$\Omega_1(\xi) = -\int_1^2 \sinh(\sqrt{\lambda}(2-r)) \sinh(\sqrt{\lambda}(\xi-1)) h(r)\, dr$$

$$+ \int_1^\xi \sinh(\sqrt{\lambda}(\xi-1)) \sinh(\sqrt{\lambda}) h(r)\, dr$$

$$= \frac{1}{2} \left[-\int_\xi^2 \cosh(\sqrt{\lambda}(1+\xi-r)) h(r)\, dr \right.$$

$$+ \int_1^2 \cosh(\sqrt{\lambda}(3-\xi-r)) h(r)\, dr$$

$$\left. - \int_1^\xi \cosh(\sqrt{\lambda}(\xi-r-1)) h(r)\, dr \right],$$

with the second equality following from the use of identities such as

$$2 \sinh(\sqrt{\lambda}(2-r)) \sinh(\sqrt{\lambda}(\xi-1))$$
$$= \cosh(\sqrt{\lambda}(2-r) + \sqrt{\lambda}(\xi-1))$$
$$- \cosh(\sqrt{\lambda}(2-r) - \sqrt{\lambda}(\xi-1)),$$

$$\Omega_2(\xi) = -\int_1^2 \cosh(\sqrt{\lambda}(2-r)) \cosh(\sqrt{\lambda}(\xi-1)) h(r)\, dr$$

$$+ \int_1^\xi \sinh(\sqrt{\lambda}(\xi-r)) \sinh(\sqrt{\lambda}) h(r)\, dr$$

$$= -\frac{1}{2} \left[\int_\xi^2 \cosh(\sqrt{\lambda}(1+\xi-r)) h(r)\, dr \right.$$

$$+ \int_1^2 \cosh(\sqrt{\lambda}(3-\xi-r)) h(r)\, dr$$

$$\left. + \int_1^\xi \cosh(\sqrt{\lambda}(\xi-r-1)) h(r)\, dr \right],$$

and

$$\Omega_3(\xi) = 2\int_1^\xi \sinh(\sqrt{\lambda}(\xi - r))\cosh(\sqrt{\lambda})h(r)\,dr$$
$$- \int_1^2 \sinh(\sqrt{\lambda}(2 - r))\cosh(\sqrt{\lambda}(\xi - 1))h(r)\,dr$$
$$- \int_1^2 \cosh(\sqrt{\lambda}(2 - r))\sinh(\sqrt{\lambda}(\xi - 1))h(r)\,dr$$
$$= -\int_1^\xi \sinh(\sqrt{\lambda}(1 - \xi + r))h(r)\,dr - \int_\xi^2 \sinh(\sqrt{\lambda}(1 + \xi - r))h(r)\,dr.$$

Note that for $\xi \in [1, 2]$, $|1 + \xi - r| \leq 1$ whenever $r \in [\xi, 2]$, and $|3 - \xi - r| \leq 1$ whenever $r \in [1, 2]$, and $|\xi - r - 1| \leq 1$ whenever $r \in [1, \xi]$. Thus by pulling the factor of $\left(-e^{\sqrt{\lambda}}T_+(\lambda)^2 + e^{-\sqrt{\lambda}}T_-(\lambda)^2\right)^{-1}$ into the integrand of Ω_1, we see that

$$\left|\frac{\lambda^2}{\det M_\lambda}\frac{\cosh^2(\lambda)}{\sqrt{\lambda}}\Omega_1(\xi)\right| \lesssim \frac{1}{\sqrt{\lambda}}\|h\|_{L^2}.$$

Arguing similarly for Ω_2 and Ω_3, we get from (17) that

$$|w_5| \lesssim \left(\frac{1}{\sqrt{|\lambda|}} + \sqrt{|\lambda|} + 1\right)\|h\|_{L^2} \lesssim \sqrt{|\lambda|}\|h\|_{L^2},$$

with the implicit constant independent of ξ. It follows that

$$\|w\|_{L^2} \lesssim \sqrt{|\lambda|}\|y\|, \quad |\lambda| \geq N,$$

for some constant $N > 0$ independent of y and, in particular,

$$\|x\| \lesssim |\lambda|^{1/2}\|y\|,$$

with the implicit constant independent of the specific y and x. □

4 Optimal Energy Decay for Classical Solutions

For the reader's convenience, we state below the version of the Borichev–Tomilov theorem used in [4].

Theorem 4 *Let* $(T(t))_{t\geq 0}$ *be a bounded* C_0*-semigroup on a Hilbert space* X *with generator* A *such that* $\sigma(A) \cap i\mathbb{R} = \emptyset$. *Then for any* $\alpha > 0$, *the following are equivalent:*

1. $\|R(is, A)\| = O(|s|^\alpha)$ as $|s| \to \infty$;
2. $\|T(t)A^{-1}\| = O(t^{-1/\alpha})$ as $t \to \infty$;
3. $\|T(t)x\| = o(t^{-1/\alpha})$ as $t \to \infty$ for all $x \in D(A)$.

Using the abstract but powerful tool above, we can convert the resolvent estimate in Theorem 3 into a rate of energy decay of classical solutions of (2), deriving the main result of the article. The rate itself will follow easily from Theorem 4 as we shall soon see, but optimality will require a little more work.

Theorem 5 *If $x \in D(A)$, then $E_x(t) = o(t^{-4})$ as $t \to \infty$. Moreover, this rate is optimal in the sense that, given any positive function r satisfying $r(t) = o(t^{-4})$ as $t \to \infty$, there exists $x \in D(A)$ such that $E_x(t) \neq o(r(t))$ as $t \to \infty$.*

Before we begin the proof, we state the following summary proposition needed to show optimality. What is stated below is more or less a collection of results from [4].

Proposition 1 *Let B be the generator of the $(S(t))_{t \geq 0}$ on the Hilbert space*

$$Z_* = \{(u, v, w) \in H^1(0, 1) \times L^2(0, 1) \times L^2(1, 3/2) : u(0) = 0\}$$

that solves the following well-posed problem:

$$\begin{cases} u_{tt}(\xi, t) = u_{\xi\xi}(\xi, t), & \xi \in (0, 1), t > 0, \\ w_t(\xi, t) = w_{\xi\xi}(\xi, t), & \xi \in (1, 3/2), t > 0, \\ u(0, t) = w(3/2, t) = 0, & t > 0, \\ u_t(1, t) = w(1, t), \quad u_\xi(1, t) = w_\xi(1, t), & t > 0, \\ u(\xi, 0) = u(\xi), \quad u_t(\xi, 0) = v(\xi), & \xi \in (0, 1), \\ w(\xi, 0) = w(\xi), & \xi \in (1, 3/2). \end{cases} \quad (18)$$

Then

$$\limsup_{|s| \to \infty} |s|^{-1/2} \|R(is, B)\| > 0. \quad (19)$$

In particular, for any positive function r satisfying $r = o(t^{-4})$ as $t \to \infty$, there exists

$$x_* \in D(B) = \{(u, v, w) \in H^2(0, 1) \times H^1(0, 1) \times H^2(1, 3/2)$$
$$: u(0) = v(0) = w(3/2) = 0,$$
$$v(1) = w(1), \; u'(1) = w'(1)\}$$

such that

$$E_{x_*}(t) = \int_0^1 |u'(\xi, t)|^2 + |v(\xi, t)|^2 \, d\xi + \int_1^{3/2} |w(\xi, t)|^2 \, d\xi \neq o(r(t))$$

as $t \to \infty$.

Proof After a rescaling of the heat component by a factor of 2, this is the same problem as what is studied in [4, Sect. 5], namely the coupled wave-heat equation that leads to the optimal resolvent bound in [4, Theorem 3.1], but with Dirichlet wave condition. The problem is well-posed, therefore, and the same resolvent estimates hold up to a constant and they remain optimal in the sense of (19). This is again proved in the same exact way as [4, Theorem 3.4] by using the argument found there based on Rouché's theorem. Note that in this case, however, $\sigma(B) \cap i\mathbb{R} = \emptyset$.

The final part of the proposition follows from (19) and is proved along the lines of [4, Remark 4(a)]. We flesh that remark out here. Assume for a contradiction that there exists a positive function r satisfying $r(t) = o(t^{-2})$ as $t \to \infty$ such that for all $x \in D(B)$, $\|S(t)x\| = o(r(t))$. Without loss of generality, r is non-increasing since we can replace r with $r_1(t) = \sup_{t \leq \tau} r(\tau)$ which also satisfies $r_1(t) = o(t^{-2})$ and $\|S(t)x\| = o(r_1(t))$ for all $x \in D(B)$. Then for all $y \in X$, there exists a constant C_y such that

$$r(t)^{-1}\|S(t)R(1, B)y\| \leq C_y, \quad t \geq 0.$$

Hence by the uniform boundedness principle, there exists $C > 0$ independent of y such that

$$\|S(t)R(1, B)\| \leq Cr(t).$$

In particular, if we adjust the constant C to be slightly larger, $m(t) < Cr(t)$, where $m(t) = \sup_{t \leq \tau} \|S(\tau)R(1, B)\|$. By [3, Proposition 1.3],

$$\|R(is, B)\| \lesssim 1 + m_*^{-1}\left(\frac{1}{2(|s|+1)}\right), \quad s \in \mathbb{R},$$

where m_*^{-1} is a right inverse of the function m, mapping $(0, m(0)]$ onto $[0, \infty)$. This contradicts (19) if $|s|^{-1/2} m_*^{-1}\left(\frac{1}{2(|s|+1)}\right) \to 0$ as $|s| \to \infty$, which we now show.

Notice first that because $t^2 Cr(t) \to 0$ as $t \to \infty$ and $(Cr)_*^{-1}(|s|) \to \infty$ as $|s| \to 0$, we have that $(Cr)_*^{-1}(|s|)^2 |s| \to 0$ as $|s| \to 0$, where $(Cr)_*^{-1}$ is a right inverse of the function Cr, mapping $(0, Cr(0)]$ onto $[0, \infty)$.

Hence

$$(Cr)_*^{-1}\left(\frac{1}{2(|s|+1)}\right)^2 \frac{1}{2(|s|+1)} \to 0$$

as $|s| \to \infty$. But since $m < Cr$ and both functions are non-increasing, it follows that $m_*^{-1} \leq (Cr)_*^{-1}$ on the interval $(0, m(0)]$ and we are done.

Remark 1 In the above proof, we can alternatively prove the simpler optimality statement that

$$\|S(t)R(1, B)\| \geq ct^{-2}, \quad t \geq 1,$$

for some constant $c > 0$ by combining [1, Theorem 4.4.14] with the fact that the specific $\lambda_n^{\pm}$ in [4, Theorem 3.4] are evenly spaced.

We finally prove the decay rate in Theorem 5 using Theorem 4 as promised and its optimality by showing that the system (18) is effectively contained within (2).

Proof of Theorem **5** By Theorem 4, we have that

$$E_x(t) = \frac{1}{2}\|T(t)x\|^2 = o(t^{-4})$$

as $t \to \infty$ for any $x \in D(A)$ since Theorem 3 gives us the rate $\|R(is, A)\| = O(|s|^{1/2})$ as $s \to \infty$.

To show optimality, assume that there exists a positive function r satisfying $r = o(t^{-4})$ as $t \to \infty$. Proposition 1 produces an $x_* = (u_*, v_*, w_*) \in D(B) \subset Z_*$ with the classical solution to (2) given by $x_*(t) = (u_*(t), v_*(t), w_*(t))$ for $t \geq 0$ and for which $E_{x_*}(t) \neq o(r(t))$ as $t \to \infty$. Define $x_1 = (u_1, v_1, w_1, \tilde{u}_1, \tilde{v}_1) \in X$ by

$$u_1(\xi) = u_*(\xi), \quad \xi \in [0, 1],$$
$$v_1(\xi) = v_*(\xi), \quad \xi \in [0, 1],$$
$$w_1(\xi) = \begin{cases} w_*(\xi), & \xi \in [1, 3/2], \\ -w_*(3-\xi), & \xi \in [3/2, 2], \end{cases}$$
$$\tilde{u}_1(\xi) = -u_*(3-\xi), \quad \xi \in [2, 3],$$
$$\tilde{v}_1(\xi) = -v_*(3-\xi), \quad \xi \in [2, 3],$$

and note that x_1 satisfies all the conditions necessary to be in $D(A)$, including the $H^2(1, 2)$ condition since on a symmetric interval around the only potentially problematic point $\xi = 3/2$, the function w_1 is the negative reflection of an H^2 function around a point at which it is 0. Moreover, the classical solution to (2) for initial data x_1 is given by $x_1(t) = (u_1(t), v_1(t), w_1(t), \tilde{u}_1(t), \tilde{v}_1(t))$ for $t \geq 0$, where

$$u_1(\xi, t) = u_*(\xi, t), \quad \xi \in [0, 1],$$
$$v_1(\xi, t) = v_*(\xi, t), \quad \xi \in [0, 1],$$
$$w_1(\xi, t) = \begin{cases} w_*(\xi, t), & \xi \in [1, 3/2], \\ -w_*(3-\xi, t), & \xi \in [3/2, 2], \end{cases}$$
$$\tilde{u}_1(\xi, t) = -u_*(3-\xi, t), \quad \xi \in [2, 3],$$
$$\tilde{v}_1(\xi, t) = -v_*(3-\xi, t), \quad \xi \in [2, 3].$$

It follows that

$$E_{x_1}(t) = 2E_{x_*}(t) \neq o(r(t))$$

as $t \to \infty$. □

5 Possible Future Directions

In this final section, we pose and comment on a few questions about possible future directions arising out of systems similar to that described by (2). The last of these questions could potentially be very interesting and not easily tractable.

Note, however, that the question likely to be asked first—what happens when the Dirichlet conditions are replaced by Neumann conditions—is easily answered. In this case, the semigroup is actually unbounded. The function $x(\xi, t) = (at, a, at)$ for any constant $a \neq 0$ solves the variant of (2) where the fourth line is changed to $u_t(0, t) = \tilde{u}_t(3, t) = 0$ for the initial condition $(0, a, 0)$, which yields an unbounded orbit of the semigroup in this case. That said, an alternative formulation involving a different state space can be chosen for the Neumann problem, one that is more physically intuitive and for which the same method as for the Dirichlet case can be applied to obtain the same rate of decay. With that out of the way, we ask the following natural two part question.

Open Question 1

1. Does the rate of energy decay remain the same up to a constant when extra wave and heat parts are added, for example, in a wave-heat-wave-heat or a wave-heat-wave-heat-wave system?
2. If the energy remains optimally bounded by Ct^{-4} for C dependent on the particular system, can we find an explicit N-formula for the multiplicative constant C_N bounding the energy of the system composed of N wave-heat pairs all coupled together?

Our first reaction at the thought of answering this question is one of hesitation, as the methods used in this article involved inverting a 4×4 matrix, and a system composed of N wave-heat pairs would require the inversion of a $(4N - 2) \times (4N - 2)$ matrix. Though we believe that the answer to the first part of the above question is affirmative, the second part would require some clever matrix tricks to avoid rather messy computations. It is notable, however, that the matrices would have 0 entries everywhere, except off of a diagonal of width at most four. So perhaps it is doable.

The idea of, perhaps inductively, obtaining a formula for C_N as above leads to the question of homogenisation, that is, the computation of a limit equation. For $N \in \mathbb{N}$ and a given smooth function f, consider the following system of mixed hyperbolic and elliptic type that was studied in [12]:

$$\begin{cases} \partial_t^2 u_N(\xi, t) - \partial_\xi^2 u_N(\xi, t) = \partial_t f(\xi, t), & \xi \in \bigcup_{j \in \{1,\dots,N\}} \left(\frac{j-1}{N}, \frac{2j-1}{2N}\right), t \in \mathbb{R}, \\ u_N(\xi, t) - \partial_\xi^2 u_N(\xi, t) = f(\xi, t), & \xi \in \bigcup_{j \in \{1,\dots,N\}} \left(\frac{2j-1}{2N}, \frac{j}{N}\right), t \in \mathbb{R}, \\ \partial_\xi u_N(0, t) = \partial_\xi u_N(1, t), & t \in \mathbb{R}, \end{cases} \quad (20)$$

subject to zero initial conditions and the requirement that the u_N and their derivatives are continuous. We use the notation ∂ for derivatives as in [12] to avoid a mess involving the subscript N. Note that requiring conditions of continuity at the junction points results in the coupling considered in [13] rather than that of [4] and (2).

Waurick showed in [12] that as $N \to \infty$, the sequence of solutions $(u_N)_{N \in \mathbb{N}}$ converges weakly in $L^2_{loc}([0, 1] \times \mathbb{R})$ to u, the solution to the limit equation

$$\frac{1}{2}\partial_t^2 u(\xi, t) + \partial_t u(\xi, t) + \frac{1}{2}u(\xi, t) - 2\partial_\xi^2 u(\xi, t) = f(\xi, t) + \partial_t f(\xi, t), \quad t \in \mathbb{R},$$

subject to zero initial conditions and Neumann boundary on both ends. Furthermore, he showed that this limit admitted exponentially stable solutions in the sense of [11]. However, when the elliptic part, $u_N(\xi, t) - \partial_\xi^2 u_N(\xi, t) = f(\xi, t)$, is replaced with the corresponding parabolic part, $\partial_t u_N(\xi, t) - \partial_\xi^2 u_N(\xi, t) = f(\xi, t)$, the limit equation becomes

$$\frac{1}{2}\partial_t^2 u(\xi, t) + \partial_t u(\xi, t) - 2\partial_\xi^2 u(\xi, t) = f(\xi, t) + \partial_t f(\xi, t), \quad t \in \mathbb{R},$$

subject again to zero initial conditions and Neumann boundary. Crucially, this limit equation is not exponentially stable, raising the following question.

Open Question 2 For the homogenised limit equation with mixed hyperbolic and parabolic parts as above, can the limit solution be posed and solved by a non-uniformly stable semigroup? □

In [8], resolvent estimates of some kind are calculated in a way that depends on N via the Gelfand transform, before a numerical analysis is conducted. How this might somehow be converted to a resolvent estimate for the limit problem itself remains an interesting unanswered question.

Acknowledgements The author thanks David Seifert and Charles Batty for helpful discussions on the topic of this article and is especially indebted to David for his feedback on several drafts of the same. Heartfelt appreciation goes out to the reviewer as well, who was extremely meticulous in spotting out errors and making suggestions for improvement. Finally, the author is also grateful to the University of Sydney for funding this work through the Barker Graduate Scholarship.

Appendix—Entries for the Cofactor Matrix C in Sect. 3

$$c_{11} = -\lambda^{3/2}[e^{\sqrt{\lambda}}T_+(\lambda) + e^{-\sqrt{\lambda}}T_-(\lambda)],$$
$$c_{12} = \lambda^2 \cosh(\lambda)(e^{\sqrt{\lambda}}T_+ - e^{-\sqrt{\lambda}}T_-),$$
$$c_{13} = -\lambda^2 \cosh(\lambda)[e^{\sqrt{\lambda}}T_+(\lambda) + e^{-\sqrt{\lambda}}T_-(\lambda)],$$
$$c_{14} = \lambda^{3/2}\cosh(\lambda),$$

$$c_{21} = -\lambda[e^{\sqrt{\lambda}}T_+(\lambda) - e^{-\sqrt{\lambda}}T_-(\lambda)],$$

$$c_{22} = -\lambda^2 \sinh(\lambda)[e^{\sqrt{\lambda}}T_+(\lambda) - e^{-\sqrt{\lambda}}T_-(\lambda)],$$

$$c_{23} = \lambda^2 \sinh(\lambda)[e^{\sqrt{\lambda}}T_+(\lambda) + e^{-\sqrt{\lambda}}T_-(\lambda)],$$

$$c_{24} = -\lambda^{3/2} \sinh(\lambda),$$

$$c_{31} = \lambda^{3/2} \cosh(\lambda),$$

$$c_{32} = \lambda^{5/2} \sinh(\lambda) \cosh(\lambda),$$

$$c_{33} = \lambda^2 \cosh^2(\lambda),$$

$$c_{34} = -\lambda^{3/2}[e^{\sqrt{\lambda}}T_+(\lambda)e^{-\sqrt{\lambda}}T_-(\lambda)],$$

$$c_{41} = -\lambda^{3/2} \sinh(\lambda),$$

$$c_{42} = -\lambda^{5/2} \sinh^2(\lambda),$$

$$c_{43} = -\lambda^2 \cosh(\lambda) \sinh(\lambda),$$

$$c_{44} = -\lambda[e^{\sqrt{\lambda}}T_+(\lambda) - e^{-\sqrt{\lambda}}T_-(\lambda)].$$

References

1. Arendt, W., Batty, C.J.K., Hieber, M., Neubrander, F.: Vector-valued Laplace transforms and Cauchy problems, 2nd edn., Monographs in Mathematics, vol. 96. Birkhäuser/Springer Basel AG, Basel (2011). MR 2798103
2. Avalos, G., Triggiani, R.: Mathematical analysis of PDE systems which govern fluid-structure interactive phenomena. Bol. Soc. Parana. Mat. (3) **25**(1–2), 17–36 (2007). MR 2379673
3. Batty, C.J.K., Duyckaerts, T.: Non-uniform stability for bounded semi-groups on Banach spaces. J. Evol. Equ. **8**(4), 765–780 (2008). MR 2460938
4. Batty, C.J.K., Paunonen, L., Seifert, D.: Optimal energy decay in a one-dimensional coupled wave-heat system. J. Evol. Equ. **16**(3), 649–664 (2016). MR 3551240
5. Batty, C.J.K., Paunonen, L., Seifert, D.: Optimal energy decay for the wave-heat system on a rectangular domain. SIAM J. Math. Anal. **51**(2), 808–819 (2019). MR 3928347
6. Borichev, A., Tomilov, Y.: Optimal polynomial decay of functions and operator semigroups. Math. Ann. **347**(2) (2010), 455–478. MR 2606945
7. Brezis, H.: Functional Analysis, Sobolev Spaces and Partial Differential Equations, Universitext. Springer, New York (2011). MR 2759829
8. Franz, S., Waurick, M.: Resolvent estimates and numerical implementation for the homogenisation of one-dimensional periodic mixed type problems. ZAMM Z. Angew. Math. Mech. **98**(7), 1284–1294 (2018). MR 3832855
9. Ng, A.C.S., Seifert, D.: Optimal energy decay in a one-dimensional wave-heat system with infinite heat part. J. Math. Anal. Appl. **482**(2), 123563 (2020). MR 4015695
10. Rozendaal, J., Seifert, D., Stahn, R.: Optimal rates of decay for operator semigroups on Hilbert spaces. Adv. Math. **346**, 359–388 (2019). MR 3910799
11. Trostorff, S.: Exponential stability for linear evolutionary equations. Asymptot. Anal. **85**(3-4), 179–197 (2013). MR 3156625
12. Waurick, M.: Stabilization via homogenization. Appl. Math. Lett. **60**, 101–107 (2016). MR 3505860
13. Zhang, X., Zuazua, E.: Polynomial decay and control of a 1-d hyperbolic-parabolic coupled system. J. Differ. Equ. **204**(2), 380–438 (2004). MR 2085542

Implicit Convolution Fokker-Planck Equations: Extended Feller Convolution

Wha-Suck Lee and Christiaan Le Roux

Abstract Fokker-Planck equations are partial differential equations in the transition function of the Markov process. In the evolution equation approach, we re-write partial differential equations as ordinary differential equations in Banach spaces. In particular, an implicit evolution equation is used to re-write the Fokker-Planck equation for a pair of discontinuous Markov processes. In this paper we consider the continuous analogue in the form of two homogeneous Markov processes intertwined by the extended Chapman-Kolmogorov equation. Abstract harmonic analysis techniques are used to extend the Feller convolution. Then the associated Fokker-Planck equations are re-written as an implicit evolution equation expressed in terms of the extended Feller convolution.

Keywords Extended Chapman-Kolmogorov equation · Intertwined homogeneous Markov processes

1 Introduction

A *dynamic boundary condition* for an evolution equation describes (mathematically) a dynamical interaction between a boundary, modelled as a physical entity, and the interior which models another physical entity. As such the boundary condition will also contain time derivatives. For non-perfect thermal contact between the interior and the boundary, the dynamic interaction can be formulated as an implicit evolution equation of the form

---

W.-S. Lee (✉)
Department of Mathematics and Applied Mathematics, North-West University, Private Bag X1290, Potchefstroom 2520, South Africa
e-mail: drwhasuck.lee@gmail.com

DSI-NRF Centre of Excellence in Mathematical and Statistical Sciences (CoE-MaSS), Johannesburg, South Africa

C. Le Roux
Department of Mathematics and Applied Mathematics, University of Pretoria, Private Bag X20, Hatfield 0028, South Africa

© Springer Nature Switzerland AG 2020
J. Banasiak et al. (eds.), *Semigroups of Operators – Theory and Applications*, Springer Proceedings in Mathematics & Statistics 325,
https://doi.org/10.1007/978-3-030-46079-2_18

$$\tfrac{d}{dt}[Bu(t)] = Au(t); \lim_{t \to 0^+}[Bu(t)] = y \in Y, \tag{1}$$

where A and B are linear operators from a Banach space X to a Banach space Y.

In [1], the classical absorbing barrier problem of a stochastic process is formulated as a dynamic boundary condition. This resulted in a pair of non-homogeneous Markov processes with a pair of distinct finite state spaces intertwined by the extended Chapman-Kolmogorov equation. The present problem of two distinct *continuous* state spaces and two *homogeneous* Markov processes models random transitions within a continuum of "life" states and from the "life" states to a continuum of "death" states (see [2, Sect. 8.1.22] for the case of a single "death" (coffin) state).

For a single homogeneous Markov process, the Feller convolution captures both the operator representation of the transition functions and expresses the Chapman-Kolmogorov equation as a commutative convolution semigroup. In particular, Feller's convolution (see [3, Chap. V.4, Eq. (4.1)]) provides the operator representation of a probability distribution, but without clear motivation. Therefore, in Sect. 2, we fit Feller's convolution into the framework of admissible homomorphisms introduced in [4, Sect. 2]: admissible homomorphisms replace probability distributions, and the product of admissible homomorphisms replaces the Feller convolution of probability distributions. Then a Feller convolution semigroup is shown to be equivalent to a star-semigroup (defined in [4, Sect. 6]).

For a pair of homogeneous Markov processes, the backward extended Chapman-Kolmogorov equation describes uni-directional transitions between two distinct states of "life" and "death". Such transitions require the notion of a *two-state space* stochastic kernel in Sect. 3 since Feller's joint conditional distribution ([3, Chap. V.9, Definition (9.3)]) is inadequate. For the application of empathy theory, we introduce the dual notion of the *forward* extended Chapman-Kolmogorov equation (Eq. (18)) by a *conjugate* stochastic kernel in Sect. 4. Then, in Sect. 5, we use the (non-commutative) extended Feller convolution to express the forward extended Chapman-Kolmogorov equation as a non-commutative star-empathy (see [4, Sect. 6]). Thus, the Riesz representation of a two-state space distribution can be regarded as an admissible homomorphism on a product test-space.

In Sect. 6, the first Markov process is a C_0-strongly continuous star-semigroup under a suitable analytic condition near the time origin. By empathy theory, the second Markov process is a strongly continuous evolution family evolving in empathy with the star-semigroup (Sect. 7). In the discrete case, strong continuity enables a Laplace transform approach to the Fokker-Planck equation of empathy pseudo-Poisson processes ([1, Corollary 4.2]). Similarly, we use the Laplace transform theory in the framework of admissible homomorphisms to construct a Fokker-Planck equation for a strongly continuous intertwined Brownian motion (Sect. 8). This is the reason for the use of admissible homomorphisms in place of distributions.

2 Convolution Semigroup as a Star Semigroup

The framework of admissible homomorphisms gives an explicit construction of Feller's intuitive representation [3] of a probability distribution, Q, as a bounded linear operator, $\mathcal{Q}$, on a test space of continuous functions. We give a brief description of admissible homomorphisms here (see [4] for full details).

Let Φ_U be the test space $\mathrm{BUC}(\mathbb{R}, \mathbb{C})$ of bounded uniformly continuous functions from the group $\mathbb{R}$ to $\mathbb{C}$. The Riesz representation of the distribution $Q\{dy\}$ is the homomorphism $Q': \Phi_U \to \mathbb{C}$ that is the expectation of f: $\langle Q', f \rangle = Q'(f) := \int_{\mathbb{R}} Q\{dy\} f(y)$ for all $f \in \Phi_U$. Since Φ_U is shift invariant, the following shift induced function is well-defined:

$$x \mapsto [Q \circledast f](x) := \langle Q', f_{-x} \rangle = \int_{\mathbb{R}} Q\{dy\} f(x+y), \tag{2}$$

where we define shift by the group element $x \in \mathbb{R}$ as the function $f_x(y) = f(y-x)$. In construct (2), we use Bobrowski's equivalent format (see [2, Definition 7.5.1]) of the Feller convolution.

The construct (2) is also well-defined for other test spaces Φ in place of Φ_U. In fact, $Q \circledast f \in \Phi_U$ for all $f \in \Phi_U$ [4, Theorem 12]. When Q' and Φ have the property $Q \circledast f \in \Phi$ for all $f \in \Phi$, we call Q' a Φ-admissible homomorphism. We denote the set of all Φ-admissible homomorphisms by $\mathscr{A}_\Phi$.

Following Feller, we define the bounded linear operator $\mathcal{Q}: \Phi \to \Phi$ by

$$\mathcal{Q} f = Q \circledast f \text{ for all } f \in \Phi. \tag{3}$$

The constructs (2) and (3) place the morpheme $Q\{dy\} \mapsto Q' \mapsto \mathcal{Q}$ (implicit in [3]) explicitly in the framework of admissible homomorphisms [4, Sect. 2], where $\mathcal{Q}$ is called the *dualism* of Q'. Construct (3) is a special case of the definition of dualism in [4, Eq. (3)]. So we write $\mathcal{Q} = \Gamma(Q')$.

By this dualism, we have a product of admissible homomorphisms [4, Eq. (4)]: if R is another distribution (function) with associated homomorphism R' and dualism $\mathcal{R}$, the product $Q' * R' \in \mathscr{A}_\Phi$ is defined by

$$\langle Q' * R', f \rangle = \langle Q', \mathcal{R} f \rangle \text{ for all } f \in \Phi. \tag{4}$$

If we define the Feller convolution of distributions as in (2) by $[Q \star R](E) = \int_{\mathbb{R}} Q\{dy\} R(E+y)$ for all Borel subsets $E \subset \mathbb{R}$, then

$$[Q \star R]' = Q' * R'; \tag{5}$$
$$\Gamma(Q' * R') = \mathcal{Q} \circ \mathcal{R}. \tag{6}$$

The following subspaces of Φ_U are possible test spaces: (a) $\Phi_0 := C_0(\mathbb{R}, \mathbb{C})$, the space of continuous functions with zero limits at $\pm\infty$; (b) $\Phi_\infty := C[\mathbb{R}, \mathbb{C}]$, the space of continuous functions with finite limits at $\pm\infty$. For all these test spaces, the following result follows by Eq. (5).

Theorem 1 *The map $Q \mapsto Q'$ is injective. Admissible homomorphisms replace distributions. Their convolution $*$ replaces the Feller convolution $\star$.*

A Markov process captures the notion of a non-deterministic time evolution. Let $\mathbf{X} = \{X_t\}_{t>0}$ be a Markov process with time-homogeneous transition function $\{Q_t(x, B)\}_{t>0}$. Then $\{Q_t(x, B)\}_{t>0}$ is *intertwined* by the Chapman-Kolmogorov equation if

$$Q_{t+s}(x, B) = \int_{y \in \mathbb{R}} Q_t(x, \{dy\}) Q_s(y, B) \text{ for all } s, t > 0. \qquad (7)$$

We deal exclusively with *homogeneous* Markov processes, that is, $\{Q_t(x, B)\}_{t>0}$ is both time- and space-homogeneous; $\{Q_t(x, B)\}_{t>0}$ is *space-homogeneous* if $Q_t(x, B) = Q_t(x + r, B + r)$ for every $r \in \mathbb{R}$ and Borel set $B \subset \mathbb{R}$. Then $\{Q_t(x, B)\}_{t>0}$ is a time continuum of distributions

$$\mathbf{Q} := \{Q_t\{dy\}\}_{t>0} = \{Q_t(0, \{dy\})\}_{t>0}$$

since $Q_t\{B - x\} = Q_t(x, B)$. We call $\mathbf{Q}$ a *distribution transition function*. By construct (3), $\mathbf{Q}$ has an operator representation $\{\mathcal{Q}_t : \Phi_U \to \Phi_U\}_{t>0}$.

Let $\mathbf{X}$ be a homogeneous Markov process intertwined by the Chapman-Kolmogorov equation. Then, in terms of the Feller convolution $\star$, the distribution transition function $\mathbf{Q}$ is a Feller convolution semigroup:

$$Q_{t+s}\{dy\} = Q_t\{dy\} \star Q_s\{dy\} \text{ for all } s, t > 0; \qquad (8)$$
$$\mathcal{Q}_{t+s} = \mathcal{Q}_t \circ \mathcal{Q}_s \text{ for all } s, t > 0. \qquad (9)$$

In the framework of admissible homomorphisms we replace the *distribution* transition function $\mathbf{Q}$ by a time continuum, $\mathfrak{q}' := \{Q'_t\}_{t>0}$, of admissible homomorphisms on Φ_U, which we call an *admissible transition function*. Then the Feller convolution semigroup is expressed as a star-semigroup (defined in [4, Sect. 6]) since the dualism of the admissible homomorphism Q'_t is precisely Feller's operator representation $\mathcal{Q}_t$ of the distribution $Q_t\{dy\}$.

Theorem 2 *Let $\mathbf{X}$ be a homogeneous Markov process intertwined by the Chapman-Kolmogorov equation (7). Then, in terms of the product $*$, the admissible transition function $\mathfrak{q}'$ is a star-semigroup:*

$$Q'_{t+s} = Q'_t * Q'_s \text{ for all } s, t > 0; \qquad (10)$$
$$\mathcal{Q}_{t+s} = \mathcal{Q}_t \circ \mathcal{Q}_s \text{ for all } s, t > 0. \qquad (11)$$

Note that the dualism $\mathcal{Q}_t$ backtracks to the homomorphism Q'_t by the point evaluation map $\theta'_0 : f \mapsto f(0)$ as follows: $\theta'_0(\mathcal{Q}_t f) = \langle Q'_t, f \rangle$ for each $f \in \Phi_U$.

3 Two-State Space Stochastic Kernels

Two distinct discrete state spaces, namely $\mathbb{N}_X := \{1, 2, ..., m\}$, the set of m "life" states, and $\mathbb{N}_{\overline{Y}} := \{\overline{1}, \overline{2}, ..., \overline{n}\}$, the set of n "death" states, result from the two-state space approach to an absorbing barrier of a Markov process [1]. The continuous analogues of $\mathbb{N}_X$ and $\mathbb{N}_{\overline{Y}}$ are the continuums of life and death states, $\mathbb{R}_X$ and $\mathbb{R}_{\overline{Y}}$, respectively. The pair of state spaces $\mathbb{R}_X$ and $\mathbb{R}_{\overline{Y}}$ are distinct copies of $\mathbb{R}$. The bar notation distinguishes a death entity from a life entity.

Consequently there are two types of transitions: (a) from a life state to another life state within $\mathbb{R}_X$ and (b) from a life state in $\mathbb{R}_X$ to a death state in $\mathbb{R}_{\overline{Y}}$. The single-state space transition (a) is described by the stochastic kernel $Q(x, B)$, which gives the probability of X transitioning from a point $x \in \mathbb{R}_X$ to a Borel set $B \subset \mathbb{R}_X$. Similarly, the uni-directional transition (b) is described by the *two-state space* stochastic kernel $R(x, \overline{B})$, which gives the probability of $\overline{Y}$ transitioning from a point $x \in \mathbb{R}_X$ to a Borel set $\overline{B} \subset \mathbb{R}_{\overline{Y}}$.

Let $Q(x, \{dy\})$ denote the distribution on $\mathbb{R}_X$. The distribution $R(x, \{d\overline{y}\})$ on $\mathbb{R}_{\overline{Y}}$ is defined by the conditional probability

$$R(x, \overline{B}) = P(\overline{Y} \in \overline{B} | X = x) := \lim_{h \to 0} \frac{P(\{\overline{Y} \in \overline{B}\} \text{ and } \{X \in [x, x+h]\})}{F([x, x+h])}, \quad (12)$$

where F is the distribution of X. As in the classical case, the two-state space stochastic kernel $R(x, \overline{B})$ plays a dual role: (a) For each $x \in \mathbb{R}_X$, $R(x, \overline{B})$ is the probability distribution $R(x, \{d\overline{y}\})$ and (b) for each Borel set $\overline{B} \subset \mathbb{R}_{\overline{Y}}$, $R(x, \overline{B})$ is the point function that maps $x \in \mathbb{R}_X$ to $P(\overline{Y} \in \overline{B} | X = x)$. Note that definition (12) differs from Feller's joint conditional distribution [3, Ch. V.9, Definition (9.3)], which treats both $[x, x+h]$ and $\overline{B}$ as subsets of a single state space, $\mathbb{R}$.

Consider a pair of *homogeneous* Markov processes $(\mathbf{X}, \mathbf{Y})$ intertwined by the backward extended Chapman-Kolmogorov equation introduced in [1]:

$$\begin{cases} Q_{t+s}(x, B) = \displaystyle\int_{y \in \mathbb{R}} Q_t(x, \{dy\}) Q_s(y, B) \text{ for all } s, t > 0; & (13a) \\ R_{t+s}(x, \overline{B}) = \displaystyle\int_{y \in \mathbb{R}} Q_t(x, \{dy\}) R_s(y, \overline{B}) \text{ for all } s, t > 0. & (13b) \end{cases}$$

We call Eq. (13b) the *two-state space backward (extended Chapman-Kolmogorov) transition equation*. We call the pair $(\mathbf{X}, \mathbf{Y})$ *homogeneous* if both transition functions in the pair $(Q_t(x, B), R_t(x, \overline{B}))_{t>0}$ are homogeneous. In that case it reduces to a pair of *distribution* transition functions

$$(\mathbf{Q}, \mathbf{R}) := (Q_t\{dy\}, R_t\{d\overline{y}\})_{t>0} = (Q_t(0, \{dy\}), R_t(0, \{d\overline{y}\}))_{t>0}.$$

Proposition 1 *Let* $(\mathbf{X}, \mathbf{Y})$ *be a pair of homogeneous Markov processes intertwined by the backward extended Chapman-Kolmogorov equation. The two-state space backward transition Eq. (13b) can be expressed in terms of* $(\mathbf{Q}, \mathbf{R})$:

$$R_{t+s}\{\overline{B}\} := R_{t+s}(0, \overline{B}) = \int_{y \in \mathbb{R}} Q_t\{dy\} R_s(0, \overline{B} - y) \text{ for all } s, t > 0. \quad (14)$$

Remark 1 Suppose that we do not make a distinction between $\mathbb{R}_X$ and $\mathbb{R}_{\overline{Y}}$ and take both $Q_t\{dy\}$ and $R_t\{d\bar{y}\}$ as measures on $\mathbb{R}$. Then by the commutativity of the Feller convolution, Eq. (14) becomes (compare with Eq. (8)):

$$R_{t+s}\{d\bar{y}\} = Q_t\{dy\} \star R_s\{d\bar{y}\} = R_s\{d\bar{y}\} \star Q_t\{dy\} \text{ for all } s, t > 0. \quad (15)$$

Then it follows that $\mathcal{R}_{t+s} = \mathcal{Q}_t \circ \mathcal{R}_s = \mathcal{R}_s \circ \mathcal{Q}_t$ for all $s, t > 0$, where $\mathcal{Q}_t$ and $\mathcal{R}_t$ are the operator representations of $Q_t\{dy\}$ and $R_t\{d\bar{y}\}$, respectively. This commutative relation contradicts the uni-directional transitions from life states to death states.

4 Forward Extended Chapman-Kolmogorov Equation

In the discrete case, a *reverse* empathy results in the operator representation of the uni-directional two-state space backward transition Eq. (13b) (see [1, Sect. 3, Theorem 3.1]). To remedy this, we introduce the notion of a conjugate stochastic kernel that results in the dual notion of the *forward* extended Chapman-Kolmogorov equation.

We now consider the continuous analogue of conjugate stochastic kernel. Let the stochastic kernels $Q_t(x, B)$ and $R_t(x, \overline{B})$ of Eqs. (13) have probability transition *density* functions $q_t(x, y)$ and $r_t(x, \bar{y})$, respectively. Then, for $t > 0$ and Borel sets $B \subset \mathbb{R}_X$, we define the *conjugate kernels* $\overline{Q}_t(y, B)$ and $\overline{R}_t(\bar{y}, B)$ by

$$\overline{Q}_t(y, B) := \int_{x \in B} q_t(x, y) dx, \ y \in \mathbb{R}_X; \quad (16)$$

$$\overline{R}_t(\bar{y}, B) := \int_{x \in B} r_t(x, \bar{y}) dx, \ \bar{y} \in \mathbb{R}_{\overline{Y}}. \quad (17)$$

The symbols $\overline{Q}_t\{dy\} = \overline{Q}_t(0, \{dy\})$ and $\overline{R}_t\{dy\} = \overline{R}_t(\bar{0}, \{dy\})$ will denote the corresponding distributions on $\mathbb{R}_X$.

Let $Q_t(x, B)$ and $R_t(x, \overline{B})$ be the stochastic kernels of Eqs. (13) with probability transition density functions $q_t(x, y)$ and $r_t(x, \bar{y})$. If $Q_t(x, B)$ ($R_t(x, \overline{B})$, resp.) is space-homogeneous, the conjugate kernel $\overline{Q}_t(y, B)$ ($\overline{R}_t(\bar{y}, B)$, resp.) is also space-homogeneous. Analogous to the discrete case, the conjugate transition functions $\overline{Q}_t(y, B)$ and $\overline{R}_t(\bar{y}, B)$ satisfy the *forward extended Chapman-Kolmogorov equation*

$$\begin{cases} \overline{Q}_{t+s}(y, B) = \displaystyle\int_{x \in \mathbb{R}} \overline{Q}_t(y, \{dx\})\overline{Q}_s(x, B) \text{ for all } s, t > 0; & (18a) \\ \overline{R}_{t+s}(\bar{y}, B) = \displaystyle\int_{x \in \mathbb{R}} \overline{R}_t(\bar{y}, \{dx\})\overline{Q}_s(x, B) \text{ for all } s, t > 0. & (18b) \end{cases}$$

This follows from an application of Fubini's theorem and the fact that $r_s(x, \bar{y})$ is the Radon-Nikodym derivative of the measure $\overline{R}_s(\bar{y}, \{dx\})$.

Theorem 3 *Let* $(\mathbf{X}, \mathbf{Y})$ *be a pair of homogeneous Markov processes with a pair of probability transition density functions* $(q_t(x, y), r_t(x, \bar{y}))_{t>0}$. *If* $(\mathbf{X}, \mathbf{Y})$ *is intertwined by the backward extended Chapman-Kolmogorov equation (13), then the corresponding pair of conjugate transition functions* $(\overline{Q}_t(y, B), \overline{R}_t(\bar{y}, B))_{t>0}$ *is intertwined by the forward extended Chapman-Kolmogorov equation (18).*

The distribution transition functions $\mathbf{Q}$ and $\mathbf{R}$ are defined on distinct spaces, $\mathbb{R}_X$ and $\mathbb{R}_{\bar{Y}}$. The conjugation operation produces a corresponding pair of distribution transition functions $(\overline{\mathbf{Q}}, \overline{\mathbf{R}}) := (\overline{Q}_t\{dy\}, \overline{R}_t\{dy\})_{t>0}$ on $\mathbb{R}_X$. We now have the following analogue of Proposition 1.

Proposition 2 *Let* $(\mathbf{X}, \mathbf{Y})$ *be a pair of homogeneous Markov processes intertwined by the backward extended Chapman-Kolmogorov equation. The two-state space forward transition Eq. (18b) can be expressed in terms of* $(\overline{\mathbf{Q}}, \overline{\mathbf{R}})$:

$$\overline{R}_{t+s}\{B\} := \overline{R}_{t+s}(\bar{0}, B) = \int_{x \in \mathbb{R}} \overline{R}_t\{dy\}\overline{Q}_s(0, B - x) \text{ for all } s, t > 0. \quad (19)$$

The problem of the distributions $\overline{Q}_t\{dy\}$ and $\overline{R}_t\{dy\}$ associated with (19) commuting as in Eq. (15) persists because their operator representations also commute as in Remark 1. Therefore we next construct the required non-commutative generalization of the Feller convolution by representing $\overline{Q}_t\{dy\}$ and $\overline{R}_t\{dy\}$ as admissible homomorphisms on a product test space.

5 Non-commutative Extended Feller Convolution

The two-state space approach forces us to consider the product space $\Phi_X \times \Phi_{\bar{Y}}$, where $\Phi_X := BUC(\mathbb{R}_X, \mathbb{C})$ and $\Phi_{\bar{Y}} := BUC(\mathbb{R}_{\bar{Y}}, \mathbb{C})$. To deal with one-parameter shifts, we consider the diagonal subgroup $G := \{(\sigma, \sigma) | \sigma \in \mathbb{R}\}$ instead of the product group $\mathbb{R} \times \mathbb{R}$. Then for each $(f, \bar{g}) \in \Phi_X \times \Phi_{\bar{Y}}$, we define a corresponding function $\varphi : G \to \mathbb{C}^2 := \mathbb{C} \times \mathbb{C}$ by $\varphi(\sigma, \sigma) = (f(\sigma), \bar{g}(\bar{\sigma}))$, where σ and $\bar{\sigma}$ denote the same numerical value. The product test space Φ_P is defined as the set of all such functions φ. The one-parameter translation operators $R^{(\sigma, \sigma)}$, $\sigma \in G$ are defined by $R^{(\sigma, \sigma)}\varphi =$

$\varphi_\sigma : (\tau, \tau) \in G \to \varphi(\tau - \sigma, \tau - \sigma) = \big(f(\tau - \sigma), \bar{g}(\tau - \sigma)\big)$. We shall henceforth denote (σ, σ) as σ, $R^{(\sigma,\sigma)}$ as R^σ and $\varphi_{(\sigma,\sigma)}$ as φ_σ.

In order to work with the product test space Φ_P instead of a single test space Φ_X, we define the following mappings to lift classical linear functionals as $\mathbb{C}^2$-linear functionals:

(a) liftings $\ell_1 : \mathbb{C} \to \mathbb{C}^2 : z \mapsto (z, 0)$ and $\ell_2 : \mathbb{C} \to \mathbb{C}^2 : z \mapsto (0, z)$;
(b) liftings $\ell_X : \Phi_X \to \Phi_P : f \mapsto (f, 0_{\bar{Y}})$ and $\ell_{\bar{Y}} : \Phi_{\bar{Y}} \to \Phi_P : f \mapsto (0_X, \bar{f})$, where $0_{\bar{Y}} \in \Phi_{\bar{Y}}$ and $0_X \in \Phi_X$ are the zero functions;
(c) the projection $\pi_X : \Phi_P \to \Phi_X : \varphi = (f, \bar{g}) \mapsto f$.

Now the dual roles of each Φ_P-admissible linear functional x'_P as homomorphisms from Φ_P to $\mathbb{C}^2$ and operators $X_P : \Phi_P \to \Phi_P$ give an algebraic structure to the class $\mathscr{A}_{\Phi_P}$ by means of an associative product: for Φ_P-admissible homomorphisms x'_P and y'_P, the product $x'_P * y'_P$ defined in [4, Eq. (4)] is given by

$$\langle x'_P * y'_P, \varphi \rangle = \langle x'_P, Y_P \varphi \rangle \text{ for all } \varphi \in \Phi_P, \tag{20}$$

where $Y_P = \Gamma(y'_P)$ is the dualism of y'_P. We shall show that $*$ is a non-commutative extension of the classical commutative Feller convolution $\star$ of Borel measures on $\mathbb{R}$. We therefore call it a *convolution product*.

Lemma 1 (a) *Let Q' be a Φ_X-admissible homomorphism with dualism $Q = \Gamma(Q')$: $\Phi_X \to \Phi_X$. Let $Q'_P := \ell_1 \circ Q' \circ \pi_X$, i.e.,*

$$\langle Q'_P, \varphi \rangle = \big(\langle Q', f \rangle, 0\big) \in \mathbb{C}^2 \text{ for all } \varphi = (f, \bar{g}) \in \Phi_P. \tag{21}$$

Then Q'_P is a Φ_P-admissible homomorphism and its dualism is $\mathcal{Q}_P = \Gamma(Q'_P) = \ell_X \circ \Gamma(Q') \circ \pi_X$, i.e., $\mathcal{Q}_P \varphi = \big(\mathcal{Q}f, 0_{\bar{Y}}\big) \in \Phi_P$ for all $\varphi = (f, \bar{g}) \in \Phi_P$.

(b) *Let R' be a Φ_X-admissible homomorphism with dualism $\mathcal{R} = \Gamma(R') : \Phi_X \to \Phi_X$. Let $R'_P := \ell_2 \circ R' \circ \pi_X$, i.e.,*

$$\langle R'_P, \varphi \rangle = \big(0, \langle R', f \rangle\big) \in \mathbb{C}^2 \text{ for all } \varphi = (f, \bar{g}) \in \Phi_P. \tag{22}$$

Then R'_P is a Φ_P-admissible homomorphism and its dualism is $\mathcal{R}_P = \Gamma(R'_P) = \ell_{\bar{Y}} \circ \Gamma(R') \circ \pi_X$, i.e., $\mathcal{R}_P \varphi = \big(0_X, \overline{\mathcal{R}f}\big) \in \Phi_P$ for all $\varphi = (f, \bar{g}) \in \Phi_P$.

Although both Q' and R' are "life" homomorphisms on Φ_X, R'_P lifts R' as a "death" homomorphism that has a "death" dualism on Φ_P.

Now we replace the intertwined *conjugate* distributions $\overline{Q}_t\{dy\}$ and $\overline{R}_t\{dy\}$ on $\mathbb{R}_X$ by the classical (admissible) linear functionals $\overline{Q}'_t(X)$ and $\overline{R}'_t(X)$ on Φ_X, respectively (see Theorem 1). By virtue of Lemma 1 we lift $\overline{Q}'_t(X)$ and $\overline{R}'_t(X)$ as the Φ_P-admissible homomorphisms

$$Q'_P(t) := \ell_1 \circ \overline{Q}'_t(X) \circ \pi_X, \quad R'_P(t) := \ell_2 \circ \overline{R}'_t(X) \circ \pi_X,$$

respectively. This produces a corresponding pair of conjugate Φ_P-admissible transition functions $(\bar{q}'_P, \bar{r}'_P) := (\overline{Q}'_P(t), \overline{R}'_P(t))_{t>0}$.

Theorem 4 *Let $(\mathbf{X}, \mathbf{Y})$ be a pair of homogeneous Markov processes intertwined by the backward extended Chapman-Kolmogorov equation (13). Then, in terms of the product $*$ defined in (20), the pair of conjugate Φ_P-admissible transition functions $(\bar{q}'_P, \bar{r}'_P)$ is a star-empathy:*

$$\begin{cases} \overline{Q}'_P(t+s) = \overline{Q}'_P(t) * \overline{Q}'_P(s) \ \text{for all } s, t > 0; & (23a) \\ \overline{R}'_P(t+s) = \overline{R}'_P(t) * \overline{Q}'_P(s) \ \text{for all } s, t > 0. & (23b) \end{cases}$$

Moreover, $\overline{Q}'_P(s) * \overline{R}'_P(t)$ is the zero homomorphism on Φ_P for all $s, t > 0$.

In view of Eqs. (23) we call $(\bar{q}'_P, \bar{r}'_P)$ the *conjugate extended Riesz representation* on Φ_P of the distribution transition functions $(\mathbf{Q}, \mathbf{R})$.

6 Conjugate Convolution Semigroup: Single Test Space

Consider a *single* homogeneous Markov process $\mathbf{X}$ with distribution transition function $\mathbf{Q} = \{Q_t\{dy\}\}_{t>0}$. Then the associated admissible transition function $q' = \{Q'_t\}_{t>0}$ is defined on a single test space. We now establish sufficient conditions for the admissible transition functional q' and its conjugate $\bar{q}' = \{\overline{Q}'_t\}_{t>0}$ to be strongly continuous. Later, in Sect. 7, we show that these results carry over to $\bar{q}'_P$, which is defined on a product test space.

From this point onwards, we assume that (i) each stochastic kernel $Q_t(x, B)$ has a transition density function $q_t(x, y)$, as in Eq. (16), and (ii) that $\mathbf{Q}$ is a *convolution semigroup* as defined by Bobrowski [2, Definition 7.6.1]: it satisfies Eq. (8) and the initial condition $\lim_{t \to 0^+} Q_t\{dy\} = \delta_0$. This analytic condition denotes the *weak* convergence of the distributions $Q_t\{dy\}$ to δ_0, the Dirac measure concentrated at 0, on the single test space Φ_U. Next we investigate how well the conjugation operation preserves the properties of $\mathbf{Q}$.

Example 1 Let $\mathbf{X}$ be the standard Brownian motion in $\mathbb{R}$. Then the distributions $Q_t\{dy\}$ and $\overline{Q}_t\{dy\}$ are both equal to the normal (Gaussian) distribution with zero mean and variance t by the symmetry of the normal transition density function. Thus both $\mathbf{Q}$ and $\overline{\mathbf{Q}}$ are convolution semigroups.

In the framework of normed admissible homomorphisms, a convolution semigroup is expressed simultaneously as a C_0-strongly continuous star-semigroup and an *isometric* operator C_0-semigroup. These properties are preserved by the conjugation operation.

Theorem 5 *Let $\mathbf{Q}$ be a convolution semigroup. Then $\overline{\mathbf{Q}}$ is a convolution semigroup: $\bar{q}'$ is a strongly continuous star-semigroup and the* dualism *transition function $\overline{\mathfrak{Q}} := \{\overline{\mathcal{Q}}_t\}_{t>0}$, where $\overline{\mathcal{Q}}_t = \Gamma(\overline{Q}'_t)$, is an operator C_0-semigroup.*

Theorem 5 holds for any test space that is a closed subspace of Φ_U, in particular the smaller Banach test space Φ_∞ used by Feller [3, Sect. IX.2]. The strongly continuous operator semigroups $\mathfrak{Q}$ and $\overline{\mathfrak{Q}}$ restricted to Φ_∞ have generators defined on dense subspaces of Φ_∞ (see [5, Theorem 11.5.2]). The *Feller* generators, A and $\overline{A}$, of $\mathfrak{Q}$ and $\overline{\mathfrak{Q}}$, respectively, are defined on the subspace $C^\infty[\mathbb{R}, \mathbb{C}]$ of infinitely differentiable functions in Φ_∞ with derivatives in Φ_∞. Thus Feller's definition [3, Sect. IX.2, Def. 4] of a generator differs slightly from that of Hille and Phillips [5].

The distribution transition function $\mathbf{Q}$ is defective in the setting of the backward extended Chapman-Kolmogorov equation. If $\mathbf{Q}$ is defective, then so also is $\overline{\mathbf{Q}}$ since $\mathbb{R} = -\mathbb{R}$.

Proposition 3 *Let $\mathbf{X}$ be a homogeneous Markov process such that $\mathbf{Q}$ is a convolution semigroup and defective. Then there exists a unique $c > 0$ and a unique distribution transition function $\mathbf{P} = \{P_t\{dy\}\}_{t>0}$ that is a convolution semigroup with proper distributions such that*

$$Q_t\{dy\} = e^{-ct} P_t\{dy\}. \tag{24}$$

The Feller generator of $\mathbf{Q}$ is $A - cI$, where A is the Feller generator of $\mathbf{P}$.

Example 2 Let $\mathbf{X}$ and $\mathbf{Q}$ satisfy the hypotheses of Proposition 3. If $\mathbf{P}$ is the transition distribution function associated with the standard Brownian motion of Example 1, then we call $\mathbf{X}$ a *defective Brownian motion* and the Feller generator $\overline{A}$ of $\overline{\mathbf{Q}}$ is given by $\overline{A} f = \frac{1}{2}\left(\frac{d^2}{dx^2} - 2c\right) f$ for all $f \in C^\infty[\mathbb{R}, \mathbb{C}]$.

7 Conjugate Convolution Semigroup: Product Test Space

Consider the pair $(\mathbf{X}, \mathbf{Y})$ of homogeneous Markov processes with a pair of distribution transition functions $(\mathbf{Q}, \mathbf{R})$ intertwined by the backward extended Chapman-Kolmogorov equation (13). The *conjugate* extended Riesz representation $(\bar{\mathfrak{q}}'_P, \bar{\mathfrak{r}}'_P)$ of $(\mathbf{Q}, \mathbf{R})$ is defined on the product test-space Φ_P.

Here the product test-space Φ_P is associated with $\Phi_\infty(X) \times \Phi_\infty(\overline{Y})$ where $\Phi_\infty(X) := C[\mathbb{R}_X, \mathbb{C}] \subset \Phi_X$ and $\Phi_\infty(\overline{Y}) := C[\mathbb{R}_{\overline{Y}}, \mathbb{C}] \subset \Phi_{\overline{Y}}$ (see Sect. 5). Under the product norm, Φ_P is a Banach space. We obtain the following product test-space analogue of Theorem 5 by setting $\theta'_0(X) := \ell_1 \circ \theta'_0 \circ \pi_X$, and the Φ_P-dualism transition functions as $\overline{\mathfrak{Q}}_P = \{\overline{\mathfrak{Q}}_P(t)\}_{t>0}$ and $\overline{\mathfrak{R}}_P := \{\overline{\mathcal{R}}_P(t)\}_{t>0}$, where $\overline{\mathfrak{Q}}_P(t) = \Gamma(\overline{Q}'_P(t))$ and $\overline{\mathcal{R}}_P(t) = \Gamma(\overline{R}'_P(t))$. To see this, note $\langle \overline{Q}'_P(t), \varphi \rangle \to \langle \theta'_0(X), \varphi \rangle$ as $t \to 0^+$ for all $\varphi \in \Phi_P$.

Proposition 4 *If $\mathbf{Q}$ is a convolution semigroup, then $\bar{\mathfrak{q}}'_P$ is strongly continuous and $\overline{\mathfrak{Q}}_P$ is an operator C_0-semigroup on Φ_P.*

By Theorem 4, $(\bar{\mathfrak{q}}'_P, \bar{\mathfrak{r}}'_P)$ is a star-empathy. A star empathy can be analyzed in the framework of *normed* admissible homomorphisms that has a full Laplace transform theory (see [4, Sect. 5, 11]) which is applicable to $\bar{q}'_P$ and $\bar{r}'_P$ because the strong continuity of $\bar{q}'_P$ ensures the strong continuity of $\bar{r}'_P$ (see the proof of [6, Theorem 2]) and the Laplace transforms of $\bar{q}'_P$ and $\bar{r}'_P$ exist for all $\lambda > 0$ by their uniform boundedness on $(0, \infty)$, which follows from $\overline{Q}_t\{dy\}$ and $\overline{R}_t\{dy\}$ being probability measures.

Lemma 2 *For all $\varphi \in \Phi_P$ and all $\lambda > 0$, the Laplace transforms of $\bar{q}'_P$ and $\bar{r}'_P$,*

$$\langle \bar{q}'_P(\lambda), \varphi \rangle := \int_0^\infty e^{-\lambda t} \langle \overline{Q}'_P(t), \varphi \rangle dt, \quad \langle \bar{r}'_P(\lambda), \varphi \rangle := \int_0^\infty e^{-\lambda t} \langle \overline{R}'_P(t), \varphi \rangle dt,$$

respectively, exist as Lebesgue integrals. The Laplace transforms $\bar{q}'_P(\lambda)$ and $\bar{r}'_P(\lambda)$ are bounded admissible homomorphisms on Φ_P.

The crucial concept of Laplace-closedness follows from the definition of a conjugate kernel (see Eq. (16)): $\bar{q}'_P$ is Laplace-closed with respect to itself (see [4, Eq. (9) & p. 210]) and $\bar{r}'_P$ is Laplace-closed with respect to $\bar{q}'_P$. Then, by [4, Theorem 11], we have the following central result.

Theorem 6 *The conjugate extended Riesz representation on Φ_P of $(\mathbf{Q}, \mathbf{R})$ satisfies the pseudo-resolvent equations*

$$\begin{cases} \bar{q}'_P(\lambda) - \bar{q}'_P(\mu) = (\mu - \lambda)\bar{q}'_P(\lambda) * \bar{q}'_P(\mu); & (25a) \\ \bar{r}'_P(\lambda) - \bar{r}'_P(\mu) = (\mu - \lambda)\bar{r}'_P(\lambda) * \bar{q}'_P(\mu) & (25b) \end{cases}$$

for all $\lambda, \mu > 0$. In addition, for all $t > 0$,

$$\bar{q}'_P(\lambda) * \overline{Q}'_P(t) = \overline{Q}'_P(t) * \bar{q}'_P(\lambda); \quad (26)$$

$$\bar{r}'_P(\lambda) * \overline{Q}'_P(t) = \overline{R}'_P(t) * \bar{q}'_P(\lambda). \quad (27)$$

For $\lambda > 0$, let $\overline{\mathcal{Q}}_P(\lambda) := \Gamma(\bar{q}'_P(\lambda))$ and $\overline{\mathcal{R}}_P(\lambda) := \Gamma(\bar{r}'_P(\lambda))$ be the dualisms of the Laplace transforms. Then $\overline{\mathcal{Q}}_P(\lambda)$ is the λ-potential operator or resolvent operator defined in [7, Sect. 1.3, Eq.(1.29)]. By Proposition 6, $\overline{\mathcal{Q}}_P(\lambda)$ and $\overline{\mathcal{R}}_P(\lambda)$ are bounded operators on the product test space Φ_P.

Example 3 Let $(\mathbf{Q}, \mathbf{R})$ be as in Proposition 4 with $\mathbf{Q}$ as in Example 2. Then the Feller generator of $\overline{\mathfrak{Q}}'_P$ is defined on $\Delta_P := C^\infty[\mathbb{R}_X, \mathbb{C}] \times \Phi_\infty(\overline{Y})$ and is given by

$$\overline{A}_P \varphi = \left(\tfrac{1}{2} f'' - cf, 0_{\overline{Y}} \right) \text{ for all } \varphi := (f, \bar{g}) \in \Delta_P. \quad (28)$$

For all $\varphi := (f, \bar{g}) \in \Delta_P$, $[\overline{\mathcal{Q}}_P(\lambda)\varphi](x, x) = \int_0^\infty e^{-(\lambda+c)t}[p_t * f](x)dt$ for all $x \in \mathbb{R}$, where $p_t(y) = \frac{1}{\sqrt{2\pi t}} \exp\left(-\frac{y^2}{2t}\right)$ is the probability density function of the standard Brownian motion (Example 1).

8 Implicit Convolution Fokker-Planck Equations

In [1], implicit Fokker-Planck equations are derived for a pair of discontinuous intertwined counting processes by means of the Laplace transform. Here the operator representations of the transition functions results in a pair of evolution families with one family evolving in empathy with the second family, which is a semigroup. Unfortunately these distribution functions are not homogeneous. Thus, the Fokker-Planck equations cannot be formulated directly in terms of the distributions. In the present setting, the homogeneity of the transition functions enables us to overcome this limitation by expressing the Fokker-Planck equation as a convolution equation. This convolution is the non-commutative extended Feller convolution introduced in Sect. 5.

Let $(\mathbf{X}, \mathbf{Y})$ be an *intertwined Brownian motion*, that is, there is another homogeneous Markov process $\mathbf{Y}$ in empathy with a convolution semigroup $\mathbf{X}$; $(\mathbf{X}, \mathbf{Y})$, $(\mathbf{Q}, \mathbf{R})$ and $(\bar{\mathsf{q}}'_P, \bar{\mathsf{r}}'_P)$ are as in Sect. 7 and $\mathbf{X}$ is a defective Brownian motion, as in Example 3. For each fixed $\varphi = (f, \bar{g}) \in \Delta_P$, define functions u_P and v_P from $(0, \infty) \times \mathbb{R}$ to $\mathbb{C}^2$ by

$$v_P(t, x) = [\overline{\mathcal{Q}}_P(t)\varphi](x, x), \quad u_P(t, x) = [\overline{\mathcal{R}}_P(t)\varphi](x, x). \tag{29}$$

Thus $v_P(t, x) = (v(t, x), 0)$, where $v(t, x) = [\overline{\mathcal{Q}}_t(X)f](x)$. By Eq. (28), v_P satisfies the Fokker-Planck equation

$$\frac{\partial v_P}{\partial t} = \overline{A}_P v_P = \begin{pmatrix} \frac{1}{2}\frac{\partial^2 v}{\partial x^2} - cv \\ 0_{\tilde{Y}} \end{pmatrix}, \tag{30}$$

i.e., v satisfies the scalar Fokker-Planck equation $\frac{\partial v}{\partial t} = \frac{1}{2}\frac{\partial^2 v}{\partial x^2} - cv$ in $(0, \infty) \times \mathbb{R}_X$. Translated back to the original homomorphisms $\overline{\mathcal{Q}}'_t(X)$, Eq. (30) becomes

$$\begin{pmatrix} \frac{\partial}{\partial t} \int_{\mathbb{R}_X} \overline{\mathcal{Q}}_t\{dy\} f(x+y) \\ 0_{\tilde{Y}} \end{pmatrix} = \begin{pmatrix} \frac{1}{2} \int_{\mathbb{R}_X} \overline{\mathcal{Q}}_t\{dy\}(\frac{\partial^2}{\partial x^2} - 2c) f(x+y) \\ 0_{\tilde{Y}} \end{pmatrix}. \tag{31}$$

Following the approach of [4, Sect. 12], we will now use Laplace transforms to show that $u_P(t, x)$ satisfies a Fokker-Planck equation in the form of an implicit (convolution) evolution equation formulated in terms of admissible homomorphisms. Therefore, we do not derive the star implicit evolution equation directly from the intertwined pseudo-resolvent $(\bar{\mathsf{q}}'_P(\lambda), \bar{\mathsf{r}}'_P(\lambda))_{\lambda>0}$. Instead, by applying the dualism mapping Γ to Eqs. (25)–(27), we obtain analogous equations for the operator-valued dualisms $(\overline{\mathcal{Q}}_P(\lambda), \overline{\mathcal{R}}_P(\lambda))_{\lambda>0}$ which are precisely the identities (4)–(7) of [8, Lemma 3]. We then assume the invertibility assumption,

$$\overline{\mathcal{R}}_P(\xi) \text{ is invertible for some } \xi > 0, \tag{32}$$

to define the domains $\Delta_X := \overline{\mathcal{Q}}_P(\lambda)[\Phi_P]$ and $\Delta_{\overline{Y}} := \overline{\mathcal{R}}_P(\lambda)[\Phi_P]$ for $\lambda > 0$. These subspaces of Φ_P do not depend on the choice of λ since $(\bar{q}'_P, \bar{r}'_P)$ is a star-empathy by Theorem 4.

The pair of generators A and B of (1) from $\Delta_{\overline{Y}}$ to Φ_P are defined by

$$B = \overline{\mathcal{Q}}_P(\lambda)[\overline{\mathcal{R}}_P(\lambda)]^{-1}, \quad A = \lambda B - [\overline{\mathcal{R}}_P(\lambda)]^{-1}.$$

A and B are independent of λ (see [8, Sect. 5]). Note that $B[\Delta_{\overline{Y}}] = \Delta_X$. Then by [8, Theorem 2.8(a), Thm. 5.2], $A = \overline{A}'_P B$, where $\overline{A}'_P$ is the Feller generator of $\overline{\mathcal{Q}}_P = \{\overline{\mathcal{Q}}_P(t)\}_{t>0}$, and for each $\varphi = (f, 0_{\overline{Y}}) \in \Delta_X \cap \Delta_P$,

$$\begin{cases} \dfrac{\partial}{\partial t}(Bu_P) = Au_P; & (33a) \\ \lim_{t \to 0^+} Bu_P(t,x) = \varphi(x,x), \ x \in \mathbb{R}. & (33b) \end{cases}$$

The implicit evolution Eq. (33) can be expressed in terms of the admissible homomorphisms $A' = \theta'_0(X) \circ A$ and $B' = \theta'_0(X) \circ B$:

$$\begin{cases} \dfrac{d}{dt}\langle B' * \overline{R}'_P(t), \varphi \rangle = \langle A' * \overline{R}'_P(t), \varphi \rangle & \text{for a.e. } t > 0; \quad (34a) \\ \lim_{t \to 0^+} \langle B' * \overline{R}'_P(t), \varphi \rangle = \langle \theta'_0(X), \varphi \rangle. & (34b) \end{cases}$$

In terms of the original homomorphisms $\overline{R}'_t(X)$, the second component of Eq. (34a) is simply $0_{\overline{Y}} = 0_{\overline{Y}}$ and the first component is

$$\frac{\partial}{\partial t}\langle B' * \overline{R}'_t(X), f_{-x} \rangle = \langle B' * \overline{R}'_t(X), \frac{1}{2}\left(\frac{\partial^2}{\partial x^2} - 2c\right) f_{-x} \rangle \quad \text{for a.e. } t > 0. \quad (35)$$

Note that $\overline{Q}'_t(X) = B' * \overline{R}'_t(X)$ on $\Delta_X \cap \Delta_P$.

Acknowledgements W.-S. Lee thanks Professor Jacek Banasiak and Professor Sanne Ter Horst for their decisive, generous and wise support. The support of the DST-NRF Centre of Excellence in Mathematical and Statistical Sciences (CoE-MaSS) towards this research is hereby acknowledged by W.-S. Lee. Opinions expressed and conclusions arrived at, are those of the authors and are not necessarily to be attributed to the CoE. Part of the work of W.-S. Lee was also funded by the Claude Leon Foundation and the NWU Postoctoral Fellowship.

References

1. Lee, W.-S., Sauer, N.: Intertwined Markov processes: the extended Chapman-Kolmogorov equation. Proc. R. Soc. Edinb. A **148**, 123–131 (2018)
2. Bobrowski, A.: Functional Analysis for Probability and Stochastic Processes. Cambridge University Press, Cambridge (2005)
3. Feller, W.: An Introduction to Probability Theory and Its Applications. Vol. II, 2nd edn. Wiley Ser. Pro. Wiley, New York (1971)
4. Lee, W.-S., Sauer, N.: Intertwined evolution operators. Semigroup Forum **94**, 204–228 (2017)
5. Hille, E., Phillips, R.S.: Functional Analysis and Semi-groups. AMS Colloq. Publ. 31, AMS (2000)
6. Sauer, N., Banasiak, J., Lee, W.-S.: Causal relations in support of implicit evolution equations. Bull. South Ural State Univ. Ser.: Math. Model. Program. Comput. Softw. **11**, 85–102 (2018)
7. Böttcher, B., Schilling, R., Wang, J.: Lévy Matters III. Lévy-Type Processes: Construction, Approximation and Sample Path Properties. Lecturer Notes in Mathematics, vol. 2099. Springer, Berlin (2013)
8. Sauer, N.: Empathy theory and the Laplace transform. In: Linear Operators. Banach Center Publications, Institute of Mathematics, Polish Academy of Sciences, Warszawa, vol. 38, pp. 325–338 (1997)

Asymptotic Properties of Stochastic Semigroups with Applications to Piecewise Deterministic Markov Processes

Katarzyna Pichór and Ryszard Rudnicki

Abstract The paper is devoted to stochastic semigroups, i.e. semigroups of linear operators on integrable functions preserving the set of densities. We present some results concerning their asymptotic stability and asymptotic decomposition. Finally we give applications to stochastic semigroups generated by piecewise deterministic Markov processes: pure jump-type processes, stochastic billiards and to biological models of gene expressions, electrical activity of a neuron, and two-phase cell cycle.

Keywords Stochastic semigroup · Piecewise deterministic Markov process · Asymptotic stability · Asymptotic decomposition · Biological models

1 Introduction

The paper is devoted to stochastic semigroups, i.e. semigroups of linear operators on integrable functions preserving the set of densities. Stochastic semigroups are generated by partial differential equations of different types and describe the behaviour of the distributions of Markov processes like diffusion processes, piecewise deterministic Markov processes (PDMPs) and hybrid stochastic processes [23]. They are also used to study ergodic properties of dynamical systems [9]. A lot of applications of stochastic semigroups and PDMPs to biological models can be found in [26].

We present some results concerning asymptotic properties of stochastic semigroups (asymptotic stability, sweeping). In particular, we show that if a uniformly continuous stochastic semigroup has a unique invariant density $f^* > 0$, then it is asymptotically stable. We recall recent results concerning asymptotic decomposition of partially integral semigroups [18, 19]. We also collect some simple conclusions

---

K. Pichór
Institute of Mathematics, University of Silesia, Bankowa 14, 40-007 Katowice, Poland
e-mail: katarzyna.pichor@us.edu.pl

R. Rudnicki (✉)
Institute of Mathematics, Polish Academy of Sciences, Bankowa 14, 40-007 Katowice, Poland
e-mail: rudnicki@us.edu.pl

© Springer Nature Switzerland AG 2020
J. Banasiak et al. (eds.), *Semigroups of Operators – Theory and Applications*,
Springer Proceedings in Mathematics & Statistics 325,
https://doi.org/10.1007/978-3-030-46079-2_19

from these general results which can be directly applied to study long-time behaviour of stochastic semigroups induced by PDMPs.

According to a non-rigorous definition by Davis [4], the class of PDMPs is a general family of stochastic models covering virtually all non-diffusion applications. More precisely, a PDMP is a continuous-time Markov process ξ_t with values in some metric space X and such that there is an increasing sequence of random times (t_n), called jump times, such that sample paths (trajectories) of ξ_t are defined in a deterministic way in each interval (t_n, t_{n+1}). If a PDMP with values in some metric space X is time-homogeneous, then it usually induces a stochastic semigroup $\{P(t)\}_{t\geq 0}$ on a space $L^1(X)$ given in the following way. If ξ_0 has a density f, then ξ_t has a density $P(t)f$.

The first application is to pure jump-type Markov processes. An example is the kangaroo process [16]. Then we study one-dimensional stochastic billiards as models of collisionless kinetic equations [15]. It is interesting that the Lebowitz and Rubinow model of cell cycle [10] can be identified with a one-dimensional stochastic billiard with symmetric reflection boundary conditions. Next example of applications is a model of electrical activity of a neuron [27]. Finally, we study a piecewise deterministic version of the two-phase cell cycle model introduced in [14]. In our case we are interested in the evolution of age and maturity density distribution in a single line of cells.

2 Stochastic Operators and Semigroups

Let a triple (X, Σ, m) be a σ-finite measure space. Denote by D the subset of the space $L^1 = L^1(X, \Sigma, m)$ which contains all densities

$$D = \{f \in L^1 \colon f \geq 0, \ \|f\| = 1\}.$$

A linear operator $P \colon L^1 \to L^1$ is called *stochastic* if $P(D) \subseteq D$.

One can define a stochastic operator by means of a *transition probability function*. We recall that $\mathscr{P}(x, \Gamma)$ is a transition probability function on (X, Σ) if $\mathscr{P}(x, \cdot)$ is a probability measure on (X, Σ) for each $x \in X$ and the function $x \mapsto \mathscr{P}(x, \Gamma)$ is measurable for each $\Gamma \in \Sigma$. Assume that $\mathscr{P}$ has the following property

$$m(\Gamma) = 0 \implies \mathscr{P}(x, \Gamma) = 0 \text{ for } m\text{-a.e. } x \text{ and } \Gamma \in \Sigma. \tag{1}$$

Then for every $f \in D$ the measure

$$\mu(\Gamma) = \int f(x) \mathscr{P}(x, \Gamma) m(dx)$$

is absolutely continuous with respect to the measure m. This fact is a simple consequence of the Radon–Nikodym theorem, which says that the measure ν is absolutely continuous with respect to the measure m iff the following implication $m(\Gamma) = 0 \Rightarrow \nu(\Gamma) = 0$ holds for all sets $\Gamma \in \Sigma$. Now, the formula $Pf = d\mu/dm$ defines a stochastic operator $P : L^1 \to L^1$. Moreover, if $P^* \colon L^\infty \to L^\infty$ is the adjoint operator of P then $P^*g(x) = \int g(y)\,\mathscr{P}(x, dy)$.

A family $\{P(t)\}_{t \geq 0}$ of linear operators on L^1 is called a *stochastic semigroup* if it is a strongly continuous semigroup (C_0-semigroup) and all operators $P(t)$ are stochastic. Let $\mathfrak{D}(A)$ be the set of all $f \in L^1$ such that there exists the limit $Af = \lim_{t \to 0^+} t^{-1}(P(t)f - f)$. The operator $A \colon \mathfrak{D}(A) \to L^1$ is called the *infinitesimal generator* (briefly the *generator*) of the semigroup $\{P(t)\}_{t \geq 0}$.

Having a stochastic operator P and a constant $\lambda > 0$ we can define a stochastic semigroup

$$P(t)f = \sum_{k=0}^{\infty} \frac{(\lambda t)^k e^{-\lambda t}}{k!} P^k f. \qquad (2)$$

Then the operator $A = -\lambda I + \lambda P$ is the generator of this semigroup. The operator A is bounded, and so the semigroup $\{P(t)\}_{t \geq 0}$ is *uniformly continuous*, i.e.,

$$\lim_{t \to t_0} \|P(t) - P(t_0)\| = 0 \text{ for } t_0 \geq 0, \qquad (3)$$

and vice versa, every uniformly continuous stochastic semigroup is of the form (2) (see [23, p. 267]).

3 Asymptotic Properties of Stochastic Semigroups

Now, we introduce some notions which characterize the asymptotic behaviour of iterates of stochastic operators P^n, $n = 0, 1, 2, \ldots$, and stochastic semigroups $\{P(t)\}_{t \geq 0}$. The iterates of stochastic operators form a discrete-time semigroup and we can use notation $P(t) = P^t$ for their powers and we formulate most of definitions and results for both types of semigroups without distinguishing them.

A stochastic semigroup $\{P(t)\}_{t \geq 0}$ is *asymptotically stable* if there exists a density f^* such that

$$\lim_{t \to \infty} \|P(t)f - f^*\| = 0 \quad \text{for} \quad f \in D. \qquad (4)$$

From (4) it follows immediately that f^* is *invariant* with respect to $\{P(t)\}_{t \geq 0}$, i.e. $P(t)f^* = f^*$ for each $t \geq 0$.

A stochastic semigroup $\{P(t)\}_{t \geq 0}$ is called *sweeping* with respect to a set $B \in \Sigma$ if for every $f \in D$

$$\lim_{t \to \infty} \int_B P(t)f(x)\,m(dx) = 0.$$

We say that a stochastic semigroup $\{P(t)\}_{t \geq 0}$ satisfies the *Foguel alternative* [9] if it is asymptotically stable or sweeping from all compact sets.

Although asymptotic properties of stochastic operators and semigroups were intensively studied (see [9, 23]), new results still appear. Our aim is to present some results concerning asymptotic decomposition of stochastic semigroups, but we begin with recalling some older theorems which precede them.

A stochastic semigroup $\{P(t)\}_{t\geq 0}$ is called *partially integral* if there exists a measurable function $q: (0, \infty) \times X \times X \to [0, \infty)$, called a *kernel*, such that

$$P(t)f(y) \geq \int_X q(t, x, y) f(x) m(dx) \tag{5}$$

for every density f and

$$\int_X \int_X q(t, x, y) m(dx) m(dy) > 0$$

for some $t > 0$. The following result was proved in [17].

Theorem 1 *Let $\{P(t)\}_{t\geq 0}$ be a continuous time partially integral stochastic semigroup. Assume that the semigroup $\{P(t)\}_{t\geq 0}$ has a unique invariant density f^*. If $f^* > 0$ a.e., then the semigroup $\{P(t)\}_{t\geq 0}$ is asymptotically stable.*

One of the classical tools to prove the existence of invariant densities is the *abstract ergodic theorem* due to Kakutani and Yosida (see [9] Theorem 5.2.1). We precede the formulation of this theorem by some definitions.

Let E be a Banach space and let E^* be the *dual space*. A sequence (x_n) of elements of E is called *weakly convergent* to $x \in E$, if $x^*(x_n) \to x^*(x)$ for each $x^* \in E^*$. The set $F \subset E$ is called *weakly precompact* if each sequence (x_n) of elements of F contains a weakly convergent subsequence in E. The prefix *pre* underlines the fact, that the limit of the subsequence belongs to E, but we do not assume that it belongs to F.

Theorem 2 *Let P be a stochastic operator on the space $L^1(X, \Sigma, m)$. If for a given $f \in D$ the sequence*

$$A_n f = \frac{1}{n} \sum_{k=0}^{n-1} P^k f$$

is weakly precompact, then the sequence $(A_n f)$ converges strongly to some $f^* \in D$ and $P f^* = f^*$.

In particular, if for some $f \in D$ there exists a $g \in L^1$ such that $A_n f \leq g$ for all $n \in \mathbb{N}$, then the sequence $\{A_n f\}$ is weakly precompact and, consequently, the operator P has an invariant density.

Now, we recall another version of the Kakutani-Yosida theorem. Let (X, Σ, μ) be a probability space. A stochastic operator P on $L^1(X, \Sigma, \mu)$ is called a *double stochastic operator* if $P\mathbf{1}_X = \mathbf{1}_X$. Let P be a double stochastic operator. Then

$$\Sigma_{\text{inv}} = \{A \in \Sigma : P^* \mathbf{1}_A = \mathbf{1}_A\}$$

is a σ-algebra and it is called the σ-algebra of invariant sets.

Theorem 3 *Let P be a double stochastic operator on $L^1(X, \Sigma, \mu)$. If $f \in L^1(X, \Sigma, \mu)$, then the sequence $A_n f$ converges in L^1 to $\bar{f} = \mathrm{E}(f|\Sigma_{\mathrm{inv}})$ and $P\bar{f} = \bar{f}$.*

The proof of Theorem 3 can be found for example in [6] (Theorem A, Chapter VII). The statement of Theorem 3 can be a little strengthened. Since P is a contraction and $P\bar{f} = \bar{f}$, we have

$$\left\| \frac{1}{n} \sum_{k=0}^{n-1} P^{k+m} f - \bar{f} \right\| = \left\| P^m \left(\frac{1}{n} \sum_{k=0}^{n-1} P^k f - \bar{f} \right) \right\| \le \left\| \frac{1}{n} \sum_{k=0}^{n-1} P^k f - \bar{f} \right\|$$

and from Theorem 3 we conclude

$$\lim_{n \to \infty} \sup_{m \in \mathbb{N}} \left\| \frac{1}{n} \sum_{k=0}^{n-1} P^{k+m} f - \bar{f} \right\| = 0. \tag{6}$$

Formula (6) allows us to prove the following result.

Theorem 4 *Let $\{P(t)\}_{t \ge 0}$ be a uniformly continuous stochastic semigroup. If the semigroup $\{P(t)\}_{t \ge 0}$ has an invariant density $f^* > 0$ a.e., then for each density h there exists an invariant density $\bar{h}$ such that $\bar{h} = \lim_{t \to \infty} P(t)h$ in L^1. If additionally, f^* is a unique invariant density, then $\lim_{t \to \infty} P(t)h = f^*$.*

We split the proof of Theorem 4 into lemmas. We begin with some result concerning the convergence of the Toeplitz type sums:

$$S(t) = \sum_{n=0}^{\infty} p_{t,n} P^n.$$

We assume that

$$p_{t,n} \ge 0, \quad \sum_{n=0}^{\infty} p_{t,n} = 1, \quad \lim_{t \to \infty} \sup_n p_{t,n} = 0 \tag{7}$$

and that there exist positive integers $\underline{n}_t < \bar{n}_t$ such that

$$\lim_{t \to \infty} \sum_{n=\underline{n}_t}^{\bar{n}_t} p_{t,n} = 1, \quad \lim_{t \to \infty} \sup_{\underline{n}_t \le n \le \bar{n}_t} \left| \frac{p_{t,n+1}}{p_{t,n}} - 1 \right| = 0. \tag{8}$$

Lemma 1 *If P is a double stochastic operator on $L^1(X, \Sigma, \mu)$ and the sequences $p_{t,n}$ satisfy conditions (7), (8), then $\lim_{t \to \infty} \|S(t)f - \mathrm{E}(f|\Sigma_{\mathrm{inv}})\| = 0$ for each function $f \in L^1(X, \Sigma, \mu)$ and $P\mathrm{E}(f|\Sigma_{\mathrm{inv}}) = \mathrm{E}(f|\Sigma_{\mathrm{inv}})$.*

Proof Let $\bar{f} = \mathrm{E}(f|\Sigma_{\mathrm{inv}})$. Fix $\varepsilon > 0$. From (6) we find n such that

$$\left\|\frac{1}{n}\sum_{k=m}^{m+n-1} P^k f - \bar{f}\right\| < \varepsilon, \quad \left\|\frac{1}{n+1}\sum_{k=m}^{m+n} P^k f - \bar{f}\right\| < \varepsilon.$$

Next, from (8) we choose t_0, such that $\sum_{n=\underline{n}_t}^{\bar{n}_t} p_{t,n} > 1 - \varepsilon$ for $t \geq t_0$. We split the interval $[\underline{n}_t, \bar{n}_t]$ into subintervals which contain n or $n+1$ positive integers. If a subinterval contains n integers, we replace the sum $s = \sum_{k=m}^{m+n-1} p_{t,k} P^k$ in the expression $S(t)$ by the sum $s' = \sum_{k=m}^{m+n-1} n^{-1} s_{t,m} P^k$, where $s_{t,m} = \sum_{k=m}^{m+n-1} p_{t,k}$. We make a similar operation if a subinterval contains $n+1$ integers. We conclude from the second part of condition (8), that $\|s - s'\| < \varepsilon s_{t,m}$ for sufficiently large t. Finally we obtain inequality $\|S(t)f - \bar{f}\| < 4\varepsilon$. □

Lemma 2 *If P is a double stochastic operator, then*

$$\lim_{t\to\infty} \|e^{t(P-I)}f - \mathrm{E}(f|\Sigma_{\mathrm{inv}})\| = 0 \quad \text{for } f \in L^1(X, \Sigma, \mu)$$

and $P\,\mathrm{E}(f|\Sigma_{\mathrm{inv}}) = \mathrm{E}(f|\Sigma_{\mathrm{inv}})$.

Proof Let $p_{t,n} = \dfrac{t^n e^{-t}}{n!}$. Then $e^{t(P-I)} = \sum_{n=0}^{\infty} p_{t,n} P^n$ and $p_{t,n} = \mathrm{Prob}(N_t = n)$, where $(N_t)_{t\geq 0}$ is a Poisson process with intensity $\lambda = 1$. In order to check that $p_{t,n}$ satisfies conditions (7) and (8), we define $\underline{n}_t = [t - t^{2/3}]$, $\bar{n}_t = [t + t^{2/3}]$. We observe that

$$\frac{\underline{n}_t - t}{\sqrt{t}} \to -\infty, \quad \frac{\bar{n}_t - t}{\sqrt{t}} \to \infty \quad \text{as } t \to \infty. \tag{9}$$

Condition (7) follows immediately from the definition of $p_{t,n}$. The first part of condition (8) follows from (9) and from the fact that $(N_t - t)/\sqrt{t}$ converges in distribution to $\mathcal{N}(0, 1)$. If $n \in [\underline{n}_t, \bar{n}_t]$ then $\left|\frac{n}{t} - 1\right| \leq t^{-1/3}$ and the formula

$$\frac{p_{t,n+1}}{p_{t,n}} = \frac{t^{n+1} e^{-t}}{(n+1)!} \cdot \frac{n!}{t^n e^{-t}} = \frac{t}{n+1}$$

implies the second part of condition (8). □

Lemma 3 *Let P be a stochastic operator on $L^1(X, \Sigma, m)$, λ be a positive constant and let $\{P(t)\}_{t\geq 0}$ be the stochastic semigroup given by (2). If $Pf^* = f^*$ for some density $f^* > 0$ a.e., then for each density h there exists the limit $\bar{h} = \lim_{t\to\infty} P(t)h$ in L^1 and $P\bar{h} = \bar{h}$. If additionally, f^* is a unique invariant density of the operator P, then $\lim_{t\to\infty} P(t)h = f^*$.*

Proof Let $\tilde{P}f = P(ff^*)/f^*$ and $\mu(dx) = f^*(x)\,m(dx)$. Then $\tilde{P}$ is a double stochastic operator on the space $L^1(X, \Sigma, \mu)$ and $P^* g = \tilde{P}^* g$ for each function $g \in L^\infty(X, \Sigma, \mu)$. From Lemma 2:

$$\lim_{t\to\infty} \|e^{\lambda t(\tilde{P}-I)}f - \mathrm{E}(f|\Sigma_{\mathrm{inv}})\|_{L^1(X,\Sigma,\mu)} = 0 \quad \text{for } f \in L^1(X,\Sigma,\mu). \tag{10}$$

Formula (10) can be written in the following way:

$$\lim_{t\to\infty} \|e^{\lambda t(P-I)}(ff^*) - f^*\mathrm{E}(f|\Sigma_{\mathrm{inv}})\|_{L^1(X,\Sigma,m)} = 0 \quad \text{for } ff^* \in L^1(X,\Sigma,m).$$

Since $\tilde{P}\mathrm{E}(f|\Sigma_{\mathrm{inv}}) = \mathrm{E}(f|\Sigma_{\mathrm{inv}})$, we have $P(f^*\mathrm{E}(f|\Sigma_{\mathrm{inv}})) = f^*\mathrm{E}(f|\Sigma_{\mathrm{inv}})$. Let $h = ff^*$ and $\bar{h} = f^*\mathrm{E}(f|\Sigma_{\mathrm{inv}})$. Then $P(t)h \to \bar{h}$ in $L^1(X,\Sigma,m)$ and $P\bar{h} = \bar{h}$. If f^* is the unique invariant density of the operator P, then the σ-algebra Σ_{inv} is trivial. Indeed, in the opposite case the function $f = f^*\mathbf{1}_A / \|f^*\mathbf{1}_A\|_m$ is another invariant density if $A \in \Sigma_{\mathrm{inv}}$ and $m(A) > 0$, $m(X \setminus A) > 0$. Since the σ-algebra Σ_{inv} is trivial, $\mathrm{E}(f|\Sigma_{\mathrm{inv}}) = \mathrm{const}$. Then $P(t)h \to cf^*$ for $h \in L^1(X,\Sigma,m)$, where $c = \int_X h(x)\, m(dx)$. If h is a density, then $P(t)h \to f^*$. $\square$

Now, we are ready to finish the proof of Theorem 4. Since $\{P(t)\}_{t\geq 0}$ is a uniformly continuous stochastic semigroup on $L^1(X,\Sigma,m)$, there exists a stochastic operator P on $L^1(X,\Sigma,m)$ and a positive constant λ such that the semigroup $\{P(t)\}_{t\geq 0}$ is given by (2). If $Pf^* = f^*$, then $P(t)f^* = f^*$ for $t \geq 0$, and vice versa, if $P(t)f^* = f^*$ for $t \geq 0$, then $Af^* = 0$, and therefore $Pf^* = f^*$. Thus, Theorem 4 is a simple consequence of Lemma 3.

4 Asymptotic Decomposition of Stochastic Semigroups

Our aim is to provide sufficient conditions for the Foguel alternative. We recall that a stochastic semigroup $\{P(t)\}_{t\geq 0}$ satisfies the Foguel alternative if it is asymptotically stable or sweeping from all compact sets. We also want to find simple sufficient conditions for asymptotic stability and sweeping of $\{P(t)\}_{t\geq 0}$.

We assume additionally that (X,ρ) is a separable metric space and $\Sigma = \mathscr{B}(X)$ is the σ-algebra of Borel subsets of X. We will consider a partially integral stochastic semigroup $\{P(t)\}_{t\geq 0}$ on the space $L^1(X,\Sigma,m)$ with a kernel q (see (5) for the definition). We assume that the kernel q satisfies the following condition:

(K) for every $x_0 \in X$ there exist an $\varepsilon > 0$, a $t > 0$, and a measurable function $\eta \geq 0$ such that $\int \eta(y)\,\mu(dy) > 0$ and

$$q(t, x, y) \geq \eta(y)\mathbf{1}_{B(x_0,\varepsilon)}(x) \quad \text{for } x, y \in X, \tag{11}$$

where $B(x_0, \varepsilon) = \{x \in X : \rho(x, x_0) < \varepsilon\}$.

Inequality (11) can be written using the transition probability function in the following way:

$$\mathscr{P}(t, x, dy) \geq \eta(y)\, m(dy) \quad \text{for } x \in B(x_0, \varepsilon). \tag{12}$$

Condition (K) is satisfied if, for example, for every point $x \in X$ there exist a $t > 0$ and an $y \in X$ such that the kernel $q(t, \cdot, \cdot)$ is continuous in a neighbourhood of (x, y) and $q(t, x, y) > 0$.

In [18] the following theorem is proved.

Theorem 5 *Let $\{P(t)\}_{t \geq 0}$ be a stochastic semigroup which satisfies (K). Then there exist an at most countable set J, a family of invariant densities $\{f_j^*\}_{j \in J}$ with disjoint supports $\{A_j\}_{j \in J}$, and a family $\{\alpha_j\}_{j \in J}$ of positive linear functionals defined on L^1 such that*

(i) *for every $j \in J$ and for every $f \in L^1$ we have*

$$\lim_{t \to \infty} \|\mathbf{1}_{A_j} P(t)f - \alpha_j(f) f_j^*\| = 0, \tag{13}$$

(ii) *if $Y = X \setminus \bigcup_{j \in J} A_j$, then for every $f \in L^1$ and for every compact set F we have*

$$\lim_{t \to \infty} \int_{F \cap Y} P(t)f(x) m(dx) = 0. \tag{14}$$

Remark 1 The *support* of a measurable function f is defined up to a set of measure zero by the formula: $\mathrm{supp}\, f = \{x \in X : f(x) \neq 0\}$.

Remark 2 It is not difficult to check (see [19]) that the sets A_j, $j \in J$, which occur in the formulation of Theorem 5, are not only disjoint but also their closures are disjoint.

Remark 3 Theorem 5 still holds if we assume that $\{P(t)\}_{t \geq 0}$ is a substochastic semigroup [19], i.e., $\|P(t)f\| \leq \|f\|$ and $P(t)f \geq 0$ for $f \geq 0$, $t \geq 0$.

Now, we present some corollaries. Directly from Theorem 5 we obtain

Corollary 1 *Assume that a stochastic semigroup $\{P(t)\}_{t \geq 0}$ satisfies condition (K) and has no invariant densities. Then $\{P(t)\}_{t \geq 0}$ is sweeping from compact sets.*

The second corollary is related to irreducible stochastic semigroups. A stochastic semigroup $\{P(t)\}_{t \geq 0}$ is *irreducible* if $\int_0^\infty P(t) f \, dt > 0$ a.e. for every density f. If the stochastic semigroup is irreducible, then it has either a unique invariant density or an invariant density does not exist.

Corollary 2 *If $\{P(t)\}_{t \geq 0}$ is a stochastic semigroup on the space $L^1(X, \Sigma, m)$ which satisfies (K) and is irreducible, then $\{P(t)\}_{t \geq 0}$ is asymptotically stable or sweeping from compact sets.*

In order to formulate the next corollary we need two auxiliary conditions. Both of them are weak versions of irreducibility and tightness.

(WI) There exists a point $x_0 \in X$ such that for each $\varepsilon > 0$ and for each density f we have

$$\int_{B(x_0,\varepsilon)} P(t)f(x)\,m(dx) > 0 \quad \text{for some } t = t(\varepsilon, f) > 0. \tag{15}$$

(T) There exists $\kappa > 0$ such that

$$\sup_{F \in \mathscr{F}} \limsup_{t \to \infty} \int_F P(t)f(x)\,m(dx) \geq \kappa \tag{16}$$

for $f \in D_0$, where D_0 is a dense subset of D and $\mathscr{F}$ is the family of all compact subsets of X.

Corollary 3 *Let $\{P(t)\}_{t \geq 0}$ be a stochastic semigroup. Assume that $\{P(t)\}_{t \geq 0}$ satisfies conditions (K), (WI), and (T). Then the semigroup $\{P(t)\}_{t \geq 0}$ is asymptotically stable.*

The proof of Corollary 3 can be found in [19]. Now, we formulate the Foguel alternative for some class of stochastic semigroups. We need an auxiliary definition. We say that a stochastic semigroup $\{P(t)\}_{t \geq 0}$ *overlaps supports* if for every $f, g \in D$ there exists $t > 0$ such that

$$m(\operatorname{supp} P(t)f \cap \operatorname{supp} P(t)g) > 0.$$

Corollary 4 *Assume that $\{P(t)\}_{t \geq 0}$ satisfies (K) and overlaps supports. Then $\{P(t)\}_{t \geq 0}$ is sweeping or $\{P(t)\}_{t \geq 0}$ has an invariant density f^* with a support A and there exists a positive linear functional α defined on $L^1(X, \Sigma, m)$ such that*

(i) *for every $f \in L^1(X, \Sigma, m)$ we have*

$$\lim_{t \to \infty} \|\mathbf{1}_A P(t)f - \alpha(f)f^*\| = 0, \tag{17}$$

(ii) *if $Y = X \setminus A$, then for every $f \in L^1(X, \Sigma, m)$ and for every compact set F we have*

$$\lim_{t \to \infty} \int_{F \cap Y} P(t)f(x)\,m(dx) = 0. \tag{18}$$

In particular, if $\{P(t)\}_{t \geq 0}$ has an invariant density f^* with the support A and $X \setminus A$ is a subset of a compact set, then $\{P(t)\}_{t \geq 0}$ is asymptotically stable.

The proof of Corollary 4 is given in [21].

5 Applications to Piecewise Deterministic Markov Processes

5.1 Asymptotic Stability of Pure Jump-Type Markov Processes

A *pure jump-type homogeneous Markov process* on a measurable space (X, Σ) is defined in the following way. Let $\psi \colon X \to [0, \infty)$ be a given measurable function and let $\mathscr{P}(x, B)$ be a given transition probability function on X. Let $t_0 = 0$ and let ξ_0 be an X-valued random variable. For each $n \geq 1$ one can choose the *n*th *jump time* t_n as a positive random variable satisfying

$$\operatorname{Prob}(t_n - t_{n-1} \leq t \mid \xi_{t_{n-1}} = x) = 1 - e^{-\psi(x)t}, \quad t \geq 0,$$

and define $\xi_t = \xi_{t_{n-1}}$ for $t_{n-1} \leq t < t_n$ and we assume that ξ_{t_n} is a random variable with values in X satisfying condition

$$\operatorname{Prob}(\xi_{t_n} \in B \mid \xi_{t_{n-1}} = x) = \mathscr{P}(x, B), \quad B \in \Sigma.$$

The infinitesimal generator L of the process $(\xi_t)_{t \geq 0}$ is of the form

$$Lf(x) = \psi(x) \int_X (f(y) - f(x)) \mathscr{P}(x, dy), \quad f \in B(X, \Sigma), \tag{19}$$

where $B(X, \Sigma)$ is the space of all bounded Σ-measurable functions. If m is a σ-finite measure on (X, Σ), and the transition probability function $\mathscr{P}$ corresponds to some stochastic operator P on $L^1(X, \Sigma, m)$ and ψ is a bounded function, then the operator

$$Af(x) = P(\psi f)(x) - \psi(x) f(x) \tag{20}$$

is an infinitesimal generator of a stochastic semigroup $\{P(t)\}_{t \geq 0}$ on $L^1(X, \Sigma, m)$ corresponding to the process $(\xi_t)_{t \geq 0}$. Let λ be such that $\lambda \geq \sup\{\psi(x) \colon x \in X\}$ and

$$\bar{P}f(x) = \lambda^{-1} \big[P(\psi f)(x) + (\lambda - \psi(x)) f(x) \big].$$

Then $Af = \lambda \bar{P} f - \lambda f$ and

$$P(t)f = \sum_{n=0}^{\infty} \frac{(\lambda t)^n e^{-\lambda t}}{n!} \bar{P}^n f = e^{\lambda t (\bar{P} - I)} f. \tag{21}$$

There is a strict relation between nonnegative fixed points of the operators P and $\bar{P}$. Both operators can be extended to all nonnegative measurable functions assuming that Pf is a pointwise limit of the sequence (Pf_n), where (f_n) is an arbitrary increasing sequence of integrable functions pointwise convergent to f. We assume that $\psi(x) > 0$. A function f^* satisfies the equation $\bar{P}f^* = f^*$ if and only if $P(\psi f^*) = \psi f^*$. If we assume additionally that $\inf_{x \in X} \psi(x) > 0$, then the existence and uniqueness of an invariant density for the operator $\bar{P}$ follows from the existence and uniqueness of an invariant density for the operator P. From Lemma 3 we obtain the following.

Corollary 5 *Assume that ψ is a positive and bounded measurable function. If there exists a positive density f^* such that $P(\psi f^*) = \psi f^*$, then for each density h there exists the limit $\bar{h} = \lim_{t\to\infty} P(t)h$ in L^1. If additionally, f^* is a unique density such that $P(\psi f^*) = \psi f^*$, then $\lim_{t\to\infty} P(t)h = f^*$. In particular, if ψ is bounded above and below and the operator P has a unique invariant density g^* and $g^* > 0$, then the semigroup $\{P(t)\}_{t\geq 0}$ is asymptotically stable.*

5.2 Collisionless Kinetic Equations

A special role in applications of PDMPs is played by a *velocity jump process*, used in descriptions of cellular or organism dispersal [16], stochastic billiards [5], cell cycle models [10, 22] and kinetic theory [12, 15].

As an example we consider a particle moving with a constant speed in a slab with finite thickness and with a partly diffusive boundary reflection. If v_t is the horizontal component of velocity and x_t is the horizontal position of particle at time t, then the model reduces to a stochastic billiard on the interval $[-a, a]$, which was studied in [15]. Between the boundary points $-a$ and a the position and velocity satisfy the equations, $x'_t = v_t$, $v'_t = 0$, and $k_-(v, \bar{v})$ and $k_+(v, \bar{v})$ are the distribution densities of velocity $\bar{v}$ after hitting the boundary $-a$ and a respectively, if v is the velocity of the particle before hitting the boundary. The evolution of distribution densities of the process $\xi_t = (x_t, v_t)$ is given by a stochastic semigroup $\{P(t)\}_{t\geq 0}$ on the space $L^1(X, \mathscr{B}(X), m)$, where $X = [-a, a] \times ([-1, 0) \cup (0, 1])$ and m is the Lebesgue measure on X. The infinitesimal generator A of the semigroup $\{P(t)\}_{t\geq 0}$ is given by the formula

$$Af(x, v) = -v\frac{\partial f}{\partial x}(x, v).$$

The functions f from the domain of the operator A satisfy the boundary conditions:

$$\bar{v}f(-a, \bar{v}) = \int_{-1}^{0} vk_-(v, \bar{v})f(-a, v)\,dv, \quad \bar{v}f(a, \bar{v}) = \int_{0}^{1} vk_+(v, \bar{v})f(a, v)\,dv.$$

Sufficient conditions for asymptotic stability of the semigroup $\{P(t)\}_{t\geq 0}$ are given in [15]. The proof is based on Theorem 2. For example, if there exist $C > 0$ and $\gamma > 0$ such that

$$k_-(v,\bar{v}) \leq C|\bar{v}|^\gamma \quad \text{for } v < 0, \bar{v} > 0 \quad \text{and} \quad k_+(v,\bar{v}) \leq C\bar{v}^\gamma \quad \text{for } v > 0, \bar{v} < 0,$$

then an invariant density exists and under some irreducibility of k_- and k_+ the semigroup $\{P(t)\}_{t\geq 0}$ is asymptotically stable. If the kernels k_- and k_+ are continuous functions and there is no invariant density then from Corollary 1 it follows that the semigroup is sweeping from compact sets. It means that

$$\lim_{t\to\infty} \int_{-\varepsilon}^{\varepsilon} \int_{-a}^{a} P(t)f(x,v)\,dx\,dv = 1 \tag{22}$$

for every density f and every $\varepsilon > 0$.

The Lebowitz and Rubinow model of cell cycle [10] can be identified with a one-dimensional stochastic billiard with symmetric reflection boundary conditions. Thus, the results from the paper [15] can be directly applied to the Lebowitz and Rubinow model. Asymptotic properties of collisionless kinetic semigroups in d-dimensional domain are studied in [12].

5.3 Gene Expression Models

Gene expression models are used to describe the process of protein production. Since the dynamics of such processes depends on the activity of genes, usually they are described by *dynamical systems with random switching*. Now, we give a short introduction to the theory of such objects.

We consider a family of deterministic processes indexed by a set $I = \{1,\ldots,k\}$. Each process is a solution of a system of differential equations defined on some subset X of the space $\mathbb{R}^d$:

$$x'(t) = b(x(t), i), \quad i \in I, \tag{23}$$

where $b\colon X \times I \to \mathbb{R}^d$ is a Lipschitz function. Temporary, elements of I are called states. We assume that for each $x_0 \in X$ there exists a unique solution $x\colon [0,\infty) \to X$ of (23) with the initial condition $x(0) = x_0$ and we denote this solution by $\pi_t^i x_0$. We assume that for each $x \in X$ there exists an intensity transition matrix $Q(x) = [q_{ij}(x)]$ between the states I. Let $q_i(x) = -q_{ii}(x)$. We choose a point $x_0 \in X$ and a state $i \in I$. We go along the trajectory $\pi_t^i x_0$, but we can leave the state i in a small time interval $[t, t+\Delta t]$ with probability

$$q_i(\pi_t^i x_0)\Delta t + o(\Delta t).$$

If we leave the state i at time t_1 then we are at the point $x_1 = \pi_{t_1}^i x_0$. We choose a new state j with probability $q_{ij}(x_1)/q_i(x_1)$. We repeat the procedure starting now from the point x_1 and the state j and continue it in subsequent steps. If $i(t)$ is the state of the system at time t then $\xi_t = (x(t), i(t))$, $t \geq 0$, is a homogeneous PDMP with values in the space $\mathbb{X} = X \times I$. The family of processes ξ_t induces a stochastic semigroup $\{P(t)\}_{t \geq 0}$ on the space $L^1(\mathbb{X}, \Sigma, m)$, where Σ is the σ-algebra of the Borel subsets of $\mathbb{X}$ and m is the product measure of the Lebesgue measure on X and the counting measure on I. Based on the paper [19] we shortly present here a general method of studying asymptotic properties of the semigroup $\{P(t)\}_{t \geq 0}$.

Condition (K) can be checked by using Lie brackets. Let $f(x)$ and $g(x)$ be two vector fields on $\mathbb{R}^d$. The *Lie bracket* $[f, g]$ is a vector field given by

$$[f, g]_j(x) = \sum_{k=1}^d \left(f_k(x) \frac{\partial g_j}{\partial x_k}(x) - g_k(x) \frac{\partial f_j}{\partial x_k}(x) \right).$$

Assume that the vector fields $b^1, \ldots, b^k$ are sufficiently smooth in a neighbourhood of a point x. We say that *Hörmander's condition* holds at x if vectors fields

$$b^2(x) - b^1(x), \ldots, b^k(x) - b^1(x), \ [b^i, b^j](x)_{1 \leq i, j \leq k}, \ [b^i, [b^j, b^l]](x)_{1 \leq i, j, l \leq k}, \ldots$$

span the space $\mathbb{R}^d$. Let $(x_0, i), (y_0, j) \in \mathbb{X}$ and assume that the trajectories of the process (ξ_t) join (y_0, j) with (x_0, i). If Hörmander's condition holds at x_0 and $q_{l'l}(x_0) > 0$ for $l \neq l'$, then condition (K) is satisfied at the state (y_0, j). The condition (WI) is satisfied at the state (x_0, i) if each state (y_0, j) we can join by the process ξ_t with some state arbitrary close to the state (x_0, i). If conditions (K) and (WI) hold and the space $\mathbb{X}$ is compact then immediately from Corollary 3 it follows that the semigroup $\{P(t)\}_{t \geq 0}$ is asymptotically stable.

Now, we consider the following gene expression model introduced by Lipniacki et al. [11] and studied in [2]. A model consists of two types of molecules: mRNA, and protein, and we assume that variables x_1, x_2 denote their concentrations. If the gene is active then mRNA transcript molecules are synthesized with a constant speed R. The protein translation proceeds with the rate $Px_1(t)$, where P is a constant. The mRNA and protein degradation rates are μ_R and μ_P, respectively. Thus the process is described by the following system

$$\begin{cases} x_1'(t) = Ri(t) - \mu_R x_1(t), \\ x_2'(t) = Px_1(t) - \mu_P x_2(t), \end{cases} \quad (24)$$

where $i(t) = 1$ if the gene is active and $i(t) = 0$ in the opposite case. We also assume that the gene is activated with rate q_0 and inactivated with rate q_1 and the rates are continuous and bounded functions of $\mathbf{x} = (x_1, x_2)$. Then the process $\xi_t = (x_1(t), x_2(t), i(t))$, $t \geq 0$, is a homogeneous PDMP.

In this model we have two dynamical systems, which correspond, respectively, to $i = 0$ and $i = 1$. Each system has a unique stationary point, either $\mathbf{x}_0 = (0, 0)$, or $\mathbf{x}_1 = \left(\frac{R}{\mu_R}, \frac{PR}{\mu_P \mu_R} \right)$, which is asymptotically stable. We assume that $q_0(\mathbf{x}_0) > 0$ and $q_1(\mathbf{x}_1) > 0$. This assumption implies that if the process starts from any point (x_1, x_2, i) then it enters the invariant set $\mathbb{X} = \left[0, \frac{R}{\mu_R} \right] \times \left[0, \frac{PR}{\mu_P \mu_R} \right] \times \{0, 1\}$ and the process visits any neighbourhood of the point $(\mathbf{x}_0, 0)$ and the point $(\mathbf{x}_1, 1)$ infinitely many times. It means that condition (WI) holds. Moreover, if we assume additionally that $q_0(\mathbf{x}_1) > 0$ or $q_1(\mathbf{x}_0) > 0$ then we can check that condition (K) holds by applying Hörmander's condition. We have $b^2(\mathbf{x}) = \mathbf{v} + b^1(\mathbf{x})$ with $\mathbf{v} = (R, 0)$. Hence

$$b^2 - b^1 = (R, 0) = \mathbf{v},$$
$$[b^1, b^2] = [b^1, \mathbf{v}] = [(-\mu_R x_1, P x_1 - \mu x_2), (R, 0)] = (R\mu_R, -RP).$$

The vectors $b^2 - b^1$ and $[b^1, b^2]$ span the space $\mathbb{R}^2$, and consequently Hörmander's condition holds. Since the set $\mathbb{X}$ is compact, the semigroup $\{P(t)\}_{t \geq 0}$ is asymptotically stable.

A more advanced model of gene expression is consider in [25]. In this model the protein production has an additional primary transcript step.

5.4 Stein's Neuronal Model

Electrical activity of a neuron is described by its *depolarization* $x(t)$. Nerve cells may be excited or inhibited through neuron's synapses. Synapses may be excitatory or inhibitory. If at time t an excitation occurs then $x(t^+) = x(t^-) + x_E$ and if an inhibition occurs, then $x(t^+) = x(t^-) - x_I$, where x_E and x_I are nonnegative constants. The jumps (excitations and inhibitions) may occur at random times with intensities $\lambda_E > 0$ and $\lambda_I > 0$, respectively. Between jumps depolarization $x(t)$ decays according to the equation $x'(t) = -\alpha x(t), \alpha > 0$. When a sufficient (threshold) level $\theta > 0$ of excitation is reached, the neuron emits an action potential (fire). This will be followed by an absolute refractory period of duration t_R, during which $x \equiv 0$ and then the process starts again.

The neural activity can be described as a piecewise deterministic Markov process in the following way. We introduce an extra 0-phase, which begins at the end of the refractory period and finishes when depolarization jumps from 0 for the first time. After a first jump we have the subthreshold phase denoted by 1. By 2 we denote the refractory phase of duration t_R. Therefore, the state space is $X = \{(0, 0)\} \cup (-\infty, \theta] \times \{1\} \cup [0, t_R] \times \{2\}$, and in the 0-phase we have only one state $(0, 0)$; if the neuron is in the 1-phase then it has a state $(x, 1)$, where x is its depolarization; and in the 2-phase it has a state $(x, 2)$, where x is the time since the moment of firing. We consider two types of jumps: when the neuron is excited or inhibited and at the end of the refractory period. Thus, we can have one or more jumps inside phase 1. Figure 1 presents a schematic diagram of the model.

Let $\xi_t = (\xi_t^1, \xi_t^2)$, $t \geq 0$, be a stochastic process with values in X describing the state of the neuron. Between jumps the process ξ_t satisfies the following system of equations

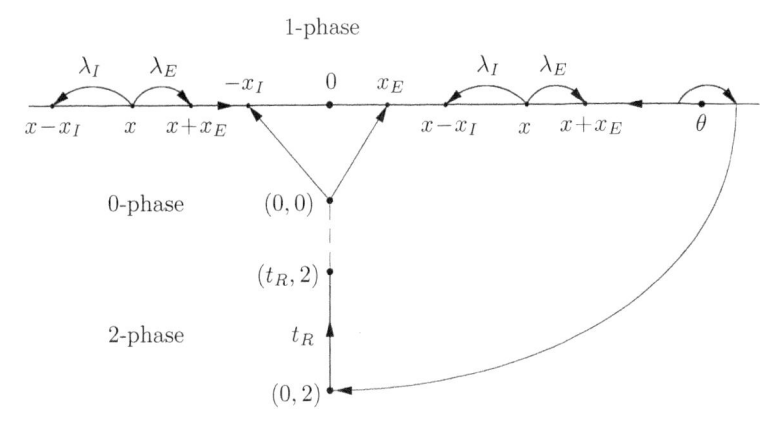

Fig. 1 A schematic diagram of the model

$$\xi_t^{1\prime} = \begin{cases} 0, & \text{if } \xi_t^2 = 0, \\ -\alpha \xi_t^1, & \text{if } \xi_t^2 = 1, \\ 1, & \text{if } \xi_t^2 = 2. \end{cases}$$

Define a measure m on the σ-algebra $\mathscr{B}(X)$ of the Borel subsets of X by $m = \delta_{(0,0)} + m_1 + m_2$, where $\delta_{(0,0)}$ is the Dirac measure at $(0, 0)$, m_1 is the Lebesgue measure on the segment $(-\infty, 0] \times \{1\}$ and m_2 is the Lebesgue measure on the segment $[0, t_R] \times \{2\}$. The family of PDMPs described above induces a stochastic semigroup $\{P(t)\}_{t \geq 0}$ on the space $L^1(X, \mathscr{B}(X), m)$.

The following result is proved in [20].

Theorem 6 *The stochastic semigroup introduced by Stein's model is asymptotically stable.*

The proof of this result is based on Corollary 3.

5.5 Two-Phase Model

We start with a short biological description of two phase-cell cycle models studied in [21]. The cell cycle is divided into the resting and proliferating phase. The duration of the resting phase is a random variable t_A which depends on the maturity m of a cell. The duration $t_B = \tau$ of the proliferating phase is constant. A cell can move from the resting phase to the proliferating phase with rate $\varphi(m)$. Cells age with unitary velocity and mature with a velocity $g_1(m)$ in the resting phase and with a velocity $g_2(m)$ in the proliferating phase. The age variable a in the proliferating phase is assumed to range from $a = 0$ at the moment of entering the proliferating phase to $a = \tau$ at the point of cytokinesis.

If the mother cell has maturity m, then the daughter cells have maturity $h(m)$.

We assume that the functions φ, h, g_1 and g_2 are sufficiently regular, $\varphi(m) = 0$ for $m \le m_P$ and $\varphi(m) > 0$ for $m > m_P$, where $m_P > 0$ is the minimum cell size when it can enter the proliferating phase. Let $Q(m) = \int_0^m \frac{\varphi(r)}{g_1(r)} dr$. We assume that $\lim_{m \to \infty} Q(m) = \infty$, which guaranties that each cell enters the proliferating phase with probability one. Let $\lambda(m) = \pi_2(-\tau, h^{-1}(m))$, where $\pi_i(t, m_0)$, $i = 1, 2$, is the solution of the equation $m'(t) = g_i(m(t))$ with the initial condition $m(0) = m_0 \ge 0$.

The above model can be treated as a piecewise deterministic Markov process. We consider a sequence of consecutive descendants of a single cell. Let s_n be a time when a cell from the nth-generation enters a proliferating phase and $t_n = s_n + \tau$ be a time of its division. If $t_{n-1} \le t < t_n$ then the state $\xi_t = (a(t), m(t), i(t))$ of the n-th cell is described by age $a(t)$, maturity $m(t)$ and the index $i(t)$, where $i = 1$ if a cell is in the resting phase and $i = 2$ if it is in the proliferating phase. Random moments $t_0, s_1, t_1, s_2, t_2, \ldots$ are called *jump times*. Between jump times the parameters change according to the following system of equations:

$$\begin{cases} a'(t) = 1, \\ m'(t) = g_{i(t)}(m(t)), \\ i'(t) = 0. \end{cases} \quad (25)$$

The process ξ_t changes at jump points according to the following rules:

$$a(s_n) = 0, \quad m(s_n) = m(s_n^-), \quad i(s_n) = 2,$$

$$a(t_n) = 0, \quad m(t_n) = h(m(t_n^-)), \quad i(t_n) = 1.$$

If $m(t_{n-1}) = m_0$ then the cumulative distribution function F of $s_n - t_{n-1}$ is given by

$$F(t) = 1 - e^{Q(m_0) - Q(\pi_1(t, m_0))}. \quad (26)$$

Then $(\xi_t)_{t \ge 0}$ is a homogeneous Markov process and it induces a stochastic semigroup $\{P(t)\}_{t \ge 0}$ on the space $L^1(X, \mathcal{B}(X), \mu)$, where

$$X = \{(a, m, 1): m \ge \pi_1(a, 0),\ a \ge 0\} \cup \{(a, m, 2): m \ge \pi_2(a, m_p),\ a \in [0, \tau]\},$$

and μ is the product of the two-dimensional Lebesgue measure and the counting measure on the set $\{1, 2\}$ (see Fig. 2).

We need two additional assumptions:

$$h(\pi_2(\tau, m)) < m \text{ for } m \ge m_P, \quad (27)$$

$$h'(\pi_2(\tau, \bar{m})) g_2(\pi_2(\tau, \bar{m})) g_1(\bar{m}) \ne g_1(h(\pi_2(\tau, \bar{m}))) g_2(\bar{m}) \quad (28)$$

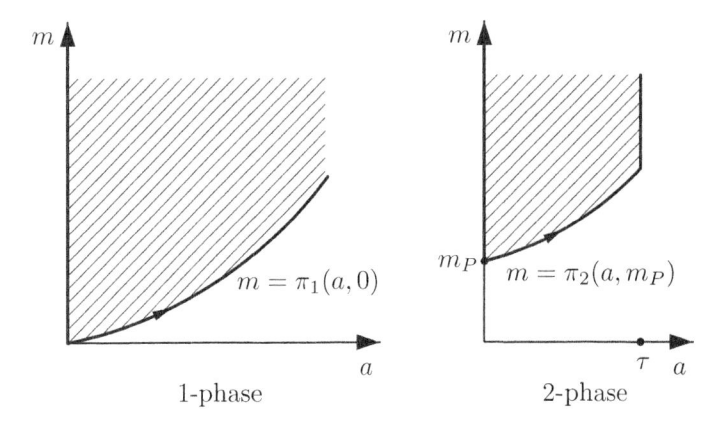

Fig. 2 The set X

for some $\bar{m} > m_P$. Condition (27) guarantees that with a positive probability each cell will have a descendant in some generation with a sufficiently small maturity. Condition (28) seems to be technical but if it is not fulfilled, the cells have *synchronous growth* and the semigroup is not asymptotically stable. In particular if $g_1 \equiv g_2$ and $h(m) = m/2$, then (28) reduces to $2g_2(m) \neq g_2(2m)$ for some $m > \pi_2(\tau, m_P)/2$. A similar condition appears in many papers concerning size-structured models [1, 3, 8, 24, 26]. The following results are proved in [21].

Theorem 7 *The semigroup* $\{P(t)\}_{t \geq 0}$ *satisfies the Foguel alternative, i.e.* $\{P(t)\}_{t \geq 0}$ *is asymptotically stable or sweeping from compact sets.*

The proof of this result is based on Theorem 1 and Corollary 1. Moreover we have the following alternative.

Corollary 6 *If* $\liminf_{m \to \infty} Q(\lambda(m)) - Q(m) > 1$ *and* $\varphi(m) \geq \varepsilon > 0$ *for sufficiently large m, then the semigroup* $\{P(t)\}_{t \geq 0}$ *is asymptotically stable. If* $Q(\lambda(m)) - Q(m) \leq 1$ *for sufficiently large m and φ is bounded, then the semigroup* $\{P(t)\}_{t \geq 0}$ *is sweeping.*

The proof of Corollary 6 is based on Corollary 4 and some results concerning asymptotic stability [7] and sweeping [13] of a stochastic operator related to a discrete time two-phase cell cycle model [28].

Acknowledgements This research was partially supported by the National Science Centre (Poland) Grant No. 2017/27/B/ST1/00100.

References

1. Banasiak, J., Pichór, K., Rudnicki, R.: Asynchronous exponential growth of a general structured population model. Acta Appl. Math. **119**, 149–166 (2012)
2. Bobrowski, A., Lipniacki, T., Pichór, K., Rudnicki, R.: Asymptotic behavior of distributions of mRNA and protein levels in a model of stochastic gene expression. J. Math. Anal. Appl. **333**, 753–769 (2007)
3. Diekmann, O., Heijmans, H.J.A.M., Thieme, H.R.: On the stability of the cell size distribution. J. Math. Biol. **19**, 227–248 (1984)
4. Davis, M.H.A.: Piecewise-deterministic Markov processes: a general class of nondiffusion stochastic models. J. Roy. Statist. Soc. Ser. B **46**, 353–388 (1984)
5. Evans, S.N.: Stochastic billiards on general tables. Ann. Appl. Probab. **11**, 419–437 (2001)
6. Foguel, S.R.: The Ergodic Theory of Markov Processes. Van Nostrand Reinhold Comp., New York (1969)
7. Gacki, H., Lasota, A.: Markov operators defined by Volterra type integrals with advanced argument. Ann. Polon. Math. **51**, 155–166 (1990)
8. Gyllenberg, M., Heijmans, H.J.A.M.: An abstract delay-differential equation modelling size dependent cell growth and division. SIAM J. Math. Anal. **18**, 74–88 (1987)
9. Lasota, A., Mackey, M.C.: Chaos, Fractals and Noise. Stochastic Aspects of Dynamics. Springer Applied Mathematical Sciences, vol. 97. New York (1994)
10. Lebowitz, J.L., Rubinow, S.L.: A theory for the age and generation time distribution of microbial population. J. Math. Biol. **1**, 17–36 (1974)
11. Lipniacki, T., Paszek, P., Marciniak-Czochra, A., Brasier, A.R., Kimmel, M.: Transcriptional stochasticity in gene expression. J. Theor. Biol. **238**, 348–367 (2006)
12. Lods, B., Mokhtar-Kharroubi, M., Rudnicki, R.: Invariant density and time asymptotics for collisionless kinetic equations with partly diffuse boundary operators. Ann. I. H. Poincaré – AN (2020). https://doi.org/10.1016/j.anihpc.2020.02.004
13. Łoskot, K., Rudnicki, R.: Sweeping of some integral operators. Bull. Pol. Ac.: Math. **37**, 229–235 (1989)
14. Mackey, M.C., Rudnicki, R.: Global stability in a delayed partial differential equation describing cellular replication. J. Math. Biol. **33**, 89–109 (1994)
15. Mokhtar-Kharroubi, M., Rudnicki, R.: On asymptotic stability and sweeping of collisionless kinetic equations. Acta Appl. Math. **147**, 19–38 (2017)
16. Othmer, H.G., Dunbar, S.R., Alt, W.: Models of dispersal in biological systems. J. Math. Biol. **26**, 263–298 (1988)
17. Pichór, K., Rudnicki, R.: Continuous Markov semigroups and stability of transport equations. J. Math. Anal. Appl. **249**, 668–685 (2000)
18. Pichór, K., Rudnicki, R.: Asymptotic decomposition of substochastic operators and semigroups. J. Math. Anal. Appl. **436**, 305–321 (2016)
19. Pichór, K., Rudnicki, R.: Asymptotic decomposition of substochastic semigroups and applications. Stochast. Dyn. **18**, 1850001 (2018)
20. Pichór, K., Rudnicki, R.: Stability of stochastic semigroups and applications to Stein's neuronal model. Discret. Contin. Dyn. Syst. B **23**, 377–385 (2018)
21. Pichór, K., Rudnicki, R.: Applications of stochastic semigroups to cell cycle models. Discret. Contin. Dyn. Syst. B **24**, 2365–2381 (2019)
22. Rotenberg, M.: Transport theory for growing cell populations. J. Theor. Biol. **103**, 181–199 (1983)
23. Rudnicki, R.: Stochastic operators and semigroups and their applications in physics and biology. In: Banasiak, J., Mokhtar-Kharroubi, M. (eds.) Evolutionary Equations with Applications in Natural Sciences, Lecture Notes in Mathematics, vol. 2126, pp. 255–318. Springer, Heidelberg (2015)
24. Rudnicki, R., Pichór, K.: Markov semigroups and stability of the cell maturation distribution. J. Biol. Syst. **8**, 69–94 (2000)

25. Rudnicki, R., Tomski, A.: On a stochastic gene expression with pre-mRNA, mRNA and protein contribution. J. Theor. Biol. **387**, 54–67 (2015)
26. Rudnicki, R., Tyran-Kamińska, M.: Piecewise Deterministic Processes in Biological Models. SpringerBriefs in Applied Sciences and Technology, Mathematical Methods. Springer, Cham, Switzerland (2017)
27. Stein, R.B.: Some models of neuronal variability. Biophys. J. **7**, 37–68 (1967)
28. Tyrcha, J.: Asymptotic stability in a generalized probabilistic/deterministic model of the cell cycle. J. Math. Biol. **26**, 465–475 (1988)

On Polynomial Stability of Coupled Partial Differential Equations in 1D

Lassi Paunonen

Abstract We study the well-posedness and asymptotic behaviour of selected PDE–PDE and PDE–ODE systems on one-dimensional spatial domains, namely a boundary coupled wave–heat system and a wave equation with a dynamic boundary condition. We prove well-posedness of the models and derive rational decay rates for the energy using an approach where the coupled systems are formulated as feedback interconnections of impedance passive regular linear systems.

Keywords Coupled PDE system · Polynomial stability · Wave equation · Strongly continuous semigroup · Systems theory · Feedback

MSC (2010) 35L05 · 35B35 · 93C05 · 47D06 · 93D15

1 Introduction

The purpose of this short paper is to discuss how an abstract "system theoretic approach" can be used in the study of stability properties of certain types of coupled linear PDE–PDE and PDE–ODE systems. In particular, several recent references have demonstrated that coupled PDE systems very often exhibit *polynomial* and the more general *non-uniform stability* [4, 7, 19], in which the energies of the classical solutions of the system decay at subexponential rates as $t \to \infty$. While polynomial stability of many specific coupled PDE systems has been proved in the literature [2, 3, 11, 14, 15, 24, 25] using a variety of powerful methods, in this paper we focus on the usage of selected abstract results from [17] establishing polynomial stability for a *class* of such coupled systems. More precisely, the results in [17] approach

---

The research is supported by the Academy of Finland Grant numbers 298182 and 310489 held by L. Paunonen.

L. Paunonen (✉)
Mathematics and Statistics, Faculty of Information Technology and Communication Sciences, Tampere University, P.O. Box 692, 33101 Tampere, Finland
e-mail: lassi.paunonen@tuni.fi

© Springer Nature Switzerland AG 2020
J. Banasiak et al. (eds.), *Semigroups of Operators – Theory and Applications*, Springer Proceedings in Mathematics & Statistics 325, https://doi.org/10.1007/978-3-030-46079-2_20

the study of stability of coupled PDE–PDE and PDE–ODE systems by considering them as abstract *systems* which form *a feedback interconnection*. We demonstrate the use of this framework by proving polynomial stability for two PDE systems, a one-dimensional "wave-heat system"

$$\rho(\xi)v_{tt}(\xi, t) = (T(\xi)v_\xi(\xi, t))_\xi, \qquad -1 < \xi < 0, \tag{1a}$$
$$w_t(\xi, t) = w_{\xi\xi}(\xi, t), \qquad 0 < \xi < 1, \tag{1b}$$
$$v_\xi(-1, t) = 0, \qquad w(1, t) = 0, \tag{1c}$$
$$v_t(0, t) = w(0, t), \qquad T(0)v_\xi(0, t) = w_\xi(0, t) \tag{1d}$$

and a wave equation with an "acoustic boundary condition"

$$\rho(\xi)v_{tt}(\xi, t) = (T(\xi)v_\xi(\xi, t))_\xi, \qquad 0 < \xi < 1, \tag{2a}$$
$$m\delta_{tt}(t) = -d\delta_t(t) - k\delta(t) - \beta v_t(1, t) \tag{2b}$$
$$v_\xi(1, t) = \delta_t(t), \qquad v_t(0, t) = 0. \tag{2c}$$

The system (1) is similar to those studied in [3, 5, 24], but with a wave part that may have spatially varying density $\rho(\cdot)$ and Young's modulus $T(\cdot)$. The system (2) is a one-dimensional analogue of wave equations used in modelling the behaviour of acoustic waves on higher dimensional spatial domains [6, 18].

The abstract system theoretic approach has been employed in several studies on stability of coupled PDEs, especially by Ammari and co-authors [1, 14], but it is still under-utilised as a technique and much of its potential remains hidden. This may be largely due to the fact that recognising particular PDE systems that fit in a given abstract framework is often less than straightforward, and formulating the particular coupled PDE system as an abstract feedback interconnection often requires some effort. The purpose of this note is to demonstrate and discuss these steps for the two coupled systems (1) and (2). In particular, our aim is to outline the general procedure and highlight the most important steps in using the results in [17] to prove polynomial stability.

Equations (1) and (2) fit into a general class of PDE–PDE and PDE–ODE systems which consist of two "abstract component systems" (see Fig. 1) with *states*, *inputs*, and *outputs* $(x(t), u(t), y(t))$ and $(x_c(t), u_c(t), y_c(t))$ satisfying the following criteria.

1. One of the systems is unstable and the other is exponentially stable.
2. Both systems are *impedance passive*, meaning that they do not contain "internal sources of energy" (see Sect. 2 for details).
3. The full coupled system is formed by a "power-preserving interconnection"

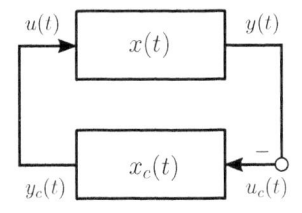

[Fig. 1: diagram of the closed-loop system with blocks $x(t)$ and $x_c(t)$, with signals $u(t)$, $y(t)$, $y_c(t)$, $u_c(t)$.]

Fig. 1 The closed-loop system

$$u(t) = y_c(t) \quad \text{and} \quad u_c(t) = -y(t) \tag{3}$$

(or alternatively $u(t) = -y_c(t)$ and $u_c(t) = y(t)$).

For such systems, the results in [17] can be used to prove polynomial or non-uniform stability of the full coupled system by verifying certain conditions on the two component systems.

In many concrete PDE–PDE and PDE–ODE systems it is fairly easy to identify the unstable component (e.g., an undamped wave or beam equation, or an ODE with a skew-adjoint system matrix) and the stable component (e.g. a heat equation or a damped wave/beam, or a stable ODE). However, making sure that the impedance passivity and the power-preserving interconnection are satisfied depends on the choices of the states, inputs, and outputs $(x(t), u(t), y(t))$ and $(x_c(t), u_c(t), y_c(t))$, as well as on the choices of the state spaces of these systems. Making the correct choices is often not straightforward, and this is precisely the process we aim to illustrate in this paper by considering the two PDE systems (1) and (2). In these example cases the component systems that make up the full coupled systems have been studied extensively in the literature, and after expressing the systems as coupled abstract systems, we will find that all the conditions required for proving the polynomial closed-loop stability are either already known for our component systems, or can be computed explicitly with minimal effort.

The general framework of *regular linear systems* used in [17] involves several technical concepts. However, many of these technicalities are only required in the proofs, and it is often not necessary to write the PDE systems under consideration in this particular form. Instead, the knowledge that such a representation exists is sufficient, and regularity has been proved in the literature for several particular types of PDE systems [8, 12, 13, 26].

The paper is organised as follows. In Sect. 2 we formulate the class of abstract systems used in the analysis in greater detail and restate a general condition for polynomial stability of abstract coupled systems from [17]. In Sects. 3 and 4 we study the coupled wave-heat system (1) and the wave Eq. (2), respectively.

If X and Y are Banach spaces and $A : X \to Y$ is a linear operator, we denote the domain of A by $D(A)$. The space of bounded linear operators from X to Y is denoted by $L(X, Y)$. If $A : X \to X$, then $\rho(A)$ denotes the resolvent set of A and the resolvent operator is $R(\lambda, A) = (\lambda - A)^{-1}$ for $\lambda \in \rho(A)$. The inner product on a Hilbert space is denoted by $\langle \cdot, \cdot \rangle$. For $T \in L(X)$ on a Hilbert space X we define $\operatorname{Re} T = \frac{1}{2}(T + T^*)$.

For two functions $f : [0, \infty) \to [0, \infty)$ and $g : [0, \infty) \to (0, \infty)$ we write $f(t) = o(g(t))$ if $f(t)/g(t) \to 0$ as $t \to \infty$.

2 Coupled Abstract Systems

In this section we will briefly summarise the most important results (in a special case related to our systems) concerning polynomial stability from [17]. Throughout the paper we consider closed-loop system consisting of two systems with $(x(t), u(t), y(t))$ and $(x_c(t), u_c(t), y_c(t))$, where $x(t) \in X$ and $x_c(t) \in X_c$ for some Hilbert spaces X and X_c, and $u(t), y(t), u_c(t), y_c(t) \in \mathbb{C}^m$ for all $t \geq 0$. Most notably we assume that the two systems are *impedance passive* in the sense that their (classical) states satisfy

$$\frac{1}{2}\frac{d}{dt}\|x(t)\|_X^2 \leq \operatorname{Re}\langle u(t), y(t)\rangle_{\mathbb{C}^m} \quad \text{and} \quad \frac{1}{2}\frac{d}{dt}\|x_c(t)\|_{X_c}^2 \leq \operatorname{Re}\langle u_c(t), y_c(t)\rangle_{\mathbb{C}^m}.$$

In addition we assume that these two systems are *regular linear systems* [23] so that their dynamics are described by

$$\dot{x}(t) = Ax(t) + Bu(t), \quad x(0) \in X, \tag{4a}$$
$$y(t) = C_\Lambda x(t) + Du(t) \tag{4b}$$

and

$$\dot{x}_c(t) = A_c x_c(t) + B_c u_c(t), \quad x_c(0) \in X_c, \tag{5a}$$
$$y_c(t) = C_{c\Lambda} x_c(t) + D_c u_c(t) \tag{5b}$$

for suitable operators $A : D(A) \subset X \to X$ and $A_c : D(A_c) \subset X_c \to X_c$ that generate strongly continuous semigroups, $D, D_c \in \mathbb{C}^{m\times m}$, and possibly unbounded operators $B, C, B_c,$ and C_c. The details of regular linear systems can be found, e.g., in [17, 21, 23]. While the detailed assumptions on the parameters of (4) and (5) are fairly technical, in this paper we will demonstrate that it is often not necessary to find the exact expressions of B, C_Λ, D and $B_c, C_{c\Lambda}, D_c$, or even A and A_c, as long as the regularity of the system under consideration have been established earlier in the literature. It should be noted that all systems on finite-dimensional spaces are regular, as are systems (4) with bounded input and output operators B and C.

The impedance passivity immediately implies that the semigroups generated by A and A_c are contractive (this follows from letting $u(t) \equiv 0$ and $u_c(t) \equiv 0$ in the definition of impedance passivity). In addition, necessarily $\operatorname{Re} D \geq 0$ and $\operatorname{Re} D_c \geq 0$, where $\operatorname{Re} T = \frac{1}{2}(T + T^*)$.

The main motivation for considering regular linear systems with possibly unbounded operators $B, C_\Lambda, B_c,$ and $C_{c\Lambda}$ is that these systems can be used to describe systems with couplings on the boundaries of the PDEs. Another great benefit of regular linear systems is that this class has a very strong feedback theory [23]. In particular, the results in [23] imply that if either $D \geq 0$ or $D_c = 0$, then the closed-loop system with state $x_e(t) = (x(t), x_c(t)) \in X \times X_c$ is associated with a contraction semigroup $T_e(t)$ (see [17] for details). In particular this implies that the closed-loop system has a well-defined solution and

$$\|x(t)\|_X^2 + \|x_c(t)\|_{X_c}^2 \leq \|x(0)\|_X^2 + \|x_c(0)\|_{X_c}^2, \qquad t \geq 0.$$

The results in [17] establish polynomial stability of the closed-loop system under the following conditions. The theorem makes use of the *transfer function* of the system $(x_c(t), u_c(t), y_c(t))$, which can either be computed using the formula $P_c(\lambda) = C_{c\Lambda}(\lambda - A_c)^{-1} B_c + D_c$ or using the Laplace transform of the original PDE system, in which case $\hat{y}_c(\lambda) = P_c(\lambda) \hat{u}_c(\lambda)$.

Proposition 1 *Let* (A, B, C_Λ, D) *and* $(A_c, B_c, C_{c\Lambda}, D_c)$ *be two impedance passive regular linear systems where A is skew-adjoint and has compact resolvent, and either $D \geq 0$ or $D_c = 0$. If the system (A, B, C_Λ, D) becomes exponentially stable with negative output feedback $u(t) = -y(t)$, and if there exists $\alpha, \eta_0 > 0$ such that*

$$\operatorname{Re} P_c(is) \geq \frac{\eta_0}{1 + |s|^\alpha} \qquad \forall s \in \mathbb{R}, \tag{6}$$

then the closed-loop system is polynomially stable so that

$$\left\| \begin{bmatrix} x(t) \\ x_c(t) \end{bmatrix} \right\|_{X \times X_c} = o(t^{-\alpha}), \qquad \text{as } t \to \infty$$

for all classical solutions of the closed-loop system.

If $0 \in \rho(A)$, *then it is sufficient that* (6) *holds for* $|s| \geq s_0$ *where* $s_0 > 0$ *is such that* $[-is_0, is_0] \subset \rho(A)$.

Proof It is shown in [17, Theorem 3.7] that if the conditions of the proposition hold, then the generator A_e of the contraction semigroup $T_e(t)$ satisfies $i\mathbb{R} \subset \rho(A_e)$ and $\|R(is, A_e)\| = M_R(1 + |s|^\alpha)$ for some constant $M_R > 0$ and for all $s \in \mathbb{R}$. The claim therefore follows from [7, Theorem 2.4]. □

In Proposition 1 the "classical solutions" of the closed-loop system refer to those solutions of the coupled PDE system that correspond to the states $x(t)$ and $x_c(t)$ for which $(x(0), x_c(0)) \in X \times X_c$ belongs to the domain of the closed-loop semigroup generator. This domain is characterised in detail in [17, Sect. 3], but unfortunately the description in terms of the operators (A, B, C_Λ, D) and $(A_c, B_c, C_{c\Lambda}, D_c)$ is typically not very illustrative. In this paper we will not discuss the properties of the classical solutions of the original coupled system in detail in general cases, but instead we will only present the existence of the classical solutions in the cases of the two PDE systems in Sects. 3 and 4. As should be expected from a correct abstract formulation, these classical solutions are precisely those solutions of the original coupled systems which satisfy the boundary conditions and for which all derivatives in the system exist in a suitable sense.

Remark 1 Since we assume $A^* = -A$, the impedance passivity of (A, B, C_Λ, D) can be used to show that the property that $u(t) = -y(t)$ stabilizes the system exponentially is equivalent to the *exact observability* of the pair (C, A) (or alternatively *exact controllability* of the pair (A, B)) [16, 20].

3 The Coupled Wave-Heat System

In this section we study the polynomial stability of the wave-heat system

$$\rho(\xi)v_{tt}(\xi, t) = (T(\xi)v_\xi(\xi, t))_\xi, \quad -1 < \xi < 0, \tag{7a}$$
$$w_t(\xi, t) = w_{\xi\xi}(\xi, t), \quad 0 < \xi < 1, \tag{7b}$$
$$v_\xi(-1, t) = 0, \quad w(1, t) = 0, \tag{7c}$$
$$v_t(0, t) = w(0, t), \quad T(0)v_\xi(0, t) = w_\xi(0, t) \tag{7d}$$

with initial conditions $v(\cdot, 0) \in H^1(-1, 0)$, $v_t(\cdot, 0) \in L^2(-1, 0)$, and $w(\cdot, 0) \in L^2(0, 1)$. Here $\rho(\cdot)$ is the mass density of the string and $T(\cdot)$ is the Young's modulus [26, Sect. 5]. The system is similar to those considered in [5, 24], but the physical parameters $\rho(\cdot)$ and $T(\cdot)$ of the wave part are allowed to be spatially varying. We assume $\rho(\cdot)$ and $T(\cdot)$ are continuously differentiable on $[-1, 0]$, and $0 < c_0 \leq \rho(\xi), T(\xi) \leq c_1$ for some constants $c_0, c_1 > 0$ and for all $\xi \in [-1, 0]$.

In this case the natural interpretation is that the unstable system is the wave equation on $(-1, 0)$, and the stable system is the heat equation on $(0, 1)$. The coupling boundary conditions (7d) at $\xi = 0$ can indeed be interpreted as a power-preserving interconnection (3) if we choose

$$u_c(t) = -w_\xi(0, t), \quad y_c(t) = w(0, t), \quad u(t) = v_t(0, t), \quad y(t) = T(0)v_\xi(0, t).$$

Only based on the coupling boundary conditions (7d) it would be possible to choose the converse roles for the inputs and the outputs. While this choice would also lead to a "wave-part" with the same properties, it turns out that the heat equation with the Dirichlet boundary input would not be a regular linear system on an L^2-space. Because of this, the above choice is more suitable.

Now that the inputs and outputs have been fixed, we continue by choosing a suitable state $x(t)$ and the space X in such a way that the wave equation can be represented as an impedance passive regular linear system. The abstract representations of wave equations are well-understood, and in particular we can achieve these properties by writing the wave part on its "energy space" $X = L^2(-1, 0) \times L^2(-1, 0)$ with the state $x(t) = (x_1(t), x_2(t))$ where

$$x_1(t) = \rho(\cdot)v_t(\cdot, t) \quad \text{(momentum distribution)},$$
$$x_2(t) = v_\xi(\cdot, t). \quad \text{(strain)}$$

In these variables the wave part becomes

$$\frac{d}{dt}\begin{bmatrix} x_1(\xi, t) \\ x_2(\xi, t) \end{bmatrix} = \begin{bmatrix} 0 & \partial_\xi \\ \partial_\xi & 0 \end{bmatrix}\begin{bmatrix} \rho(\xi)^{-1}x_1(\xi, t) \\ T(\xi)x_2(\xi, t) \end{bmatrix}$$
$$\rho(0)^{-1}x_1(0, t) = u(t), \quad T(-1)x_2(-1, t) = 0,$$
$$y(t) = T(0)x_2(0, t).$$

If we define the norm on the state space $X = L^2(-1, 0) \times L^2(-1, 0)$ by

$$\|x(t)\|_X^2 = \int_{-1}^{0} \left[\rho(\xi)^{-1}|x_1(\xi, t)|^2 + T(\xi)|x_2(\xi, t)|^2 \right] d\xi,$$

then the total energy of the wave part is given by [22, Ex. 1.6]

$$E_x(t) = \frac{1}{2}\|x(t)\|_X^2$$

for every classical solution $x(t)$.

The system operator A is chosen to be $A = \begin{bmatrix} 0 & \partial_\xi \\ \partial_\xi & 0 \end{bmatrix}\begin{bmatrix} \rho(\cdot)^{-1} & 0 \\ 0 & T(\cdot) \end{bmatrix}$ with

$$D(A) = \{ (x_1, x_2) \in H^1(-1, 0) \times H^1(-1, 0) \mid x_2(-1) = x_1(0) = 0 \}.$$

It has been shown in [26, Sect. 5] that the wave equation is a regular linear system. In addition, it is impedance passive since every classical state $x(t) = (x_1(t), x_2(t))$ satisfies

$$\frac{1}{2}\frac{d}{dt}\|x(t)\|_X^2 = \text{Re}\langle \dot{x}(t), x(t)\rangle_X$$
$$= \text{Re}\left[\int_{-1}^{0} (T(\xi)v_\xi(\xi, t))_\xi \overline{v_t(\xi, t)} + T(\xi)v_\xi(\xi, t)\overline{(v_t(\xi, t))_\xi} d\xi\right]$$
$$= \text{Re}\left[T(\xi)v_\xi(\xi, t)\overline{v_t(\xi, t)}\right]_{\xi=-1}^{0} = \text{Re}\left(T(0)v_\xi(0, t)\overline{v_t(0, t)}\right) = \text{Re}\, u(t)\overline{y(t)}.$$

The system operator A of the wave part is skew-adjoint and has compact resolvent by [22, Theorem 4.2(iv)]. Finally, the wave part is stabilised exponentially with negative output feedback $u(t) = -\kappa y(t)$ as shown in [22, Ex. 5.21] (see also [9]).

The stable part of (7) consisting of the heat equation is given by

$$w_t(\xi, t) = w_{\xi\xi}(\xi, t), \qquad 0 < \xi < 1,$$
$$w_\xi(0, t) = -u_c(t), \qquad w(1, t) = 0$$
$$y_c(t) = w(0, t).$$

This simple PDE system can be formulated as an abstract linear system on $X_c = L^2(0, 1)$ by choosing the state $x_c(t) = w(\cdot, t)$ and $A_c = \partial_{\xi\xi}$ with domain

$$D(A_c) = \{ x_c \in H^2(0, 1) \mid x'_c(0) = x_c(1) = 0 \}.$$

The input and output operators can be chosen such that $B_c u_c = -\delta_0(\cdot) u_c$ for $u_c \in \mathbb{C}$ and $C_c x_c = x_c(0) \in \mathbb{C}$ for all $x_c(\cdot) \in D(A_c)$. With these choices we have from [21, Proposition 6.5] that the heat system is regular. If we choose the standard L^2-norm on $X_c = L^2(0, 1)$, then the heat system is also impedance passive system, since every classical state $x_c(t)$ satisfies (using $w(1, t) = 0$)

$$\frac{1}{2}\frac{d}{dt}\|x_c(t)\|_{L^2}^2 = \operatorname{Re}\langle \dot{x}_c(t), x_c(t)\rangle_{L^2} = \operatorname{Re}\int_0^1 w_{\xi\xi}(\xi, t)\overline{w(\xi, t)}d\xi$$

$$= \operatorname{Re}\left[w_\xi(\xi, t)\overline{w(\xi, t)}\right]_{\xi=0}^1 - \operatorname{Re}\int_0^1 w_\xi(\xi, t)\overline{w_\xi(\xi, t)}d\xi$$

$$\leq \operatorname{Re}(-w_\xi(0, t))\overline{w(0, t)} = \operatorname{Re} u_c(t)\overline{y_c(t)}.$$

The following proposition generalises the main results of [5, 24] to the case where the wave part is allowed to have spatially varying parameters $\rho(\cdot)$ and $T(\cdot)$.

Proposition 2 *For all initial conditions*

$$v(\cdot, 0) \in H^2(-1, 0), \quad v_t(\cdot, 0) \in H^1(-1, 0), \quad \text{and} \quad w(\cdot, 0) \in H^2(0, 1)$$

which satisfy the boundary conditions of (7) at $t = 0$, the system (7) has a solution $(v(\cdot, \cdot), w(\cdot, \cdot))$ which satisfies the boundary conditions for all $t \geq 0$ and

$$v(\cdot, \cdot) \in C(0, \infty; H^2(-1, 0)) \cap C^2(0, \infty; L^2(-1, 0)),$$
$$w(\cdot, \cdot) \in C(0, \infty; H^2(0, 1)) \cap C^1(0, \infty; L^2(0, 1)).$$

The energy

$$E_{tot}(t) = \frac{1}{2}\int_{-1}^0 \rho(\xi)|v_t(\xi, t)|^2 + T(\xi)|v_\xi(\xi, t)|^2 d\xi + \frac{1}{2}\int_0^1 |w(\xi, t)|^2 d\xi$$

of every such classical solution of (7) satisfies

$$E_{tot}(t) = o(t^{-4}).$$

Proof We will not present the details in this paper, but it follows from the definition of the systems (A, B, C_Λ, D) and $(A_c, B_c, C_{c\Lambda}, D_c)$ that with the given assumptions the initial state $(x(0), x_c(0))$ belongs to the domain $D(A_e)$ of the generator A_e of the closed-loop semigroup on $X \times X_c$. Because of this the existence and the stated properties follow from the property that the classical solution of the closed-loop system satisfies $(x(t), x_c(t)) \in C(0, \infty; D(A_e)) \cap C^1(0, \infty; X \times X_c)$. Thus

$$v_\xi(\cdot, \cdot) \in C(0, \infty; H^1(-1, 0)) \cap C^1(0, \infty; L^2(-1, 0))$$
$$v_t(\cdot, \cdot) \in C(0, \infty; H^1(-1, 0)) \cap C^1(0, \infty; L^2(-1, 0))$$
$$w(\cdot, \cdot) \in C(0, \infty; H^2(0, 1)) \cap C^1(0, \infty; L^2(0, 1)),$$

which in particular implies the first claim.

We have $E_{tot}(t) = \frac{1}{2}\|x(t)\|^2 + \frac{1}{2}\|x_c(t)\|^2$, and therefore the decay rate can be deduced from Proposition 1 if we can show that the conditions are satisfied for $\alpha = 1/2$. The system operator A of the wave part is skew-adjoint and has compact resolvent by [22, Theorem 4.2]. It is further shown in [22, Ex. 5.21] that this kind of a system is stabilized with negative output feedback $u(t) = -y(t)$. Finally, we will show that $D_c = 0$ and derive lower bound of the form (6) for $\mathrm{Re}\, P_c(is)$ with $\alpha = 1/2$. For $\lambda \in \rho(A)$ and $u_c \in \mathbb{C}$ we have that $P_c(\lambda)u_c = y_c$ where $y_c \in \mathbb{C}$ is such that [10, Sect. 1]

$$\lambda w(\xi) = w_{\xi\xi}(\xi), \qquad \xi \in (0, 1)$$
$$-w_\xi(0) = u_c, \qquad w(1) = 0$$
$$y_c = w(0).$$

The solution $w(\xi)$ of this ODE is $w(\xi) = \frac{\sinh(\sqrt{\lambda}(1-\xi))}{\sqrt{\lambda}\cosh(\sqrt{\lambda})}u_c$, and therefore

$$P_c(\lambda)u_c = y_c = w(0) = \frac{\sinh(\sqrt{\lambda})}{\sqrt{\lambda}\cosh(\sqrt{\lambda})}u_c = \frac{\tanh(\sqrt{\lambda})}{\sqrt{\lambda}}u_c.$$

For regular linear systems $D_c = \lim_{\lambda \to \infty} P_c(\lambda)$, and since $\tanh(\sqrt{\lambda})$ is uniformly bounded for $\lambda > 0$, we have $D_c = 0$. A direct computation also shows that

$$\mathrm{Re}\, P_c(is) = \frac{1}{2\sqrt{2}\sqrt{|s|}}\frac{\sinh(\sqrt{2|s|}) + \sin(\sqrt{2|s|})}{\cosh(\sqrt{2|s|})}.$$

Since $\mathrm{Re}\, P_c(is)$ is bounded and nonzero for $-\pi/2 \le s \le \pi/2$, and $\mathrm{Re}\, P_c(is) \ge 0.4|s|^{-1/2}$ for all $|s| \ge \pi/2$, the estimate (6) holds for $\alpha = 1/2$ and for some $\eta_0 > 0$. Because of this, Proposition 1 implies that for all classical solutions of the closed-loop system we have

$$E_{tot}(t) = \frac{1}{2} \left\| \begin{bmatrix} x(t) \\ x_c(t) \end{bmatrix} \right\|^2_{X \times X_c} = o(t^{-4}).$$

□

4 Wave Equation with an Acoustic Boundary Condition

In this section we consider a one-dimensional wave equation with an "acoustic boundary condition",

$$\rho(\xi)v_{tt}(\xi,t) = (T(\xi)v_\xi(\xi,t))_\xi, \qquad 0 < \xi < 1, \tag{8a}$$
$$m\delta_{tt}(t) = -d\delta_t(t) - k\delta(t) - \beta v_t(1,t) \tag{8b}$$
$$v_\xi(1,t) = \delta_t(t), \qquad v_t(0,t) = 0. \tag{8c}$$

This PDE–ODE system is similar to those studied on multi-dimensional domains in [6, 18]. In particular, polynomial decay of energy was shown in the article [18] for these types of models under geometric constraints on the boundary conditions.

We again allow the physical parameters $\rho(\cdot)$ and $T(\cdot)$ to depend on the spatial variable. The functions $\rho(\cdot)$ and $T(\cdot)$ satisfy the same assumptions as in Sect. 3, and $m > 0$, $d > 0$, and $k > 0$ are the mass, the damping coefficient and the spring coefficient of the ODE (8b) at $\xi = 1$ [6].

We will prove polynomial decay of the energy of the system (8) by writing it as a power-preserving interconnection between two impedance passive systems—an infinite-dimensional one and a finite-dimensional one. We begin by investigating the dynamic boundary condition (8b). This is a second order ordinary differential equation with state $\delta(t)$, and the term $-\beta v_t(1,t)$ acts as an external input in this equation. On the other hand, the derivative $\delta_t(t)$ determines the boundary condition (8c) of the wave equation, and can therefore be considered as an output of this ODE. If we again consider $(x(t), u(t), y(t))$ to be the wave equation and $(x_c(t), u_c(t), y_c(t))$ to describe the ODE at $\xi = 1$, then the above analysis indicates that the inputs and outputs of the component systems could be chosen as

$$y(t) = -u_c(t) \qquad \leftrightarrow \qquad \beta v_t(1,t) = -u_c(t)$$
$$u(t) = y_c(t) \qquad \leftrightarrow \qquad v_\xi(1,t) = \delta_t(t)$$

However, we need to be careful in the choices of the coefficients of the inputs in order to achieve impedance passivity of the component systems. Making of the appropriate choices is demonstrated in the following.

We can again write the wave equation on its energy space $X = L^2(0,1) \times L^2(0,1)$ in the variables $x_1(\xi,t) = \rho(\xi)v_t(\xi,t)$ and $x_2(\xi,t) = v_\xi(\xi,t)$ and with the norm

$$\|x(t)\|_X^2 = \int_0^1 \left[\rho(\xi)^{-1}|x_1(\xi,t)|^2 + T(\xi)|x_2(\xi,t)|^2\right]d\xi.$$

The system operator A of the wave part is $A = \begin{bmatrix} 0 & \partial_\xi \\ \partial_\xi & 0 \end{bmatrix} \begin{bmatrix} \rho(\cdot)^{-1} & 0 \\ 0 & T(\cdot) \end{bmatrix}$, now with domain

$$D(A) = \{(x_1, x_2) \in H^1(0,1) \times H^1(0,1) \mid x_1(0) = x_2(1) = 0\}.$$

A direct computation analogous to the one in Sect. 3 shows that

$$\frac{1}{2}\frac{d}{dt}\|x(t)\|_X^2 = \operatorname{Re}\langle \dot{x}(t), x(t)\rangle_X = \operatorname{Re}\left(T(1)v_\xi(1,t)\overline{v_t(1,t)}\right)$$

and thus in order to achieve impedance passivity for the wave part, we should choose the input and output of the wave part as $u(t) = T(1)v_\xi(1,t)$ and $y(t) = v_t(1,t)$. These choices and the requirement for the power-preserving interconnection also fix the coefficients of the inputs and outputs of the finite-dimensional system. In particular, we have $-\beta v_t(1,t) = -\beta y(t) =: \beta u_c(t)$, and necessarily $y_c(t) = u(t) = T(1)v_\xi(1,t) = T(1)\dot{\delta}_t(t)$. The full finite-dimensional system thus becomes

$$\frac{d}{dt}\begin{bmatrix}\delta(t)\\ \dot{\delta}(t)\end{bmatrix} = \begin{bmatrix} 0 & 1 \\ -k/m & -d/m \end{bmatrix}\begin{bmatrix}\delta(t)\\ \dot{\delta}(t)\end{bmatrix} + \begin{bmatrix} 0 \\ \beta \end{bmatrix} u_c(t)$$

$$y_c(t) = [0, T(1)]\begin{bmatrix}\delta(t)\\ \dot{\delta}(t)\end{bmatrix}.$$

We can define the system on $X_c = \mathbb{C}^2$ with matrices (A_c, B_c, C_c, D_c) chosen as

$$A_c = \begin{bmatrix} 0 & 1 \\ -k/m & -d/m \end{bmatrix}, \quad B_c = \begin{bmatrix} 0 \\ \beta \end{bmatrix}, \quad C_c = [0, T(1)], \quad D_c = 0.$$

For this system the impedance passivity can be achieved with a suitable choice of the norm of $X_c = \mathbb{C}^2$. Indeed, if we take a norm $\|(z_1, z_2)^T\|_{X_c}^2 = c_1|z_1|^2 + c_2|z_2|^2$ for some $c_1, c_2 > 0$, we can compute

$$\frac{1}{2}\frac{d}{dt}\|x_c(t)\|_{X_c}^2 = \operatorname{Re}\langle \dot{x}_c(t), x_c(t)\rangle_{X_c}$$

$$= c_1 \operatorname{Re}\dot{\delta}(t)\overline{\delta(t)} + c_2 \operatorname{Re}\left(-k\delta(t)/m - d\dot{\delta}(t)/m + \beta u_c(t)\right)\overline{\dot{\delta}(t)}$$

$$\leq (c_1 - c_2 k/m)\operatorname{Re}\dot{\delta}(t)\overline{\delta(t)} + \beta T(1)^{-1}c_2 \operatorname{Re} u_c(t)\overline{T(1)\dot{\delta}(t)}$$

which is equal to $\operatorname{Re} u_c(t)\overline{y_c(t)}$ if we choose $c_2 = T(1)/\beta$ and $c_1 = c_2 k/m$.

The following proposition establishes the polynomial decay of energy for the system (8).

Proposition 3 *For all initial conditions*

$$v(\cdot, 0) \in H^2(0, 1), \quad v_t(\cdot, 0) \in H^1(0, 1), \quad \delta(0), \delta_t(0) \in \mathbb{R}$$

which satisfy the boundary conditions of (8) at $t = 0$, the system (8) has a solution $(v(\cdot, \cdot), \delta(\cdot))$ which satisfies the boundary conditions for all $t \geq 0$ and

$$v(\cdot, \cdot) \in C(0, \infty; H^2(0, 1)) \cap C^2(0, \infty; L^2(0, 1)), \quad \delta(\cdot) \in C^2(0, \infty; \mathbb{C})$$

The energy

$$E_{tot}(t) = \frac{1}{2} \int_0^1 \rho(\xi)|v_t(\xi, t)|^2 + T(\xi)|v_\xi(\xi, t)|^2 d\xi$$
$$+ \frac{T(1)}{2\beta m} \left(k\delta(t)^2 + m\delta_t(1, t)^2\right)$$

of every such classical solution of (8) satisfies

$$E_{tot}(t) = o(t^{-1}).$$

Proof The definitions of the systems (A, B, C_Λ, D) and $(A_c, B_c, C_{c\Lambda}, D_c)$ again imply that with the given assumptions the initial state $(x(0), x_c(0))$ belongs to the domain $D(A_e)$ of the generator A_e of the closed-loop semigroup on $X \times X_c$. Thus the closed-loop system has a classical solution such that $(x(t), x_c(t)) \in C(0, \infty; D(A_e)) \cap C^1(0, \infty; X \times X_c)$. This also implies that the boundary conditions are satisfied for all $t \geq 0$, and

$$v_\xi(\cdot, \cdot) \in C(0, \infty; H^1(-1, 0)) \cap C^1(0, \infty; L^2(-1, 0))$$
$$v_t(\cdot, \cdot) \in C(0, \infty; H^1(-1, 0)) \cap C^1(0, \infty; L^2(-1, 0)),$$
$$\delta_t(\cdot, \cdot) \in C^1(0, \infty; \mathbb{C}),$$

which implies the first claim.

The system operator A of the wave part is again skew-adjoint with compact resolvent by [22, Theorem 4.2] and analogously as in [22, Ex. 5.21] we can show that it is stabilized exponentially with negative output feedback $u(t) = -y(t)$. Moreover, it is easy to show that $0 \in \rho(A)$.

We have $D_c = 0$ for the finite-dimensional system. The transfer function $P_c(is) = C_c R(is, A_c) B_c$ can be computed explicitly as

$$P_c(is) = T(1)\beta m \cdot \frac{ds^2 + is(ks - ms^3)}{(ms^2 - k)^2 + d^2 s^2},$$

and in particular we have

$$\operatorname{Re} P_c(is) = T(1)\beta m \cdot \frac{ds^2}{m^2 s^4 + (d^2 - 2km)s^2 + k^2}.$$

The transfer function is equal to zero at $s = 0$, but for $\alpha = 2$ and for any $s_0 > 0$ there exists a constant $\eta_0 > 0$ such that the estimate (6) holds for all $|s| \geq s_0$. Since $E_{tot}(t) = \frac{1}{2}\|x(t)\|_X^2 + \frac{1}{2}\|x_c(t)\|_{X_c}^2$, the claim now follows from Proposition 1. $\square$

References

1. Abbas, Z., Ammari, K., Mercier, D.: Remarks on stabilization of second-order evolution equations by unbounded dynamic feedbacks. J. Evol. Equ. **16**(1), 95–130 (2016)
2. Ammari, K., Mercier, D., Régnier, V., Valein, J.: Spectral analysis and stabilization of a chain of serially connected Euler-Bernoulli beams and strings. Commun. Pure Appl. Anal. **11**(2), 785–807 (2012)
3. Avalos, G., Lasiecka, I., Triggiani, R.: Heat-wave interaction in 2–3 dimensions: optimal rational decay rate. J. Math. Anal. Appl. **437**(2), 782–815 (2016)
4. Batty, C.J.K., Duyckaerts, T.: Non-uniform stability for bounded semi-groups on Banach spaces. J. Evol. Equ. **8**, 765–780 (2008)
5. Batty, C.J.K., Paunonen, L., Seifert, D.: Optimal energy decay in a one-dimensional coupled wave-heat system. J. Evol. Equ. **16**(3), 649–664 (2016)
6. Beale, J.T.: Spectral properties of an acoustic boundary condition. Indiana Univ. Math. J. **25**(9), 895–917 (1976)
7. Borichev, A., Tomilov, Y.: Optimal polynomial decay of functions and operator semigroups. Math. Ann. **347**(2), 455–478 (2010)
8. Byrnes, C.I., Gilliam, D.S., Shubov, V.I., Weiss, G.: Regular linear systems governed by a boundary controlled heat equation. J. Dyn. Control Syst. **8**(3), 341–370 (2002)
9. Cox, S., Zuazua, E.: The rate at which energy decays in a string damped at one end. Indiana Univ. Math. J. **44**(2), 545–573 (1995)
10. Curtain, R., Morris, K.: Transfer functions of distributed parameter systems: a tutorial. Automatica J. IFAC **45**(5), 1101–1116 (2009)
11. Duyckaerts, T.: Optimal decay rates of the energy of a hyperbolic-parabolic system coupled by an interface. Asymptot. Anal. **51**(1), 17–45 (2007)
12. Guo, B.Z., Shao, Z.C.: Regularity of an Euler-Bernoulli equation with Neumann control and collocated observation. J. Dyn. Control Syst. **12**(3), 405–418 (2006)
13. Guo, B.Z., Zhang, Z.X.: On the well-posedness and regularity of the wave equation with variable coefficients. ESAIM Control Optim. Calc. Var. **13**(4), 776–792 (2007)
14. Hassi, E.M.A.B., Ammari, K., Boulite, S., Maniar, L.: Stability of abstract thermo-elastic semigroups. J. Math. Anal. Appl. **435**(2), 1021–1035 (2016)
15. Mercier, D., Nicaise, S., Sammoury, M.A., Wehbe, A.: Indirect stability of the wave equation with a dynamic boundary control. Math. Nachr. **291**(7), 1114–1146 (2018)
16. Miller, L.: Resolvent conditions for the control of unitary groups and their approximations. J. Spectr. Theory **2**(1), 1–55 (2012)
17. Paunonen, L.: Stability and robust regulation of passive linear systems. SIAM J. Control Optim. **57**(6), 3827–3856 (2019)
18. Muñoz Rivera, J.E., Qin, Y.: Polynomial decay for the energy with an acoustic boundary condition. Appl. Math. Lett. **16**(2), 249–256 (2003)
19. Rozendaal, J., Seifert, D., Stahn, R.: Optimal rates of decay for operator semigroups on Hilbert spaces. Adv. Math. **346**, 359–388 (2019)
20. Tucsnak, M., Weiss, G.: Observation and Control for Operator Semigroups. Birkhäuser Basel (2009)

21. Tucsnak, M., Weiss, G.: Well-posed systems–The LTI case and beyond. Automatica J. IFAC **50**(7), 1757–1779 (2014)
22. Villegas, J.: A port-Hamiltonian approach to distributed parameter systems. Ph.D. thesis, Universiteit Twente, Twente, The Netherlands (2007)
23. Weiss, G.: Regular linear systems with feedback. Math. Control Signals Syst. **7**(1), 23–57 (1994)
24. Zhang, X., Zuazua, E.: Polynomial decay and control of a 1-d hyperbolic-parabolic coupled system. J. Differ. Equ. **204**(2), 380–438 (2004)
25. Zhang, X., Zuazua, E.: Long-time behavior of a coupled heat-wave system arising in fluid-structure interaction. Arch. Ration. Mech. Anal. **184**(1), 49–120 (2007)
26. Zwart, H., Le Gorrec, Y., Maschke, B., Villegas, J.: Well-posedness and regularity of hyperbolic boundary control systems on a one-dimensional spatial domain. ESAIM Control Optim. Calc. Var. **16**(4), 1077–1093 (2010)

Degenerate Nonlinear Semigroups of Operators and Their Applications

Ksenia V. Vasiuchkova, Natalia A. Manakova, and Georgy A. Sviridyuk

Abstract In this paper, we construct the conditions for the existence of a degenerate nonlinear resolving semigroup of shift operators for a semilinear Sobolev type equation. Based on the phase space method, we find the conditions for the existence of solutions to the Cauchy problem for a semilinear Sobolev type equation. The solutions are defined only on a simple Banach manifold (quasistationary trajectories). Also, we find the conditions under which the solution can be continued in time. The obtained abstract results are illustrated by the Cauchy–Dirichlet problem for the generalized filtration Boussinesq equation.

Keywords Semilinear Sobolev type equations · Cauchy problem · Resolving nonlinear semigroup of operator · Additive "noises" · Generalized Boussinesq filtration equations

1 Introduction

From the middle of the 20th century, the theory of (semi)groups of operators is one of the main research tools to study the Cauchy problem

$$u(0) = u_0 \qquad (1)$$

for abstract operator-differential equations

K. V. Vasiuchkova (✉) · N. A. Manakova · G. A. Sviridyuk
Department of Mathematical Physics Equations, South Ural State University,
76, Lenin ave, Chelyabinsk 454080, Russian Federation
e-mail: vasiuchkovakv@susu.ru

N. A. Manakova
e-mail: manakovana@susu.ru

G. A. Sviridyuk
e-mail: sviridyuk@susu.ru

$$\dot{u} = A(u). \tag{2}$$

One of the methods to find analytical solutions to problem (1), (2) is to construct resolving (semi)groups of operators [1]. In the case of linear equation (2), the theory of (semi)groups of operators was proposed and developed by E. Hille, R. S. Phillips, K. Yosida, W. Feller, and others, and was widely used to study various initial boundary-value problems for equations of mathematical physics [2–8].

The study of nonlinear semigroups of operators was initiated by the study of nonlinear parabolic equations. Among the first works in this field we note the papers [9–12] that made important contribution to the development of the theory. The papers [9, 10, 13] obtain the conditions for the existence of a solution to problem (1), (2) in the case of nonlinear dissipative operator A, acting in Hilbert space $\mathbf{H}$, when the image of operator $I - A$ coincides with whole initial space $\mathbf{H}$. Also, the papers [9, 10, 13] construct the semigroups of nonlinear shift operators. The main research method is the construction of approximate solutions to the initial problem as solutions to the Cauchy problem for the operator equation, where the operator in the right-hand side is dissipative and satisfies the Lipschitz condition. The work [9] was the first to obtain the necessary and sufficient conditions for the existence of a generator of resolving semigroup of nonlinear operators (an analogous of Hille–Yosida–Phillips theorem in the linear case) in the case of multi-valued nonlinear dissipative operators. Various applications to nonlinear evolution equations and quasilinear first order equations were investigated in the works [14–16].

In natural science and technology, an extensive class of processes and phenomena can be modelled by partial differential equations or systems of partial differential equations. Here the equations are not resolved with respect to the highest time derivative, and have the form

$$L\dot{u} = M(u), \quad \ker L \neq \{0\}. \tag{3}$$

Such equations are usually called Sobolev type equations [17–20]. For the first time, this term was introduced by Showalter [21]. Problem (1), (3) is difficult to study. The main reason is that the problem is unsolvable in principle for an arbitrary initial value u_0, even if u_0 belongs to a dense lineal of the initial Banach space $\mathscr{U}$. Therefore, it is relevant to search and describe the structure of set of initial value condition (1), for which problem (1), (3) has a solution. One of the successful methods to study linear ($M \in Cl(\mathscr{U}; \mathscr{F})$) problem (1), (3) is the theory of degenerate resolving (semi)groups of operators, which was proposed in [22], and was developed in [23–25]. The papers [22, 23] construct analytical (semi)groups of resolving operators with kernels, and obtain the conditions under which the image of resolving (semi)group coincides with the phase space of the corresponding equation defined on the subspace. Other successful approaches to study degenerate equations are presented in [21, 26].

The paper [27] was the first to propose the phase space method, which is one of the effective methods to study the Cauchy problem for the semilinear Sobolev type equations ($M(u) = M_1 u + M_2(u)$). A phase space is a closure of the set of all admissible initial values u_0 that are the vectors for which there exists the unique (local) solution to the Cauchy problem. Then, the vector u_0 is chosen such that the following two conditions hold in the neighborhood $\mathscr{O}_{u_0}$ of u_0. First, the phase space **M** is a smooth Banach manifold (phase manifold). Second, singular (i.e. with condition $\ker L \neq \{0\}$) Eq. (3) is reduced to the regular equation

$$\dot{u} = F(u), \qquad (4)$$

where F is a smooth section of the tangent bundle $T\mathscr{O}_{u_0}$. This method allows to find solutions to a wide class of problems [28–30]. Also, we note another important concept, which considers quasistationary (semi)trajectories [31] that pass through the point u_0, and pointwise belong to the phase space **M**. Moreover, any stationary trajectory of Eq. (3) is quasistationary, but the converse is wrong. Note that the phase space of equation (3) can lie on a smooth manifold having singularities such as Whitney k-folds [32].

The review [33] gives the conditions under which the phase space of equation (3) is a simple Banach C^∞-manifold, and there exists the unique quasistationary (semi)trajectory of Eq. (3) that passes through the point u_0. Recall that a Banach C^∞-manifold is called simple if its atlas is equivalent to the atlas containing the unigue map. In particular, the sufficient conditions for the simplicity of the phase space of equation (3) are obtained in the case of the s-monotone, p-coercive operator $(-M_2)$ and the Fredholm operator L. Note that Eq. (3) with Showalter–Sidorov initial value condition can have several solutions, if the phase space is not simple [34].

The purpose of the article is to find the conditions for the existence of a non-local solution to Cauchy problem (1), (3). In Paragraph 1, we use the phase space method to construct the quasistationary semitrajectories, i.e. the solutions to Eq. (3) defined on the phase manifold **M**. Also, we show that the solution can be continued in time on $(0, +\infty)$ in the case of the dissipative operator M. In Paragraph 2, we construct a non-linear resolving semigroup of shift operators of Eq. (3) defined on **M**. The obtained abstract results are illustrated by the Cauchy–Dirichlet problem for the generalized filtration Boussinesq equation [35]. The paper [36] was the first to obtain the conditions for the existence of a local solution to the Showalter–Sidorov–Dirichlet problem. Also, the paper [36] shows that the phase space of the generalized filtration Boussinesq equation is a smooth Banach manifold. We obtain a priori estimates in order to prove that the solution can be continued in time on $(0, +\infty)$, and to construct the resolving semigroup of operators of Boussinesq equation.

We use the methods developed in Paragraph 1 and Paragraph 2 in order to find the solutions to the stochastic semilinear Sobolev type equation

$$L \overset{o}{\eta} = M(\eta), \qquad (5)$$

endowed with the weakened (in the sense of S.G. Krein) Cauchy condition

$$\lim_{t \to 0+} (\eta(t) - \eta_0) = 0. \qquad (6)$$

Here $\eta = \eta(t)$ is the desired stochastic process, η_0 is a given random variable, and the symbol $\overset{o}{\eta}$ denotes the Nelson–Gliklikh derivative of the stochastic process $\eta = \eta(t)$ [37]. In order to study the Cauchy problem, we construct the spaces of K-"noises", i.e. the spaces of stochastic K-processes that are almost surely differentiable in the sense of Nelson–Gliklikh. This approach is based on the paper [38]. Note that this approach allows to transfer on the stochastic case the methods of functional analysis that are applied in the deterministic case [20, 24, 25]. Further, we construct the quasistationary semitrajectories of Eq. (5) that pass through the point $\eta_0(\omega)$ for each fixed $\omega \in \Omega$, where $\Omega \equiv (\Omega, \mathscr{A}, \mathbf{P})$ is a complete probability space. We use the quasistationary semitrajectories in order to construct the stochastic nonlinear resolving semigroup of operators defined on the phase manifold $\mathbf{M}$.

2 Phase Space Quasistationary Semitrajectories

Let $\mathbf{H} = (\mathbf{H}; \langle \cdot, \cdot \rangle)$ be a real separable Hilbert space identified with its conjugate space; $(\mathscr{H}, \mathscr{H}^*)$ and $(\mathscr{B}, \mathscr{B}^*)$ be dual (with respect to the duality $\langle \cdot, \cdot \rangle$) pairs of reflexive Banach spaces such that the embeddings

$$\mathscr{B} \hookrightarrow \mathscr{H} \hookrightarrow \mathbf{H} \hookrightarrow \mathscr{H}^* \hookrightarrow \mathscr{B}^* \tag{7}$$

are dense and continuous. Let $L \in \mathscr{L}(\mathscr{H}; \mathscr{H}^*)$ be a linear, continuous, self-adjoint nonnegatively defined Fredholm operator, and the operator $M \in C^k(\mathscr{B}; \mathscr{B}^*), k \geq 1$.

In view of the self-adjointness and Fredholm property of the operator L we identify $\mathscr{H} \supset \ker L \equiv \operatorname{coker} L \subset \mathscr{H}^*$, this implies $\mathscr{H}^* = \operatorname{coker} L \oplus \operatorname{im} L$. Denote by $\overline{\operatorname{im} L}$ closure im L in topology $\mathscr{B}^*$, then $\mathscr{B}^* = \operatorname{coker} L \oplus \overline{\operatorname{im} L}$. Denote by Q projector $\mathscr{B}^*$ along coker L on $\overline{\operatorname{im} L}$. Then, if $u = u(t), t \in (0, T]$ is a solution of Eq. (3), then the solution belongs to the set

$$\mathbf{M} = \begin{cases} \{u \in \mathscr{B} : (I - Q)M(u) = 0\}, & \text{if } \ker L \neq \{0\}; \\ \mathscr{B}, & \text{if } \ker L = \{0\}. \end{cases} \tag{8}$$

Consider the set

$$\operatorname{coim} L = \{u \in \mathscr{H} : \langle u, v \rangle = 0 \quad \forall v \in \ker L \setminus \{0\}\}.$$

In view of embeddings (7), the space $\mathscr{B}$ can be represented as a direct sum $\mathscr{B} = \mathscr{B}^0 \oplus \mathscr{B}^1$, where $\ker L \equiv \mathscr{B}^0$, coim $L \cap \mathscr{B} \equiv \mathscr{B}^1$. Denote by P the projector on $\mathscr{B}^1$ along $\mathscr{B}^0$.

Definition 1 A vector-function $u \in C^k((0, T]; \mathscr{B})$, $k \in \mathbf{N} \cup \{\infty\}$, is called *a solution to Eq.* (3), satisfies the equation for some $T \in \mathbf{R}_+$. A solution $u = u(t)$ to Eq. (3) is called a *solution to Cauchy problem* (1), (3), if $u = u(t)$ also satisfies initial value condition (1).

Remark 1 A little unlike the standard [31], a vector function $u = u(t)$ is called a *quasistationary semitrajectory of Eq.* (3), if $u = u(t)$ is a solution to this equation and belongs to the set **M**, i.e. $u(t) \in \mathbf{M}$ for all $t \in (0, T]$. If a quasistationary semitrajectory of Eq. (3) satisfies condition (1), then the semitrajectory is called a quasistationary semitrajectory of Eq. (3) that pass through the point u_0.

Definition 2 [28] Suppose that the point $u_0 \in \mathbf{M}$ and $u_0^1 = Pu_0 \in \mathscr{B}^1$. A set **M** is *a Banach C^k-manifold* at the point u_0, if there exist the neighborhoods $\mathscr{O}_0^{\mathbf{M}} \subset \mathbf{M}$ and $\mathscr{O}_0^1 \subset \mathscr{B}^1$ of the the points u_0 and u_0^1 respectively, and C^k-diffeomorphism $\delta : \mathscr{O}_0^1 \to \mathscr{O}_0^{\mathbf{M}}$ such that δ^{-1} is equal to the restriction of projector P on $\mathscr{O}_0^{\mathbf{M}}$. A set **M** is called *Banach C^k-manifold modelled by the subspace $\mathscr{B}^1$*, if **M** is a Banach C^k-manifold at each of its points. A connected Banach C^k-manifold is called *simple*, if any atlas of the manifold is equivalent to the atlas containing the unique map.

Theorem 1 *Suppose that the set **M** is a simple Banach C^k-manifold at the point u_0. Then there exists the unique quasistationary semitrajectory of Eq. (3) that pass through the point u_0.*

Proof In view of the representation of the space $\mathscr{B} = \mathscr{B}^0 \oplus \mathscr{B}^1$ and the existence of the projector P of the space $\mathscr{B}$ on $\mathscr{B}^1$ along $\mathscr{B}^0$, any vector $u = u^0 + u^1$, where $(I - P)u = u^0 \in \mathscr{B}^0$, and $Pu = u^1 \in \mathscr{B}^1$. Denote by L_1 a restriction of the operator L on the subspace $\mathscr{B}^1$. Then, in view of the linearity of the operator L, we obtain

$$Lu = L(u^0 + u^1) = L_1 u^1,$$

and Eq. (3) is equivalent to the system of equations

$$\begin{aligned} \dot{u}^1 &= L_1^{-1} QM(\delta(u^1)), \quad u = \delta(u^1), \\ 0 &= (I - Q)M(u), \end{aligned} \quad (9)$$

where L_1^{-1} is inverse operator to the operator L.

Let $\mathscr{O}_0^{\mathbf{M}}$ and $\mathscr{O}_0^1$ be neighborhoods of the points $u_0 \in \mathbf{M}$ and $u_0^1 = Pu_0$, respectively. Since the set **M** is a Banach C^k-manifold at the point u_0, then there exists a diffeomorphism $\delta \in C^k(\mathscr{O}_0^1; \mathscr{O}_0^{\mathbf{M}})$. Denote by δ'_{u^1} the Frechet derivative of the operator δ at the point $u^1 \in \mathscr{O}_0^1$, then the operator $\delta'_{u^1} \in \mathscr{L}(\mathscr{B}^1; T_u \mathbf{M})$ for each fixed point $u^1 \in \mathscr{O}_0^1$. Here $T_u \mathbf{M}$ is the tangent space at the point $u = \delta(u^1)$. Let us act on the left by the operator δ'_{u^1} on both sides of the first equation of system (9):

$$\delta'_{u^1} \dot{u}^1 = \delta'_{u^1} L_1^{-1} QM(\delta(u^1)), \quad (10)$$

as

$$\delta'_{u^1} \dot{u}^1 = \frac{d}{dt} \delta(u^1) = \dot{u},$$

then

$$\dot{u} = F(u),$$

where operator $F = \delta'_{u^1} L_1^{-1} QM : u \to T_u\mathbf{M}$. Then the operator $F \in C^k(\mathcal{O}_0^{\mathbf{M}}; T_0\mathbf{M})$, where $T_0\mathbf{M}$ is a restriction of the tangent bundle $T\mathbf{M}$ on $\mathcal{O}_0^{\mathbf{M}}$. In view of the classical theorem of the existence and uniqueness of solution to the Cauchy problem [39], we obtain the simple local solvability of the problem

$$\dot{u} = F(u), \quad u(0) = u_0. \tag{11}$$

Let us show that the solution to problem (11) is a quasistationary semitrajectory of Eq. (3). Let $u = u(t)$ be the solution to problem (11). Since $u = \delta(u^1)$, then $u(t) \in \mathbf{M}$ for each $t \in (0, T)$. In addition, the vector function $u^1(t) = Pu(t)$ satisfies Eq. (3), since $\delta'_{u^1} \in \mathcal{L}(\mathcal{B}^1; T_u\mathbf{M})$ is a toplinear isomorphism. So, if $u = u(t)$ is a solution of problem (11), then $u = u(t)$ is the quasistationary semitrajectory of Eq. (3) that pass through the point u_0. Thus, the Theorem is proved.

Definition 3 The operator $M : \mathcal{B} \to \mathcal{B}^*$ is called dissipative, if

$$\langle M(u) - M(v), u - v \rangle \leq 0 \quad \forall u, v \in \mathcal{B}.$$

Note that the dissipativity of the operator M is equivalent to the monotonicity of the operator $(-M)$. According to the paper [33], the s-monotonicity (i.e. $\langle -M'_u v, v \rangle \geq 0 \ \forall u, v \in \mathcal{B}$) of the operator $(-M)$ implies the strict monotonicity of the operator, and, consequently, the dissipativity of the operator.

Consider the stationary equation

$$M(v) = 0 \tag{12}$$

and suppose that there exists a solution to Eq. (12). Further, let the operator M be dissipative. Introduce the norm $|u|^2 = \langle Lu, u \rangle$ in coim L. In view of Courant principle, this norm is equivalent to the norm induced from the overspace $\mathbf{H}$. Then, it follows from Eq. (3) that

$$\frac{1}{2}\frac{d}{dt}|u(t)|^2 = \frac{1}{2}\frac{d}{dt}|u(t) - v|^2 = \langle M(u) - M(v), u - v \rangle \leq 0,$$

where v is a solution to Eq. (12), which in turn is a stationary solution to Eq. (3). Hence, in view of the uniqueness of solution to Cauchy problem (11), we obtain that the solution can be continued on the interval $(0, +\infty)$. Then the following theorem is correct.

Theorem 2 *Suppose that the set u_0 set $\mathbf{M}$ is a simple Banach C^k-manifold at the point u_0, there exists a solution to Eq. (12), and the operator M is dissipative. Then there exists the unique quasistationary semitrajectory $u \in C^k((0, +\infty); \mathcal{B})$ of Eq. (3) that pass through the point u_0.*

Definition 4 A set $\mathbf{M} \subset \mathcal{B}$ is called a *phase space* of Eq. (3), if
(i) any solution $u = u(t)$ to Eq. (3) belongs to the set $\mathbf{M}$ pointwise, i.e. $u(t) \in \mathbf{M}$, $t \in (0, +\infty)$;

(ii)) for any $u_0 \in \mathbf{M}$ there exists the unique solution to problem (1), (3).

Remark 2 If the set $\mathbf{M}$ is a simple Banach C^k-manifold, then $\mathbf{M}$ coincides with the phase space of equation (3).

3 Semigroups of Operators

Let $(\mathscr{V}, <\cdot,\cdot>)$ be a Hilbert space. Consider a family of (in the general case, nonlinear) operators $V^t(\cdot): \text{dom } V^t \subset \mathscr{V} \to \mathscr{V} \ \forall t \geq 0$ with a common domain $D(V) = \text{dom } V^t \subset \mathscr{V} \ \forall t \geq 0$. Denote by $C(\mathscr{V})$ the set of continuous operators.

Definition 5 A mapping $V^\bullet \in C^k([0, +\infty); C(\mathscr{V}))$ is called a semigroup of operators, if
$$V^{t+s}(u) = V^t(V^s(u)) \ \forall s, t \geq 0 \ \forall u \in \mathscr{V}.$$

We identify the semigroup $V^\bullet$ with its graph $\{V^t : t \in \mathbf{R}_+\}$.

Definition 6 A semigroup $\{V^t : t \in \mathbf{R}_+\}$ is called contractive, if
$$\|V^t(u) - V^t(v)\|_\mathscr{V} \leq \|u - v\|_\mathscr{V} \ \forall t \geq 0 \ \forall u, v \in \mathscr{V}.$$

Definition 7 An operator $A : \text{dom } A \subset \mathscr{V} \to \mathscr{V}$ is called an *infinitesimal generator* of the semigroup $\{V^t : t \in \mathbf{R}_+\}$, if the condition
$$A(u) = \lim_{t \to 0+} \frac{V^t(u) - u}{t}, \tag{13}$$
holds, where dom A consists of all $u \in \mathscr{V}$ such that (13) is satisfied.

Let us construct a semigroup of resolving operators of Eq. (3). Define a shift operator $U^t(u_0) \equiv u(t)$, where $u(t)$ is a solution to problem (11). Then $\{U^t : t \in \mathbf{R}_+\}$ forms a nonlinear semigroup of shift operators with common domain $D(U) = \mathbf{M} \subset \mathscr{B}$. The infinitesimal generator of the semigroup of operators $\{U^t : t \in \mathbf{R}_+\}$ is operator $F = \delta'_{u^1} L_1^{-1} QM : u \to T_u \mathbf{M}$, defined in Theorem 1. Since the operator M is dissipative and the operator L_1 is positive defined, then the operator F is dissipative, and, therefore, the constructed semigroup is contractive. Indeed,

$$< \frac{U^t(u) - u}{t} - \frac{U^t(v) - v}{t}, u - v >_\mathbf{H} = \frac{1}{t} < U^t(u) - U^t(v), u - v >_\mathbf{H} - \|u - v\|_\mathbf{H}^2 \leq$$
$$\leq \frac{1}{t} \|u - v\|_\mathbf{H} \left(\|U^t(u) - U^t(v)\|_\mathbf{H} - \|u - v\|_\mathbf{H} \right) \leq 0,$$

then
$$\|U^t(u) - U^t(v)\|_\mathbf{H} \leq \|u - v\|_\mathbf{H}.$$

This completes the proof following Theorem.

Theorem 3 *Suppose that the set* **M** *is a simple Banach C^k-manifold at point u_0, and there exists a solution to Eq. (12), and the operator M is dissipative. Then there exists a resolving semigroup of operators $\{U^t : t \in \mathbf{R}_+\}$ of Eq. (3), defined on the manifold* **M**.

Remark 3 Similarly to the linear case [23], a semigroup of operators $\{U^t : t \in \mathbf{R}_+\}$ of Eq. (3) is called *degenerate*, since the semigroup is given not on the whole space, but on some set **M** having the structure of a simple Banach manifold.

4 General Mathematical Filtration Boussinesq Model

Let $\Omega \subset \mathbf{R}^n$ be a bounded domain with a smooth boundary of class C^∞. In the cylinder $\Omega \times \mathbf{R}_+$ consider Dirichlet problem

$$u(s,t) = 0, \ (s,t) \in \partial\Omega \times \mathbf{R}_+ \tag{14}$$

for the generalized filtration Boussinesq equation [35, 36]

$$(\lambda - \Delta)u_t = \Delta(|u|^{p-2}u), \ p \geq 2. \tag{15}$$

Equation (15) is the most interesting particular case of the equation obtained by Dzektser [35]. Here the desired function $u = u(s,t)$ corresponds to the potential of speed of movement of the free surface of the filtered liquid; the parameter $\lambda \in \mathbf{R}$ characterizes the medium, and this parameter λ can take negative values.

Let $\mathbf{H} = W_2^{-1}(\Omega)$, $\mathscr{H} = L_2(\Omega)$, $\mathscr{B} = L_p(\Omega)$ (all functional spaces are defined on domain Ω). Note that there exists the dense and continuous embedding $\overset{\circ}{W}{}_2^1(\Omega) \hookrightarrow L_q(\Omega)$ for $p \geq \frac{2n}{n+2}$, therefore $L_p(\Omega) \hookrightarrow W_2^{-1}(\Omega)$, where $\frac{1}{p} + \frac{1}{q} = 1$. In **H**, define the scalar product by the formula

$$\langle x, y \rangle = \int_\Omega x\tilde{y}\,ds \ \forall x, y \in \mathbf{H}, \tag{16}$$

where $\tilde{y}$ is the generalized solution to the homogeneous Dirichlet problem for Laplace operator $(-\Delta)$ in the domain Ω. Let $\mathscr{B}^* = (L_p(\Omega))^*$ and $\mathscr{H}^* = (L_2(\Omega))^*$, where $(L_p(\Omega))^*$ is conjugate space with respect to duality (16). For thus defined $\mathscr{H}^*$ and $\mathscr{B}^*$ there exist dense and continuous embeddings (7). Define the operators L and M as follows:

$$\langle Lu, v \rangle = \int_\Omega (\lambda u\tilde{v} + uv)\,ds, \ u, v \in \mathscr{H};$$

$$\langle M(u), v \rangle = -\int_\Omega |u|^{p-2}uv\,ds, \ u, v \in \mathscr{B}.$$

Let $\{\varphi_k\}$ be the sequence of eigenfunctions of the homogeneous Dirichlet problem for Laplace operator $(-\Delta)$ in the domain Ω, and $\{\lambda_k\}$ be the corresponding sequence of eigenvalues numbered in non-increasing order taking into account the multiplicity.

Lemma 1 (i) *For all $\lambda \geq -\lambda_1$ the operator $L \in \mathscr{L}(\mathscr{H}; \mathscr{H}^*)$ is self-adjoint, Fredholm, and non-negatively defined, and the orthonormal family $\{\varphi_k\}$ of its functions is total in the space $\mathscr{H}$.*

(ii) *Operator $M \in C^1(\mathscr{B}; \mathscr{B}^*)$ is dissipative and p-coercive.*

Proof Statement (i) is a classical result. As for statement (ii), the Frechet derivative of the operator M at the point $u \in \mathscr{B}$ is determined by the formula

$$|\langle M'_u v, w \rangle| = (p-1) \left| \int_\Omega |u|^{p-2} vw \, ds \right| \leq \text{const} \|u\|_{L_p(\Omega)}^{p-2} \|v\|_{L_p(\Omega)} \|w\|_{L_p(\Omega)}.$$

Hence, the operator $M \in C^1(\mathscr{B}; \mathscr{B}^*)$. The dissipativity of the operator M is a consequence of the s-monotonicity of the operator $(-M)$ [36]:

$$\langle -M'_u v, v \rangle = (p-1) \int_\Omega |u|^{p-2} v^2 \, du > 0, \; u, v \in L_p(\Omega), \; u, v \neq 0.$$

In addition, the operator $(-M)$ is p-coercive (a.g. $< M(x), x > \geq C_M \|x\|^p$ and $\|M(x)\|_* \leq C^M \|x\|^{p-1} \; \forall x \in \mathscr{B}$) [36]:

$$\langle -M(u), u \rangle = \int_\Omega |u|^p \, ds = \|u\|_{L_p(\Omega)}^p,$$

$$|\langle -M(u), v \rangle| \leq \int_\Omega |u|^{p-1} |v| \, ds \leq \|u\|_{L_p(\Omega)}^{p-1} \|v\|_{L_p(\Omega)}.$$

Hence, the operator $(-M)$ is strong coercive. So, Lemma is proved.

If $\lambda \geq -\lambda_1$

$$\ker L = \begin{cases} \{0\}, \; if \; \lambda > -\lambda_1; \\ \text{span}\{\varphi_1\}, \; if \; \lambda = -\lambda_1. \end{cases}$$

Therefore

$$\text{im } L = \begin{cases} \mathscr{H}^*, \; if \; \lambda > -\lambda_1; \\ \{u \in \mathscr{H}^* : \langle u, \varphi_1 \rangle = 0\}, \; if \; \lambda = -\lambda_1, \end{cases}$$

$$\text{coim } L = \begin{cases} \mathscr{H}, \; if \; \lambda > -\lambda_1; \\ \{u \in \mathscr{H} : \langle u, \varphi_1 \rangle = 0\}, \; if \; \lambda = -\lambda_1. \end{cases}$$

Hence, the projectors

$$P = Q = \begin{cases} \mathbf{I}, \ \lambda > -\lambda_1; \\ \mathbf{I} - \langle \cdot, \varphi_1 \rangle, \ \lambda = -\lambda_1. \end{cases}$$

Construct the set

$$\mathbf{M} = \begin{cases} \mathscr{B}, \ if \ \lambda > -\lambda_1; \\ \{u \in \mathscr{B} : \int_\Omega |u|^{p-2} u \varphi_1 \, ds = 0\}, \ if \ \lambda = -\lambda_1. \end{cases}$$

Theorem 4 *Suppose that $p \geq \frac{2n}{n+2}$, $\lambda \geq -\lambda_1$. Then*
  (i) the set $\mathbf{M}$ is a simple Banach C^1-manifold modelled by the space coim $L \cap \mathscr{B}$;
  (ii) $\forall u_0 \in \mathbf{M}$ there exists the unique solution $u \in C^k((0, +\infty); \mathbf{M})$ to problem (1), (14), (15).

Proof Statement (i) was obtained in [32]. Statement (ii) is a consequence of Theorem 2 and Lemma 1.

Define the shift operator $U^t(u_0) \equiv u(t)$, where $u(t)$ is a solution to problem (1), (14), (15). Then $\{U^t : t \in \mathbf{R}_+\}$ forms a nonlinear semigroup of operators with domain $D(U) = \mathbf{M}$.

Theorem 5 *Suppose that $p \geq \frac{2n}{n+2}$, $\lambda \geq -\lambda_1$. Then there exists a resolving semigroup of contractive operators $\{U^t : t \in \mathbf{R}_+\}$ of Eq. (15) defined on the manifold $\mathbf{M}$.*

Proof Obviously, $u \equiv 0$ is a stationary solution to Eq. (15). Statement of the theorem is a consequence of Theorem 3 and Lemma 1.

5 Stochastic K-Processes. Phase Space

Consider a complete probability space $\Omega \equiv (\Omega, \mathscr{A}, \mathbf{P})$ and the set of real numbers $\mathbf{R}$ endowed with a Borel σ-algebra. According to [38], a measurable mapping $\xi : \Omega \to \mathbf{R}$ is called a *random variable*. The set of random variables having zero expectations (i.e. $\mathbf{E}\xi = 0$) and finite variances (i.e. $\mathbf{D}\xi < +\infty$) forms Hilbert space $\mathbf{L_2}$ with the scalar product $(\xi_1, \xi_2) = \mathbf{E}\xi_1\xi_2$, where $\mathbf{E}, \mathbf{D}$ is the expectation and variance of the random variable, respectively. Denote by $\mathscr{A}_0$ the σ-subalgebra of σ-algebra $\mathscr{A}$. Construct the space $\mathbf{L_2^0}$ of random variables that are measurable with respect to $\mathscr{A}_0$, then $\mathbf{L_2^0}$ is a subspace of the space $\mathbf{L_2}$. Suppose that $\Pi : \mathbf{L_2} \to \mathbf{L_2^0}$ is the orthoprojector, and $\xi \in \mathbf{L_2}$, then $\Pi\xi$ is called the *conditional expectation* of the random variable ξ and is denoted by $\mathbf{E}(\xi|\mathscr{A}_0)$.

Let $\mathscr{I} \subset \mathbf{R}$ be a set. Consider two mappings: $f : \mathscr{I} \to \mathbf{L_2}$, that each $t \in \mathscr{I}$, associates with a random variable $\xi \in \mathbf{L_2}$, and $g : \mathbf{L_2} \times \Omega \to \mathbf{R}$, that each pair (ξ, ω) associates with a point $\xi(\omega) \in \mathbf{R}$. A mapping $\eta : \mathscr{I} \times \Omega \to \mathbf{R}$, of the form $\eta = \eta(t, \omega) = g(f(t), \omega)$ is called an *(one-dimensional) random process*. According to [38], a random process η is called *continuous*, if almost surely all its trajectories are continuous. Denote by $\mathbf{CL_2}$ the set of continuous random processes, which forms a Banach space. An example of the continuous process is one-dimensional Wiener process $\beta = \beta(t)$, which can be represented as [38]

$$\beta(t) = \sum_{k=0}^{\infty} \xi_k \sin \frac{\pi}{2}(2k+1)t, \qquad (17)$$

where ξ_k are uncorrelated Gaussian variables, $\mathbf{E}\xi_k = 0$, $\mathbf{D}\xi_k = [\frac{\pi}{2}(2k+1)]^{-2}$.

Fix $\eta \in \mathbf{CL_2}$ and $t \in \mathscr{I}$ and denote by $\mathscr{N}_t^\eta$ the σ-algebra generated by the random variable $\eta(t)$. Denote $\mathbf{E}_t^\eta = \mathbf{E}(\cdot|\mathscr{N}_t^\eta)$.

Definition 8 (i) Suppose that $\eta \in \mathbf{CL_2}$. A random variable

$$D\eta(t,\cdot) = \lim_{\Delta t \to 0+} \mathbf{E}_t^\eta \left(\frac{\eta(t+\Delta t, \cdot) - \eta(t, \cdot)}{\Delta t} \right)$$

$$\left(D_*\eta(t,\cdot) = \lim_{\Delta t \to 0+} \mathbf{E}_t^\eta \left(\frac{\eta(t, \cdot) - \eta(t-\Delta t, \cdot)}{\Delta t} \right) \right)$$

is called a *mean derivative on the right* $D\eta(t,\cdot)$ *(on the left* $D_*\eta(t,\cdot)$*) of a random process* η *at the point* $t \in (\varepsilon, \tau)$, if the limit exists in the sense of a uniform metric on $\mathbf{R}$. A random process η is called *mean differentiable on the right (on the left)* on $\mathscr{I}$, if there exists the mean derivative on the right (on the left) at each point $t \in \mathscr{I}$.

(ii) Let the random process $\eta \in \mathbf{CL_2}$ be mean differentiable on the right and on the left on $\mathscr{I}$. The derivative $\overset{o}{\eta} = D_S\eta = \frac{1}{2}(D + D_*)\eta$ is called the symmetric mean derivative.

Remark 4 Futher, the symmetric mean derivative is called the *Nelson–Gliklikh derivative*. Denote the lth Nelson–Gliklikh derivative of the random process η by $\overset{o\,(l)}{\eta}$, $l \in \mathbf{N}$. Note that the Nelson–Gliklikh derivative coincides with the classical derivative, if $\eta(t)$ is a deterministic function. In the case of the one-dimensional Wiener process $\beta = \beta(t)$, the following statements are correct [37]:

(i) $\overset{o}{\beta}(t) = \frac{\beta(t)}{2t}$ for all $t \in \mathbf{R}_+$;

(ii) $\overset{o\,(l)}{\beta}(t) = (-1)^{l-1} \cdot \prod_{i=1}^{l-1}(2i-1) \cdot \frac{\beta(t)}{(2t)^l}$, $l \in \mathbf{N}$, $l \geq 2$.

Consider the space of "noises" $\mathbf{C}^l\mathbf{L_2}$, $l \in \mathbf{N}$, i.e. the space of random processes from $\mathbf{CL_2}$, whose trajectories are almost surely differentiable by Nelson–Gliklikh on $\mathscr{I}$ up to the order l inclusive.

Consider a real separable Hilbert space $(\mathscr{V}, <\cdot,\cdot>)$ with orthonormal basis $\{\varphi_k\}$. Each element $u \in \mathscr{V}$ can be represented as $u = \sum_{k=1}^{\infty} u_k \varphi_k = \sum_{k=1}^{\infty} <u, \varphi_k> \varphi_k$. Next, choose a monotonely decreasing numerical sequence $K = \{\mu_k\}$ such that $\sum_{k=1}^{\infty} \mu_k^2 <$

$+\infty$. Consider a sequence of random variables $\{\xi_k\} \subset \mathbf{L}_2$, such that $\sum_{k=1}^{\infty} \mu_k^2 \mathbf{D}\xi_k < +\infty$. Denote by $\mathscr{V}_K \mathbf{L}_2$ the Hilbert space of *random K-variables* having the form $\xi = \sum_{k=1}^{\infty} \mu_k \xi_k \varphi_k$. Moreover, a random K-variable $\xi \in \mathscr{V}_K \mathbf{L}_2$ exists, if, for example, $\mathbf{D}\xi_k < \text{const } \forall k$. Note that space $\mathscr{V}_K \mathbf{L}_2$ is a Hilbert space with scalar product $(\xi^1, \xi^2) = \sum_{k=1}^{\infty} \mu_k^2 \mathbf{E}\xi_k^1 \xi_k^2$. Consider a sequence of random processes $\{\eta_k\} \subset \mathbf{CL}_2$ and define $\mathscr{V}$-*valued continuous stochastic K-process*

$$\eta(t) = \sum_{k=1}^{\infty} \mu_k \eta_k(t) \varphi_k \tag{18}$$

if series (18) converges uniformly by the norm $\mathscr{V}_K \mathbf{L}_2$ on any compact set in $\mathscr{I}$. Consider the Nelson–Gliklikh derivatives of random K-process

$$\overset{o\,(l)}{\eta}(t) = \sum_{k=1}^{\infty} \mu_k \overset{o\,(l)}{\eta_k}(t) \varphi_k$$

on the assumption that there exist the Nelson–Gliklikh derivatives up to the order l inclusive in the right-hand side, and all series converge uniformly according to the norm $\mathscr{V}_K \mathbf{L}_2$ on any compact from $\mathscr{I}$. Next, consider the space $\mathbf{C}(\mathscr{I}; \mathscr{V}_K \mathbf{L}_2)$ of continuous stochastic K-processes and the space $\mathbf{C}^l(\mathscr{I}; \mathscr{V}_K \mathbf{L}_2)$ of stochastic K-processes whose trajectories are almost surely continuously differentiable by Nelson–Gliklikh up to the order $l \in \mathbf{N}$ inclusive. An example of a K-process from the space $\mathbf{C}^l(\mathscr{I}; \mathscr{V}_K \mathbf{L}_2)$ is a Wiener K-process [38] $W_K(t) = \sum_{k=1}^{\infty} \mu_k \beta_k(t) \varphi_k$, where $\{\beta_k\} \subset \mathbf{C}^l \mathbf{L}_2$ is a sequence of one-dimensional Wiener processes (Brownian motions on $\mathscr{I}$). Note that the space $\mathbf{C}^l(\mathscr{I}; \mathscr{V}_K \mathbf{L}_2)$ will call *the space of K-"noises"*.

Let us extend the constructions from Paragraph 1 to the stochastic case. By analogy with Paragraph 1, consider a real separable Hilbert space $\mathbf{H} = (\mathbf{H}; \langle \cdot, \cdot \rangle)$ identified with its conjugate space, and dual pairs of reflexive Banach spaces $(\mathscr{H}, \mathscr{H}^*)$ and $(\mathscr{B}, \mathscr{B}^*)$, such that embeddings (7) are dense and continuous. Let an operator $L \in \mathscr{L}(\mathscr{H}; \mathscr{H}^*)$ be linear, continuous, self-adjoint, non-negative defined Fredholm operator, and an operator $M \in C^k(\mathscr{B}; \mathscr{B}^*)$, $k \geq 1$, be dissipative. In space $\mathbf{H}$ choose an orthonormal basis $\{\varphi_k\}$ so that $\text{span}\{\varphi_1, \varphi_2, \ldots, \varphi_l\} = \ker L$, $\dim \ker L = l$ and the following condition holds:

$$\{\varphi_k\} \subset \mathscr{B}. \tag{19}$$

We use the space $\mathbf{H}$ in order to construct the space of $\mathbf{H}$-valued random K-variables $\mathbf{H}_K \mathbf{L}_2$ as a completion of the linear span of random K-variables $\xi = \sum_{k=1}^{\infty} \mu_k \xi_k \varphi_k$. Taking into account that the operator L is self-adjoint and Fredholm, we identify $\mathbf{H} \supset \ker L \equiv \text{coker } L \subset \mathbf{H}^*$ and, similarly, construct the space $\mathbf{H}_K^* \mathbf{L}_2$

according to the corresponding orthonormal basis. We use the subspace ker L in order to construct the subspace $[\ker L]_K \mathbf{L}_2 \subset \mathbf{H}_K \mathbf{L}_2$ and, similarly, the subspace $[\operatorname{coker} L]_K \mathbf{L}_2 \subset \mathbf{H}_K^* \mathbf{L}_2$. Taking into account that embeddings (7) are continuous and dense, we construct the spaces $\mathscr{H}_K^* \mathbf{L}_2 = [\operatorname{coker} L]_K \mathbf{L}_2 \oplus [\operatorname{im} L]_K \mathbf{L}_2$ and $\mathscr{B}_K^* \mathbf{L}_2 = [\operatorname{coker} L]_K \mathbf{L}_2 \oplus \overline{[\operatorname{im} L]}_K \mathbf{L}_2$.

We use the subspace $\operatorname{coim} L \subset \mathscr{H}$ in order to construct the subspace $[\operatorname{coim} L]_K \mathbf{L}_2$ such that the space $\mathscr{H}_K \mathbf{L}_2 = [\ker L]_K \mathbf{L}_2 \oplus [\operatorname{coim} L]_K \mathbf{L}_2$. Denote $[\ker L]_K \mathbf{L}_2 \equiv \mathscr{B}_K^0 \mathbf{L}_2$ such that the space $\operatorname{coim} L \cap \mathscr{B}$ in order to construct the set $\mathscr{B}_K^1 \mathbf{L}_2$, then $\mathscr{B}_K \mathbf{L}_2 = \mathscr{B}_K^0 \mathbf{L}_2 \oplus \mathscr{B}_K^1 \mathbf{L}_2$. The following lemma is correct, since the operator L is self-adjoint and Fredholm.

Lemma 2 *(i) Let operator $L \in \mathscr{L}(\mathscr{H}; \mathscr{H}^*)$ be a linear, continuous, self-adjoint, non-negatively defined Fredholm operator, then the operator $L \in \mathscr{L}(\mathscr{H}_K \mathbf{L}_2; \mathscr{H}_K^* \mathbf{L}_2)$, and*

$$\mathbf{H}_K \mathbf{L}_2 \supset [\ker L]_K \mathbf{L}_2 \equiv [\operatorname{coker} L]_K \mathbf{L}_2 \subset \mathbf{H}_K^* \mathbf{L}_2$$

if

$$\mathbf{H} \supset \ker L \equiv \operatorname{coker} L \subset \mathbf{H}^*.$$

(ii) There exists a projector Q of the space $\mathscr{B}_K^* \mathbf{L}_2$ on $\overline{[\operatorname{im} L]}_K \mathbf{L}_2$ along $[\operatorname{coker} L]_K \mathbf{L}_2$.

(iii) There exists a projector P of the space $\mathscr{B}_K \mathbf{L}_2$ on $\mathscr{B}_K^1 \mathbf{L}_2$ along $\mathscr{B}_K^0 \mathbf{L}_2$.

Suppose that $\mathscr{I} \equiv (0, +\infty)$. We use the space $\mathbf{H}$ in order to construct the spaces of K-"noises" spaces $\mathbf{C}^k(\mathscr{I}; \mathbf{H}_K \mathbf{L}_2)$ and $\mathbf{C}^k(\mathscr{I}; \mathscr{B}_K \mathbf{L}_2)$, $k \in \mathbf{N}$. Next, we extend the constructions from Paragraph 1 to the stochastic case. Consider the stochastic Sobolev type equation

$$L \overset{o}{\eta} = M(\eta). \tag{20}$$

A solution to Eq. (20) is a stochastic K-process. Stochastic K-processes $\eta = \eta(t)$ and $\zeta = \zeta(t)$ are considered to be equal, if almost surely each trajectory of one of the processes coincides with a trajectory of other process.

Definition 9 A stochastic K-process $\eta \in C^1(\mathscr{I}; \mathscr{B}_K \mathbf{L}_2)$ is called a *solution to Eq.* (20), if almost surely all trajectories of η satisfy Eq. (20) for all $t \in \mathscr{I}$. A solution $\eta = \eta(t)$ to Eq. (20) that satisfies the initial value condition

$$\lim_{t \to 0+} (\eta(t) - \eta_0) = 0 \tag{21}$$

is called a *solution to Cauchy problem* (20), (21), if the solution satisfies condition (21) for some random K-variable $\eta_0 \in \mathscr{B}_K \mathbf{L}_2$.

Fix $\omega \in \Omega$. Let $\eta = \eta(t), t \in \mathscr{I}$ be a solution to Eq. (20), then η belongs to the set

$$\mathbf{M} = \begin{cases} \{\eta \in \mathscr{B}_K \mathbf{L}_2 : (\mathbf{I} - Q) M(\eta) = 0\}, & \text{if } \ker L \neq \{0\}; \\ \mathscr{B}_K \mathbf{L}_2, & \text{if } \ker L = \{0\}. \end{cases} \tag{22}$$

Theorem 6 *Suppose that the set* **M** *is a simple Banach C^k-manifold at the point $\eta_0 \in \mathscr{B}_K \mathbf{L}_2$, there exists a solution to the equation*

$$M(\eta) = 0,$$

and the operator M is dissipative. Then there exists a resolving semigroup of operators $\{\Xi^t : t \in \mathbf{R}_+\}$ of Eq. (20) defined on the manifold **M**.

Next, we consider the Dirichlet problem

$$\eta(s, t) = 0, \ (s, t) \in \partial \Omega \times \mathbf{R}_+ \qquad (23)$$

for the stochastic Boussinesq equation

$$(\lambda - \Delta) \overset{o}{\eta} = \Delta(|\eta|^{p-2}\eta), \ p \geq 2. \qquad (24)$$

Define the shift operator $\Xi^t(\eta_0) \equiv \eta(t)$, where $\eta(t)$ is the solution of problem (21), (23), (24). Then $\{\Xi^t : t \in \mathbf{R}_+\}$ forms a nonlinear semigroup of operators with domain $D(\Xi) = \mathbf{M}$.

Theorem 7 *Suppose that $p \geq \frac{2n}{n+2}$, $\lambda \geq -\lambda_1$. Exists a resolving semigroup of compressing operators $\{\Xi^t : t \in \mathbf{R}_+\}$ of Eq. (24) defined on the manifold* **M**.

References

1. Clement, P., Heihmans, H.J.A.M.: One-Parameter Semigroups. North-Holland, Amsterdam (1987)
2. Engel, K., Nagel, R.: One-Parameter Semigroups for Linear Evolution Equations. Springer, N.Y. (2000)
3. Goldstein, J.: Semigroups of Linear Operators and Applications. Oxford University Press, N.Y. (1985)
4. Lunardi, A.: Analytic Semigroups and Optimal Regularity in Parabolic Problems. Birkhauser, Basel (1995)
5. Batkai, A., Piazzera, S.: Semigroups for Delay Equations. A.K. Peters, Canada (2005)
6. Vrabie, I.: C_0-Semigroups and Applications. North-Holland, Amsterdam (2003)
7. Favini, A., Tanabe H.: Regularity results and solution semigroups for retarded functional differential equations. Bull. South Ural State Univ. Ser.: Math. Model. Program. Comput. Softw. **10**(1), 48–69 (2017). https://doi.org/10.14529/mmp170104 (in Russian)
8. Banasiak, J., Bobrowski, A.: A semigroup related to a convex combination of boundary conditions obtained as a result of averaging other semigroups. J. Evol. Equ. **15**(1), 223–237 (2015)
9. Komura, Y.: Nonlinear semi-groups in Hilbert space. J. Math. Soc. Jpn. **19**(4), 493–507 (1967)
10. Kato, T.: Nonlinear semigroups and evolution equations. J. Math. Soc. Jpn. **19**(4), 508–520 (1967)
11. Brezis, H.: Operateurs Maximaux Monotones et Semigroupes de Contractions Dans Les Espaces de Hubert. North-Holland, Amsterdam (1974)
12. Crandall, M.G.: An Introduction to Evolution Governed by Accretive Operators. Academic Press, N.Y. (1976)

13. Dubinskii, Yu.A: Nonlinear elliptic and parabolic equations. J. Soviet Math. **12**(5), 475–554 (1979). (in Russian)
14. Miyadera, I.: Nonlinear Semigroups. American Mathematical Society (1992)
15. Aulbach, B.: Nguyen van Minh: nonlinear semigroups and the existence and stability of solutions of semilinear nonautonomous evolution equations. Abstract Appl. Anal. **1**(4), 351–380 (1996)
16. Belleni-Morante, A., McBride, A.: Applied Nonlinear Semigroups. Wiley, Chichester (1998)
17. Al'shin, A.B., Korpusov, M.O., Sveshnikov, A.G.: Blow-up in Nonlinear Sobolev Type Equations. Walter de Gruyter & Co., Berlin (2011)
18. Zamyshlyaeva, A.A.: The higher-order Sobolev-type models. Bull. South Ural State Univ. Ser.: Math. Model. Program. Comput. Softw. **7**(2), 5–28 (2014). https://doi.org/10.14529/mmp140201 (in Russian)
19. Keller, A.V., Zagrebina, S.A.: Some generalizations of the Showalter–Sidorov problem for Sobolev-type models. Bull. South Ural State Univ. Ser.: Math. Model. Program. Comput. Softw. **8**(2), 5–23 (2015). https://doi.org/10.14529/mmp150201 (in Russian)
20. Favini, A., Sviridyuk, G., Sagadeeva, M.: Linear Sobolev type equations with relatively p-radial operators in space of "noises". Mediter. J. Math. **12**(6), 4607–4621 (2016). https://doi.org/10.1007/s00009-016-0765-x
21. Showalter, R.E.: The Sobolev equation. Appl. Anal. **5**(1), 15–22 (1975). https://doi.org/10.1080/00036817508839103
22. Sviridyuk, G.A.: On the general theory of operator semigroups. Russian Math. Surv. **49**(4), 45–74 (1994). https://doi.org/10.1070/RM1994v049n04ABEH002390
23. Sviridyuk, G.A., Fedorov, V.E.: Linear Sobolev Type Equations and Degenerate Semigroups of Operators. VSP, Utrecht, Boston, Koln, Tokyo (2003)
24. Favini, A., Sviridyuk, G.A., Zamyshlyaeva, A.A.: One class of Sobolev type equations of higher order with additive "white noise". Commun. Pure Appl. Anal. **15**(1), 185–196 (2016). https://doi.org/10.3934/cpaa.2016.15.185
25. Favini, A., Zagrebina, S.A., Sviridyuk, G.A.: Multipoint initial-final value problems for dynamical Sobolev-type equations in the space of noises. Electron. J. Differ. Equ. **2018**(128), 1–10 (2018). https://doi.org/10.3934/cpaa.2016.15.185
26. Favini, A., Yagi, A.: Degenerate Differential Equations in Banach Spaces. Marcel Dekker Inc., N.Y. (1999)
27. Sviridyuk, G.A.: The manifold of solutions of an operator singular pseudoparabolic equation. Proce. USSR Acad. Sci. **289**(6), 1315–1318 (1986). (in Russian)
28. Sviridyuk, G.A., Kazak, V.O.: The phase space of an initial-boundary value problem for the Hoff equation. Math. Notes **71**(1–2), 262–266 (2002). https://doi.org/10.1023/A:1013919 50060533
29. Kondyukov, A.O., Sukacheva, T.G.: Phase space of the initial-boundary value problem for the Oskolkov system of highest order. Bull. South Ural State Univ. Ser.: Math. Model. Program. Comput. Softw. **11**(4), 67–77 (2018). https://doi.org/10.14529/mmp180405
30. Sviridyuk, G.A., Manakova, N.A.: Phase space of the Cauchy-Dirichlet problem for the Oskolkov equation of nonlinear filtration. Russian Math. **47**(9), 33–38 (2003)
31. Sviridyuk, G.A.: Quasistationary trajectories of semilinear dynamical equations of Sobolev type. Russian Acad. Sci. Izvestiya Math. **42**(3), 601–614 (1994)
32. Bokareva, T.A., Sviridyuk, G.A.: Whitney folds of the phase spaces of some semilinear equations of Sobolev type. Math. Notes **55**(3–4), 237–242 (1994). https://doi.org/10.1007/BF02110776
33. Manakova, N.A., Sviridyuk, G.A.: Nonclassical equations of mathematical physics. phase space of semilinear Sobolev type equations. Bull. South Ural State Univ. Ser.: Math. Mech. Phys. **8**(3), 31–51 (2016). https://doi.org/10.14529/mmph160304 (in Russian)
34. Manakova, N.A., Gavrilova, O.V.: About nonuniqueness of solutions of the Showalter–Sidorov problem for one mathematical model of nerve impulse spread in membrane. Bull. South Ural State Univ. Ser.: Math. Model. Program. Comput. Softw. **11**(4), 161–168 (2018). https://doi.org/10.14529/mmp180413

35. Dzektser, E.S.: The generalization of the equations of motion of groundwater. Proc. USSR Acad. Sci. **5**, 1031–1033 (1972). (In Russian)
36. Sviridyuk, G.A., Semenova, I.N.: Solvability of an inhomogeneous problem for the generalized Boussinesq filtration equations. Differ. Equ. **24**(9), 1065–1069 (1988)
37. Gliklikh, Y.E.: Investigation of Leontieff type equations with white noise protect by the methods of mean derivatives of stochastic processes. Bull. South Ural State Univ. Ser.: Math. Model. Program. Comput. Softw. (13), 24–34 (2012)
38. Sviridyuk G.A., Manakova N.A.: Dynamic models of Sobolev type with the Showalter–Sidorov condition and additive "noises". Bull. South Ural State Univ. Ser.: Math. Model. Program. Comput. Softw. **7**(1), 90–103 (2014). (in Russian)
39. Leng, S.: Introduction to Differentiable Manifolds. Springer, N.Y. (2002)

Sharp Interior and Boundary Regularity of the SMGTJ-Equation with Dirichlet or Neumann Boundary Control

Roberto Triggiani

Abstract We consider the third order (in time) linear equation known as SMGTJ-equation, as defined on a multidimensional bounded domain and subject to either Dirichlet or Neumann boundary control. We then establish corresponding sharp interior and boundary regularity results.

Keywords SMGTJ - Equation · Boundary Regularity

1 Introduction: The SMGTJ Equation

Let Ω be a bounded domain in $\mathbb{R}^3$ with sufficiently smooth boundary $\Gamma = \partial\Omega$, as specified below. In this paper, we consider the problem of sharp interior and boundary **regularity** of the linear version of a third order (in time) PDE **with non-homogeneous term on the boundary**. Both Dirichlet and Neumann boundary terms will be considered. The equation, which should be called SMGTJ [for G. G. Stokes (1851), F. K. Moore and W. E. Gibson (1960), P. A. Thompson (1972) and P. M. Jordan (2004)], see [12, 13, 33, 37, 39], arises in a variety of physical contexts such as: effects of the radiation of heat on the propagation of sound; propagation of disturbances in a gas subject to relaxation effects; behavior of viscoelastic materials; propagation of acoustic waves, etc. In particular, if in classical models in nonlinear acoustics (Kuznetsov equation, Westervelt equation, Kokhlov–Zobolotskaya–Kuznetsov equation), one replaces the Fourier Law for the heat flux with the Maxwell-Cattaneo Law (to avoid the paradox of infinite speed of propagation), one obtains a third order in time PDE, whose linear part is the one considered in the present paper; that is [12, 13]

$$\tau\psi_{ttt} + \psi_{tt} - c^2\Delta\psi - b\Delta\psi_t = 0 \quad \text{in } (0, T] \times \Omega, \tag{1.1}$$

R. Triggiani (✉)
Department of Mathematical Sciences, University of Memphis, Memphis, TN 38152, USA
e-mail: rtrggani@memphis.edu

where $\tau > 0$, $b > 0$, $c^2 > 0$ are fixed constants, whose physical meaning is not relevant here. See [34].

We are taking Ω in $\mathbb{R}^3$, as this is the physically significant setting. However, the mathematical analysis works on any $\mathbb{R}^d$, $d = 1, 2, \ldots$ in the Dirichlet case and in $\mathbb{R}^d$, $d = 2, 3, \ldots$ in the Neumann case.

Part A: Dirichlet Case

2 Linear Third Order SMGTJ-Equation with Non-homogeneous Dirichlet Boundary Term

Henceforth we shall take $\tau = 1$ in (1.1) w.l.o.g.

If the linear third order equation (1.1) is written in terms of the pressure, then Dirichlet non-homogeneous boundary terms are appropriate [14]. We then consider the following mixed problem in the unknown $y(t, x)$:

$$\begin{cases} y_{ttt} + \alpha y_{tt} - c^2 \Delta y - b \Delta y_t = 0 & \text{in } Q = (0, T] \times \Omega \quad (2.1a) \\ y|_{t=0} = y_0; \quad y_t|_{t=0} = y_1; \quad y_{tt}|_{t=0} = y_2 & \text{in } \Omega \quad (2.1b) \\ y|_\Sigma = g & \text{in } \Sigma = (0, T] \times \Gamma \quad (2.1c) \end{cases}$$

2.1 Case $g \equiv 0$.

A rather comprehensive study of this case was carried out in [34] in the constant coefficient case via semigroup/functional analytic techniques, and in [15, 16] in the variable coefficient case via energy methods. Here we shall only report a subset of these results which are relevant to the present paper. Define the positive self adjoint operator on $H = L^2(\Omega)$:

$$Af = -\Delta f, \qquad \mathcal{D}(A) = H^2(\Omega) \cap H_0^1(\Omega), \qquad (2.2)$$

so that problem (2.1a), (2.1b) and (2.1c) (with $g = 0$) can be re-written abstractly as

$$u_{ttt} + \alpha u_{tt} + c^2 Au + b A u_t = 0 \quad \text{on } H = L^2(\Omega), \qquad (2.3)$$

along with I.C. u_0, u_1, u_2. We re-write it as a first order problem as

$$\frac{d}{dt} \begin{bmatrix} u \\ u_t \\ u_{tt} \end{bmatrix} = G \begin{bmatrix} u \\ u_t \\ u_{tt} \end{bmatrix}, \quad G = \begin{bmatrix} 0 & I & 0 \\ 0 & 0 & I \\ -c^2 A & -bA & -\alpha I \end{bmatrix}. \qquad (2.4)$$

Introduce the following spaces:

$$U_0 \equiv H \times H \times H \tag{2.5a}$$

$$U_1 \equiv \mathcal{D}(A^{\frac{1}{2}}) \times \mathcal{D}(A^{\frac{1}{2}}) \times H; \qquad U_2 \equiv \mathcal{D}(A) \times \mathcal{D}(A) \times \mathcal{D}(A^{\frac{1}{2}}) \tag{2.5b}$$

$$U_3 \equiv \mathcal{D}(A) \times \mathcal{D}(A^{\frac{1}{2}}) \times H; \qquad U_4 \equiv \mathcal{D}(A^{\frac{3}{2}}) \times \mathcal{D}(A) \times \mathcal{D}(A^{\frac{1}{2}}) \tag{2.5c}$$

Theorem 2.1 ([34, Sect. 2]) *The operator G in (2.4) generates a s.c. group e^{Gt} on each of the spaces U_1, U_2, U_3, U_4 with appropriate domains so that*

$$\begin{bmatrix} u(t) \\ u_t(t) \\ u_{tt}(t) \end{bmatrix} = e^{Gt} \begin{bmatrix} u_0 \\ u_1 \\ u_2 \end{bmatrix} \in C([0, T]; U_i), \quad i = 1, 2, 3, 4 \tag{2.6}$$

for $[u_0, u_1, u_2] \in U_i$, $i = 1, 2, 3, 4$. Below we shall emphasize the case U_3, whereby then

$$G : U_3 \supset \mathcal{D}(G) = \mathcal{D}(A) \times \mathcal{D}(A) \times \mathcal{D}(A^{\frac{1}{2}}) \longrightarrow U_3 \tag{2.4 bis}$$

The group generation property points out that the third order equation (2.3) has a 'hyperbolic' character. In fact, as in [34], rewrite (2.3) as

$$(u_t + \alpha u)_{tt} + bA\left(\frac{c^2}{b}u + u_t\right) = 0. \tag{2.7}$$

This suggests introducing a new variable, as in [34]

$$\text{either } z = \frac{c^2}{b}u + u_t, \qquad \text{or else } \xi = u_t + \alpha u \tag{2.8}$$

(i) Thus,

$$\text{If } \alpha = \frac{c^2}{b}, \text{ then } (2.7) \Longrightarrow z_{tt} + bAz = 0 \ (z = \xi), \tag{2.9}$$

the pure abstract wave equation.
(ii) Otherwise,

$$z = \frac{c^2}{b}u + u_t = (\alpha u + u_t) - \gamma u, \qquad \gamma = \alpha - \frac{c^2}{b} \tag{2.10}$$

$$(u_t + \alpha u)_{tt} = z_{tt} + \gamma u_{tt} = z_{tt} + \gamma \left(z - \frac{c^2}{b}u\right)_t. \tag{2.11}$$

Substituting (2.8), (2.10), (2.11) in (2.7) leads to the following hyperbolic-dominated system

$$\begin{cases} z_{tt} = -bAz - \gamma z_t + \gamma \frac{c^2}{b} z - \gamma \left(\frac{c^2}{b}\right)^2 u & \text{(2.12a)} \\ u_t = -\frac{c^2}{b} u + z & \text{(2.12b)} \end{cases}$$

(model #2 in [34, Sect. 2]) coupling the hyperbolic z-equation with the scalar ODE in u.

2.2 Case $y_0 = 0$, $y_1 = 0$, $y_2 = 0$, $g \neq 0$.

In this case, we seek to obtain sharp regularity of the map

$$g \longrightarrow \left\{ y, y_t, y_{tt}, \left. \frac{\partial y}{\partial \nu}\right|_\Sigma \right\} \tag{2.13}$$

from the Dirichlet boundary datum g to the interior solution $\{y, y_t, y_{tt}\}$ and the Neumann boundary trace $\left.\frac{\partial y}{\partial \nu}\right|_\Sigma$ of problem (2.1a), (2.1b) and (2.1c). Our main result in the present Part A (Dirichlet) is the following

Theorem 2.2 *With reference to problem (2.1a)–(2.1c) with zero I.C., we have the following interior regularity results:*

$g \in L^2(0, T; L^2(\Gamma)) \cap C([0, \epsilon); L^2(\Gamma))$, $g(0) = 0$, $\epsilon > 0$ *small*

$$\implies \begin{cases} y \in C([0, T]; L^2(\Omega)) & \text{(2.14a)} \\ y_t \in C([0, T]; [\mathcal{D}(A^{\frac{1}{2}})]' = H^{-1}(\Omega)), & \text{(2.14b)} \\ y_{tt} \in L^2(0, T; [\mathcal{D}(A)]') & \text{(2.14c)} \\ \left.\frac{\partial y}{\partial \nu}\right|_\Sigma \in H^{-1}(\Sigma)) & \text{(2.14d)} \end{cases}$$

For (2.14a) and (2.14b), see (3.20) and (3.21), respectively. For (2.14c), see (3.31a). Finally for (2.14d) see (4.1) of Theorem 4.1. Moreover,

$$\begin{cases} g \in C([0, T]; L^2(\Gamma)) \\ g(0) = 0 \end{cases} \implies y_{tt} \in C([0, T]; [\mathcal{D}(A)]'), \tag{2.14e}$$

see (3.31b), all the maps being continuous.

Remark 2.1 The results of Theorem 2.2 should be compared with the following results for general second order hyperbolic equations, even with variable coefficients, which we however report only for the canonical wave equation.

Theorem 2.3 ([23, 25, 26]) *Consider the following mixed problem, where Ω is a bounded domain in $\mathbb{R}^d$, $d \geq 1$, with sufficiently smooth boundary Γ:*

$$\begin{cases} w_{tt} = \Delta w & \text{in } Q = (0, T] \times \Omega & (2.15a) \\ w|_{t=0} = 0; \ w_t|_{t=0} = 0 & \text{in } \Omega & (2.15b) \\ w|_\Sigma = g & \text{in } \Sigma = (0, T] \times \Gamma & (2.15c) \end{cases}$$

Then

$$\begin{cases} g \to \left\{ w, w_t, w_{tt}, \dfrac{\partial w}{\partial \nu}\Big|_\Sigma \right\} \text{ continuously} \\ L^2(0, T; L^2(\Gamma)) \to C\left([0, T]; L^2(\Omega) \times H^{-1}(\Omega) \times H^{-2}(\Omega)\right) \times H^{-1}(\Sigma), \end{cases}$$
(2.16)

where $H^{-1}(\Sigma) = $ dual of $\{h \in H_0^1(\Sigma)\}$ i.e. with $h(\cdot, 0) = 0$ and $h(\cdot, T) = 0$ on Γ (but actually, $h(\cdot, T) = 0$ is not needed).

Indeed, our proof of Theorem 2.2 in Sect. 3 (interior regularity) and Sect. 4 (boundary regularity) will critically be based on Theorem 2.3. This result is reported also in [31, Chap. 10, Sect. 5]. The proof of Theorem 2.3 is by PDE-techniques, either directly [25, 26], or much more conveniently, by duality [23].

In fact, consider the following problem, dual of problem (2.15a)–(2.15c)

$$\begin{cases} \phi_{tt} = \Delta \phi + f & \text{in } Q & (2.17a) \\ \phi|_{t=T} = \phi_0; \ \phi_t|_{t=T} = \phi_1 & \text{in } \Omega & (2.17b) \\ \phi|_\Sigma = 0 & \text{in } \Sigma & (2.17c) \end{cases}$$

Theorem 2.4 ([26], [23, Lemma 2.1, p. 154]) *The following (sharp, hidden) trace regularity holds true*

$$\int_0^T \int_\Gamma \left(\frac{\partial \phi}{\partial \nu}\right)^2 d\Sigma = \mathcal{O}_T\left(\|\{\phi_0, \phi_1\}\|^2_{H_0^1(\Omega) \times L^2(\Omega)} + \|f\|^2_{L^1(0,T;L^2(\Omega))}\right). \quad (2.18)$$

Since [23] it has been ascertained that a most convenient roadmap is to first show (by PDE-techniques) Theorem 2.4 and then obtain Theorem 2.3 on $\{w, w_t, w_{tt}\}$ by duality.

Problem (2.12a), (2.12b) can be likewise re-written in an abstract form as

$$\frac{d}{dt}\begin{bmatrix} z \\ z_t \\ u \end{bmatrix} = \mathbb{A}\begin{bmatrix} z \\ z_t \\ u \end{bmatrix}, \quad \mathbb{A} = \begin{bmatrix} 0 & I & 0 \\ -bA + \gamma\frac{c^2}{b}I & -\gamma I & -\gamma\left(\frac{c^2}{b}\right)I \\ I & 0 & -\frac{c^2}{b}I \end{bmatrix}. \quad (2.19)$$

where $\mathbb{A}$ likewise generates a s.c. group $e^{\mathbb{A}t}$ on each of the following spaces [34, Sect. 2]

$$H_0 = \mathcal{D}(A^{\frac{1}{2}}) \times H \times H \tag{2.20a}$$

$$H_1 \equiv \mathcal{D}(A^{\frac{1}{2}}) \times H \times \mathcal{D}(A^{\frac{1}{2}}); \qquad H_2 \equiv \mathcal{D}(A) \times \mathcal{D}(A^{\frac{1}{2}}) \times \mathcal{D}(A) \tag{2.20b}$$

$$H_3 \equiv \mathcal{D}(A^{\frac{1}{2}}) \times H \times \mathcal{D}(A); \qquad H_4 \equiv \mathcal{D}(A) \times \mathcal{D}(A^{\frac{1}{2}}) \times \mathcal{D}(A^{\frac{3}{2}}) \tag{2.20c}$$

The H_i spaces for $\mathbb{A}$ are the perfect counterpart of the spaces U_i for G, $i = 0, \ldots, 4$. One has $e^{Gt} = M e^{\mathbb{A}t} M^{-1}$ where the operator M and its inverse M^{-1} are given explicitly in [34].

The constant $\gamma = \alpha - \frac{c^2}{b}$ plays a critical role in the stability of the s.c. group e^{Gt} on U_i, equivalently of the s.c. group $e^{\mathbb{A}t}$ on H_i. Indeed, e^{Gt} is uniformly stable on each U_i (with a sharp explicit decay rate) if and only if $\gamma > 0$. The case $\gamma = 0$, see (2.9) corresponds to the point spectrum $\sigma_p(\mathbb{A})$ of $\mathbb{A}$ being on the imaginary axis, while the point $-\frac{c^2}{b}$ is in its continuous spectrum [34]. Paper [4] claims that if $\gamma < 0$, and at least in the 1-D case, the boundary homogeneous Eq. (1.1) admits a chotic and topologically mixing semigroup on Banach spaces of Herzog's type. General criteria for hypercyclic and chaotic semigroups were given in [7] and further extended in [3] with applications in [2].

2.3 Proof of Theorem 2.2: Preliminary Analysis

We introduce, as usual, the Dirichlet map

$$Dg = \varphi \iff \{\Delta\varphi = 0 \text{ in } \Omega, \ \varphi|_\Gamma = g\}. \tag{2.21a}$$

$$D: L^2(\Gamma) \to H^{\frac{1}{2}}(\Omega) \subset H^{\frac{1}{2}-2\epsilon}(\Omega) = \mathcal{D}(A^{\frac{1}{4}-\epsilon}), \text{ or } A^{\frac{1}{4}-\epsilon}D \in \mathcal{L}\left(L^2(\Gamma); L^2(\Omega)\right) \tag{2.21b}$$

by elliptic theory [24, 25, 41], with $\epsilon > 0$ arbitrary. One cannot take $\epsilon = 0$, see [31, Remark 3.1.4, p. 186]. At first we shall take $g \in H^1(0, T; L^2(\Gamma))$, so that $g_t \in L^2(0, T; L^2(\Gamma)) = L^2(\Sigma)$. We next return to Eq. (2.1a) and re-write it, as usual [25, 41], [31, Appendix 3B, pp. 420–424], via (2.21a) as

$$y_{ttt} + \alpha y_{tt} - c^2 \Delta(y - Dg) - b\Delta(y - Dg)_t = 0 \quad \text{in } Q \tag{2.22}$$

or abstractly, via (2.2), as

$$y_{ttt} + \alpha y_{tt} + c^2 A(y - Dg) + bA(y - Dg)_t = 0 \quad \text{in } H. \tag{2.23}$$

Extending, as usual [25, 31], the original operator A in (2.2): $L^2(\Omega) \supset \mathcal{D}(A) \to L^2(\Omega)$ to $A_e : L^2(\Omega) \to [\mathcal{D}(A^*)]' = [\mathcal{D}(A)]'$; duality [ ]' w.r.t. $H = L^2(\Omega)$ by isomorphism, and retaining the symbol A also for A_e for such extension, we re-write Eq. (2.23) as

$$(y_t + \alpha y)_{tt} + bA\left(\frac{c^2}{b}y + y_t\right) = c^2 ADg + bADg_t \in [\mathcal{D}(A)]'. \tag{2.24}$$

Setting as in (2.10)

$$z = \frac{c^2}{b}y + y_t = (\alpha y + y_t) - \gamma y, \qquad \gamma = \alpha - \frac{c^2}{b} \tag{2.25}$$

and proceeding as in going from (2.10) to (2.12a), (2.12b), we re-write problem (2.1a), (2.1b) and (2.1c) as the following hyperbolic-dominated system

$$\begin{cases} z_{tt} = -bAz - \gamma z_t + \gamma \frac{c^2}{b}z - \gamma\left(\frac{c^2}{b}\right)^2 y + c^2 ADg + bADg_t \in [\mathcal{D}(A)]' & (2.26a) \\ y_t = -\frac{c^2}{b}y + z & (2.26b) \end{cases}$$

along with the I.C. (we are taking $y_0 = 0$, $y_1 = 0$, $y_2 = 0$)

$$z_0 = \frac{c^2}{b}y_0 + y_1 = 0, \qquad z_1 = \frac{c^2}{b}y_1 + y_2 = 0. \tag{2.26c}$$

3 First Proof of Theorem 2.2 (Interior Regularity): Direct Method

Step 1 The coupling $\gamma\left(\frac{c^2}{b}\right)^2 y = \gamma\left(\frac{c^2}{b}\right)^2 \int_0^t e^{-\frac{c^2}{b}(t-\tau)}z(\tau)d\tau$ between the hyperbolic z-dynamics in (2.26a) and the ODE y-equation in (2.26b) is a mild (lower order) integral term. Thus, essentially w.l.o.g., we may take at first

$$\gamma = 0, \quad \text{i.e. } \alpha = \frac{c^2}{b}, \tag{3.1}$$

see (2.9), to simplify the computations. This will not affect the sought-after regularity of the map in (2.9). The terms z_t, z that by taking $\gamma = 0$ disappear are benign terms for the argument that follows. Thus, we obtain the simplified problem

$$\begin{cases} z_{tt} = -bAz + c^2ADg + bADg_t \in [\mathcal{D}(A)]' & (3.2a) \\ y_t = -\dfrac{c^2}{b} y + z & (3.2b) \end{cases}$$

along with zero I.C., where now under the (essentially w.l.o.g.) assumption (3.1), the z-problem is uncoupled; that is, explicitly, in PDE-form

$$\begin{cases} z_{tt} = b\Delta z & \text{in } Q = (0, T] \times \Omega & (3.3a) \\ z\big|_{t=0} = 0; \quad z_t\big|_{t=0} = 0 & \text{in } \Omega & (3.3b) \\ z\big|_\Sigma = \dfrac{c^2}{b} g + g_t & \text{in } \Sigma = (0, T] \times \Gamma & (3.3c) \\ & & (3.3d) \end{cases}$$

$$\begin{cases} y_t = -\dfrac{c^2}{b} y + z & (3.4a) \\ y\big|_{t=0} = 0 & (3.4b) \end{cases}$$

Orientation Thus, under the (essentially benign) assumption (3.1), the crux of our proof consists in applying to the z-wave equation, either as a mixed problem as in (3.3a)–(3.3c), or else in the abstract form (3.2a), the optimal regularity results (at present for the solution $\{z, z_t\}$ in the interior) from Theorem 2.3 of Remark 2.1 and then use these results for z to obtain corresponding results for y, via (3.4a), (3.4b). In carrying out this strategy, the challenge we face is that we seek to reduce the assumption of regularity of the Dirichlet boundary term $g \in H^1(0, T; L^2(\Gamma))$ to a sort of 'minimal' level, as the term g_t is not present in the original problem (2.1a), (2.1b) and (2.1c), but is sneaked in at the level of the technical step in (2.23). Of course $g \in H^1(0, T; L^2(\Gamma))$ allows one to invoke the results of Theorem 2.3 at once and thus obtain a preliminary (conservative) result: the map

$$g \in H^1(0, T; L^2(\Gamma)) \;\to\; \{z, z_t\} \in C([0, T]; L^2(\Omega) \times H^{-1}(\Omega)) \qquad (3.5a)$$

continuously. From here, it then readily follows by (3.4a), (3.4b); as $y_0 = 0$, $z_0 = 0$:

$$y(t) = \int_0^t e^{-\frac{c^2}{b}(t-\tau)} z(\tau) d\tau \in C([0,T]; L^2(\Omega)) \tag{3.5b}$$

$$y_t(t) = \int_0^t e^{-\frac{c^2}{b}(t-\tau)} z_t(\tau) d\tau \in C([0,T]; H^{-1}(\Omega)) \tag{3.5c}$$

continuously. We note for future reference that for the positive self-adjoint operator A in (2.2) we have

$$\mathcal{D}(A^{\frac{1}{2}}) \equiv H_0^1(\Omega), \quad \text{hence } [\mathcal{D}(A^{\frac{1}{2}})]' \equiv H^{-1}(\Omega) \tag{3.6}$$

(norm-equivalence) [23, Eq. (3.1), p. 17]. Our goal is precisely to refine the regularity results in (3.5a)–(3.5c), i.e. by eventually dropping the H^1-regularity of g in time, via a limit approximation argument.

Step 2 To this end, we shall use critically representation formulas of solutions of second order hyperbolic equations with (presently) Dirichlet non-homogeneous terms, such as the z-problem (3.3a)–(3.3c), by use of cosine/sine operators. Such formulas—(3.8a), (3.9a) below—for **boundary non-homogeneous** second order PDEs were first introduced in [41] in 1977 and used extensively since, e.g. in [25–28], [23, Sect. 3], etc. The author most gratefully acknowledges to have learnt cosine operator theory (originally, for **boundary homogeneous** problems) from Sova [36], Kisynski [17–22], Da Prato–Giusti [6]. This theory for boundary non-homogeneous problems was later collected in Fattorini [10, 11]. For convenience and easy reference, we shall recall critical results as needed in our present development in Appendix A. The (negative self-adjoint) operator $-A : L^2(\Omega) \supset \mathcal{D}(A) \to L^2(\Omega)$ generates a s.c. (self-adjoint) cosine operator $\mathcal{C}(t)$ with corresponding sine operator $\mathcal{S}(t)x = \int_0^t \mathcal{C}(\tau) x d\tau$. As reported in Appendix A after [25, 26, 31, 41], the representation formulae for the solution of the Dirichlet-boundary problem (3.3a)–(3.3c), or its abstract version (3.2a) with henceforth

$$b = 1, \quad c = 1, \quad \text{w.l.o.g.} \tag{3.7}$$

are given by ($\mathcal{S}(0) = 0$) [23, p. 172]

$$z(t) = A \int_0^t \mathcal{S}(t-\tau) D g(\tau) d\tau + A \int_0^t \mathcal{S}(t-\tau) D g_t(\tau) d\tau \tag{3.8a}$$

$$= z^{(1)}(t) + z^{(2)}(t) \tag{3.8b}$$

$$z_t(t) = A \int_0^t \mathcal{C}(t-\tau)Dg(\tau)d\tau + A \int_0^t \mathcal{C}(t-\tau)Dg_t(\tau)d\tau \quad (3.9a)$$

$$= z_t^{(1)}(t) + z_t^{(2)}(t). \quad (3.9b)$$

Step 3 We now invoke the optimal regularity theory of the wave (or, generally, 2nd order hyperbolic equations), this time with Dirichlet boundary term $g \in L^2(0, T; L^2(\Gamma))$ and obtain, continuously (Theorem 2.3):

$$g \in L^2(0, T; L^2(\Gamma)) \implies \begin{cases} z^{(1)}(t) = A \int_0^t \mathcal{S}(t-\tau)Dg(\tau)d\tau \in C([0,T]; L^2(\Omega)) & (3.10a) \\ z_t^{(1)}(t) = A \int_0^t \mathcal{C}(t-\tau)Dg(\tau)d\tau \in C([0,T]; H^{-1}(\Omega)) & (3.10b) \end{cases}$$

Step 4 Next, let with $\epsilon > 0$ arbitrarily small

$$g \in L^2(0, T; L^2(\Gamma)) \cap C([0, \epsilon]; L^2(\Gamma)), \quad g(0) = 0. \quad (3.11)$$

Integrating by parts ($\mathcal{S}(0) = 0$) we obtain from (3.8a), (3.8b)

$$z^{(2)}(t) = A \int_0^t \mathcal{S}(t-\tau)Dg_t(\tau)d\tau = \left[A\mathcal{S}(t-\tau)Dg(\tau)\right]_{\tau=0}^{\tau=t} - A \int_0^t \mathcal{C}(t-\tau)Dg(\tau)d\tau \quad (3.12)$$

$$= \cancel{A\mathcal{S}(0)Dg(t)} - \cancel{A\mathcal{S}(t)Dg(0)} - A \int_0^t \mathcal{C}(t-\tau)Dg(\tau)d\tau$$

or, as in (3.10b):

$$\begin{cases} z^{(2)}(t) = -A \int_0^t \mathcal{C}(t-\tau)Dg(\tau)d\tau = -z_t^{(1)}(t) \in C([0, T]; H^{-1}(\Omega)) \\ \text{for } g \text{ as in (3.11), continuously.} \end{cases} \quad (3.13)$$

Thus, by (3.10a) and (3.13) used in (3.8b), we obtain

$$\begin{cases} z(t) = z^{(1)}(t) + z^{(2)}(t) \in C([0, T]; H^{-1}(\Omega)) \\ \text{for } g \text{ as in (3.11), continuously.} \end{cases} \quad (3.14)$$

Step 5 Next, returning to the y-equation in (2.26b) [with (3.7) w.l.o.g.] and z given by (3.14) we obtain as in (3.5b), (3.5c)

$$y(t) = \int_0^t e^{-(t-\tau)} z(\tau) d\tau = \int_0^t e^{-(t-\tau)} z^{(1)}(\tau) d\tau + \int_0^t e^{-(t-\tau)} z^{(2)}(\tau) d\tau \quad (3.15a)$$

$$= y^{(1)}(t) + y^{(2)}(t). \quad (3.15b)$$

Returning to (3.10a) for $z^{(1)}$, we obtain conservatively

$$\begin{cases} y^{(1)}(t) = \int_0^t e^{-(t-\tau)} z^{(1)}(\tau) d\tau \in C([0, T]; L^2(\Omega)) \\ \text{for } g \in L^2(0, T; L^2(\Gamma)), \text{ continuously.} \end{cases} \quad (3.16)$$

Next, with $z^{(2)} = -z_t^{(1)}(t)$ as given by (3.13), we compute

$$y^{(2)}(t) = \int_0^t e^{-(t-\tau)} z^{(2)}(\tau) d\tau = -\int_0^t e^{-(t-\tau)} z_\tau^{(1)}(\tau) d\tau \quad (3.17)$$

$$= -\left[e^{-(t-\tau)} z^{(1)}(\tau) \right]_{\tau=0}^{\tau=t} + \int_0^t e^{-(t-\tau)} z^{(1)}(\tau) d\tau \quad (3.18)$$

$$\begin{cases} y^{(2)}(t) = -z^{(1)}(t) + \cancel{e^{-t} z^{(1)}(0)} + \int_0^t e^{-(t-\tau)} z^{(1)}(\tau) d\tau \in C([0, T]; L^2(\Omega)) \\ \text{for } g \text{ as in (3.11), continuously.} \end{cases}$$
$$(3.19)$$

as the expression of $z^{(2)} = -z_t^{(1)}(t)$ in (3.13) has such a constraint. As to the regularity noted in (3.19), we invoke (3.10a) for the term $z^{(1)}(t)$, while the same regularity holds true for the second convolution term, this time conservatively.

We conclude by (3.16) on $y^{(1)}$ and (3.19) on $y^{(2)}$ that

$$\begin{cases} y(t) = y^{(1)}(t) + y^{(2)}(t) = -z^{(1)}(t) + 2 \int_0^t e^{-(t-\tau)} z^{(1)}(\tau) d\tau \in C([0, T]; L^2(\Omega)) \\ \text{for } g \text{ as in (3.11), continuously.} \end{cases}$$
$$(3.20)$$

Then (3.20) shows the first result in (2.14a) of Theorem 2.2.

Step 6 Next, with (2.25) = (3.4a), invoke (3.14) for z and (3.20) for y and obtain via (3.6):

$$\begin{cases} y_t = -\dfrac{c^2}{b} y + z \in C\left([0, T]; H^{-1}(\Omega) = [\mathcal{D}(A^{\frac{1}{2}})]'\right) \\ \text{for } g \text{ as in (3.11), continuously.} \end{cases} \quad (3.21)$$

Then (3.21) shows the second result in (2.14b) of Theorem 2.2.
Step 7 We finally need to establish the regularity of y_{tt}. This will be obtained via (3.4a) from

$$y_{tt} = -\frac{c^2}{b} y_t + z_t. \tag{3.22}$$

Thus, we need to establish the regularity of z_t, as that of y_t is given by (3.21). But $z_t = z_t^{(1)} + z_t^{(2)}$ by (3.9b), where the regularity of $z_t^{(1)} = -z^{(2)}(t)$ by (3.13) was already established in (3.10b).

Step 8 We seek the regularity of $z_t^{(2)}$ from its representation formula in (3.9a), (3.9b). We compute from (3.9b), with g as in (3.11)

$$z_t^{(2)}(t) = A \int_0^t \mathcal{C}(t-\tau) D g_t(\tau) d\tau \quad \text{(by parts, recalling (A.2))} \tag{3.23}$$

$$= \left[A\mathcal{C}(t-\tau) Dg(\tau) \right]_{\tau=0}^{\tau=t} + A \int_0^t A\mathcal{S}(t-\tau) Dg(\tau) d\tau \tag{3.24}$$

$$= ADg(t) - A\mathcal{C}(t)Dg(0) + AA \int_0^t \mathcal{S}(t-\tau) Dg(\tau) d\tau \tag{3.25}$$

$$= ADg(t) + Az^{(1)}(t) \tag{3.26}$$

recalling $\mathcal{C}(0) = I$ and (3.10a). Thus by (3.10a)

$$g \in L^2(0, T; L^2(\Gamma)) \Longrightarrow Az^{(1)}(t) = AA \int_0^t \mathcal{S}(t-\tau) Dg(\tau) d\tau \in C([0, T]; [\mathcal{D}(A)]') \tag{3.27}$$

continuously. Moreover, by (2.21b) with any $\epsilon_1 > 0$

$$g \in L^2(0, T; L^2(\Gamma)) \Longrightarrow ADg = A^{\frac{3}{4}+\epsilon_1}(A^{\frac{1}{4}-\epsilon_1}D)g \in L^2(0, T; [\mathcal{D}(A^{\frac{3}{4}+\epsilon_1})]')$$
$$\subset L^2(0, T; [\mathcal{D}(A)]') \tag{3.28a}$$

$$g \in C([0, T]; L^2(\Gamma)) \Longrightarrow ADg = A^{\frac{3}{4}+\epsilon_1}(A^{\frac{1}{4}-\epsilon_1}D)g \in C([0, T]; [\mathcal{D}(A^{\frac{3}{4}+\epsilon_1})]')$$
$$\subset C([0, T]; [\mathcal{D}(A)]') \tag{3.28b}$$

continuously. By (3.26), (3.28a) and (3.27)

$$\begin{cases} g \in L^2(0, T; L^2(\Gamma)) \cap C([0, \epsilon); L^2(\Gamma)) \\ g(0) = 0, \end{cases} \Longrightarrow z_t^{(2)}(t) = ADg + Az^{(1)} \in L^2(0, T; [\mathcal{D}(A)]')$$
$$\tag{3.29a}$$

continuously, as well as by (3.26), (3.28b) and (3.27)

$$\begin{cases} g \in C([0, T]; L^2(\Gamma)) \\ g(0) = 0, \end{cases} \implies z_t^{(2)}(t) = ADg + Az^{(1)} \in C([0, T]; [\mathcal{D}(A)]') \tag{3.29b}$$

continuously.

Step 9 In turn, combining (3.10b) for $z_t^{(1)}$ with (3.29a) for $z_t^{(2)}$ in (3.9b), we conclude

$$\begin{cases} g \in L^2(0, T; L^2(\Gamma)) \cap C([0, \epsilon); L^2(\Gamma)) \\ g(0) = 0, \end{cases} \implies z_t = z_t^{(1)} + z_t^{(2)} \in L^2(0, T; [\mathcal{D}(A)]') \tag{3.30a}$$

continuously, as well as by (3.29b) on $z_t^{(2)}$ this time

$$\begin{cases} g \in C([0, T]; L^2(\Gamma)) \\ g(0) = 0, \end{cases} \implies z_t = z_t^{(1)} + z_t^{(2)} \in C([0, T]; [\mathcal{D}(A)]') \tag{3.30b}$$

continuously. Finally, from (3.22), recalling the regularity of y_t in (2.14b) and the regularity of z_t in (3.30a), we obtain

$$\begin{cases} g \in L^2(0, T; L^2(\Gamma)) \cap C([0, \epsilon); L^2(\Gamma)) \\ g(0) = 0, \end{cases} \implies y_{tt} = -\frac{c^2}{b} y_t + z_t \in L^2(0, T; [\mathcal{D}(A)]') \tag{3.31a}$$

continuously, as well as by (2.14b) and (3.30b)

$$\begin{cases} g \in C([0, T]; L^2(\Gamma)) \\ g(0) = 0, \end{cases} \implies y_{tt} = -\frac{c^2}{b} y_t + z_t \in C([0, T]; [\mathcal{D}(A)]') \tag{3.31b}$$

continuously. Then (3.31b) shows the result in (2.14c) and (2.14e) of Theorem 2.2.

4 Proof of Theorem 2.2: Regularity of the Boundary Trace $\left.\dfrac{\partial y}{\partial \nu}\right|_\Sigma$

In this section we shall establish the boundary regularity (2.14d) of Theorem 2.2. It is here repeated for convenience.

Theorem 4.1 *With reference to problem (2.1a)–(2.1c), we have*

$$\begin{cases} g \in L^2(0, T; L^2(\Gamma)) \cap C([0, \epsilon); L^2(\Gamma)) \\ g(0) = 0 \end{cases} \implies \left.\frac{\partial y}{\partial \nu}\right|_\Sigma \in H^{-1}(\Sigma) \qquad (4.1)$$

continuously. Here $H^{-1}(\Sigma) =$ dual of $\{h \in H_0^1(\Sigma)\}$ i.e. with $h(\cdot, 0) = 0$ and $h(\cdot, T) = 0$ on Γ (but actually, $h(\cdot, T) = 0$ is not needed.)

Proof **Step 1** We return to the solution representation formula (3.8b) complemented by (3.13), with g as in (3.11), in particular $g(0) = 0$:

$$z(t) = z^{(1)}(t) + z^{(2)}(t) = z^{(1)}(t) - z_t^{(1)}(t) \qquad (4.2)$$

$$= A \int_0^t \mathcal{S}(t-\tau) Dg(\tau)d\tau - A \int_0^t \mathcal{C}(t-\tau) Dg_t(\tau)d\tau. \qquad (4.3)$$

recalling (3.10a), (3.10b). We next invoke critically the boundary regularity results in [23, Eq. (2.14) of Theorem 2.3, p. 153], recalled in Theorem 2.3, Eq. (2.16) of Remark 2.1 and also (A.5):

$$g \in L^2(0, T; L^2(\Gamma)) \implies \left.\frac{\partial z^{(1)}}{\partial \nu}\right|_\Sigma = -D^* A A \int_0^t \mathcal{S}(t-\tau) Dg(\tau) d\tau \in H^{-1}(\Sigma). \qquad (4.4)$$

Step 2 We now return to (3.20), with g as in (3.11) (and $b = 1, c = 1$, w.l.o.g. as in (3.7))

$$y(t) = -z^{(1)}(t) + 2\int_0^t e^{-(t-\tau)} z^{(1)}(\tau) d\tau \qquad (4.5)$$

where then, recalling (4.4), we obtain

$$\left.\frac{\partial y}{\partial \nu}\right|_\Sigma = -\left.\frac{\partial z^{(1)}}{\partial \nu}\right|_\Sigma + 2\int_0^t e^{-(t-\tau)} \frac{\partial z^{(1)}}{\partial \nu}(\tau) d\tau \in H^{-1}(\Sigma) \qquad (4.6)$$

for g as in (3.11), continuously. Then (4.6) establishes Theorem 4.1. □

5 A Boundary Trace Result for the u-Problem (2.3)

The optimal interior (and boundary) regularity results for second order hyperbolic equations with Dirichlet boundary terms were proven directly in [25, 26] and by duality in [23] via Theorem 2.4. This latter paper then set up the road map to obtain optimal interior (and boundary) regularity results for a variety of other dynamics such as: Schrödinger equations with Dirichlet boundary term, plate-like equations with certain boundary terms, all by duality. Dual results (boundary traces) are of interest in themselves. In this paper, we shall also show an independent dual result, which will be used in Sect. 6 in reproving (essentially) the main Theorem 2.2. The dual result will be given in Sect. 6. In this section, we provide an independent boundary trace result for a problem which is not the dual problem of the original non-homogeneous problem (2.1a)–(2.1c), but which is very closely related to the actual dual problem. This will be presented in Sect. 6. The independent boundary trace results of the present and next sections for the SMGTJ mixed problem will be critically based on a duality result of second order hyperbolic equation, namely Theorem 2.4. In the present section, the object of our interest is problem (2.3), re-written in PDE-form as

$$\begin{cases} u_{ttt} + \alpha u_{tt} - c^2 \Delta u - b \Delta u_t = 0 & \text{in } Q = (0, T] \times \Omega & (5.1a) \\ u\big|_{t=T} = u_0; \ u_t\big|_{t=T} = u_1; \ u_{tt}\big|_{t=T} = u_2 & \text{in } \Omega & (5.1b) \\ u\big|_\Sigma = 0 & \text{in } \Sigma = (0, T] \times \Gamma & (5.1c) \end{cases}$$

$$\text{abstractly} \qquad u_{ttt} + \alpha u_{tt} + c^2 A u + b A u_t = 0 \qquad (5.1d)$$

with I.C. at $t = T$ (i.e. backward in time). We shall select the I.C. in the space U_3 defined in (2.5c)

$$\{u_0, u_1, u_2\} \in U_3 \equiv \mathcal{D}(A) \times \mathcal{D}(A^{\frac{1}{2}}) \times H. \qquad (5.2)$$

Accordingly, in view of Theorem 2.1, we have

$$\begin{bmatrix} u_0 \\ u_1 \\ u_2 \end{bmatrix} \in U_3 \rightarrow \begin{bmatrix} u(t) \\ u_t(t) \\ u_{tt}(t) \end{bmatrix} = e^{G(T-t)} \begin{bmatrix} u_0 \\ u_1 \\ u_2 \end{bmatrix} \in C([0, T]; U_3) \qquad (5.3)$$

continuously, where e^{Gt} is a s.c. group on U_3 with infinitesimal generator G defined in (2.4). Explicitly,

$$[u_0, u_1, u_2] \in U_3 = \mathcal{D}(A) \times \mathcal{D}(A^{\frac{1}{2}}) \times L^2(\Omega) \Longrightarrow$$

$$[u, u_t, u_{tt}] \in C([0, T]; \mathcal{D}(A) \times \mathcal{D}(A^{\frac{1}{2}}) \times L^2(\Omega))$$
$$= C([0, T]; [H^2(\Omega) \cap H_0^1(\Omega)] \times H_0^1(\Omega) \times L^2(\Omega)), \qquad (5.4)$$

continuously, to be invoked repeatedly below. The key boundary trace result of the present section is the following Theorem.

Theorem 5.1 *With reference to problem (5.1a)–(5.1b), (5.2), the following trace estimate holds true*

$$\int_0^T \int_\Gamma \left(\frac{\partial u_t}{\partial \nu}\right)^2 d\Sigma = \mathcal{O}_T\left(\|\{u_0, u_1, u_2\}\|_{U_3}^2\right) \tag{5.5}$$

where $\mathcal{O}$ denotes a constant depending on T, $T > 0$ arbitrary, as well as the equation coefficients.

Remark 5.1 Estimate (5.5) is an illustration of a sharp, so called hidden regularity result, in line with the phenomenon first discovered in [23, 26] for second order hyperbolic equations, and reported (for the wave equation) in Theorem 2.4. The term $u_t \in C([0, T]; H_0^1(\Omega))$ optimally in the interior, and yet $\frac{\partial u_t}{\partial \nu} \in L^2(0, T; L^2(\Gamma))$, i.e. it possesses '$\frac{1}{2}$' space regularity better than a formal application of trace theory by reducing the time regularity from C to L^2.

Proof of Theorem 5.1. Step 1 Rewrite Eq. (5.1a) as in (2.7)

$$(u_t + \alpha u)_{tt} - b\Delta\left(\frac{c^2}{b}u + u_t\right) = 0 \quad \text{in } Q \tag{5.6}$$

and introduce now the variable ξ in (2.8) recalling (5.4)

$$\xi = \alpha u + u_t \in C([0, T]; H_0^1(\Omega)), \text{ so that } \frac{c^2}{b}u + u_t = \xi - \gamma u, \quad \gamma = \alpha - \frac{c^2}{b}. \tag{5.7}$$

Rewrite problem (5.1a)–(5.1c) accordingly as

$$\begin{cases} \xi_{tt} - b\Delta\xi + b\gamma\Delta u = 0 & \text{in } Q & (5.8a) \\ \xi\big|_{t=T} = \xi_0 = \alpha u_0 + u_1; \quad \xi_t\big|_{t=T} = \xi_1 = \alpha u_1 + u_2 & \text{in } \Omega & (5.8b) \\ \xi\big|_{\Sigma} \equiv 0 & \text{in } \Sigma & (5.8c) \end{cases}$$

$$\text{abstractly} \quad \xi_{tt} + bA\xi - b\gamma Au = 0 \tag{5.8d}$$

with I.C. at $t = T$. For problem (5.8a)–(5.8c), the key sharp/hidden regularity result is

Theorem 5.2 With reference to problem (5.8a)–(5.8c), we have

$$\int_0^T \int_\Gamma \left(\frac{\partial \xi}{\partial \nu}\right)^2 d\Sigma = \mathcal{O}_T\left(\|\{u_0, u_1, u_2\}\|_{U_3}^2\right), \quad U_3 = \mathcal{D}(A) \times \mathcal{D}(A^{\frac{1}{2}}) \times L^2(\Omega) \tag{5.9}$$

where $\mathcal{O}_T$ denotes a constant depending on T, $T > 0$ arbitrary.

Remark 5.2 Estimate (5.9) is again an illustration of a sharp, 'hidden' regularity result, as $\xi \in C([0, T]; H_0^1(\Omega))$ optimally in the interior, by (5.7), and thus (5.9) does not follow by a formal application of trace theory.

Proof of Theorem 5.2 The proof is critically based on the approach and results of [23] reproduced also in [31, Sect. 10.5.10, p. 958]. Let, as usual, $h(x) = [h_1(x), \ldots, h_d(x)] \in C^2(\overline{\Omega})$ be a vector field, such that $h|_\Gamma = \nu =$ outward unit normal vector. Let $H(x) = \left[\dfrac{\partial h_i}{\partial x_j}\right], i, j = 1, \ldots, d$ be the usual Jacobian matrix. The multiplier method, with multiplier $h \cdot \nabla \xi$, applied to Eq. (5.8a), say with $b = 1$ w.l.o.g., gives the usual identity [31, Eq. (10.5.10.5), p. 959]

$$\int_\Sigma \frac{\partial \xi}{\partial \nu} h \cdot \nabla \xi d\Sigma + \frac{1}{2} \int_\Sigma \xi_t^2 h \cdot \nu d\Sigma - \frac{1}{2} \int_\Sigma |\nabla \xi|^2 h \cdot \nu d\Sigma$$
$$= \int_Q H \nabla \xi \cdot \nabla \xi dQ + \frac{1}{2} \int_Q \xi_t^2 \operatorname{div} h \, dQ - \frac{1}{2} \int_Q |\nabla \xi|^2 \operatorname{div} h \, dQ$$
$$+ b\gamma \int_Q \Delta u h \cdot \nabla \xi dQ + \left[(\xi_t, h \cdot \nabla \xi)_\Omega\right]_0^T, \quad (5.10)$$

where from (5.4), (5.7), (5.8c)

$$\begin{cases} \xi_t \equiv 0 \text{ on } \Sigma; \quad h \cdot \nabla \xi = \dfrac{\partial \xi}{\partial \nu} h \cdot \nu = \dfrac{\partial \xi}{\partial \nu} \text{ on } \Sigma; \quad |\nabla \xi|^2 = \left(\dfrac{\partial \xi}{\partial \nu}\right)^2 \text{ on } \Sigma \\ \xi = \alpha u + u_t \in C([0, T]; H_0^1(\Omega)); \quad |\nabla \xi| \in C([0, T]; L^2(\Omega)) \\ \xi_t \in C([0, T]; L^2(\Omega)); \quad \Delta u \in C([0, T]; L^2(\Omega)) \end{cases}$$
(5.11)

continuously on $\{u_0, u_1, u_2\} \in U_3$. Thus, as usual via (5.11), identity (5.10) reduces to

$$\int_0^T \int_\Gamma \left(\frac{\partial \xi}{\partial \nu}\right)^2 d\Sigma = \mathcal{O}_T\left(\|\{u_0, u_1, u_2\}\|_{U_3}^2\right) \quad (5.12)$$

and Theorem 5.2 is established. In effect, we could get estimate (5.12) at once by applying Theorem 2.4 directly, with Δu as given by (5.11). The above computations based on (5.10) give a flavor of the proof of Theorem 2.4 given in [23, Lemma 2.1, p. 154].

Step 2 From (5.7), we estimate

$$\begin{cases} \left|\dfrac{\partial u_t}{\partial \nu}\right| - |\alpha| \left|\dfrac{\partial u}{\partial \nu}\right| \leq \left|\dfrac{\partial u_t}{\partial \nu} + \alpha \dfrac{\partial u}{\partial \nu}\right| = \left|\dfrac{\partial \xi}{\partial \nu}\right| & (5.13a) \\ \dfrac{1}{2}\left|\dfrac{\partial u_t}{\partial \nu}\right|^2 - \alpha^2 \left|\dfrac{\partial u}{\partial \nu}\right|^2 \leq \left|\dfrac{\partial \xi}{\partial \nu}\right|^2 & (5.13b) \end{cases}$$

which used in (5.12) yields

$$\int_0^T \int_\Gamma \left(\frac{\partial u_t}{\partial \nu}\right)^2 d\Sigma \leq \mathcal{O}\left(\|\{u_0, u_1, u_2\}\|_{U_3}^2\right) + 2\alpha^2 \int_0^T \int_\Gamma \left(\frac{\partial u}{\partial \nu}\right)^2 d\Sigma. \quad (5.14)$$

Step 3 Via $u \in C([0, T]; H^2(\Omega))$ from (5.4), we obtain directly via trace theory

$$\int_0^T \int_\Gamma \left(\frac{\partial u}{\partial \nu}\right)^2 d\Sigma \leq C_T \left(\|\{u_0, u_1, u_2\}\|_{U_3}^2\right) \quad (5.15)$$

which inserted in (5.14) yields

$$\int_0^T \int_\Gamma \left(\frac{\partial u_t}{\partial \nu}\right)^2 d\Sigma = \mathcal{O}_T \left(\|\{u_0, u_1, u_2\}\|_{U_3}^2\right) \quad (5.16)$$

as desired, and Theorem 5.1 is proved.

6 A Dual Result of the Non-homogeneous Dirichlet Problem (2.1a)–(2.1c)

It turns out that the dual problem of the boundary non-homogeneous Dirichlet problem (2.1a)–(2.1c) is actually the problem

$$\begin{cases} v_{ttt} - \alpha v_{tt} + c^2 \Delta v - b \Delta v_t = 0 & \text{in } Q & (6.1a) \\ v|_{t=T} = v_0; \quad v_t|_{t=T} = v_1; \quad v_{tt}|_{t=T} = v_2 & \text{in } \Omega & (6.1b) \\ v|_\Sigma = 0 & \text{in } \Sigma & (6.1c) \end{cases}$$

abstractly $\quad v_{ttt} - \alpha v_{tt} - c^2 A v + b A v_t = 0 \quad (6.2)$

along with I.C. at $t = T$. This is very closely related to problem (5.1a)–(5.1c), or its abstract version (5.1d), as the present considerations attest. Write (6.1a) as in (2.7) = (5.6).

$$(v_t - \alpha v)_{tt} - b\Delta \left(-\frac{c^2}{b} v + v_t\right) = 0 \text{ in } Q. \quad (6.3)$$

Set, similarly to (2.8)

$$\eta = v_t - \alpha v \quad \text{so that } \zeta = -\frac{c^2}{b} v + v_t = \eta + \gamma v, \quad \gamma = \alpha - \frac{c^2}{b} \quad (6.4)$$

and then problem (6.1a)–(6.1c) reduces to

$$\begin{cases} \eta_{tt} - b\Delta\eta - b\gamma\Delta v = 0 & \text{in } Q & (6.5a) \\ \eta|_{t=T} = \eta_0 = v_1 - \alpha v_0; \quad \eta_t|_{t=T} = \eta_1 = v_2 - \alpha v_1 & \text{in } \Omega & (6.5b) \\ \eta|_\Sigma = 0 & \text{in } \Sigma & (6.5c) \end{cases}$$

$$\text{abstractly} \qquad \eta_{tt} + bA\eta + b\gamma Av = 0 \qquad (6.5d)$$

with I.C. at $t = T$, to be compared with the ξ-problem (5.8a)–(5.8c) and its abstract version (5.8d). Moreover, in terms of the variable ζ in (6.4), we have, with $\gamma = \alpha - \frac{c^2}{b}$:

$$\zeta = -\frac{c^2}{b}v + v_t = (-\alpha v + v_t) + \gamma v = \eta + \gamma v \qquad (6.6)$$

$$(v_t - \alpha v)_{tt} = \zeta_{tt} - \gamma v_{tt} = \zeta_{tt} - \gamma \left(\zeta - \frac{c^2}{b}v\right)_t = \zeta_{tt} - \gamma \zeta_t - \gamma \frac{c^2}{b}v_t \qquad (6.7)$$

$$= \zeta_{tt} - \gamma \zeta_t - \gamma \frac{c^2}{b}\zeta - \gamma \left(\frac{c^2}{b}\right)^2 v. \qquad (6.8)$$

Hence (6.3) $(v_t - \alpha v)_{tt} + bA\left(-\frac{c^2}{b}v + v_t\right) = 0$ is rewritten by (6.6)–(6.8) as

$$\begin{cases} \zeta_{tt} = -bA\zeta + \gamma \zeta_t + \gamma \frac{c^2}{b}\zeta + \gamma \left(\frac{c^2}{b}\right)^2 v & (6.9a) \\ v_t = \frac{c^2}{b}v + \zeta, & (6.9b) \end{cases}$$

to be compared with the z-problem (2.12a), (2.12b). The ζ-problem (6.9a), (6.9b) and the z-problem in (2.12a), (2.12b) differ only by innocuous changes of signs. Thus, the well-posedness theory developed in [34] and recalled in Theorem 2.1 for the z-problem in (2.12a), (2.12b) applies also to the ζ-problem (6.9a), (6.9b) over a finite time interval. In particular, the implication on the original v-problem (6.1a)–(6.1c)—counterpart of the u-problem in Theorem 2.1 is the following Theorem.

Theorem 6.1 *With reference to problem (6.1a)–(6.1c) we have*

$$\{v_0, v_1, v_2\} \in U_3 \equiv \mathcal{D}(A) \times \mathcal{D}(A^{\frac{1}{2}}) \times H \implies \{v, v_t, v_{tt}\} \in C([0, T]; U_3) \quad (6.10)$$

as a s.c. group on U_3.

Returning to the v-problem (6.1a)–(6.1c), we obtain the following Theorem.

Theorem 6.2 *With reference to the v-problem (6.1a)–(6.1c), abstractly (6.2), we have*

$$\int_0^T \int_\Gamma \left(\frac{\partial v_t}{\partial \nu}\right)^2 d\Sigma = \mathcal{O}_T\left(\|\{v_0, v_1, v_2\}\|_{U_3}^2\right), \quad U_3 = \mathcal{D}(A) \times \mathcal{D}(A^{\frac{1}{2}}) \times H.$$
(6.11)

The proof is exactly the same as that of Theorem 5.1. The first step is showing

Theorem 6.3 *With reference to the η-problem (6.5a)–(6.5d), we have*

$$\int_0^T \int_\Gamma \left(\frac{\partial \eta}{\partial \nu}\right)^2 d\Sigma = \mathcal{O}\left(\|\{v_0, v_1, v_2\}\|_{U_3}^2\right).$$
(6.12)

The proof of Theorem 6.3 is exactly the same as that of Theorem 5.2 (the difference in sign $+b\gamma$ in (5.8a) versus $-b\gamma$ in (6.5a) is irrelevant, under the common property $\Delta u, \Delta v \in C([0,T]; H)$. Also, apply at once Theorem 2.4.

Next, by duality on the trace result in Theorem 6.2 for the v-problem (6.1a)–(6.1c), we shall re-obtain (in a slightly weaker form) the basic interior regularity result of Theorem 2.2 for $\{y, y_t, y_{tt}\}$. While the proof of Theorem 2.2 was 'direct', the proof of Theorem 6.4 is 'by duality'.

Theorem 6.4 *With reference to the Dirichlet problem (2.1a)–(2.1c), we have*

$$\begin{cases} g \in C([0,T]; L^2(\Gamma)) \\ g(0) = 0, \end{cases} \implies \{y, y_t, y_{tt}\} \in C([0,T]; H \times [\mathcal{D}(A^{\frac{1}{2}})]' \times [\mathcal{D}(A)]')$$
(6.13)

continuously.

Proof **Step 1** We shall first establish the following

Proposition 6.5 *With reference to the Dirichlet problem (2.1a)–(2.1c), we have, for each $0 < t \leq T$:*

$$\begin{cases} g \in C([0,T]; L^2(\Gamma)) \\ g(0) = 0, \end{cases} \implies \{y(t), y_t(t), y_{tt}(t)\} \in H \times [\mathcal{D}(A^{\frac{1}{2}})]' \times [\mathcal{D}(A)]' \quad (6.14a)$$

$$\implies \int_0^t [y(\tau), y_t(\tau), y_{tt}(\tau)] d\tau \in H \times [\mathcal{D}(A^{\frac{1}{2}})]' \times [\mathcal{D}(A)]' \quad (6.14b)$$

continuously.

Proof of Proposition 6.5. **Step (i)** By Theorem 6.1 and 6.2, we have

$$\{v_0, v_1, v_2\} \in U_3 = \mathcal{D}(A) \times \mathcal{D}(A^{\frac{1}{2}}) \times H \tag{6.15a}$$

$$\implies \begin{cases} \{v, v_t, v_{tt}\} \in C([0, T]; U_3 = \mathcal{D}(A) \times \mathcal{D}(A^{\frac{1}{2}}) \times H) \\ \text{AND} \\ \dfrac{\partial v_t}{\partial \nu} \in L^2(0, T; L^2(\Gamma)) \end{cases} \tag{6.15b}$$

continuously.

Step (ii) We now invoke the duality identity (B.4) in Appendix B, which we rewrite here for convenience for a generic t, $0 < t \leq T$:

$$\langle y_{tt}(t) + \alpha y_t(t), v_0 \rangle - \langle y_t(t) + \alpha y(t), v_1 \rangle + \langle y(t), v_2 \rangle$$
$$- b \langle y(t), \Delta v_0 \rangle \tag{6.16}$$
$$= - \left\langle c^2 g + b g_t, \dfrac{\partial v}{\partial \nu} \right\rangle_{L^2(0, t; L^2(\Gamma))}.$$

With $g \in L^2(0, t; L^2(\Gamma))$ and under (6.15a), (6.15b) and hence $\dfrac{\partial v}{\partial \nu} \in C([0, t]; H^{\frac{1}{2}}(\Gamma))$, we have regarding the first term on the RHS of (6.16):

$$\left\langle g, \dfrac{\partial v}{\partial \nu} \right\rangle_{L^2(0, t; L^2(\Gamma))} = \int_0^t \int_\Gamma g \dfrac{\partial v}{\partial \nu} d\Gamma \, dt < \infty. \tag{6.17}$$

Step (iii) Under hypothesis (6.14a) for g and under (6.15a), (6.15b) for the adjoint v-problem, we compute the last term on the RHS of (6.16) by parts as follows, for $0 < t \leq T$

$$\int_0^t \left(g_t(\tau), \dfrac{\partial v(\tau)}{\partial \nu} \right)_{L^2(\Gamma)} d\tau = \left[\left(g(\tau), \dfrac{\partial v(\tau)}{\partial \nu} \right)_{L^2(\Gamma)} \right]_{\tau=0}^{\tau=t}$$
$$- \int_0^t \left(g(\tau), \dfrac{\partial v_t(\tau)}{\partial \nu} \right)_{L^2(\Gamma)} d\tau \tag{6.18}$$
$$= \left(g(t), \dfrac{\partial v(t)}{\partial \nu} \right)_{L^2(\Gamma)} - \left(g(0), \dfrac{\partial v(0)}{\partial \nu} \right)_{L^2(\Gamma)}$$
$$- \int_0^t \left(g(\tau), \dfrac{\partial v_t(\tau)}{\partial \nu} \right)_{L^2(\Gamma)} d\tau \tag{6.19}$$

With $g \in C([0, T]; L^2(\Gamma))$ as in (6.14a) and $\dfrac{\partial v}{\partial \nu} \in C([0, t]; H^{\frac{1}{2}}(\Gamma))$, we have that the first term on the RHS of (6.19) is well-defined (conservatively). Notice that it can also be re-written via (A.5) as

$$\left(g(t), \frac{\partial v(t)}{\partial \nu}\right)_{L^2(\Gamma)} = \left(g(t), -D^* A v(t)\right)_{L^2(\Gamma)} \tag{6.20}$$
$$= -(ADg(t), v(t))_{L^2(\Gamma)} \text{ (well-defined)}$$

with $v \in C([0, T]; \mathcal{D}(A))$ by (6.10) and $ADg \in C([0, T]; [\mathcal{D}(A)]')$ conservatively by (2.21b) and (6.14a). Then by (6.19), (6.20)

$$\left\langle g_t, \frac{\partial v}{\partial \nu}\right\rangle_{L^2(0,t;L^2(\Gamma))} = -(ADg(t), v(t))_{L^2(\Omega)} - \left\langle g, \frac{\partial v_t}{\partial \nu}\right\rangle_{L^2(0,t;L^2(\Gamma))} \tag{6.21}$$
(well-defined, by (6.20) and (6.15b)),

continuously with respect to g as in (6.14a) and $\{v_0, v_1, v_2\}$ as in (6.15a), (6.15b). Thus, the RHS of identity (6.17) is well-defined for g as in (6.14a) and $\{v_0, v_1, v_2\} \in U_3$ as in (6.15a), (6.15b), critically because of Theorem 6.2. Next we turn to the LHS of identity (6.16) and then the argument displayed in (B.15) of Appendix B, yields (6.14a), for any $0 < t \leq T$.

Specifically, for each t, $0 < t \leq T$ via duality pairing:

$$\left.\begin{array}{l} v_2 \in H \to y(t) \in H \\ v_1 \in \mathcal{D}(A^{\frac{1}{2}}) \to y_t(t) + \alpha y(t) \in [\mathcal{D}(A^{\frac{1}{2}})]' \end{array}\right\} \implies y_t(t) \in \left[\mathcal{D}(A^{\frac{1}{2}})\right]' \quad \begin{array}{l}(6.22)\\(6.23)\end{array}$$

$$v_0 \in \mathcal{D}(A) \to y_{tt}(t) + \alpha y_t(t) \in [\mathcal{D}(A)]' \implies y_{tt}(t) \in [\mathcal{D}(A)]' \tag{6.24}$$

Thus, (6.23)–(6.24) imply $\{y(t), y_t(t), y_{tt}(t)\} \in H \times [\mathcal{D}(A^{\frac{1}{2}})]' \times [\mathcal{D}(A)]'$. The same argument then gives

$$\int_0^t Y(\tau) d\tau \in H \times [\mathcal{D}(A^{\frac{1}{2}})]' \times [\mathcal{D}(A)]', \text{ for any } 0 < t \leq T, \tag{6.25}$$

where we have set $Y(t) = [y(t), y_t(t), y_{tt}(t)]$. Proposition 6.5 is proved.

Step 2 In light of (6.14b)=(6.25), we are in the same situation as in [23, Corollary 3.2, p. 173], whereby then the map

$$t \to \int_0^t [y(\tau), y_t(\tau), y_{tt}(\tau)] d\tau \tag{6.26}$$

is continuous $[0, T] \to H \times [\mathcal{D}(A^{\frac{1}{2}})]' \times [\mathcal{D}(A)]'$. Theorem 6.4 is proved. □

Part B: Neumann Case

7 Linear Third Order SMGTJ-Equation with Non-homogeneous Neumann Boundary Term

Likewise we shall take $\tau = 1$ in (1.1) w.l.o.g. If the linear third order equation (1.1) is written in terms of scalar velocity potential (where pressure $= k\dfrac{d}{dt}$ (velocity potential)), then the Neumann non-homogeneous boundary terms are appropriate [14]. In this part B, we shall consider the following mixed problem in the unknown $y(t, x)$:

$$\begin{cases} y_{ttt} + \alpha y_{tt} - c^2 \Delta y - b \Delta y_t = 0 & \text{in } Q = (0, T] \times \Omega & (7.1a) \\ y|_{t=0} = y_0; \; y_t|_{t=0} = y_1; \; y_{tt}|_{t=0} = y_2 & \text{in } \Omega & (7.1b) \\ \left.\dfrac{\partial y}{\partial \nu}\right|_\Sigma = g & \text{in } \Sigma = (0, T] \times \Gamma & (7.1c) \end{cases}$$

7.1 Case $g = 0$.

This case is the perfect counterpart of Sect. 2.1 in the Dirichlet case, under the present setting whereby $H = L^2(\Omega)/\mathbb{R}$ and $-A$ is now the Neumann Laplacian

$$Af = -\Delta f, \quad \mathcal{D}(A) = \left\{ h \in H^2(\Omega) : \dfrac{\partial h}{\partial \nu} = 0 \text{ on } \Gamma \right\}. \tag{7.2}$$

A is likewise strictly positive self-adjoint on H so that the fractional powers $A^\theta, 0 < \theta < 1$, are well defined on H. Thus, Equation (2.3) through (2.12a), (2.12b) still hold true now including Theorem 2.1, with $-A$ the Neumann Laplacian in (7.2), rather than the Dirichlet Laplacian as in Sect. 2.1.

7.2 Case $y_0 = 0, y_1 = 0, y_2 = 0, g \neq 0$

In this case, we seek to obtain sharp regularity of the map

$$g \to \{y, y_t, y_{tt}; y|_\Sigma\}$$

from the Neumann boundary datum g to the interior solution $\{y, y_t, y_{tt}\}$ and the Dirichlet boundary trace $y|_\Sigma$. As in Part A, our analysis of the above question will critically fall on the regularity of the wave equation (or a more generally of second order hyperbolic equations) under Neumann boundary control. Unlike the Dirichlet boundary control case invoked critically in Part A, the Neumann boundary control case has two peculiarities: (i) it is dimension dependent (the case $d = 1$ is markedly more regular [31, Sect. 9.8.4, Theorem 9.8.4.1, p. 859]; (ii) it is geometry dependent [29, 30, 38] (reported in [31, p. 739]). Accordingly, set throughout Part B (Neumann)

$$\widehat{\alpha} = \beta = \frac{2}{3} \quad \text{for a general sufficiently smooth domain} \Omega \text{ in } \mathbb{R}^d, d \geq 2 \quad (7.3a)$$

$$\widehat{\alpha} = \beta = \frac{3}{4} \quad \text{for a parallelepiped in } \mathbb{R}^d, d \geq 2 \quad (7.3b)$$

In reference [30] the parameter $\widehat{\alpha}$ refers to interior regularity while β refer to boundary regularity. As we shall invoke a number of results from [30], clarity requires that we keep both of them.

Theorem 7.1 *With reference to problem (7.1a)–(7.1c), we have the following regularity results*

$$g \in L^2(0, T; L^2(\Gamma)) \cap C([0, \epsilon); L^2(\Gamma)), \quad g(0) = 0, \ \epsilon > 0 \text{ small}$$

$$\implies \begin{cases} y \in C([0, T]; H^{\widehat{\alpha}}(\Omega) = \mathcal{D}(A^{\frac{\widehat{\alpha}}{2}})) & (7.4a) \\ y_t \in C([0, T]; H^{\widehat{\alpha}-1}(\Omega) = [\mathcal{D}(A^{\frac{1-\widehat{\alpha}}{2}})]'), & (7.4b) \\ y_{tt} \in L^2(0, T; H^{\widehat{\alpha}-2}(\Omega) = [\mathcal{D}(A^{1-\frac{\widehat{\alpha}}{2}})]') & (7.4c) \\ y|_\Sigma \in H^{2\widehat{\alpha}-1}(\Sigma)). & (7.4d) \end{cases}$$

$$\begin{cases} g \in C([0, T]; L^2(\Gamma)) \\ g(0) = 0 \end{cases} \implies y_{tt} \in C([0, T]; H^{\widehat{\alpha}-2}(\Omega) \equiv [\mathcal{D}(A^{1-\frac{\widehat{\alpha}}{2}})]'). \quad (7.4e)$$

Remark 7.1 The result of Theorem 7.1 should be compared with the following results for general second order hyperbolic equations, even with space-variable coefficients, which we however report only for the canonical wave equation. Thus, let Ω be an open bounded domain in $\mathbb{R}^d, d \geq 2$, with smooth boundary Γ. (For $d = 1$, sharper regularity results hold true [31, Sect. 9.8.4, Theorem 9.8.4.1, p. 859])

Theorem 7.2 ([29, 30, 38]) *Consider the mixed problem*

$$\begin{cases} w_{tt} = \Delta w & \text{in } Q = (0, T] \times \Omega & (7.5a) \\ w|_{t=0} = 0; \quad w_t|_{t=0} = 0 & \text{in } \Omega & (7.5b) \\ \dfrac{\partial w}{\partial \nu}\bigg|_{\Sigma} = g & \text{in } \Sigma = (0, T] \times \Gamma. & (7.5c) \end{cases}$$

Recall the constants $\widehat{\alpha}, \beta$ *from* (7.3a), (7.3b). *With reference to problem* (7.1a)–(7.1c), *we have the following interior regularity*

$$g \in L^2(0, T; L^2(\Gamma)) \Longrightarrow \begin{cases} w \in C([0, T]; H^{\widehat{\alpha}}(\Omega) = \mathcal{D}(A^{\frac{\widehat{\alpha}}{2}})) & (7.6a) \\ w_t \in C([0, T]; H^{\widehat{\alpha}-1}(\Omega) = [\mathcal{D}(A^{\frac{1-\widehat{\alpha}}{2}})]') & (7.6b) \end{cases}$$

duality w.r.t. $H = L^2(\Omega)/\mathbb{R}$, *as well as the independent boundary (trace) regularity*

$$g \in L^2(0, T; L^2(\Gamma)) \Longrightarrow w|_{\Sigma} \in H^{2\widehat{\alpha}-1}(\Sigma) \qquad (7.7)$$

continuously.

Next, consider the following problem, dual of problem (7.5a)–(7.5c)

$$\begin{cases} \phi_{tt} = \Delta \phi + f & \text{in } Q & (7.8a) \\ \phi|_{t=T} = \phi_0; \quad \phi_t|_{t=T} = \phi_1 & \text{in } \Omega & (7.8b) \\ \dfrac{\partial \phi}{\partial \nu}\bigg|_{\Sigma} = 0 & \text{in } \Sigma. & (7.8c) \end{cases}$$

Theorem 7.3 ([30, Theorem B(1), p. 118, proved in (2.10) Theorem 2.0, p. 123]) *With reference to problem* (7.8a)–(7.8c), *let* $\phi_0 = \phi_1 = 0$, $f \in L^2(Q)$. *Then, with* β *in* (7.3a), (7.3b):

$$\|\phi|_{\Sigma}\|_{H^{\beta}(\Sigma)} = \mathcal{O}_T\left(\|f\|^2_{L^2(Q)}\right). \qquad (7.9)$$

Theorem 7.4 ([30, Theorem E with $\theta = 0$, p. 119]) *With reference to problem* (7.8a)–(7.8c), *let*

$$\{\phi_0, \phi_1\} \in H^{1-\widehat{\alpha}}(\Omega) \times [H^{\widehat{\alpha}}(\Omega)]', \quad f \in [H^{\widehat{\alpha}}(Q)]'. \qquad (7.10a)$$

Then we have the following (sharp hidden) trace regularity

$$\|\phi|_{\Gamma}\|_{L^2(\Sigma)} = \mathcal{O}\left(\|\{\phi_0, \phi_1\}\|^2_{H^{1-\widehat{\alpha}}(\Omega) \times [H^{\widehat{\alpha}}(\Omega)]'} + \|f\|_{[H^{\widehat{\alpha}}(Q)]'} \cdot\right), \qquad (7.10b)$$

Proof **Step 1** The case $\phi_0 = \phi_1 = 0$ is contained in [30, Theorem E with $\theta = 0$, p. 119].

Step 2 Let now $f \equiv 0$. This case is not explicitly contained in [30], but it can be deduced by interpolation from two results in [30].

First we have

$$\{\phi_0, \phi_1\} \in H^1(\Omega) \times L^2(\Omega) \implies \phi|_\Sigma \in H^\beta(\Sigma) \tag{7.11}$$

continuously. This is [30, Theorem C(1), p. 118]. It is proved as [30, Theorem 7.1, (7.1), (7.2), p. 158]. Next we have

$$\{\phi_0, \phi_1\} \in L^2(\Omega) \times [H^1(\Omega)]' \implies \phi|_\Gamma \in H^{\widehat{\alpha}-1}(\Sigma) \tag{7.12}$$

continuously. This is [30, Theorem 8.3, (8.7) and (8.8), p. 162].

Finally, by (complex) interpolation between statement (7.11) and statement (7.12) to obtain with $0 < \theta < 1$:

$$\left[H^1(\Omega), L^2(\Omega)\right]_{1-\theta} = H^\theta(\Omega); \quad \text{or} \quad \left[H^1(\Omega), L^2(\Omega)\right]_{\widehat{\alpha}} = H^{1-\widehat{\alpha}}(\Omega); \tag{7.13}$$

for ϕ_0; next

$$\left[L^2(\Omega), [H^1(\Omega)]'\right]_{1-\theta} = [H^{1-\theta}(\Omega)]'; \quad \text{or} \quad \left[L^2(\Omega, [H^1(\Omega)]')\right]_{\widehat{\alpha}} = [H^{\widehat{\alpha}}(\Omega)]'; \tag{7.14}$$

for ϕ_1; finally

$$\left[H^\beta(\Sigma), H^{\widehat{\alpha}-1}(\Sigma)\right]_{1-\theta} = H^{\beta\theta + (1-\theta)(\widehat{\alpha}-1)}(\Sigma) = L^2(\Sigma), \tag{7.15}$$

where $\beta\theta + (1-\theta)(\widehat{\alpha} - 1) = 0$ with $\widehat{\alpha} = \beta$ for $\theta = 1 - \widehat{\alpha}$. Thus,

$$\left[H^\beta(\Sigma), H^{\widehat{\alpha}-1}(\Sigma)\right]_{\widehat{\alpha}} = L^2(\Sigma). \tag{7.16}$$

Then Eqs. (7.13)–(7.16) conclude the interpolation argument and the case $f \equiv 0$ is also proved. Theorem 7.4 is established. □

7.3 Proof of Theorem 7.1: Preliminary Analysis

We introduce, as usual [26, 31, 41], the Neumann map

$$Nh = \phi \iff \left\{\Delta \phi = 0 \text{ in } \Omega, \; \left.\frac{\partial \phi}{\partial \nu}\right|_\Gamma = h, \; \phi \in L^2(\Omega)/\mathbb{R}\right\} \tag{7.17a}$$

$$N : L^2(\Gamma) \to H^{\frac{3}{2}}(\Omega) \subset H^{\frac{3}{2}-2\epsilon}(\Omega) = \mathcal{D}(A^{\frac{3}{4}-\epsilon}), \text{ or } A^{\frac{3}{4}-\epsilon}N \in \mathcal{L}(L^2(\Gamma); L^2(\Omega)) \tag{7.17b}$$

for any $\epsilon > 0$. At first we shall take $g \in H^1(0, T; L^2(\Gamma))$, so that $g_t \in L^2(0, T; L^2(\Gamma)) \equiv L^2(\Sigma)$. We next return to Eq. (7.1a) and re-write it, as usual via (7.17a), as

$$y_{ttt} + \alpha y_{tt} - c^2 \Delta(y - Ng) - b\Delta(y - Ng)_t = 0 \quad \text{in } Q \qquad (7.18)$$

or abstractly, via (7.2), as

$$y_{ttt} + \alpha y_{tt} + c^2 A(y - Ng) + bA(y - Ng)_t = 0 \quad \text{in } H = L^2(\Omega)/\mathbb{R}. \qquad (7.19)$$

Extending, as usual, the original operator A in (7.2) $H \supset \mathcal{D}(A) \to H$ to $A_e : H \to [\mathcal{D}(A^*)]' = [\mathcal{D}(A)]'$, duality w.r.t. H by isomorphism, and retaining the symbol A for such an extension, we re-write Eq. (7.19) as

$$(y_t + \alpha y)_{tt} + bA\left(\frac{c^2}{b}y + y_t\right) = c^2 ANg + bANg_t \in [\mathcal{D}(A)]'. \qquad (7.20)$$

See [31, Vol. 1, pp. 420–424; Vol. 2, p. 1061]. Setting, as in (2.10)

$$z = \frac{c^2}{b}y + y_t = (\alpha y + y_t) - \gamma y, \qquad \gamma = \alpha - \frac{c^2}{b} \qquad (7.21)$$

and proceeding as in going from (2.10) to (2.26a), (2.26b) or (2.23) to (2.26a), (2.26b), we re-write problem (7.1a)–(7.1c) as the following hyperbolic-dominated system

$$\begin{cases} z_{tt} = -bAz - \gamma z_t + \gamma\frac{c^2}{b}z - \gamma\left(\frac{c^2}{b}\right)^2 y + c^2 ANg + bANg_t \in [\mathcal{D}(A)]' & (7.22a) \\ y_t = -\frac{c^2}{b}y + z & (7.22b) \end{cases}$$

along with the I.C. (we are taking $y_0 = 0, y_1 = 0, y_2 = 0$):

$$z_0 = \frac{c^2}{b}y_0 + y_1 = 0, \qquad z_1 = \frac{c^2}{b}y_1 + y_2 = 0 \qquad (7.22c)$$

8 First Proof of Theorem 7.1 (Interior Regularity): Direct Proof

Step 1 (same strategy as in Step 1 of Sect. 3) The coupling between the hyperbolic z-dynamics in (7.22a) and the ODE y-equation in (7.22b) is mild (lower order), as the coupling $\gamma \left(\frac{c^2}{b}\right)^2 y = \gamma \left(\frac{c^2}{b}\right)^2 \int_0^t e^{-\frac{c^2}{b}(t-\tau)} z(\tau) d\tau$, an integral operator. Thus, essentially w.l.o.g., we may take at first

$$\gamma = 0, \quad \text{i.e. } \alpha = \frac{c^2}{b}, \tag{8.1}$$

see (7.21), to simplify the computations. This will not affect the sought-after regularity of the map in (7.4a)–(7.4e). Thus, we obtain the simplified problem

$$\begin{cases} z_{tt} = -bAz + c^2 ADg + bADg_t \in [\mathcal{D}(A)]' & (8.2a) \\ y_t = -\frac{c^2}{b} y + z & (8.2b) \end{cases}$$

along with zero I.C., where now under the (essentially w.l.o.g.) assumption (8.1), the z-problem is uncoupled; that is, explicitly, in PDE-form

$$\begin{cases} z_{tt} = b\Delta z & \text{in } Q = (0, T] \times \Omega & (8.3a) \\ z|_{t=0} = 0; \quad z_t|_{t=0} = 0 & \text{in } \Omega & (8.3b) \\ z|_\Sigma = \frac{c^2}{b} g + g_t & \text{in } \Sigma = (0, T] \times \Gamma & (8.3c) \end{cases}$$

$$\begin{cases} y_t = -\frac{c^2}{b} y + z & (8.4a) \\ y|_{t=0} = 0 & (8.4b) \end{cases}$$

Orientation Thus, under the (essentially benign) assumption (8.1), the crux of our proof consists in applying to the wave equation, either as a mixed problem as in (8.3a)–(8.3c), or else in the abstract form (8.2a), the optimal regularity results (at present for the solution $\{z, z_t\}$ in the interior from [29, 30, 38] reported in Theorem 7.2 for convenience, and then use these results for z to obtain corresponding results for y, via (8.4a), (8.4b). In carrying out this strategy, the challenge is that we seek to reduce the assumption of regularity of the Dirichlet boundary term $g \in H^1(0, T; L^2(\Gamma))$ to a sort of 'minimal' level, as the term g_t is not present in the original problem (7.1a), (7.1b) and (7.1c), but is sneaked in at the level of the technical step in (7.19). Of course $g \in H^1(0, T; L^2(\Gamma))$ allows one to invoke the results of Theorem 7.2 at once and thus obtain a preliminary conservative result: the map

$$g \in H^1(0, T; L^2(\Gamma)) \quad \to \quad \{z, z_t\} \in C([0, T]; H^{\widehat{\alpha}}(\Omega) \times H^{\widehat{\alpha}-1}(\Omega)) \quad (8.5a)$$

continuously. From here, it then readily follows via (8.4a), (8.4b)

$$y(t) = \int_0^t e^{-\frac{c^2}{b}(t-\tau)} z(\tau) d\tau \in C([0, T]; H^{\widehat{\alpha}}(\Omega)) \quad (8.5b)$$

$$y_t(t) = \int_0^t e^{-\frac{c^2}{b}(t-\tau)} z_t(\tau) d\tau \in C([0, T]; H^{\widehat{\alpha}-1}(\Omega)) \quad (8.5c)$$

continuously, differentiating (8.4a) in t, and using $y_t|_{t=0} = 0$. Our goal is precisely to refine the regularity result (8.5a).

Step 2 To this end, we shall use critically two main results: (i) the sharp (interior and boundary) regularity theory of Theorem 7.2 obtained by purely PDE-techniques such as pseudo-differential operators and micro-local analysis, and (ii) representation formulas to express (but not to obtain from) such results. For convenience and easy reference, we shall provide in Appendix A a short account of these results as needed in our present development. The (negative self-adjoint) operator $-A : H \supset \mathcal{D}(A) \to H$ generates a s.c. (self-adjoint) cosine operator $\mathcal{C}(t)$ with corresponding sine operator $\mathcal{S}(t)x = \int_0^t \mathcal{C}(\tau)x d\tau$. As reported in Appendix A, the representation formulae of problem (8.3a)–(8.3c), or its abstract version (8.2a) with henceforth

$$b = 1, \quad c = 1, \quad \text{w.l.o.g.} \quad (8.6)$$

are given by

$$z(t) = A \int_0^t \mathcal{S}(t - \tau) N g(\tau) d\tau + A \int_0^t \mathcal{S}(t - \tau) N g_t(\tau) d\tau \quad (8.7a)$$

$$= z^{(1)}(t) + z^{(2)}(t) \quad (8.7b)$$

$$z_t(t) = A \int_0^t \mathcal{C}(t - \tau) N g(\tau) d\tau + A \int_0^t \mathcal{C}(t - \tau) N g_t(\tau) d\tau \quad (8.8a)$$

$$= z_t^{(1)}(t) + z_t^{(2)}(t) \quad (8.8b)$$

Step 3 We now invoke the sharp regularity theory of Theorem 7.2 of the wave (in fact, generally, of 2nd order hyperbolic equations), this time with Neumann boundary term $g \in L^2(0, T; L^2(\Gamma))$ and obtain, continuously:

$$g \in L^2(0, T; L^2(\Gamma)) \Longrightarrow \begin{cases} z^{(1)}(t) = A \int_0^t \mathcal{S}(t-\tau)Ng(\tau)d\tau \in C([0, T]; H^{\widehat{\alpha}}(\Omega) = \mathcal{D}(A^{\frac{\widehat{\alpha}}{2}})) & (8.9a) \\ z_t^{(1)}(t) = A \int_0^t \mathcal{C}(t-\tau)Ng(\tau)d\tau \in C([0, T]; H^{\widehat{\alpha}-1}(\Omega) = [\mathcal{D}(A^{\frac{1-\widehat{\alpha}}{2}})]') & (8.9b) \end{cases}$$

Step 4 Next, let with $\epsilon > 0$ arbitrarily small

$$g \in L^2(0, T; L^2(\Gamma)) \cap C([0, \epsilon); L^2(\Gamma)), \quad g(0) = 0. \tag{8.10}$$

Integrating by parts ($\mathcal{S}(0) = 0$) we obtain from (8.7a), (8.7b)

$$z^{(2)}(t) = A \int_0^t \mathcal{S}(t-\tau)Ng_t(\tau)d\tau = \left[A\mathcal{S}(t-\tau)Ng(\tau) \right]_{\tau=0}^{\tau=t} - A \int_0^t \mathcal{C}(t-\tau)Ng(\tau)d\tau \tag{8.11}$$

$$= \cancel{A\mathcal{S}(0)Ng(t)} - \cancel{A\mathcal{S}(t)Ng(0)} - A \int_0^t \mathcal{C}(t-\tau)Ng(\tau)d\tau \tag{8.12a}$$

or, as in (8.9b):

$$\begin{cases} z^{(2)}(t) = -A \int_0^t \mathcal{C}(t-\tau)Ng(\tau)d\tau = -z_t^{(1)}(t) \in C([0, T]; H^{\widehat{\alpha}-1}(\Omega) = [\mathcal{D}(A^{\frac{1-\widehat{\alpha}}{2}})]') \\ \text{for } g \text{ as in (8.10), continuously.} \end{cases} \tag{8.12b}$$

Thus, by (8.9b) and (8.12b) used in (8.7b), we obtain

$$\begin{cases} z(t) = z^{(1)}(t) + z^{(2)}(t) \in C([0, T]; H^{\widehat{\alpha}-1}(\Omega) = [\mathcal{D}(A^{\frac{1-\widehat{\alpha}}{2}})]') \\ \text{for } g \text{ as in (8.10), continuously.} \end{cases} \tag{8.13}$$

Step 5 Next, returning to the y-equation in (8.2b) [with (8.6) w.l.o.g.] and z given by (8.7b) we obtain

$$y(t) = \int_0^t e^{-(t-\tau)} z(\tau)d\tau = \int_0^t e^{-(t-\tau)} z^{(1)}(\tau)d\tau + \int_0^t e^{-(t-\tau)} z^{(2)}(\tau)d\tau \tag{8.14a}$$

$$= y^{(1)}(t) + y^{(2)}(t) \tag{8.14b}$$

as $y_0 = 0$. Returning to (8.9a), we obtain conservatively

Sharp Interior and Boundary Regularity of the SMGTJ-Equation ...

$$\begin{cases} y^{(1)}(t) = \int_0^t e^{-(t-\tau)} z^{(1)}(\tau) d\tau \in C([0, T]; H^{\widehat{\alpha}}(\Omega) = \mathcal{D}(A^{\frac{\widehat{\alpha}}{2}})) \\ \text{for } g \in L^2(0, T; L^2(\Gamma)), \text{ continuously.} \end{cases} \quad (8.15)$$

Next, with $z^{(2)} = -z_t^{(1)}(t)$ as given by (8.12b), we compute via (8.14a), (8.14b)

$$y^{(2)}(t) = \int_0^t e^{-(t-\tau)} z^{(2)}(\tau) d\tau = -\int_0^t e^{-(t-\tau)} z_t^{(1)}(\tau) d\tau \quad (8.16)$$

$$= -\left[e^{-(t-\tau)} z^{(1)}(\tau) \right]_{\tau=0}^{\tau=t} + \int_0^t e^{-(t-\tau)} z^{(1)}(\tau) d\tau \quad (8.17)$$

$$\begin{cases} y^{(2)}(t) = -z^{(1)}(t) + e^{-t} z^{(1)}(0) + \int_0^t e^{-(t-\tau)} z^{(1)}(\tau) d\tau \in C([0, T]; H^{\widehat{\alpha}}(\Omega) = \mathcal{D}(A^{\frac{\widehat{\alpha}}{2}})) \\ \text{valid for } g \text{ as in (8.10), continuously.} \end{cases} \quad (8.18)$$

as the expression of $z^{(2)}(t) = -z_t^{(1)}(t)$ in (8.12b) has such a constraint. As to the regularity noted in (8.18), we invoke (8.9a) for the term $z^{(1)}(t)$, while the same regularity holds true for the second convolution term, this time conservatively. We conclude by (8.15) on $y^{(1)}$ and (8.18) on $y^{(2)}$ that

$$\begin{cases} y(t) = y^{(1)}(t) + y^{(2)}(t) = -z^{(1)}(t) + 2\int_0^t e^{-(t-\tau)} z^{(1)}(\tau) d\tau \in C([0, T]; H^{\widehat{\alpha}}(\Omega) = \mathcal{D}(A^{\frac{\widehat{\alpha}}{2}})) \\ \text{for } g \text{ as in (8.10), continuously.} \end{cases} \quad (8.19)$$

Then (8.19) shows the first result in (7.4a) of Theorem 7.1.

Step 6 Next, with (8.4a), invoke (8.13) for z and (8.19) for y and obtain via (8.4a)

$$\begin{cases} y_t = -\dfrac{c^2}{b} y + z \in C([0, T]; H^{\widehat{\alpha}-1}(\Omega) = [\mathcal{D}(A^{\frac{1-\widehat{\alpha}}{2}})]') \\ \text{for } g \text{ as in (8.10), continuously.} \end{cases} \quad (8.20)$$

Then (8.20) shows the second result in (7.4b) of Theorem 7.1.

Step 7 We finally need to establish the regularity of y_{tt}. This will be obtained from

$$y_{tt} = -\frac{c^2}{b} y_t + z_t. \quad (8.21)$$

Thus, we need to establish the regularity of z_t, as the regularity of y_t is given by (8.20). But $z_t = z_t^{(1)} + z_t^{(2)}$, where the regularity of $z_t^{(1)} = -z^{(2)}(t)$ was already established in (8.12b).

Step 8 We seek the regularity of $z_t^{(2)}$ from its representation formula in (8.8a). We compute from (8.8a)

$$z_t^{(2)}(t) = A \int_0^t \mathcal{C}(t-\tau) N g_t(\tau) d\tau \quad \text{(by parts, recalling (A.2))} \tag{8.22}$$

$$= \left[A\mathcal{C}(t-\tau) N g(\tau) \right]_{\tau=0}^{\tau=t} + A \int_0^t A\mathcal{S}(t-\tau) N g(\tau) d\tau \tag{8.23}$$

$$= ANg(t) - \overline{A\mathcal{C}(t)Ng(0)} + AA \int_0^t \mathcal{S}(t-\tau) N g(\tau) d\tau \tag{8.24}$$

$$= ANg(t) + Az^{(1)}(t) \tag{8.25}$$

recalling $\mathcal{C}(0) = I$ and (8.9a). Thus, again by (8.9a)

$$g \in L^2(0, T; L^2(\Gamma)) \implies Az^{(1)}(t) = AA \int_0^t \mathcal{S}(t-\tau) N g(\tau) d\tau \in C([0, T]; [\mathcal{D}(A^{1-\frac{\hat{\alpha}}{2}})]')$$

(8.26)

continuously. Moreover, by (7.17b)

$$g \in L^2(0, T; L^2(\Gamma)) \implies ANg = A^{\frac{1}{4}+\epsilon_1}(A^{\frac{3}{4}-\epsilon_1}N)g \in L^2(0, T; [\mathcal{D}(A^{\frac{1}{4}+\epsilon_1})]')$$
$$\subset L^2(0, T; [\mathcal{D}(A^{1-\frac{\hat{\alpha}}{2}})]')$$

(8.27a)

as well as

$$g \in C([0, T]; L^2(\Gamma)) \implies ANg = A^{\frac{1}{4}+\epsilon_1}(A^{\frac{3}{4}-\epsilon_1}N)g \in C([0, T]; [\mathcal{D}(A^{\frac{1}{4}+\epsilon_1})]')$$
$$\subset C([0, T]; [\mathcal{D}(A^{1-\frac{\hat{\alpha}}{2}})]')$$

(8.27b)

continuously, since $\hat{\alpha} = \frac{2}{3}$ or $\frac{3}{4}$ by (7.3a), (7.3b), so $1 - \frac{\hat{\alpha}}{2} = \frac{2}{3}$ or $\frac{5}{8}$ and so $1 - \frac{\hat{\alpha}}{2} > \frac{1}{4} + \epsilon$. Combining (8.26) and (8.27a) in (8.26), we obtain

$$\begin{cases} g \in L^2(0, T; L^2(\Gamma)) \cap C([0, \epsilon); L^2(\Gamma)) \\ g(0) = 0, \end{cases} \implies z_t^{(2)} = ANg + Az^{(1)} \in L^2(0, T; [\mathcal{D}(A^{1-\frac{\hat{\alpha}}{2}})]')$$

(8.28a)

as well as, via (8.26) and (8.27b)

$$\begin{cases} g \in C([0,T]; L^2(\Gamma)) \\ g(0) = 0, \end{cases} \Longrightarrow z_t^{(2)} = ANg + Az^{(1)} \in C([0,T]; [\mathcal{D}(A^{1-\frac{\hat{\alpha}}{2}})]')$$
(8.28b)

continuously.

Step 9 In turn, combining (8.9b) for $z_t^{(1)}$ with (8.28a) for $z_t^{(2)}$ in (8.8b), we conclude

$$\begin{cases} g \in L^2(0,T; L^2(\Gamma)) \cap C([0,\epsilon); L^2(\Gamma)) \\ g(0) = 0, \end{cases} \Longrightarrow z_t = z_t^{(1)} + z_t^{(2)} \in L^2(0,T; [\mathcal{D}(A^{1-\frac{\hat{\alpha}}{2}})]')$$
(8.29a)

as well as, via (8.28b) for $z_t^{(2)}$

$$\begin{cases} g \in C([0,T]; L^2(\Gamma)) \\ g(0) = 0, \end{cases} \Longrightarrow z_t = z_t^{(1)} + z_t^{(2)} \in C([0,T]; [\mathcal{D}(A^{1-\frac{\hat{\alpha}}{2}})]') \quad (8.29\text{b})$$

since $\frac{1-\hat{\alpha}}{2} < 1 - \frac{\hat{\alpha}}{2}$.

Step 10 Combining (8.20) on y_t with (8.29a) for z_t, we conclude via (8.21)

$$\begin{cases} g \in L^2(0,T; L^2(\Gamma)) \cap C([0,\epsilon); L^2(\Gamma)) \\ g(0) = 0, \end{cases} \Longrightarrow y_{tt} = -\frac{c^2}{b} y_t + z_t \in L^2(0,T; [\mathcal{D}(A^{1-\frac{\hat{\alpha}}{2}})]')$$
(8.30a)

as well as, via (8.29b) for z_t

$$\begin{cases} g \in C([0,T]; L^2(\Gamma)) \\ g(0) = 0, \end{cases} \Longrightarrow y_{tt} = -\frac{c^2}{b} y_t + z_t \in C([0,T]; [\mathcal{D}(A^{1-\frac{\hat{\alpha}}{2}})]')$$
(8.30b)

continuously. Then (8.30a), (8.30b) shows the interior regularity results in (7.4c) and (7.4d). Theorem 7.1 is proved, except for the boundary regularity statement (7.4c), which will be established in Sect. 9.

9 Proof of Theorem 7.1: Regularity of the Boundary Trace $y|_\Sigma$

In this section we shall establish the boundary regularity (7.4c) of Theorem 7.1. It is here repeated for convenience.

Theorem 9.1 *With reference to problem (7.1a)–(7.1c), we have*

$$\begin{cases} g \in L^2(0,T; L^2(\Gamma)) \cap C([0,\epsilon); L^2(\Gamma)) \\ g(0) = 0 \end{cases} \Longrightarrow y|_\Sigma \in H^{2\hat{\alpha}-1}(\Sigma)) \quad (9.1)$$

continuously.

Proof **Step 1** We return to the solution formula (8.7a) complemented by (8.12b), with g as in (8.10), i.e. as on the LHS of (9.1):

$$z(t) = z^{(1)}(t) + z^{(2)}(t) = z^{(1)}(t) - z_t^{(1)}(t) \tag{9.2}$$

$$= A \int_0^t \mathcal{S}(t - \tau) N g(\tau) d\tau - A \int_0^t \mathcal{C}(t - \tau) N g(\tau) d\tau. \tag{9.3}$$

We next invoke critically the boundary regularity results reported in (7.7) of Theorem 7.2 as well as (A.4), (A.6):

$$g \in L^2(0, T; L^2(\Gamma)) \Longrightarrow z^{(1)}\big|_\Sigma = (N^*A)A \int_0^t \mathcal{S}(t - \tau) N g(\tau) d\tau \in H^{2\hat{\alpha}-1}(\Sigma). \tag{9.4}$$

Step 2 We return to (8.19), with g as in (8.10)

$$y(t) = -z^{(1)}(t) + 2 \int_0^t e^{-(t-\tau)} z^{(1)}(\tau) d\tau \tag{9.5}$$

where then, recalling (9.4), we obtain

$$y\big|_\Sigma = -z^{(1)}\big|_\Sigma + 2 \int_0^t e^{-(t-\tau)} z^{(1)}\big|_\Sigma (\tau) d\tau \in H^{2\hat{\alpha}-1}(\Sigma) \tag{9.6}$$

continuously. Then (9.6) proves Theorem 9.1, hence (7.4c) of Theorem 7.1. □

10 A Boundary Trace Result for the u-Problem

In this section we consider problem (7.1a)–(7.1c) with $g \equiv 0$, rewritten in PDE-form as

$$\begin{cases} u_{ttt} + \alpha u_{tt} - c^2 \Delta u - b \Delta u_t = 0 & \text{in } Q = (0, T] \times \Omega \quad (10.1a) \\ u\big|_{t=T} = u_0; \quad u_t\big|_{t=T} = u_1; \quad u_{tt}\big|_{t=T} = u_2 & \text{in } \Omega \quad (10.1b) \\ \dfrac{\partial u}{\partial \nu}\bigg|_\Sigma = 0 & \text{in } \Sigma = (0, T] \times \Gamma \quad (10.1c) \end{cases}$$

abstractly, with A as in (7.2)

$$u_{ttt} + \alpha u_{tt} + c^2 A u + b A u_t = 0 \tag{10.1d}$$

with I.C. at $t = T$ (i.e. backward in time). This is the counterpart of problem (5.1a)–(5.1d) in the corresponding Dirichlet case. In this section, we shall consider two spaces for the I.C., see (2.5c) for U_3

$$\{u_0, u_1, u_2\} \in U_3 \equiv \mathcal{D}(A) \times \mathcal{D}(A^{\frac{1}{2}}) \times H \tag{10.2}$$

$$\{u_0, u_1, u_2\} \in U_5 \equiv \mathcal{D}(A^{1-\frac{\hat{\alpha}}{2}}) \times \mathcal{D}(A^{\frac{1-\hat{\alpha}}{2}}) \times \left[\mathcal{D}(A^{\frac{\hat{\alpha}}{2}})\right]' \tag{10.3a}$$

$$= \begin{cases} \mathcal{D}(A^{\frac{2}{3}}) \times \mathcal{D}(A^{\frac{1}{6}}) \times \left[\mathcal{D}(A^{\frac{1}{3}})\right]', & \alpha = \frac{2}{3} \tag{10.3b} \\ \mathcal{D}(A^{\frac{5}{8}}) \times \mathcal{D}(A^{\frac{1}{8}}) \times \left[\mathcal{D}(A^{\frac{3}{8}})\right]'. & \alpha = \frac{3}{4} \tag{10.3c} \end{cases}$$

recalling (7.3a), (7.3b). Notice that the regularity of the components of U_5 is reduced by $\mathcal{D}(A^{\frac{1}{2}})$ from u_0 to u_1 to u_2, in line with the spaces U_3 or U_4 in (2.5c).

Accordingly, in view of Theorem 2.1, we have

$$\begin{bmatrix} u_0 \\ u_1 \\ u_2 \end{bmatrix} \in U_i \to \begin{bmatrix} u(t) \\ u_t(t) \\ u_{tt}(t) \end{bmatrix} = e^{G(T-t)} \begin{bmatrix} u_0 \\ u_1 \\ u_2 \end{bmatrix} \in C([0, T]; U_i) \tag{10.4}$$

continuously, for the solution of (10.1a)–(10.1d), where e^{Gt} is a s.c. group on U_i, with infinitesimal generator G defined in (2.4), with A as in (7.2) in the Neumann case.

Theorem 10.1 *With reference to problem (10.1a)–(10.1d) and to (10.2), the following trace estimates hold true:*

(a)
$$\|u_t\|^2_{H^\beta(\Sigma)} = \mathcal{O}_T \left(\|\{u_0, u_1, u_2\}\|^2_{U_3} \right), \tag{10.5}$$

where $\beta = \dfrac{2}{3}$ or $\beta = \dfrac{3}{4}$ is defined in (7.3a), (7.3b) and where $\mathcal{O}_T$ denotes a constant depending on $\Sigma = \Gamma \times (0, T]$ and the equation coefficients, but not on U_3.

(b)
$$\int_0^T \int_\Gamma |u_t|^2 d\Gamma dt = \mathcal{O}_T \left(\|\{u_0, u_1, u_2\}\|^2_{U_5} \right). \tag{10.6}$$

Remark 10.1 The above two trace regularity results do not follow from the interior regularity. They express a 'hidden' regularity property. As to (10.5), the interior regularity of u_t is $u_t \in C([0,T]; \mathcal{D}(A^{\frac{1}{2}})) \equiv H^1(\Omega))$ by (10.4), (10.2), hence $u_t|_\Gamma \in C([0,T]; H^{1/2}(\Gamma))$ by trace theory. Instead (10.5) yields, in particular, $u_t|_\Sigma \in L^2(0,T; H^\beta(\Gamma))$, with $\beta - \frac{1}{2} = \frac{1}{6}$ or $\frac{1}{4}$ stronger in space regularity by (7.3a), (7.3b), for $\beta = \frac{2}{3}$ or $\beta = \frac{3}{4}$. Likewise, as to (10.6), one has the interior regularity $u_t \in C([0,T]; \mathcal{D}(A^{\frac{1}{6}})) \equiv H^{\frac{1}{3}}(\Omega))$ say by (10.3b) for $\widehat{\alpha} = \frac{2}{3}$, which does not yield the trace regularity $u_t|_\Sigma \in L^2(0,T; L^2(\Gamma))$ of (10.6).

Proof of Theorem 10.1: Step 1 Rewrite (10.1a) as in (2.7)

$$(u_t + \alpha u)_{tt} - b\Delta\left(\frac{c^2}{b}u + u_t\right) = 0 \quad \text{in } Q \tag{10.7}$$

and introduce the new variable ξ in (2.8)

$$\xi = \alpha u + u_t \in \begin{cases} C([0,T]; \mathcal{D}(A^{\frac{1}{2}})) & \text{for } \{u_0, u_1, u_2\} \in U_3 & (10.8a) \\ C([0,T]; \mathcal{D}(A^{\frac{1}{6}})) & \text{for } \{u_0, u_1, u_2\} \in U_5, \ \widehat{\alpha} = \frac{2}{3} & (10.8b) \end{cases}$$

in the less regular case $\widehat{\alpha} = \frac{2}{3}$ in (10.3b), as it follows from (10.4) along with (10.2) or (10.3a) respectively,

$$\frac{c^2}{b}u + u_t = \xi - \gamma u, \qquad \gamma = \alpha - \frac{c^2}{b}. \tag{10.8c}$$

Rewrite problem (10.1a)–(10.1c) accordingly as

$$\begin{cases} \xi_{tt} - b\Delta\xi + b\gamma\Delta u = 0 & \text{in } Q & (10.9a) \\ \xi|_{t=T} = \xi_0 = \alpha u_0 + u_1; \ \xi_t|_{t=T} = \xi_1 = \alpha u_1 + u_2 & \text{in } \Omega & (10.9b) \\ \left.\dfrac{\partial\xi}{\partial\nu}\right|_\Sigma \equiv 0 & \text{in } \Sigma & (10.9c) \end{cases}$$

and set

$$\xi = \xi^{(1)} + \xi^{(2)} \tag{10.10}$$

$$\begin{cases} \xi_{tt}^{(1)} - b\Delta\xi^{(1)} + b\gamma\Delta u = 0 \\ \xi^{(1)}\big|_{t=T} = 0, \quad \xi_t^{(1)}\big|_{t=T} = 0; \\ \dfrac{\partial \xi^{(1)}}{\partial \nu}\bigg|_\Sigma = 0 \end{cases} \quad \begin{cases} \xi_{tt}^{(2)} - b\Delta\xi^{(2)} = 0 & \text{in } Q \quad (10.11\text{a}) \\ \xi^{(2)}\big|_{t=T} = \xi_0, \quad \xi_t^{(2)}\big|_{t=T} = \xi_1 & \text{in } Q \quad (10.11\text{b}) \\ \dfrac{\partial \xi^{(2)}}{\partial \nu}\bigg|_\Sigma = 0 & \text{in } Q \quad (10.11\text{c}) \end{cases}$$

Step 2

Theorem 10.2 *The following trace estimates hold true (recall β in (7.3a), (7.3b)):*

(a)
$$\|\xi|_\Sigma\|^2_{H^\beta(\Sigma)} = \mathcal{O}_T\left(\|\{u_0, u_1, u_2\}\|^2_{U_3}\right), \tag{10.12}$$

(b)
$$\begin{cases} \|\xi|_\Sigma\|^2_{L^2(\Sigma)} = \mathcal{O}_T\left(\|\{u_0, u_1, u_2\}\|^2_{U_5}\right) \\ U_5 \text{ given by (10.3a)–(10.3c)} \end{cases} \tag{10.13}$$

Proof of Theorem 10.2

(a) With $\{u_0, u_1, u_2\} \in U_3$ in (10.2), we have

$$\Delta u \in C([0, T]; L^2(\Omega)), \tag{10.14}$$

continuously, by (10.4), $i = 3$. We next invoke [30, Theorem B(1), p. 118] (recalled in Theorem 7.3), as applied to the $\xi^{(1)}$-problem in (10.11a)–(10.11c), with RHS as in (10.14) and obtain

$$\xi^{(1)}\big|_\Sigma \in H^\beta(\Sigma), \text{ continuously in } U_3. \tag{10.15}$$

Next, with

$$\xi^{(2)}\big|_{t=T} = \xi_0 = \alpha u_0 + u_1 \in \mathcal{D}(A^{\frac{1}{2}}) \equiv H^1(\Omega); \quad \xi_t^{(2)}\big|_{t=T} = \xi_1 = \alpha u_1 + u_2 \in L^2(\Omega), \tag{10.16}$$

we invoke this time [30, Theorem C(1), p 118] (recalled in the (7.11) of Theorem 7.4), as applied to the $\xi^{(2)}$-problem in (10.11a)–(10.11c), with I.C. as in (10.16) and obtain

$$\xi^{(2)}\big|_\Sigma \in H^\beta(\Sigma), \tag{10.17}$$

continuously in $\{\xi_0, \xi_1\} \in H^1(\Omega) \times L^2(\Omega)$, hence continuously in U_3. Combining (10.15) and (10.17) in (10.10), we obtain (10.12) and part (a) of Theorem 10.2 is proved.

(b) Now with $\{u_0, u_1, u_2\} \in U_5$ in (10.3a), we have $u \in C\left([0, T]; H^{2-\widehat{\alpha}}(\Omega) \equiv \mathcal{D}(A^{1-\frac{\widehat{\alpha}}{2}})\right)$ by (10.4) and hence continuously

$$\Delta u = Au \in C\left([0, T]; \left[H^{\widehat{\alpha}}(\Omega)\right]'\right), \quad \widehat{\alpha} = \frac{2}{3} \text{ or } \widehat{\alpha} = \frac{3}{4} \text{ as in (7.3a)–(7.3b).} \tag{10.18}$$

We next invoke [30, Theorem E, with $\theta = 0$, p. 119] (recalled in the Theorem 7.4), as applied to the $\xi^{(1)}$-problem in (10.11a)–(10.11c), with RHS as in (10.18) and zero I.C. and obtain

$$\xi^{(1)}\big|_{\Sigma} \in L^2(\Sigma), \text{ continuously in } U_5. \tag{10.19}$$

Next, with

$$\xi^{(2)}\big|_{t=T} = \xi_0 = \alpha u_0 + u_1 \in H^{1-\widehat{\alpha}}(\Omega) = \mathcal{D}(A^{\frac{1-\widehat{\alpha}}{2}}), \quad \left[\mathcal{D}(A^{\frac{1}{6}}) \equiv H^{\frac{1}{3}}(\Omega) \text{ for } \widehat{\alpha} = \frac{2}{3}\right] \tag{10.20a}$$

$$\xi_t^{(2)}\big|_{t=T} = \xi_1 = \alpha u_1 + u_2 \in \left[H^{\widehat{\alpha}}(\Omega)\right]' \equiv \left[\mathcal{D}(A^{\frac{\widehat{\alpha}}{2}})\right]', \quad \left[\left[\mathcal{D}(A^{\frac{1}{3}})\right]' \equiv \left[H^{\frac{2}{3}}(\Omega)\right]' \text{ for } \widehat{\alpha} = \frac{2}{3}\right] \tag{10.20b}$$

we obtain by Theorem 7.4, Eq. (7.10b) [with $f \equiv 0$]

$$\xi^{(2)}\big|_{\Sigma} \in L^2(\Sigma) \tag{10.21}$$

continuously in $\{\xi_0, \xi_1\} \in H^{1-\widehat{\alpha}}(\Omega) \times \left[H^{\widehat{\alpha}}(\Omega)\right]'$, hence continuously in U_5. Combining (10.19) with (10.21) in (10.10), we obtain (10.13) and part 10.2 of Theorem 10.2 is proved. □

Step 3 (continuing the proof of Theorem 10.1b)

(a) Thus, by Theorem 10.2a, we have for $\{u_0, u_1, u_2\} \in U_3$ and β in (7.3a), (7.3b)

$$\xi\big|_{\Sigma} = \alpha u + u_t\big|_{\Sigma} \in H^{\beta}(\Sigma) \tag{10.22}$$

while $u \in C([0, T]; \mathcal{D}(A)) \subset C([0, T]; H^2(\Omega))$ by (10.4), (10.2) implies

$$u\big|_{\Sigma} \in C([0, T]; H^{\frac{3}{2}}(\Omega)) \subset H^{\beta}(\Sigma) \tag{10.23}$$

continuously in U_3. Then, (10.22) and (10.23) imply $u_t\big|_{\Sigma} \in H^{\beta}(\Sigma)$ and (10.5) of Theorem 10.1a is proved.

(b) Similarly, by Theorem 10.2b, we have for $\{u_0, u_1, u_2\} \in U_5$,

$$\xi\big|_{\Sigma} = \alpha u + u_t\big|_{\Sigma} \in L^2(\Sigma) \tag{10.24}$$

continuously, while $u \in C([0, T]; H^{2-\widehat{\alpha}}(\Omega))$ by (10.4), (10.3a) implies

$$u\big|_\Sigma \in C([0, T]; H^{\frac{3}{2}-\widehat{\alpha}}(\Gamma)) \subset L^2(\Sigma) \tag{10.25}$$

continuously in U_5. Then, (10.24) and (10.25) imply $u_t\big|_\Sigma \in L^2(\Sigma)$ and (10.6) of Theorem 10.1b is established. □

11 A Dual Result of the Neumann Problem (7.1a)–(7.1c)

The present section is the Neumann counterpart of the Dirichlet Sect. 6. Therefore, it will simply list the counterpart results. The dual problem of the boundary non-homogeneous Neumann problem (7.1a)–(7.1c) is the problem

$$\begin{cases} v_{ttt} - \alpha v_{tt} + c^2 \Delta v - b\Delta v_t = 0 & \text{in } Q \quad (11.1a) \\ v\big|_{t=T} = v_0; \quad v_t\big|_{t=T} = v_1; \quad v_{tt}\big|_{t=T} = v_2 & \text{in } \Omega \quad (11.1b) \\ \dfrac{\partial v}{\partial \nu}\bigg|_\Sigma = 0 & \text{in } \Sigma \quad (11.1c) \end{cases}$$

$$\text{abstractly} \quad v_{ttt} - \alpha v_{tt} - c^2 A v + b A v_t = 0 \tag{11.2}$$

with $-A$ the Neumann Laplician in (7.2). This is very closely related to problem (10.1a)–(10.1c), or its abstract version (10.1d) for reasons similar to those given in Sect. 6. The relevant results are as follows

Theorem 11.1 *With reference to the v-problem (11.1a)–(11.1c), abstractly (11.2), we have*

$$\{v_0, v_1, v_2\} \in U_i \implies \{v, v_t, v_{tt}\} \in C([0, T]; U_i), \quad i = 3, 5 \tag{11.3a}$$

$$U_3 = \mathcal{D}(A) \times \mathcal{D}(A^{\frac{1}{2}}) \times H, \quad U_5 = \mathcal{D}(A^{1-\frac{\widehat{\alpha}}{2}}) \times \mathcal{D}(A^{\frac{1-\widehat{\alpha}}{2}}) \times \left[\mathcal{D}(A^{\frac{\widehat{\alpha}}{2}})\right]'. \tag{11.3b}$$

This is the counterpart of Theorem 6.1 in the Dirichlet case, at least for $i = 3$.

Theorem 11.2 *With reference to problem (11.1a)–(11.1c), we have*

(a)

$$\|v_t\|^2_{H^\beta(\Sigma)} = \mathcal{O}_T\left(\|\{v_0, v_1, v_2\}\|^2_{U_3}\right), \tag{11.4}$$

where β is defined in (7.3a), (7.3b).

(b)

$$\int_0^T \int_\Gamma |v_t|^2 d\Gamma dt = \mathcal{O}_T\left(\|\{v_0, v_1, v_2\}\|_{U_5}^2\right). \tag{11.5}$$

This is the counterpart of Theorem 10.1. It is a sharp, hidden trace regularity result.

Next, by duality on the trace result (11.5) for the v-problem (11.1a)–(11.1c), we shall re-obtain (in a slightly weaker form) the basic interior regularity result of Theorem 7.1.

Theorem 11.3 *With reference to problem (7.1a)–(7.1c), we have*

$$\begin{cases} g \in C([0,T]; L^2(\Gamma)) \\ g(0) = 0 \end{cases} \implies \{y, y_t, y_{tt}\} \in C\left([0,T]; \mathcal{D}(A^{\frac{\hat{\alpha}}{2}}) \times [\mathcal{D}(A^{\frac{1-\hat{\alpha}}{2}})]' \times [\mathcal{D}(A^{1-\frac{\hat{\alpha}}{2}})]'\right), \tag{11.6}$$

continuously.

Proof The proof is the conceptual counterpart of Theorem 6.4 in the Dirichlet case, subject to further technicalities proper of the Neumann problem.

Step 1 We shall first establish the following.

Proposition 11.4 *With reference to the Neumann problem (7.1a)–(7.1c), we have, for each $0 < t \leq T$:*

$$\begin{cases} g \in C([0,T]; L^2(\Gamma)) \\ g(0) = 0, \end{cases} \implies \{y(t), y_t(t), y_{tt}(t)\} \in \mathcal{D}(A^{\frac{\hat{\alpha}}{2}}) \times [\mathcal{D}(A^{\frac{1-\hat{\alpha}}{2}})]' \times [\mathcal{D}(A^{1-\frac{\hat{\alpha}}{2}})]' \tag{11.7a}$$

$$\implies \int_0^t [y(\tau), y_t(\tau), y_{tt}(\tau)] d\tau \in \mathcal{D}(A^{\frac{\hat{\alpha}}{2}}) \times [\mathcal{D}(A^{\frac{1-\hat{\alpha}}{2}})]' \times [\mathcal{D}(A^{1-\frac{\hat{\alpha}}{2}})]' \tag{11.7b}$$

continuously.

Proof of Proposition 11.4. It is based by duality on Theorem 11.2b, Eq. (11.5), counterpart of estimate (10.6).

Step (i) By Theorem 11.1 and 11.2b, we have

$$\{v_0, v_1, v_2\} \in U_5 = \mathcal{D}(A^{1-\frac{\hat{\alpha}}{2}}) \times \mathcal{D}(A^{\frac{1-\hat{\alpha}}{2}}) \times \left[\mathcal{D}(A^{\frac{\hat{\alpha}}{2}})\right]' \tag{11.8a}$$

$$\implies \begin{cases} \{v, v_t, v_{tt}\} \in C([0,T]; U_5) \\ \text{AND} \\ v_t|_\Gamma \in L^2(0,T; L^2(\Gamma)) \end{cases} \tag{11.8b}$$

continuously.

Step (ii) We now invoke the duality identity (B.5) in Appendix B, Neumann case, and obtain for a generic t, $0 < t \leq T$ (notation denotes duality pairing):

$$\langle y_{tt}(t) + \alpha y_t(t), v_0 \rangle - \langle y_t(t) + \alpha y(t), v_1 \rangle + \langle y(t), v_2 \rangle - b \langle y(t), \Delta v_0 \rangle$$
$$= \langle c^2 g + b g_t, v \rangle_{L^2(0,t;L^2(\Gamma))}. \quad (11.9)$$

With $g \in L^2(0, t; L^2(\Gamma))$ and under (11.8a), (11.8b) and hence $v \in C([0, T]; H^{2-\widehat{\alpha}}(\Omega))$ and $v|_\Sigma \in C([0, T]; H^{\frac{3}{2}-\widehat{\alpha}}(\Gamma))$ [say $v|_\Sigma \in C([0, t]; H^{\frac{4}{3}-\frac{1}{2}}(\Gamma)) = H^{\frac{5}{6}}(\Gamma))$ for $\widehat{\alpha} = \frac{2}{3}$], we have regarding the first term on the RHS of (11.9):

$$\langle g, v|_\Gamma \rangle_{L^2(0,t;L^2(\Gamma))} = \int_0^t \int_\Gamma g v|_\Gamma d\Gamma dt < \infty. \quad (11.10)$$

Step (iii) Under hypothesis (11.7a), (11.7b) and under (11.8a), (11.8b) for the adjoint v-problem, we compute the last term on the RHS of (11.9) by parts as follows, for $0 < t \leq T$

$$\int_0^t \left(g_t(\tau), v|_\Gamma(\tau) \right)_{L^2(\Gamma)} d\tau = \left[\left(g(\tau), v|_\Gamma(\tau) \right)_{L^2(\Gamma)} \right]_{\tau=0}^{\tau=t}$$
$$- \int_0^t \left(g(\tau), v_t|_\Gamma(\tau) \right)_{L^2(\Gamma)} d\tau$$
$$= \left(g(t), v|_\Gamma(t) \right)_{L^2(\Gamma)} - \cancel{\left(g(0), v|_\Gamma(0) \right)_{L^2(\Gamma)}}$$
$$- \int_0^t \left(g(\tau), v_t|_\Gamma(\tau) \right)_{L^2(\Gamma)} d\tau \quad (11.11)$$

With $g \in C([0, T]; L^2(\Gamma))$ as in (11.7a), (11.7b) and $v|_\Sigma \in C([0, t]; H^{\frac{3}{2}-\widehat{\alpha}}(\Gamma))$, we have that the first term on the RHS of (11.11) is well-defined (conservatively). Notice that it can also be re-written as

$$\left(g(t), v_t|_\Gamma \right)_{L^2(\Gamma)} = \left(g(t), N^* A v(t) \right)_{L^2(\Gamma)}$$
$$= (A N g(t), v(t))_{L^2(\Gamma)} \text{ (well-defined)} \quad (11.12)$$

with $v \in C([0, T]; \mathcal{D}(A^{1-\frac{\widehat{\alpha}}{2}}))$ by (11.8a) and $A N g \in C([0, T]; [\mathcal{D}(A^{\frac{1}{4}+\epsilon_1})]') \subset C([0, T]; [\mathcal{D}(A^{1-\frac{\widehat{\alpha}}{2}})]')$. Here we have invoked (7.17b) for N, (11.7a) for g as well as $\frac{1}{4} + \epsilon_1 < 1 + \frac{\widehat{\alpha}}{2}$. Then by (11.11) and (11.12),

$$\langle g_t, v|_\Gamma \rangle_{L^2(0,t;L^2(\Gamma))} = (A N g(t), v(t))_{L^2(\Omega)} - \langle g, v_t|_\Gamma \rangle_{L^2(0,t;L^2(\Gamma))} \quad (11.13)$$
(well-defined, by (11.12) and (11.8b)),

continuously, with respect to g as in (11.7a), (11.7b) and $\{v_0, v_1, v_2\}$ as in (11.8a), (11.8b). Thus, the RHS of identity (11.9) is well-defined for such g in (11.7a), (11.7b) and $\{v_0, v_1, v_2\} \in U_5$, critically because of Theorem 11.2b, Eq. (11.5), used in (11.8b). Next we turn to the LHS of identity (11.9) and then the duality argument below—the counterpart of the argument in B.8 or in (6.22)–(6.24) in the Dirichlet case—yields (11.7a) also in the present Neumann case. Specifically, for each $t, 0 < t \leq T$, via duality pairing in the LHS of identity (11.9), we obtain recalling $\{v_0, v_1, v_2\} \in U_5$ in (10.3a):

$$\left.\begin{array}{l} v_2 \in [\mathcal{D}(A^{\frac{\widehat{\alpha}}{2}})]' \to y(t) \in \mathcal{D}(A^{\frac{\alpha}{2}}) \\ v_1 \in \mathcal{D}(A^{\frac{1-\widehat{\alpha}}{2}}) \to y_t(t) + \alpha y(t) \in [\mathcal{D}(A^{\frac{1-\widehat{\alpha}}{2}})]' \end{array}\right\} \Longrightarrow y_t(t) \in \left[\mathcal{D}(A^{\frac{1-\widehat{\alpha}}{2}})\right]' \quad \begin{array}{r}(11.14)\\(11.15)\end{array}$$

$$v_0 \in \mathcal{D}(A^{1-\frac{\widehat{\alpha}}{2}}) \Longrightarrow y_{tt}(t) + \alpha y_t(t) \in [\mathcal{D}(A^{1-\frac{\widehat{\alpha}}{2}})]' \quad (11.16)$$

$$\Longrightarrow y_{tt}(t) \in [\mathcal{D}(A^{1-\frac{\widehat{\alpha}}{2}})]' \quad (11.17)$$

since $\dfrac{1-\widehat{\alpha}}{2} < 1 - \dfrac{\widehat{\alpha}}{2}$, and hence $\mathcal{D}(A^{1-\frac{\widehat{\alpha}}{2}}) \subset \mathcal{D}(A^{\frac{1-\widehat{\alpha}}{2}})$, hence $[\mathcal{D}(A^{\frac{1-\widehat{\alpha}}{2}})]' \subset [\mathcal{D}(A^{1-\frac{\widehat{\alpha}}{2}})]'$. Thus, conclusions (11.14), (11.15), (11.17) establish statement (11.7a). A similar argument gives (11.7b) and Proposition 11.4 proved.

Step 2 In light of (11.7b), we apply [23, Corollary 3.2, p. 173] to obtain that the map (as in (6.26))

$$t \to \int_0^t [y(\tau), y_t(\tau), y_{tt}(\tau)] d\tau$$

is continuous $[0, T] \to \mathcal{D}(A^{\frac{\widehat{\alpha}}{2}}) \times [\mathcal{D}(A^{\frac{1-\widehat{\alpha}}{2}})]' \times [\mathcal{D}(A^{1-\frac{\widehat{\alpha}}{2}})]'$. Theorem 11.3 is proved. $\square$

Acknowledgements The results of the present paper were obtained in Spring 2016, when the author was an invited visitor of the NSF-sponsored Institute of Mathematics and its Applications (IMA), University of Minnesota, Minneapolis, on the occasion of the year-long thematic program "Control Theory and its Applications", September 1, 2015–June 30, 2016. They were presented by the author at the the following conferences venues: (i) Optimal Control for Evolutionary PDEs, Cortona, Italy, June 20–24, 2016; (ii) Semigroups of Operators: Theory and Applications, Kazimierz Dolny, Poland, September 30-October 5, 2018; (iii) Colloquium at Florida International University, November 2018; (iv) XVIII Workshop on Partial Differential Equations, Petropolis, Brazil, September 2019. The author wishes to thank the IMA for its most efficient hospitality and ideal working conditions. The author is most pleased to join the other participants in celebrating the 85th birthday of Professor Jan Kisynski, whom the author was privileged to meet, and receive much math information from, in the Fall of 1974. Many happy returns to Professor Kisynski!
Research partially supported by the National Science Foundation under Grant DMS 1713506.

Appendix A

1. **Cosine Operators**. While we refer to standard work [6, 10, 11, 17, 22, 36, 40] etc for the topic of cosine operator theory on Banach space, we include here only a few results which are used and invoked in the text with reference to a Hilbert space H ($H = L^2(\Omega)$ in Part A; $H = L^2(\Omega)/\mathbb{R}$ in Part B). In line with the text, we let $(-A)$ be the (strictly positive) self-adjoint infinitesimal generator of a strongly continuous (self-adjoint) cosine operator family $\mathcal{C}(t)$ with sine operator $\mathcal{S}(t)x = \int_0^t \mathcal{C}(\tau)x d\tau$, $x \in H$, with $A^{\frac{1}{2}}\mathcal{S}(t)$ strongly continuous:

$$\mathcal{S}(t-\tau) = \mathcal{S}(t)\mathcal{C}(\tau) - \mathcal{C}(t)\mathcal{S}(\tau) \qquad \text{(A.1a)}$$

$$\mathcal{C}(t-\tau) = \mathcal{C}(t)\mathcal{C}(\tau) - A\mathcal{S}(t)\mathcal{S}(\tau), \quad \tau, t \in \mathbb{R} \qquad \text{(A.1b)}$$

We have

$$\frac{d^2\mathcal{C}(t)x}{dt^2} = -A\mathcal{C}(t)x, \, x \in \mathcal{D}(A); \qquad \frac{d\mathcal{C}(t)x}{dt} = -A\mathcal{S}(t)x, \, x \in \mathcal{D}(A^{\frac{1}{2}}), \quad \text{(A.2)}$$

$\mathcal{C}(t)$ is even on H, $\mathcal{C}(0) = I$; $\mathcal{S}(t)$ is odd on H, $\mathcal{S}(0) = 0$. The above formulae (A.2) on H with $H \supset \mathcal{D}(A) \to H$ can be extended to $[\mathcal{D}(A)]'$ with A now the extension $A_e : H \to [\mathcal{D}(A)]'$, which we still denote by A.

2. **Representation formulae of non-homogeneous boundary control for wave (second order hyperbolic) equations** [25–27, 31, 41],[23, Sect. 3]
   Dirichlet case We return to the Dirichlet non-homogeneous w-problem in (2.15a)–(2.15c). Let D be the Dirichlet map in (2.21a), (2.21b) and $(-A)$ be the Dirichlet Laplacian in (2.2). Then

$$w(t) = A \int_0^t \mathcal{S}(t-\tau)Dg(\tau)d\tau; \quad w_t(t) = A \int_0^t \mathcal{C}(t-\tau)Dg(\tau)d\tau. \quad \text{(A.3)}$$

   Neumann case We now return to the Neumann non-homogeneous w-problem in (7.5a)–(7.5c). Let N be the Neumann map in (7.17a), (7.17b) and $(-A)$ be the Neumann Laplacian in (7.2). Then

$$w(t) = A \int_0^t \mathcal{S}(t-\tau)Ng(\tau)d\tau; \quad w_t(t) = A \int_0^t \mathcal{C}(t-\tau)Ng(\tau)d\tau. \quad \text{(A.4)}$$

3. **Operator formulae for traces**
   Let $(-A)$ be the Dirichlet Laplacian in (2.2) and D the Dirichlet map in (2.21a), (2.21b). Then [41], [29, p. 181]

$$D^*A^*\phi = -\frac{\partial \phi}{\partial \nu}, \quad \phi \in \mathcal{D}(A), \qquad \text{(A.5)}$$

which can be extended to all $\phi \in H^{\frac{3}{2}+\epsilon}(\Omega) \cap H_0^1(\Omega)$, $\epsilon > 0$.
Let now $(-A)$ be the Neumann Laplacian in (7.2) and N the Neumann map in (7.17a), (7.17b). Then [41], [29, p. 196]

$$N^* A^* \phi = \phi \big|_\Gamma, \quad \phi \in \mathcal{D}(A), \tag{A.6}$$

which can be extended to all $\phi \in H^{\frac{3}{2}+\epsilon}(\Omega) \cap H_0^1(\Omega)$, $\epsilon > 0$, with $\frac{\partial \phi}{\partial \nu}\big|_\Gamma = 0$.

Appendix B The Dual Problem of the Boundary Non-homogeneous Problem (2.1a)–(2.1c). A PDE-Approach

In this Appendix we consider the following two problems:

Problem #1 (2.1a)–(2.1c)

$$\begin{cases} y_{ttt} + \alpha y_{tt} - c^2 \Delta y - b \Delta y_t = 0 & \text{in } Q = (0, T] \times \Omega \quad \text{(B.1a)} \\ y\big|_{t=0} = y_0; \quad y_t\big|_{t=0} = y_1; \quad y_{tt}\big|_{t=0} = y_2 & \text{in } \Omega \quad \text{(B.1b)} \\ \text{and either} \\ \text{Dirichlet-control } y\big|_\Sigma = g & \text{in } \Sigma = (0, T] \times \Gamma \quad \text{(B.1c)} \\ \text{or else} \\ \text{Neumann-control } \frac{\partial y}{\partial \nu}\bigg|_\Sigma = g & \text{in } \Sigma. \quad \text{(B.1d)} \end{cases}$$

Problem #2 With $T > 0$ arbitrary,

$$\begin{cases} v_{ttt} - \alpha v_{tt} + c^2 \Delta v - b \Delta v_t = 0 & \text{in } Q \quad \text{(B.2a)} \\ v\big|_{t=T} = v_0; \quad v_t\big|_{t=T} = v_1; \quad v_{tt}\big|_{t=T} = v_2 & \text{in } \Omega \quad \text{(B.2b)} \\ \text{and either} \\ \text{Dirichlet homogeneous B.C. } v\big|_\Sigma \equiv 0 & \text{in } \Sigma \quad \text{(B.2c)} \\ \text{or else} \\ \text{Neumann homogeneous B.C. } \frac{\partial v}{\partial \nu}\bigg|_\Sigma = 0 & \text{in } \Sigma. \quad \text{(B.2d)} \end{cases}$$

The v-problem (B.2a)–(B.2d) is dual to the y-problem (B.1a)–(B.1d) for zero I.C.: $y_0 = y_1 = y_2 = 0$, in the sense specified below

Theorem B.1 *(i) Under the appropriate regularity assumptions on the data: $\{y_0, y_1, y_2\}$, g, and $\{v_0, v_1, v_2\}$—to be made explicit below—the following identity holds true, where $\langle\,,\,\rangle_\Omega$ denotes the duality pairing with respect to $H = L^2(\Omega)$ and $\langle\,,\,\rangle_\Gamma$ denotes the duality pairing with respect to $L^2(\Gamma)$:*

$$\langle y_{tt}(T) + \alpha y_t(T), v_0 \rangle_\Omega - \langle y_t(T) + \alpha y(T), v_1 \rangle_\Omega + \langle y(T), v_2 \rangle_\Omega - b \langle y(T), \Delta v_0 \rangle_\Omega$$

$$+ \langle y_0, -v_{tt}(0) + \alpha v_t(0) + b\Delta v(0) \rangle_\Omega + \langle y_1, v_t(0) - \alpha v(0) \rangle_\Omega - \langle y_2, v(0) \rangle_\Omega$$

$$- \left\langle c^2 \frac{\partial y}{\partial \nu} + b \frac{\partial y_t}{\partial \nu}, v \right\rangle_{L^2(\Sigma)} + \left\langle c^2 y + b y_t, \frac{\partial v}{\partial \nu} \right\rangle_{L^2(\Sigma)} = 0. \qquad (\text{B.3})$$

(ii) Consider the Dirichlet non-homogeneous condition (B.1c) with zero I.C. $y_0 = y_1 = y_2 = 0$, coupled with the corresponding homogeneous Dirichlet condition (B.2c). Then identity (B.3) specializes to

$$\langle y_{tt}(T) + \alpha y_t(T), v_0 \rangle_\Omega - \langle y_t(T) + \alpha y(T), v_1 \rangle_\Omega + \langle y(T), v_2 \rangle_\Omega - b \langle y(T), \Delta v_0 \rangle_\Omega$$
$$= -\left\langle c^2 g + b g_t, \frac{\partial v}{\partial \nu} \right\rangle_{L^2(0,T;L^2(\Gamma))}. \qquad (\text{B.4})$$

(iii) Consider the Neumann non-homogeneous condition (B.1d) with zero I.C. $y_0 = y_1 = y_2 = 0$, coupled with the corresponding homogeneous Neumann condition (B.2d). Then identity (B.3) specializes to

$$\langle y_{tt}(T) + \alpha y_t(T), v_0 \rangle_\Omega - \langle y_t(T) + \alpha y(T), v_1 \rangle_\Omega + \langle y(T), v_2 \rangle_\Omega$$
$$- b \langle y(T), \Delta v_0 \rangle_\Omega$$
$$= \langle c^2 g + b g_t, v \rangle_{L^2(0,T;L^2(\Gamma))}. \qquad (\text{B.5})$$

Proof **Step 1** Multiply (B.1a) by v and integrate by parts. We obtain:

$$① = \int_\Omega \int_0^T y_{ttt} v \, dt d\Omega = \langle y_{tt}(T), v(T) \rangle_\Omega - \langle y_{tt}(0), v(0) \rangle_\Omega - \langle y_t(T), v_t(T) \rangle_\Omega$$

$$+ \langle y_t(0), v_t(0) \rangle_\Omega + \langle y(T), v_{tt}(T) \rangle_\Omega - \langle y(0), v_{tt}(0) \rangle_\Omega - \int_\Omega \int_0^T y v_{ttt} \, dQ \qquad (\text{B.6})$$

$$② = \int_\Omega \int_0^T y_{tt} v \, dt d\Omega = \langle y_t(T), v(T) \rangle_\Omega - \langle y_t(0), v(0) \rangle_\Omega - \langle y(T), v_t(T) \rangle_\Omega$$

$$+ \langle y(0), v_t(0) \rangle_\Omega + \int_\Omega \int_0^T y v_{tt} \, dQ \qquad (\text{B.7})$$

$$③ = \int_0^T \int_\Omega \Delta y v \, d\Omega dt = \int_0^T \left[\int_\Omega y \Delta v d\Omega + \int_\Gamma \frac{\partial y}{\partial \nu} v d\Gamma - \int_\Gamma y \frac{\partial v}{\partial \nu} d\Gamma \right] dt \qquad (\text{B.8})$$

$$④ = \int_0^T \int_\Omega \Delta y_t v \, d\Omega \, dt = \int_0^T \left[\int_\Omega y_t \Delta v \, d\Omega + \int_\Gamma \frac{\partial y_t}{\partial \nu} v \, d\Gamma - \int_\Gamma y_t \frac{\partial v}{\partial \nu} d\Gamma \right] dt \quad \text{(B.9)}$$

Step 2 We sum up: $① + \alpha② - c^2③ - b④ = 0$ and obtain

$$\langle y_{tt}(T) + \alpha y_t(T), v(T) \rangle_\Omega - \langle y_t(T) + \alpha y(T), v_t(T) \rangle_\Omega + \langle y(T), v_{tt}(T) \rangle_\Omega - b \langle y(T), \Delta v(T) \rangle_\Omega$$
$$+ \langle y(0), -v_{tt}(0) + \alpha v_t(0) + b\Delta v(0) \rangle_\Omega + \langle y_1, v_t(0) - \alpha v(0) \rangle_\Omega - \langle y_2, v(0) \rangle_\Omega$$
$$- \int_Q y \left[v_{ttt} - \alpha v_{tt} + c^2 \Delta v - b \Delta v_t \right] dQ - \left\langle c^2 \frac{\partial y}{\partial \nu} + b \frac{\partial y_t}{\partial \nu}, v \right\rangle_{L^2(\Sigma)} + \left\langle c^2 y + b y_t, \frac{\partial v}{\partial \nu} \right\rangle_{L^2(\Sigma)} = 0.$$
(B.10)

Step 3 The $\int_Q$-term in (B.10) vanishes because of (B.2a). Next, we use the I.C. in (B.2b) for the v-problem at $t = T$, and identity (B.10) reduces to (B.3). Part (i) is proved.

Step 4 In the Dirichlet case, use $y\big|_\Sigma = 0$ in (B.1c) and $v\big|_\Sigma \equiv g$ in (B.2c). Then, identity (B.3) reduces to identity (B.4).

Step 5 In the Neumann case, use $\dfrac{\partial y}{\partial \nu}\bigg|_\Sigma = g$ in (B.1d) and $\dfrac{\partial v}{\partial \nu}\bigg|_\Sigma \equiv 0$ in (B.2d). Then identity (B.3) reduces to identity (B.5).

The next is a preliminary result.

Corollary B.2 *With reference to the Dirichlet-Problem # 1 in (B.1a)–(B.1d) with I.C. $y_0 = y_1 = y_2 = 0$ and corresponding Dirichlet Problem # 2 in (B.2a)–(B.2d), assume*

$$g \in H^1(0, T_1; L^2(\Gamma)), \quad \{v_0, v_1, v_2\} \in U_3 = \mathcal{D}(A) \times \mathcal{D}(A^{\frac{1}{2}}) \times H. \quad \text{(B.11)}$$

Then, for any t, $0 < t \leq T_1$:

$$y(t), y_t(t), y_{tt}(t) \in H \times [\mathcal{D}(A^{\frac{1}{2}})]' \times [\mathcal{D}(A)]'. \quad \text{(B.12)}$$

Proof We have Theorem 6.1 for any $0 < T_1 < \infty$:

$$\{v_0, v_1, v_2\} \in U_3 \implies \{v, v_t, v_{tt}\} \in C\left([0, T_1]; U_3 = \mathcal{D}(A) \times \mathcal{D}(A^{\frac{1}{2}}) \times H\right) \quad \text{(B.13)}$$

so that just by trace theory

$$\frac{\partial v}{\partial \nu} \in C\left([0, T]; H^{\frac{1}{2}}(\Gamma))\right) \quad \text{(B.14)}$$

Thus, the RHS of identity (B.4) is well defined by (B.11), (B.14), finite on any finite time interval. Here T is an arbitrary point $0 < T \leq T_1$. We then focus on the LHS of identity (B.4) to make sure that each term is well defined as a duality pairing w.r.t. $H = L^2(\Omega)$. We obtain

$$\left.\begin{array}{r} v_2 = v_{tt}(T) \in H \implies y(T) \in H \\ v_1 = v_t(T) \in \mathcal{D}(A^{\frac{1}{2}}) \implies y_t(T) + \alpha y(T) \in [\mathcal{D}(A^{\frac{1}{2}})]' \end{array}\right\} \implies y_t(T) \in [\mathcal{D}(A^{\frac{1}{2}})]' \right\} \implies y_{tt}(T) \in [\mathcal{D}(A)]'.$$
$$v_0 = v(T) \in \mathcal{D}(A) \implies y_{tt}(T) + \alpha y_t(T) \in [\mathcal{D}(A)]'$$
(B.15)

This takes care of the first three therms on the LHS of (B.4). Notice then that the fourth term $\langle y(T), \Delta v_0 \rangle_\Omega$ is likewise automatically well-posed with $v_0 \in \mathcal{D}(A)$, $\Delta v_0 \in H = L^2(\Omega)$, $y(T) \in H$. $\square$

References

1. Alves, M., Buriol, C., Ferreira, M., Rivera, J.M., Sepulveda, M., Vera, O.: Asymptotic behaviour for the vibrations modeled by the standard linear solid model with a thermal effect. JMAA **399**, 472–479 (2013)
2. Banasiak, J.: Chaos in Kolmogorov systems with proliferation-general criteria and applications. JMAA **378**, 89–97 (2011)
3. Banasiak, J., Moszynski, M.: A generalization of Desch-Schappacher-Webb criteria for chaos. Discrete Contin. Dyn. Syst. **12**(5), 959–972 (2005)
4. Conejero, J.A., Lizama, C., Ródenas Escribá, F.: Chaotic behaviour of the solutions of the Moore-Gibson-Thompson equation. Appl. Math. Inf. Sci. **9**(N5), 2233–2238 (2015)
5. Christov, I.: Private Communication
6. Da Prato, G., Giusti, E.: Una caratterizzazione dei generatori di funzioni coseno astratte. Bollettino dell'Unione Matematica Italiana **22**, 357–362 (1967). (in Italian)
7. Desch, W., Schappacher, W., Webb, G.: Hypercyclic and chaotic semigroups of linear operators. Ergod. Theory Dyn. Syst. **17**, 793–819 (1997)
8. Fattorini, H.O..: Ordinary differential equations in linear topological spaces, I. J. Differ. Equ. **5**(1), 72–105
9. H. O. Fattorini, Ordinary differential equations in linear topological spaces, II, *J. Diff. Eqns*, 6(1), 50-70
10. Fattorini, H.O.: Second Order Linear Differential Equations in Banach Spaces. North-Holland, Amsterdam (1985)
11. Fattorini, H.O.: The Cauchy Problem. Encyclopedia of Mathematics and its Applications, p. 636. Addison-Wesley (1983)
12. Jordan, P.: An analytical study of Kuznetsov's equation: diffusive solitons, shock formation, and solution bifurcation. Phys. Lett. A **326**, 77–84 (2004)
13. Jordan, P.: Nonlinear acoustic phenomena in viscous thermally relaxing fluids: Shock bifurcation and the emergence of diffusive solitons. J. Acoustic Soc. Am. **124**(4), 2491–2491 (2008)
14. Jordan, P.: Private Communication
15. Kaltenbacher, B., Lasiecka, I., Marchand, R.: Wellposedness and exponential decay rates for the Moore-Gibson-Thompson equation arising in high intensity ultrasound. Control Cybern. **40**, 971–988 (2011)
16. Kaltenbacher, B., Lasiecka, I., Pospieszalska, M.: Well-posedness and exponential decay of the energy in the nonlinear Jordan-Moore-Gibson-Thompson equation arising in high intensity ultrasound. Math. Methods Appl. Sci. **22**(11) (2012)
17. Kisyński, Sur: les équations différentielles dans les espaces de Banach. Bull. Acad. Polon. Sci. Sér. Sci. Math. Astr. Plys. **7**, 381–385 (1959). (in French)
18. Kisyński, J.: On second order cauchy's problem in a Banach space. Bull. Acad. Polon. Sci. Sér. Sci. Math. Astr. Plys. **18**(7), 371–374 (1970)

19. Kisyński, J.: On operator-valued solutions of d'Alembert's functional equation. I Colloquium Mathematicum **23**, 107–114 (1971)
20. Kisyński, J.: On cosine operator functions and one-parameter groups of operators. Studia Mathematica **44**, 93–105 (1972)
21. Kisyński, J.: On operator-valued solutions of d'Alembert's functional equation. II Studia Mathematica **42**, 43–66 (1972)
22. Kisyński, J.: Semi-groups of operators and some of their applications to partial differential equations. In: Control Theory and Topics in Functional Analysis (Internat. Sem., Internat. Centre Theoret. Phys., Trieste, 1974), Vienna, International Atomic Energy Agency, vol. 3, pp. 305–405 (1976)
23. Lasiecka, I., Lions, J.-L., Triggiani, R.: Nonhomogeneous boundary value problems for second order hyperbolic operators. J. Math. Pures Appl. **65**, 149–192 (1986)
24. Lions, J.L., Magenes, E.: Nonhomogeneous Boundary Value Problems and Applications I. Springer, Berlin (1972)
25. Lasiecka, I., Triggiani, R.: A cosine operator approach to modeling $L_2(0, T; L_2(\Omega))$ boundary input hyperbolic equations. Appl. Math. Optimiz. **7**, 35–83 (1981)
26. Lasiecka, I., Triggiani, R.: Regularity of hyperbolic equations under $L_2(0, T; L_2(\Omega))$-Dirichlet boundary terms. Appl. Math. Optimiz. **10**, 275–286 (1983)
27. Lasiecka, I., Triggiani, R.: Feedback semigroups and cosine operators for boundary feedback parabolic and hyperbolic equations. J. Differ. Equ. **47**, 246–272 (1983)
28. Lasiecka, I., Triggiani, R.: Trace regularity of the solutions of the wave equation with homogeneous Neumann boundary conditions and data supported away from the boundary. J. Math. Anal. Appl. **141**(1), 49–71 (1989)
29. Lasiecka, I., Triggiani, R.: Sharp regularity for mixed second order hyperbolic equations of Neumann type. Part I: The L_2-boundary case. Ann. Mat. Pura Appl. (4) 157, 285–367 (1990)
30. Lasiecka, I., Triggiani, R.: Regularity theory of hyperbolic equations with non-homogeneous Neumann boundary conditions. II. General boundary data. J. Differ. Equ. **94**, 112–164 (1991)
31. Lasiecka, I., Triggiani, R.: Control theory for partial differential equations: continuous and approximation theories, Vol I: abstract parabolic systems (P. 644); Vol II: Abstract hyperbolic systems over a finite time horizon (P. 422). In: Encyclopedia of Mathematics and Its Applications Series. Cambridge University Press, Cambridge (2000)
32. Lasiecka, I., Triggiani, R.: Exact controllability of the wave equation with Neumann boundary control. Appl. Math. Optimiz. **19**(1), 243–290 (1989)
33. Moore, F.K., Gibson, W.E.: Propagation of weak disturbances in a gas subject to relaxation effects. J. Aero/Space Sci. **27**, 117–127 (1960)
34. Marchand, R., McDevitt, T., Triggiani, R.: An abstract semigroup approach to the third-order MGT equation arising in high-intensity ultrasound: Structural decomposition, spectral analysis, exponential stability. Math. Methods Appl. Sci. **35**, 1896–1929 (2012)
35. Nelson, S., Triggiani, R.: Analytic properties of cosine operators. Proceed. AMS **74**, 101–104 (1978)
36. Sova, M.: Cosine Operator Functions. Rozprawy Matematyczne, vol. 49 (1966)
37. Stokes, G.G.: An examination of the possible effect of the radiation of heat on the propagation of sound. Philos. Mag. Ser. **1**(4), 305–317 (1851)
38. Tataru, D.: On the regularity of boundary traces for the wave equation. Annali della Scuola Normale Superiore di Pisa-Classe di Scienze (4) 26, 185-206 (1998)
39. Thompson, P.A.: Compressible-Fluid Dynamics. McGraw-Hill, New York (1972)
40. Travis, C.C., Webb, G.: Second order differential equations in Banach space. In: Lakshmikantham, V. (ed.) Nonlinear Equations in Abstract Spaces, pp. 331–361. Academic Press, London (1978)
41. Triggiani, R.: A Cosine Operator Approach to Modeling Boundary Input Problems for Hyperbolic Systems. Lecture Notes in Control and Information Sciences, vol. 6, pp. 380–390. Springer, Berlin (1978)
42. Triggiani, R.: Exact boundary controllability on $L^2(\Omega) \times H^{-1}(\Omega)$ of the wave equation with Dirichlet boundary control acting on a portion of the boundary $\partial\Omega$, and related problems. Appl. Math. Optimiz. **18**, 241–277 (1988)

Inverse Problem for the Boussinesq – Love Mathematical Model

Alyona A. Zamyshlyaeva and Aleksandr V. Lut

Abstract The work is devoted to the investigation of the inverse problem for the Boussinesq – Love equation with additional conditions. Such an equation is a Sobolev type equation and models a longitudinal vibrations in a thin elastic rod. We apply the previously obtained results for an abstract problem. The original mathematical model is reduced to the Cauchy problem for abstract Sobolev type equation of the second order sufficient conditions for the unique solvability of the problem under study are obtained.

Keywords Sobolev type equation · Mathematical model · Boussinesq – Love equation · Inverse problem

1 Introduction

Let $\Omega \subset \mathbb{R}^n$ be a bounded domain with a boundary $\partial \Omega$ of class C^∞. In the cylinder $\Omega \times [0; T]$ consider the Boussinesq – Love equation

$$(\lambda - \Delta)v_{tt} = \alpha(\Delta - \lambda')v_t + \beta(\Delta - \lambda'')v + fq, \tag{1}$$

with initial conditions

$$v(x, 0) = v_0(x), \ v_t(x, 0) = v_1(x), \tag{2}$$

boundary condition

$$v(x, t)|_{\partial \Omega} = 0 \tag{3}$$

A. A. Zamyshlyaeva · A. V. Lut (✉)
South Ural State University, Lenina, 76, Chelyabinsk, Russia
e-mail: lutav@susu.ru

A. A. Zamyshlyaeva
e-mail: zamyshliaevaaa@susu.ru

and overdetermination condition

$$\int_\Omega v(x,t)K(x)dx = \Phi(t), \tag{4}$$

where $K(x)$ is a given function in $L_2(\Omega)$.

Equation (1) describes the longitudinal vibration in the elastic rod, taking into account the inertia and under external load. Conditions (2) set the initial displacement and initial speed, respectively, and (3) sets the value at the boundaries. The overdetermination condition (4) arises when, in addition to finding the function u, one needs to restore part of external load q.

Problems (1)–(4) can be reduced to a second-order Sobolev type equation

$$Av''(t) = B_1 v'(t) + B_0 v(t) + \chi(t)q(t), \ t \in [0, T], \tag{5}$$

with conditions

$$v(0) = v_0, \ v'(0) = v_1, \tag{6}$$

$$Cv(t) = \Psi(t), \tag{7}$$

where $\mathscr{U}$, $\mathscr{F}$, $\mathscr{Y}$ are Banach spaces, operators A, B_1, $B_0 \in \mathscr{L}(\mathscr{U}; \mathscr{F})$, ker $A \neq \{0\}$, $C \in \mathscr{L}(\mathscr{U}; \mathscr{Y})$, the functions $\chi : [0, T] \to \mathscr{L}(\mathscr{Y}; \mathscr{F})$, $\Psi : [0, T] \to \mathscr{Y}$.

The study of Sobolev type equations is done by many researchers [1–7]. In these works, the results for the equations of the first [1–5, 8], the second [7] and higher orders [6] are obtained. There was constructed a solution for a stochastic first-order Sobolev type equation with an (L, p)-radial operator M based on the deterministic results obtained earlier [1]. Optimal control for semilinear Sobolev type models with s-monotone and p-coercive operators was studied in [3]. The p-sectoriality condition was introduced and studied in [4], the sufficient conditions for the existence of solutions for a first order Sobolev type equation in terms of analytical semigroups were presented. In [6] the sufficient conditions for the existence and uniqueness of a strong solution to an initial-final problem for an abstract Sobolev type equation of high order with relatively polynomially A-bounded operator pencil are obtained. Moreover, Sobolev type equations find their application in mathematical modeling of various processes and phenomena [5, 7]. In addition to this type of equations, there are various others, for example, Leontief type systems [8], which can be treated using the theory of Sobolev type equations.

In this paper, the inverse problem for the mathematical model (1)–(4), which will be reduced to (5)–(7) using the results of [7], is considered. Different types of inverse problems were studied in [2, 7, 9–12]. The solvability of the inverse problem for the equation of propagation of longitudinal waves of the first order was investigated in [2]. The inverse problem for a second-order hyperbolic equation with integral conditions was studied in [11]. A theorem on the solvability of the coefficient inverse problem for a fourth-order linear partial differential equation is presented in [9]. In the case of high order differential equations, the condition of existence and uniqueness of the solution can be found in [10, 12].

The paper consists of 4 sections and a list of references. The first section contains the statement of the problem with the historiography of the issue. The second contains auxiliary results obtained in [13]. In the third section the results for the abstract Sobolev type equation of the second order are presented. In the fourth, the main result of the paper is obtained, namely, the theorem on the existence and uniqueness of the solution of the inverse problem for the Boussinesq – Love mathematical model (1)–(4) is presented.

2 Preliminary Information

In this section we introduce the results obtained in the research of higher order Sobolev type equations [13]. Denote by $\vec{B}$ the pencil of operators B_1, B_0.

Definition 1 The pencil $\vec{B}$ is called polynomially A-bounded if

$$\exists\, a \in \mathbb{R}_+ \ \forall \mu \in \mathbb{C} \ (|\mu| > a) \Rightarrow (R^A_\mu(\vec{B}) \in \mathcal{L}(\mathcal{F};\mathcal{U})).$$

Here $R^A_\mu(\vec{B}) = (\mu^2 A - \mu B_1 - B_0)^{-1}$ denotes an A-resolvent of the pencil $(\vec{B})$.

Introduce an important condition:

$$\int_\gamma R^A_\mu(\vec{B}) d\mu \equiv \mathbb{O}, \tag{A}$$

where $\gamma = \{\mu \in \mathbb{C} : |\mu| = r > a\}$.

Lemma 1 *Let the pencil $\vec{B}$ be polynomially A-bounded and condition (A) be fulfilled. Then the operators*

$$P = \frac{1}{2\pi i} \int_\gamma R^A_\mu(\vec{B}) \mu A d\mu \in \mathcal{L}(\mathcal{U}), \quad Q = \frac{1}{2\pi i} \int_\gamma \mu A R^A_\mu(\vec{B}) d\mu \in \mathcal{L}(\mathcal{F})$$

are projectors.

Put $\mathcal{U}^0 = \ker P$, $\mathcal{F}^0 = \ker Q$, $\mathcal{U}^1 = \mathrm{im}\, P$, $\mathcal{F}^1 = \mathrm{im}\, Q$. From the previous Lemma it follows that $\mathcal{U} = \mathcal{U}^0 \oplus \mathcal{U}^1$, $\mathcal{F} = \mathcal{F}^0 \oplus \mathcal{F}^1$. Let $A^k(B_l^k)$ denote the restriction of the operator $A(B_l)$ onto $\mathcal{U}^k$, $k = 0, 1$; $l = 0, 1$.

Theorem 1 *Let the pencil $\vec{B}$ be polynomially A-bounded and condition (A) be fulfilled. Then the actions of the operators split:*
  (i) $A^k \in \mathcal{L}(\mathcal{U}^k; \mathcal{F}^k)$, $k = 0, 1$;

(ii) $B_l^k \in \mathscr{L}(\mathscr{U}^k; \mathscr{F}^k)$, $k = 0, 1$, $l = 0, 1$;
(iii) there exists an operator $(A^1)^{-1} \in \mathscr{L}(\mathscr{F}^1; \mathscr{U}^1)$;
(iv) there exists an operator $(B_0^0)^{-1} \in \mathscr{L}(\mathscr{F}^0; \mathscr{U}^0)$.

Definition 2 Define the family of operators $\{K_q^1, K_q^2\}$ as follows:
$K_0^1 = \mathbb{O}$, $K_0^2 = \mathbb{I}$,
$K_{q+1}^1 = K_q^2 H_0$, $K_{q+1}^2 = K_q^1 - K_q^2 H_1$, $q = 1, 2, \ldots$
where $H_0 = (B_0^0)^{-1} A^0$; $H_1 = (B_0^0)^{-1} B_1^0$.

Definition 3 The point ∞ is called
(i) a removable singular point of the A-resolvent of pencil $\vec{B}$, if $K_1^1 \equiv \mathbb{O}$, $K_1^2 \equiv \mathbb{O}$;
(ii) a pole of order $p \in \mathbb{N}$ of the A-resolvent of pencil $\vec{B}$, if $K_p^1 \neq \mathbb{O}$ or $K_p^2 \neq \mathbb{O}$, but $K_{p+1}^1 \equiv \mathbb{O}$, $K_{p+1}^2 \equiv \mathbb{O}$;
(iii) an essentially singular point of the A-resolvent of the pencil $\vec{B}$, if $K_k^2 \neq \mathbb{O}$ for any $k \in \mathbb{N}$.

3 Abstract Problem

This section presents the results [7] obtained in the study of the abstract problem (5)–(7).

Let the pencil $\vec{B}$ be polynomially A-bounded and condition (A) be fulfilled, then $v(t)$ can be represented as $v(t) = Pv(t) + (I - P)v(t) = u(t) + w(t)$. Suppose that $\mathscr{U}^0 \subset \ker C$. Then, by virtue of Theorem 1 and Lemma 1, problem (5)–(7) is equivalent to the problem of finding the functions $u \in C^2([0, T]; \mathscr{U}^1)$, $w \in C^2([0, T]; \mathscr{U}^0)$, $q \in C^1([0, T]; \mathscr{Y})$ from the relations

$$u''(t) = S_1 u'(t) + S_0 u(t) + (A^1)^{-1} Q\chi(t)q(t), \tag{8}$$

$$u(0) = u_0, \quad u'(0) = u_1, \tag{9}$$

$$Cu(t) = \Psi(t) \equiv Cv(t), \tag{10}$$

$$H_0 w''(t) = H_1 w'(t) + w(t) + (B_0^0)^{-1}(I - Q)\chi(t)q(t), \tag{11}$$

$$w(0) = w_0, \quad w'(0) = w_1, \tag{12}$$

where $S_1 = (A^1)^{-1} B_1^1$, $S_0 = (A^1)^{-1} B_0^1$, $u_0 = Pv_0$, $u_1 = Pv_1$, $w_0 = (I - P)v_0$, $w_1 = (I - P)v_1$, $t \in [0, T]$. The inverse problem (8)–(10) is called regular, and problem (11), (12) is called singular.

The following theorem on the existence and uniqueness of the solution was obtained in [7].

Theorem 2 *Let the pencil $\vec{B}$ be polynomially A-bounded and condition (A) be fulfilled, moreover, the ∞ be a pole of order $p \in \mathbb{N}_0$ of the A-resolvent of the pencil $\vec{B}$, operator $C \in \mathscr{L}(\mathscr{U}; \mathscr{Y})$, $\mathscr{U}^0 \subset \ker C$, $\chi \in C^{p+2}([0, T]; \mathscr{L}(\mathscr{Y}; \mathscr{F}))$, $\Psi \in C^{p+4}([0, T]; \mathscr{Y})$, for any $t \in [0, T]$ operator $C(A^1)^{-1}Q\chi$ be invertible, with $(C(A^1)^{-1}Q\chi)^{-1} \in C^{p+2}([0, T]; \mathscr{L}(\mathscr{Y}))$, the condition $Cu_1 = \Psi'(0)$ be satisfied at some initial value $u_1 \in \mathscr{U}$, and the initial values $w_k = (I - P)v_k \in \mathscr{U}^0$ satisfy*

$$w_k = -\sum_{j=0}^{p} K_j^2(B_0^0)^{-1} \frac{d^{j+k}}{dt^{j+k}}\left[(I - Q)(\chi(0)q(0))\right], \quad k = 0, 1.$$

Then there exists a unique solution (v, q) of inverse problem (5)–(7), where $q \in C^{p+2}([0, T]; \mathscr{Y})$, $v = u + w$, whence $u \in C^2([0, T]; \mathscr{U}^1)$ is the solution of (8)–(10) and the function $w \in C^2([0, T]; \mathscr{U}^0)$ is a solution of (11), (12) given by

$$w(t) = -\sum_{j=0}^{p} K_j^2(B_0^0)^{-1} \frac{d^j}{dt^j}\left[(I - Q)(\chi(t)q(t))\right].$$

4 The Boussinesq – Love Mathematical Model

Reduce the Boussinesq – Love equation (1) with conditions (2)–(4) to problem (5)–(7), for this we put

$$\mathscr{U} = \{u \in W_q^{l+2}(\Omega) : u(x) = 0, x \in \partial\Omega\}, \quad \mathscr{F} = W_q^l(\Omega), \quad \mathscr{Y} = \mathscr{F},$$

where $W_q^l(\Omega)$ are Sobolev spaces $2 \leq q < \infty$, $l = 0, 1, \ldots$. Set the operators $A = (\lambda - \Delta)$, $B_1 = \alpha(\Delta - \lambda')$, $B_0 = \beta(\Delta - \lambda'')$, $Cv = \int_\Omega v(x)K(x)dx$ and the functions $\Psi(t) = \Phi(t)$, $\chi(t)$ is multiplication by $f(x, t)$. For any $l \in \{0\} \cup \mathbb{N}$ operators $C \in \mathscr{L}(\mathscr{U}; \mathscr{Y})$, $A, B_1, B_0 \in \mathscr{L}(\mathscr{U}; \mathscr{F})$.

Denote by $\sigma(\Delta)$ the spectrum of the homogeneous problem (3) for the Laplace operator Δ in Ω. The spectrum $\sigma(\Omega)$ is negative, discrete and condenses only to $-\infty$. Denote by $\{\lambda_k\}$ the set of eigenvalues numbered in non-increasing order with multiplicity, and by $\{\varphi_k\}$ the family of corresponding eigenfunctions orthonormal with respect to the inner product $< \cdot, \cdot >$ in $L^2(\Omega)$.

Since $\{\varphi_k\} \subset C^\infty(\Omega)$, then

$$\mu^2 A - \mu B_1 - B_0 = \sum_{k=1}^{\infty} |(\lambda - \lambda_k)\mu^2 + \alpha(\lambda' - \lambda_k)\mu + \beta(\lambda'' - \lambda_k)| < \cdot, \varphi_k > \varphi_k.$$

Lemma 2 *Let one of the following conditions be fulfilled:*
  (i) $\lambda \notin \sigma(\Delta)$;
  (ii) $(\lambda \in \sigma(\Delta)) \wedge (\lambda \neq \lambda')$;
  (iii) $(\lambda \in \sigma(\Delta)) \wedge (\lambda = \lambda') \wedge (\lambda \neq \lambda'')$.
Then the pencil $\vec{B} = (B_1, B_0)$ is polynomially A-bounded.

Proof In case (i) A-spectrum of the pencil $\vec{B}$ $\sigma^A(\vec{B}) = \{\mu_k^{1,2} : k \in \mathbb{N}\}$, where $\mu_k^{1,2}$ are the roots of the equation

$$(\lambda - \lambda_k)\mu^2 + \alpha(\lambda' - \lambda_k)\mu + \beta(\lambda'' - \lambda_k) = 0. \tag{13}$$

In case (ii) $\sigma^A(\vec{B}) = \{\mu_{l,k}^{1,2} : k \in \mathbb{N}\}$, where $\mu_{l,k}^{1,2}$ are the roots of the Eq. (13) with $\lambda = \lambda_l$. In case (iii) $\sigma^A(\vec{B}) = \{\mu_{l,k}^{1,2} : k \in \mathbb{N}, k \neq l\}$. ▷

Remark 1 It is easy to see that in the case when $(\lambda \in \sigma(\Delta)) \wedge (\lambda = \lambda' = \lambda'')$ the pencil $\vec{B}$ is not polynomially A-bounded.

Now check condition (A). In case (i), there exists an operator $A^{-1} \in \mathcal{L}(\mathcal{F}; \mathcal{U})$, therefore, by virtue of [13], condition (A) is satisfied. In case (ii)

$$\frac{1}{2\pi i} \int_\gamma \sum_{k=1}^\infty \frac{<\cdot, \varphi_k > \varphi_k d\mu}{(\lambda - \lambda_k)\mu^2 + \alpha(\lambda' - \lambda_k)\mu + \beta(\lambda'' - \lambda_k)} = \sum_{\lambda = \lambda_k} \frac{<\cdot, \varphi_k >}{\alpha(\lambda' - \lambda_k)} \varphi_k \neq \mathbb{O},$$

that is, (A) does not hold, so this case is excluded from further considerations. In case (iii) condition (A) holds.

Lemma 3 [13] *Let one of the conditions (i) or (iii) of Lemma 2 be fulfilled, then ∞ is a removable singular point A of the resolvent of the pencil $\vec{B}$.*

Remark 2 Note that if condition (ii) of Lemma 2 is satisfied, then ∞ is an essentially singular point A of the resolvent of the pencil $\vec{B}$.

Construct the projectors. In case (i) of Lemma 2 $P = \mathbb{I}$ and $Q = \mathbb{I}$, in case (iii) of Lemma 2

$$P = \mathbb{I} - \sum_{\lambda = \lambda_k} < \cdot, \varphi_k > \varphi_k,$$

and the projector Q has the same form, but is defined on the space $\mathcal{F}$.

Theorem 3 *Let one of the conditions (i) or (iii) of Lemma 2 be fulfilled. K, u_0, $u_1 \in \mathcal{U}^1$, $f \in C^2([0, T]; \mathcal{L}(\mathcal{Y}, \mathcal{F}))$, $\Phi \in C^4([0, T]; \mathcal{Y})$, $\sum_{\lambda \neq \lambda_k} \frac{<f(\cdot,t)q(t), K>}{\lambda - \lambda_k} \neq 0$, the condition $\int_\Omega v_1(x)K(x)dx = \Phi'(0)$ be satisfied at some initial value $v_1 \in \mathcal{U}^1$, and the initial values $w_k = (I - P)v_k \in \mathcal{U}^0$ satisfy*

$$\left< v_0 + \frac{f(\cdot,t)q(0)}{\beta(\lambda_k - \lambda'')}, \varphi_k \right> = 0 \; for \; k : \lambda_k = \lambda,$$

$$\left< v_1 + \frac{f_t(\cdot,t)q(0) + f(\cdot,t)q'(0)}{\beta(\lambda_k - \lambda'')}, \varphi_k \right> = 0 \; for \; k : \lambda_k = \lambda.$$

Then there exists a unique solution (v, q) of inverse problem (1)–(4), where $q \in C^2([0, T]; \mathcal{Y})$, $v = u + w$, whence $u \in C^2([0, T]; \mathcal{U}^1)$ is the solution of (8)–(10) and the function $w \in C^2([0, T]; \mathcal{U}^0)$ is a solution of (11), (12) given by

$$w(t) = \sum_{\lambda = \lambda_k} \left< \frac{f(\cdot,t)q(t)}{\beta(\lambda_k - \lambda'')}, \varphi_k \right> \varphi_k. \tag{14}$$

Proof By virtue of Lemma 2, we get that the pencil $\vec{B}$ is polynomially A-bounded and condition (A) is fulfilled, moreover, the ∞ is a removable singular point of the A-resolvent of the pencil $\vec{B}$.

Since $K \in \mathcal{U}^1$, then $\mathcal{U}^0 \subset ker\, C$. For $y \in \mathcal{Y}$ due to the orthonormality of the system of eigenfunctions in $L_2(\Omega)$

$$C(A^1)^{-1} Qy = \left(\sum_{\lambda \neq \lambda_k} \frac{<f(\cdot,t), \varphi_k><\varphi_k, K>}{\lambda - \lambda_k} \right) y = \left(\sum_{\lambda \neq \lambda_k} \frac{<f(\cdot,t), K>}{\lambda - \lambda_k} \right) y.$$

This operator is reversible in $\mathcal{Y}$ when

$$\sum_{\lambda \neq \lambda_k} \frac{<f(\cdot,t), K>}{\lambda - \lambda_k} \neq 0$$

and the inverse operator is continuously differentiable by t due to the condition on the function χ.

Thus, all the conditions of Theorem 2 are satisfied, then there exists a unique solution (v, q) of inverse problem (1)–(4), where $q \in C^2([0, T]; \mathcal{Y})$, $v = u + w$, whence $u \in C^2([0, T]; \mathcal{U}^1)$ is the solution of (8)–(10) and the function $w \in C^2([0, T]; \mathcal{U}^0)$ is a solution of (11), (12) given by (14). ▷

Acknowledgements The reported study was funded by RFBR, project number 19-31-90137.

References

1. Favini, A., Sviridyuk, G.A., Sagadeeva, M.A.: Linear Sobolev type equations with relatively p-radial operators in space of "Noises". Mediter. J. Math. **13**(6), 4607–4621 (2016)
2. Kozhanov, A.I., Namsaraeva, G.V.: Linear inverse problems for a class of equations of Sobolev type. Chelyabinsk Phys. Math. J. **3**(2), 153–171 (2018) (in Russian)
3. Manakova, N.A.: Mathematical models and optimal control of the filtration and deformation processes. Bull. South Ural State Univ. Ser.: Math. Model. Program. Comput. Softw. **8**(3), 5–24 (2015) (in Russian)
4. Sviridyuk, G.A., Fedorov, V.E.: On the identities of analytic semigroups of operators with kernels. Siberian Math. J. **39**(3), 522–533 (1998)
5. Sviridyuk, G.A., Zagrebina, S.A., Konkina A.S.: The Oskolkov equations on the geometric graphs as a mathematical model of the traffic flow. Bull. South Ural State Univ. Ser.: Math. Model. Program. Comput. Softw. **8**(3), 148–154 (2015) (in Russian)
6. Zamyshlyaeva, A.A., Tsyplenkova, O.N., Bychkov, E.V.: Optimal control of solutions to the initial-final problem for the Sobolev type equation of higher order. J. Comput. Eng. Math. **3**(2), 57–67 (2016)
7. Zamyshliaeva, A.A., Lut, A.V.: Inverse problem for Sobolev type mathematical models. Bull. South Ural State Univ. Ser.: Math. Model. Program. Comput. Softw. **12**(2), 25–36 (2019)
8. Shestakov, A.L., Keller, A.V., Sviridyuk, G.A.: The theory of optimal measurements. J. Comput. Eng. Math. **13**(1), 3–16 (2014)
9. Asylbekov, T.D., Chamashev, M.K.: Coefficient inverse problem for a fourth-order linear partial differential equation. News Tomsk Polytech. Univ. Georesource Engi. **317**(2), 22–25 (2010) (in Russian)
10. Kozhanov, A.I., Telesheva, L.A.: Nonlinear inverse problems with integral overdetermination for nonstationary differential equations of high order. Bull. South Ural State Univ. Ser.: Math. Model. Program. Comput. Softw. **10**(2), 24–37 (2017) (in Russian)
11. Megraliev, Ya.T., Sattorov, A.Kh.: An inverse boundary value problem with integral conditions for a second order hyperbolic equation. Rep. Acad. Sci. Republic of Tajikistan. **53**(4), 248–256 (2010) (in Russian)
12. Yuldashev, T.K., Seredkina, A.I.: Inverse problem for quazilinear partial integro-differential equations of higher order. Bull. Samara State Tech. Univ. Ser.: Phys. Math. **3**(32), 46–55 (2013) (in Russian)
13. Zamyshlyaeva, A.A., Bychkov, E.V.: The cauchy problem for the Sobolev type equation of higher order. Bull. South Ural State Univ. Ser.: Math. Model. Program. Comput. Softw. **11**(1), 5–14 (2018)

Optimal Control of Solutions to Showalter–Sidorov Problem for a High Order Sobolev Type Equation with Additive "Noise"

Alyona A. Zamyshlyaeva and Olga N. Tsyplenkova

Abstract In this paper, the problem of optimal control of solutions to the Showalter–Sidorov problem for a high-order Sobolev type equation with additive "noise" is investigated. The existence and uniqueness of a strong solution to the Showalter–Sidorov problem for this equation are proved. Sufficient conditions for the existence and uniqueness of an optimal control of such solutions are obtained. For this, we built the space of "noises". For the differentiation of additive "noise", we use the derivative of a stochastic process in the sense of Nelson–Gliklikh.

Keywords Sobolev type equations · Showalter–Sidorov problem · Additive "noise" · Strong solutions · Optimal control

Introduction

Recently, research on Sobolev type equations has expanded considerably. The incomplete Sobolev type equation

$$Av^{(n)} = Bv + f \qquad (1)$$

with the assumption ker$A \neq \{0\}$ has been studied in different aspects for $n \geq 1$ [1–6]. Here the operators A, B are linear and continuous, acting from Banach space $\mathfrak{V}$ to $\mathfrak{G}$, absolute term $f = f(t)$ models the external force.

The lack of Eq. (1) with the deterministic absolute term is that, in natural experiments, the system is subject to random perturbation, for example in the form of white

---

A. A. Zamyshlyaeva (✉)
Department of Applied Mathematics and Programming, South Ural State University,
76, Lenin ave, Chelyabinsk 454080, Russian Federation
e-mail: zamyshliaevaaa@susu.ru

O. N. Tsyplenkova
Department of Mathematical Physics Equations, South Ural State University,
76, Lenin ave, Chelyabinsk 454080, Russian Federation
e-mail: tcyplenkovaon@susu.ru

© Springer Nature Switzerland AG 2020
J. Banasiak et al. (eds.), *Semigroups of Operators – Theory and Applications*,
Springer Proceedings in Mathematics & Statistics 325,
https://doi.org/10.1007/978-3-030-46079-2_24

noise. Currently, stochastic ordinary differential equations with various additive random processes are being actively studied [7].

The first results concerning stochastic Sobolev type equations of the first order can be found in [8]. They are based on the extension of the Ito–Stratonovich–Skorokhod method to partial differential equations [9]. In [10] there was studied a stochastic Sobolev type equation of higher order

$$A \overset{o}{\eta}{}^{(n)} = B\eta + w, \tag{2}$$

where w is the stochastic process. It is required to find the random process $\eta(t)$, satisfying (in some sense) Eq. (2) and the initial conditions

$$\overset{o}{\eta}{}^{(m)}(0) = \xi_m, \ m = 0, 1, \ldots, n-1, \tag{3}$$

where ξ_m are given random variables.

At first, w was understood as white noise, which is a generalized derivative of the Wiener process. Later, a new approach to the investigation of Eq. (2) appeared [11] and is being actively developed [12, 13], where "white noise" means the Nelson–Gliklikh derivative of the Wiener process.

Of particular interest is the optimal control problem. Consider the stochastic Sobolev type equation

$$A \overset{o}{\eta}{}^{(n)} = B\eta + w + Cu, \tag{4}$$

where $\eta = \eta(t)$ is a stochastic process, $\overset{o}{\eta}$ is the Nelson–Gliklikh derivative [14] of process η, $w = w(t)$ is a stochastic process that responds for external influence; u is unknown control function from the Hilbert space $\mathfrak{U}$ of controls, operator $C \in \mathscr{L}(\mathfrak{U}; \mathfrak{G})$.

Supply (4) with initial Showalter–Sidorov condition

$$P\left(\overset{o}{\eta}{}^{(m)}(0) - \xi_m\right) = 0, \ m = 0, \ldots, n-1. \tag{5}$$

We investigate the optimal control problem: search pair $(\hat{\eta}, \hat{u})$, where $\hat{\eta}$ is a solution to problem (4), (5), and the control $\hat{u}$ belongs to $\mathfrak{U}_{ad} \subset \mathfrak{U}$, and satisfies the relation

$$J(\hat{\eta}, \hat{u}) = \min_{(\eta, u)} J(\eta, u). \tag{6}$$

Here $J(\eta, u)$ is some specially constructed penalty functional and $\mathfrak{U}_{ad}$ is a closed convex set in the Hilbert space $\mathfrak{U}$ of controls.

1 The Spaces of "Noises". Stochastic K-Processes. Phase Space

Let $\Omega \equiv (\Omega, \mathscr{A}, \mathbf{P})$ be a complete probability space, $\mathbb{R}$ be the set of real numbers endowed with the Borel σ-algebra. A measurable mapping $\xi : \Omega \to \mathbb{R}$ is called *a random variable*. The set of random variables having zero expectation ($\mathbf{E}\xi = 0$) and finite variance forms a Hilbert space $\mathbf{L_2}$ with inner product $(\xi_1, \xi_2) = \mathbf{E}\xi_1\xi_2$. Let $\mathscr{A}_0$ be a σ-subalgebra of σ-algebra $\mathscr{A}$. Construct subspace $\mathbf{L_2^0} \subset \mathbf{L_2}$ of random variables measurable with respect to $\mathscr{A}_0$. Denote the orthoprojector by $\Pi : \mathbf{L_2} \to \mathbf{L_2^0}$. Let $\xi \in \mathbf{L_2}$, then $\Pi\xi$ is called *a conditional expectation* of the random variable ξ, and is denoted by $\mathbf{E}(\xi|\mathscr{A}_0)$.

Consider a set $\mathfrak{I} \subset \mathbb{R}$ and the following two mappings. The first one, $f : \mathfrak{I} \to \mathbf{L_2}$, associates to each $t \in \mathfrak{I}$ the random variable $\xi \in \mathbf{L_2}$. The second one, $g : \mathbf{L_2} \times \Omega \to \mathbb{R}$, associates to each pair (ξ, ω) the point $\xi(\omega) \in \mathbb{R}$. The mapping $\eta : \mathbb{R} \times \Omega \to \mathbb{R}$ having form $\eta = \eta(t, \omega) = g(f(t), \omega)$, where f and g are defined above, is called a *stochastic process*. Therefore, the stochastic process $\eta = \eta(t, \cdot)$ is a random variable for each fixed $t \in \mathfrak{I}$, i.e. $\eta(t, \cdot) \in \mathbf{L_2}$, and $\eta = \eta(\cdot, \omega)$ is called *a (sample) path* for each fixed $\omega \in \Omega$. The stochastic process η is called *continuous*, if all its paths are almost sure continuous (i.e. for almost all $\omega \in \Omega$ the paths $\eta(\cdot, \omega)$ are continuous). The set of continuous stochastic processes forms a Banach space, which is denoted by $\mathbf{C}(\mathfrak{I}, \mathbf{L_2})$. Fix $\eta \in \mathbf{C}(\mathfrak{I}, \mathbf{L_2})$ and $t \in \mathfrak{I}$, and denote by $\mathscr{N}_t^\eta$ the σ-algebra generated by the random variable $\eta(t)$. For brevity, $\mathbf{E}_t^\eta = \mathbf{E}(\cdot|\mathscr{N}_t^\eta)$.

Definition 1 Let $\eta \in \mathbf{C}(\mathfrak{I}, \mathbf{L_2})$. A random process

$$\overset{o}{\eta} = \frac{1}{2}\left(\lim_{\Delta t \to 0+} \mathbf{E}_t^\eta\left(\frac{\eta(t+\Delta t, \cdot) - \eta(t, \cdot)}{\Delta t}\right) + \lim_{\Delta t \to 0+} \mathbf{E}_t^\eta\left(\frac{\eta(t, \cdot) - \eta(t-\Delta t, \cdot)}{\Delta t}\right)\right)$$

is called a *Nelson–Gliklikh derivative* $\overset{o}{\eta}$ *of the stochastic process* η at point $t \in \mathfrak{I}$, if the limits exist in the sense of the uniform metric on $\mathbb{R}$.

Let $\mathbf{C}^l(\mathfrak{I}, L_2)$, $l \in \mathbb{N}$, be a space of stochastic processes almost sure differentiable in the sense of the Nelson–Gliklikh derivative on $\mathfrak{I}$ up to order l inclusively. The spaces $\mathbf{C}^l(\mathfrak{I}, L_2)$ are called *the spaces of differentiable "noises"*. Let $\mathfrak{I} = \{0\} \cup \mathbb{R}_+$, then a well-known example [15, 16] of a vector in the space $\mathbf{C}^l(\mathfrak{I}, L_2)$ is given by a stochastic process that describes the Brownian motion in Einstein–Smoluchowski model

$$\beta(t) = \sum_{k=0}^{\infty} \xi_k \sin\frac{\pi}{2}(2k+1)t,$$

where the independent random variables $\xi_k \in \mathbf{L_2}$ are such that the variances $D\xi_k = [\frac{\pi}{2}(2k+1)]^{-2}$, $k \in \{0\} \cup \mathbb{N}$. As shown in [14], $\overset{o}{\beta}(t) = \frac{\beta(t)}{2t}$, $t \in \mathbb{R}_+$.

Now let $\mathfrak{V}$ be a real separable Hilbert space with orthonormal basis $\{\varphi_k\}$. Denote by $\mathfrak{V}_\mathbf{K} \mathbf{L}_2$ the Hilbert space, which is a completion of the linear span of *random variables*

$$\eta = \sum_{k=1}^{\infty} \sqrt{\lambda_k} \xi_k \varphi_k$$

by the norm

$$\|\eta\|_{\mathfrak{V}}^2 = \sum_{k=1}^{\infty} \lambda_k \mathbf{D}\xi_k.$$

Here the sequence $\mathbf{K} = \{\lambda_k\} \subset \mathbb{R}_+$ is such that $\sum_{k=1}^{\infty} \lambda_k < +\infty$, $\{\xi_k\} \subset \mathbf{L}_2$ is a sequence of random variables. The elements of $\mathfrak{V}_\mathbf{K} \mathbf{L}_2$ will be called random **K**-variables. Note that for existence of a random **K**-variable $\eta \in \mathfrak{V}_\mathbf{K} \mathbf{L}_2$ it is enough to consider a sequence of random variables $\{\xi_k\} \subset \mathbf{L}_2$ having uniformly bounded variances, i.e. $\mathbf{D}\xi_k \leq const$, $k \in \mathbb{N}$.

Next, consider interval $\mathfrak{J} = (\varepsilon, \tau) \subset \mathbb{R}$. Mapping $\eta : (\varepsilon, \tau) \to \mathfrak{V}_\mathbf{K} \mathbf{L}_2$ given by

$$\eta(t) = \sum_{k=1}^{\infty} \sqrt{\lambda_k} \xi_k(t) \varphi_k,$$

where the sequence $\{\xi_k\} \subset \mathbf{C}(\mathfrak{J}, L_2)$, is called a $\mathfrak{V}$-*valued continuous stochastic* **K**-*process*, if the series on the right-hand side converges uniformly on any compact in $\mathfrak{J}$ in the norm $\|\cdot\|_{\mathfrak{V}}$, and paths of process $\eta = \eta(t)$ are almost sure continuous. Continuous stochastic **K**-process $\eta = \eta(t)$ is called *continuously Nelson–Gliklikh differentiable on* $\mathfrak{J}$, if the series

$$\overset{o}{\eta}(t) = \sum_{k=1}^{\infty} \sqrt{\lambda_k} \overset{o}{\xi}_k(t) \varphi_k$$

converges uniformly on any compact in $\mathfrak{J}$ in the norm $\|\cdot\|_{\mathfrak{V}}$, and paths of process $\overset{o}{\eta} = \overset{o}{\eta}(t)$ are almost sure continuous. Let $\mathbf{C}(\mathfrak{J}, \mathfrak{V}_\mathbf{K} \mathbf{L}_2)$ be a space of continuous stochastic **K**-processes, and $\mathbf{C}^l(\mathfrak{J}, \mathfrak{V}_\mathbf{K} \mathbf{L}_2)$ be a space of continuously differentiable up to order $l \in \mathbb{N}$ stochastic **K**-processes. An example of a stochastic **K**-process, which is continuously differentiable up to any order $l \in \mathbb{N}$ inclusively, is a Wiener **K**-process [15, 16]

$$W_\mathbf{K}(t) = \sum_{k=1}^{\infty} \sqrt{\lambda_k} \beta_k(t) \varphi_k,$$

where $\{\beta_k\} \subset \mathbf{C}^l(\mathfrak{I}, L_2)$ is a sequence of Brownian motions on $\mathbb{R}_+$. Similarly, if $\mathfrak{G}$ is a real separable Hilbert space with orthonormal basis $\{\varphi_k\}$, the spaces $\mathbf{C}(\mathfrak{I}, \mathfrak{G}_\mathbf{K} L_2)$ and $\mathbf{C}^l(\mathfrak{I}, \mathfrak{G}_\mathbf{K} L_2)$, $l \in \mathbb{N}$, are constructed. Note also that spaces $\mathbf{C}^l(\mathfrak{I}, L_2)$, $\mathbf{C}(\mathfrak{I}, \mathfrak{V}_\mathbf{K} L_2)$ and $\mathbf{C}^l(\mathfrak{I}, \mathfrak{G}_\mathbf{K} L_2)$, $l \in \mathbb{N}$, are called *the spaces of differentiable* $\mathbf{K}$-*"noises"* [15].

2 Stochastic Sobolev Type Equations of High Order with Relatively p-Bounded Operators

Let the operators $A, B \in \mathscr{L}(\mathfrak{V}_\mathbf{K} L_2, \mathfrak{G}_\mathbf{K} L_2)$. Following [5], introduce the A-resolvent set $\rho^A(B) = \{\mu \in \mathbb{C} : (\mu A - B)^{-1} \in \mathscr{L}(\mathfrak{G}_\mathbf{K} L_2, \mathfrak{V}_\mathbf{K} L_2)\}$ and the A-spectrum $\sigma^A(B) = \mathbb{C} \setminus \rho^A(B)$ of operator B. The operator-functions of variable $\mu (\mu A - B)^{-1}$, $R^A_\mu(B) = (\mu A - B)^{-1} A$, $L^A_\mu(B) = A(\mu A - B)^{-1}$ with the domain $\rho^A(B)$ are correspondingly called the A-resolvent, the right and the left A-resolvents of the operator B. If the set $\sigma^A(B)$ is bounded $(\exists a > 0 : (|\mu| < a) \Rightarrow \mu \in \sigma^A(B))$ then the operator B is called (A, σ)-bounded.

Let the operator B be (A, σ)-bounded, $p \in \{0\} \cup \mathbb{N}$.

Construct the set $\sigma^A_n(B) = \{\mu \in \mathbb{C} : \mu^n \in \sigma^A(B)\}$; it is compact in $\mathbb{C}$ due to the compactness of the A-spectrum $\sigma^A(B)$ of operator B. Take the contour $\gamma = \{\mu \in \mathbb{C} : |\mu| = r, r^n > a\}$ that bounds the domain containing the points of $\sigma^A_n(B)$ and construct the projectors

$$P = \frac{1}{2\pi i} \int_\gamma \mu^{n-1} R^A_{\mu^n}(B) d\mu \in \mathscr{L}(\mathfrak{V}_\mathbf{K} L_2), \quad Q = \frac{1}{2\pi i} \int_\gamma \mu^{n-1} L^A_{\mu^n}(B) d\mu \in \mathscr{L}(\mathfrak{G}_\mathbf{K} L_2).$$

Here, $R^A_{\mu^n}(B) = (\mu^n A - B)^{-1} A$ and $L^A_{\mu^n}(B) = A(\mu A - B)^{-1}$. Put $\mathfrak{V}^0_\mathbf{K} L_2 (\mathfrak{V}^1_\mathbf{K} L_2)$ = ker P(im P), $\mathfrak{G}^0_\mathbf{K} L_2 (\mathfrak{G}^1_\mathbf{K} L_2)$ = ker Q(im Q). Thus, the spaces $\mathfrak{V}_\mathbf{K} L_2$ and $\mathfrak{G}_\mathbf{K} L_2$ since P and Q are projectors, can be decomposed into direct sums $\mathfrak{V}_\mathbf{K} L_2 = \mathfrak{V}^0_\mathbf{K} L_2 \oplus \mathfrak{V}^1_\mathbf{K} L_2$ and $\mathfrak{G}_\mathbf{K} L_2 = \mathfrak{G}^0_\mathbf{K} L_2 \oplus \mathfrak{G}^1_\mathbf{K} L_2$, whereas $\mathfrak{V}^0_\mathbf{K} L_2 \supset$ ker A. By $A_k(B_k)$ define the restriction of operator $A(B)$ onto $\mathfrak{V}^k_\mathbf{K} L_2$, $k = 0, 1$.

Lemma 1 *The operators $A_k, B_k \in \mathscr{L}(\mathfrak{V}^k_\mathbf{K} L_2; \mathfrak{G}^k_\mathbf{K} L_2)$, $k = 0, 1$; moreover, there exist the operators $B_0^{-1} \in \mathscr{L}(\mathfrak{G}^0_\mathbf{K} L_2; \mathfrak{V}^0_\mathbf{K} L_2)$ and $A_1^{-1} \in \mathscr{L}(\mathfrak{G}^1_\mathbf{K} L_2; \mathfrak{V}^1_\mathbf{K} L_2)$.*

Construct the operators $H = B_0^{-1} A_0 \in \mathscr{L}(\mathfrak{V}^0_\mathbf{K} L_2)$, $S = A_1^{-1} B_1 \in \mathscr{L}(\mathfrak{V}^1_\mathbf{K} L_2)$.

The (A, σ)-bounded operator B is called (A, p)-*bounded*, $p \in \{0\} \cup \mathbb{N}$, if ∞ is a removable singular point (i.e. $H \equiv \mathbb{O}$, $p = 0$) or the pole of order $p \in \mathbb{N}$ (i.e. $H^p \neq \mathbb{O}$, $H^{p+1} \equiv \mathbb{O}$) of the A-resolvent $(\mu A - B)^{-1}$ of operator B.

Consider the linear stochastic Sobolev type equation of higher order (2), where the absolute term w will be indicated later. Supplement (2) with the initial Showalter–Sidorov condition (5) which is the generalization of the condition [3]

$$A \overset{o(m)}{\eta}(0) = A\xi_m, \quad m = 0, \ldots, n-1,$$

and has advantages over the Cauchy condition (3) in the case of Sobolev type equations. In addition to (5), we will consider *the weakened* (in the sense of S.G. Krein) *Showalter–Sidorov condition*

$$\lim_{t \to 0+} P \left(\overset{o\,(m)}{\eta}(t) - \xi_m \right) = 0, \quad m = 0, \ldots, n-1. \tag{7}$$

The K-random process $\eta \in \mathbf{C}^n(\mathfrak{I}, \mathfrak{G}_{\mathbf{K}}\mathbf{L}_2)$ is called *a classical solution of equation* (2), if a.s. all its trajectories satisfy Eq. (2) for some K-random process $w \in \mathbf{C}(\mathfrak{I}, \mathfrak{G}_{\mathbf{K}}\mathbf{L}_2)$. The solution $\eta = \eta(t)$ of (2) is called *the classical solution* of problem (2), (7) if a.s. condition (7) is also fulfilled. The classical solutions of the problems (2), (5) and (2), (3) are defined analogously.

Consider firstly problem (3) for the homogeneous equation

$$A \overset{o\,(n)}{\eta} = B\eta. \tag{8}$$

In this case (and only in this case) consider $\mathfrak{I} = \mathbb{R}$.

Definition 2 The mapping $V \in C^\infty(\mathbb{R}; \mathscr{L}(\mathfrak{V}_{\mathbf{K}}\mathbf{L}_2))$ is called *a propagator* of Eq. (8), if for all $v \in \mathfrak{V}_{\mathbf{K}}\mathbf{L}_2$ the vector-function $\eta(t) = V(t)v$ is a solution of (8).

Theorem 1 *Let the operator B be (A, σ)-bounded. Then, the operator-functions*

$$V_m(t) = \frac{1}{2\pi i} \int_\gamma \mu^{n-m-1}(\mu^n A - B) A e^{\mu t} d\mu,$$

where $m = 0, 1, \ldots, n-1$ and the integral is understood in the sense of Riemann, define the propagators of Eq. (8).

Lemma 2 $V_m \in C^\infty(\mathbb{R}; \mathscr{L}(\mathfrak{V}_{\mathbf{K}}\mathbf{L}_2; \mathfrak{V}^1_{\mathbf{K}}\mathbf{L}_2))$, $(V_m(t))_t^{(l)} = V_{m+l}(t)$, *where* $m = 0, 1, \ldots, n-1$, $l = 0, 1, \ldots, m$; $(V_m(t))_t^{(l)}\big|_{t=0} = \mathbb{O}$ *for* $m \neq l$ *and* $(V_m(t))_t^{(m)}\big|_{t=0} = P$ *is the projector in* $\mathfrak{V}_{\mathbf{K}}\mathbf{L}_2$ *on* $\mathfrak{V}^1_{\mathbf{K}}\mathbf{L}_2$ *along* $\mathfrak{V}^0_{\mathbf{K}}\mathbf{L}_2$.

Definition 3 The set $\mathfrak{P} \subset \mathfrak{V}_{\mathbf{K}}\mathbf{L}_2$ is called *the phase space* of equation (8) if
(i) a.s. every trajectory of the solution $\eta = \eta(t)$ lies in $\mathfrak{P}$ pointwise, i.e. $\eta(t) \in \mathfrak{P}$ a.s. for all $t \in \mathbb{R}$;
(ii) for all random variables $\xi_m \in L_2(\Omega; \mathfrak{P})$, $m = 0, 1, \ldots, n-1$, there exists a unique solution $\eta \in \mathbf{C}^n_K \mathbf{L}_2$ of (3), (8).

Theorem 2 *Let the operator B be (A, p)-bounded, $p \in \{0\} \cup \mathbb{N}$. Then the subspace $\mathfrak{V}^1_{\mathbf{K}}\mathbf{L}_2$ is the phase space of equation (8).*

Proof In fact, due to Lemma 1, Eq. (8) can be reduced to the equivalent system

$$H \overset{o\,0(n)}{\eta} = \eta^0, \quad \overset{o\,0(n)}{\eta} = S\eta^1, \tag{9}$$

where $\eta^0 = (\mathbb{I} - P)\eta$, $\eta^1 = P\eta$. After applying the Nelson–Gliklikh differentiation n times to the first equation in (9) and using operator H on it, we consecutively obtain

$$0 = H^{p+1} \overset{o(n(p+1))}{\eta} = \cdots = H^2 \overset{o(2n)}{\eta} = \cdots = H \overset{o(n)}{\eta} = \eta^0. \tag{10}$$

Thus, the condition (i) of Definition 3 is true. To prove the fulfillment of the condition (ii), note that if $\xi_m \in \mathfrak{V}_\mathbf{K}^1 \mathbf{L_2}$, $m = 0, 1, \ldots, n-1$, then there exists a unique solution of (3), (9) and it is given by $\eta^1 = \eta^1(t) = \sum_{m=0}^{n-1} V_m(t)\xi_m$. Then the unique solution of (3), (8) for $\xi_m \in \mathfrak{V}_\mathbf{K}^1 \mathbf{L_2}$, $m = 0, 1, \ldots, n-1$, is given by $\eta(t) = \eta^0(t) + \eta^1(t) = \sum_{m=0}^{n-1} V_m(t)\xi_m$. The proof of the theorem is complete.

Corollary 1 *Under the conditions of Theorem 2 the solution of (3), (8) is the Gaussian K-random process if the random variables ξ_m, $m = 0, 1, \ldots, n-1$, are Gaussian.*

Lemma 3 *Let the operator B be (A, p)-bounded, $p \in \{0\} \cup \mathbb{N}$. Then for all independent random variables $\xi_m \in \mathfrak{V}_\mathbf{K} \mathbf{L_2}$, $m = 0, 1, \ldots, n-1$, there exists a.s. a unique solution $\eta \in \mathbf{C}_K^\infty \mathbf{L_2}$ of (5), (8), represented in the form $\eta(t) = \sum_{m=0}^{n-1} V_m(t)\xi_m$, $t \in \mathbb{R}$. If in addition ξ_m, $m = 0, 1, \ldots, n-1$ take values only in $\mathfrak{V}_\mathbf{K}^1 \mathbf{L_2}$, then this solution is a unique solution of (3), (8).*

Go back to Eq. (2) and notice that now $\mathfrak{J} = [0, \tau)$. Let the K-random process $w = w(t)$, $t \in [0, \tau)$ be such that

$$(\mathbb{I} - Q)w \in \mathbf{C}^{n(p+1)}(\mathfrak{J}, \mathfrak{G}_\mathbf{K} \mathbf{L_2}) \text{ and } Qw \in \mathbf{C}(\mathfrak{J}, \mathfrak{G}_\mathbf{K} \mathbf{L_2}), \tag{11}$$

then the K-random process

$$\eta(t) = -\sum_{q=0}^{p} H^q B_0^{-1}(\mathbb{I} - Q)N \overset{o(qn)}{w}(t) + \int_0^t V_{n-1}(t-s) A_1^{-1} QNw(s) ds \tag{12}$$

is a unique classical solution of (5), (2) with $\xi_m \in \mathfrak{V}_\mathbf{K}^0 \mathbf{L_2}$, $m = 0, \ldots, n-1$.

Theorem 3 *Let the operator B be (A, p)-bounded, $p \in \{0\} \cup \mathbb{N}$. For any K-random process $w = w(t)$ satisfying (11), and for all independent random variables $\xi_m \in \mathfrak{V}_\mathbf{K} \mathbf{L_2}$, $m = 0, 1, \ldots, n-1$, independent with w, there exists a.s. a unique solution $\eta \in \mathbf{C}^n(\mathfrak{J}, \mathfrak{G}_\mathbf{K} \mathbf{L_2})$ of (2), (5), represented in the form*

$$\eta(t) = \sum_{m=0}^{n-1} V_m(t)\xi_m - \sum_{q=0}^{p} H^q B_0^{-1}(\mathbb{I} - Q) \overset{o(qn)}{w}(t) + \int_0^t V_{n-1}(t-s) A_1^{-1} Qw(s) ds. \tag{13}$$

However, "white noise" $w(t) = \overset{o}{W}_K(t) = (2t)^{-1} W_K(t)$ does not satisfy condition (11), so it cannot stand on the right-hand side of (2). One approach to solving this problem is proposed in [8, 17]. To use this approach, convert the second term on the right-hand side of (12) as follows:

$$\int_\varepsilon^t V_{n-1}(t-s) A_1^{-1} Q \overset{o}{W}_K(s) ds = -V_{n-1}(t-\varepsilon) A_1^{-1} Q W_K(\varepsilon)$$
$$+ \int_\varepsilon^t \tfrac{d}{dt} V_{n-1}(t-s) A_1^{-1} W_K(s) ds = -V_{n-1}(t-\varepsilon) A_1^{-1} Q W_K(\varepsilon) \qquad (14)$$
$$+ \int_\varepsilon^t V_{n-2}(t-s) A_1^{-1} W_K(s) ds.$$

This integration by parts makes sense for any $\varepsilon \in (0, t)$, $t \in \mathbb{R}_+$ due to definition of the Nelson–Gliklikh derivative. Letting $\varepsilon \to 0$ in (14) we get

$$\int_0^t V_{n-1}(t-s) A_1^{-1} Q \overset{o}{W}_K(s) ds = \int_0^t V_{n-2}(t-s) A_1^{-1} W_K(s) ds. \qquad (15)$$

Corollary 2 *Let the operator B be (A, p)-bounded, $p \in \{0\} \cup \mathbb{N}$, $W_k \in \mathbf{C}(\mathfrak{J}, \mathfrak{G}_K^1 \mathbf{L}_2)$. Let $\mathfrak{J} \subset \mathbb{R}_+$. For all independent random variables $\xi_m \in \mathfrak{V}_K \mathbf{L}_2$, $m = 0, 1, \ldots, n-1$, independent from W_K, there exists a.s. a unique solution $\eta \in \mathbf{C}^n(\mathfrak{J}, \mathfrak{G}_K \mathbf{L}_2)$ of the problem (5) for the equation*

$$A \overset{o\,(n)}{\eta} = B\eta + \overset{o}{W}_K, \qquad (16)$$

given by

$$\eta(t) = \sum_{m=0}^{n-1} V_m(t) \xi_m + \int_0^t V_{n-2}(t-s) A_1^{-1} W_K(s) ds.$$

Theorem 4 *Let the operator B be (A, p)-bounded, $p \in \{0\} \cup \mathbb{N}$. For all random variables $\xi_m \in \mathfrak{V}_K \mathbf{L}_2$, independent from W_K, there exists a.s. unique solution $\eta = \eta(t)$ of (7), (15) given by*

$$\eta(t) = \sum_{m=0}^{n-1} V_m(t) \xi_m + \int_0^t V_{n-2}(t-s) A_1^{-1} Q W_K(s) ds - \sum_{q=0}^{p} H^q B_0^{-1}(\mathbb{I} - Q) \overset{o\,(qn+1)}{W_K}(t).$$

3 Strong Solutions

Let $L_2(\mathfrak{J}; \mathfrak{V}_\mathbf{K}\mathbf{L_2})$ be a space of stochastic processes whose paths are square-integrable on $\mathfrak{J}$.

Definition 4 A vector function

$$\eta \in H^n(\mathfrak{V}_\mathbf{K}\mathbf{L_2}) = \{\eta \in L_2(\mathfrak{J}; \mathfrak{V}_\mathbf{K}\mathbf{L_2}) : \overset{o}{\eta}{}^{(n)} \in L_2(\mathfrak{J}; \mathfrak{V}_\mathbf{K}\mathbf{L_2})\}$$

is called a *strong solution* of equation (2), if it a.s. turns the equation to identity almost everywhere on interval $(0, \tau)$. A strong solution $\eta = \eta(t)$ of Eq. (2) is called a *strong solution to problem* (2), (5) if condition (5) a.s. holds.

This is well defined by virtue of the continuity of the embedding $H^n(\mathfrak{V}_\mathbf{K}\mathbf{L_2}) \hookrightarrow C^{n-1}(\mathfrak{J}; \mathfrak{V}_\mathbf{K}\mathbf{L_2})$. The term "strong solution" has been introduced to distinguish a solution of equation (3.1) in this sense from the solution (13), which is usually said to be "classical". Note that the classical solution (13) is also a strong solution to problem (2), (5).

Let us construct the spaces

$$H^{np+n}(\mathfrak{G}_\mathbf{K}\mathbf{L_2}) = \{v \in L_2(\mathfrak{J}; \mathfrak{G}_\mathbf{K}\mathbf{L_2}) : \overset{o}{v}{}^{(np+n)} \in L_2(\mathfrak{J}; \mathfrak{G}_\mathbf{K}\mathbf{L_2}), p \in \{0\} \cup \mathbb{N}\}.$$

The space $H^{np+n}(\mathfrak{G}_\mathbf{K}\mathbf{L_2})$ is a Hilbert space with inner product

$$[v, w] = \sum_{q=0}^{np+n} \int_0^\tau \langle v^{(q)}, w^{(q)} \rangle_{\mathfrak{G}_\mathbf{K}\mathbf{L_2}} dt.$$

Let $w \in H^{np+n}(\mathfrak{G}_\mathbf{K}\mathbf{L_2})$. Introduce the operators

$$A_1 w(t) = -\sum_{q=0}^{p} H^q B_0^{-1}(\mathbb{I} - Q) \overset{o}{w}{}^{(qn)}(t),$$

$$A_2 w(t) = \int_0^t V_{n-1}(t-s) A_1^{-1} Q w(s) ds, \, t \in (0, \tau)$$

and the function

$$k(t) = \sum_{m=0}^{n-1} V_m(t) \xi_m.$$

Lemma 4 *Let the operator B be (A, p)-bounded, $p \in \{0\} \cup \mathbb{N}$. Then*
(i) $A_1 \in \mathscr{L}(H^{np+n}(\mathfrak{G}_\mathbf{K}\mathbf{L_2}); H^n(\mathfrak{V}_\mathbf{K}\mathbf{L_2}))$;
(ii) *for arbitrary* $\xi_m \in \mathfrak{V}_\mathbf{K}\mathbf{L_2}, m = \overline{0, n-1}$ *the vector function* $k \in C^n([0, \tau);$ $\mathfrak{V}_\mathbf{K}\mathbf{L_2})$;
(iii) $A_2 \in \mathscr{L}(H^{np+n}(\mathfrak{G}_\mathbf{K}\mathbf{L_2}); H^n(\mathfrak{V}_\mathbf{K}\mathbf{L_2}))$.

Theorem 5 *Let the operator B be (A,p)-bounded, $p \in \{0\} \cup \mathbb{N}$. For any K-random process $w = w(t)$ satisfying (11), and for all independent random variables $\xi_m \in \mathfrak{V}_\mathbf{K} \mathbf{L}_2$, $m = 0, 1, \ldots, n-1$, independent from w, there exists a.s. a unique strong solution to problem (2), (5).*

4 Optimal Control

Let $\mathfrak{U}$ be a real separable Hilbert space with orthonormal basis φ_k. Consider the Showalter–Sidorov problem (5) for linear inhomogeneous Sobolev type equation with additive "noise" (4).

Introduce the control space

$$\overset{o}{H}{}^{np+n}(\mathfrak{U}_\mathbf{K} \mathbf{L}_2) = \{u \in L_2(0, \tau; \mathfrak{U}_\mathbf{K} \mathbf{L}_2) : u^{(np+n)} \in L_2(0, \tau; \mathfrak{U}_\mathbf{K} \mathbf{L}_2), u^{(q)}(0) = 0 \text{ a.s.}, q = \overline{0, p}\},$$

$p \in \{0\} \cup \mathbb{N}$. It is a Hilbert space with inner product

$$[v, w] = \sum_{q=0}^{np+n} \int_0^\tau \langle v^{(q)}, w^{(q)} \rangle_{\mathfrak{U}_\mathbf{K} \mathbf{L}_2} dt.$$

In the space $\overset{o}{H}{}^{np+n}(\mathfrak{U}_\mathbf{K} \mathbf{L}_2)$ we single out a closed convex subset $\overset{o}{H}{}^{np+n}_\partial(\mathfrak{U}_\mathbf{K} \mathbf{L}_2)$, which will be called *the set of admissible controls*.

Definition 5 A vector function $\hat{u} \in \overset{o}{H}{}^{np+n}_\partial(\mathfrak{U}_\mathbf{K} \mathbf{L}_2)$ is called an *optimal control of solutions to problem* (4), (5), if relation (6) holds.

We need to prove the existence of a unique control $\hat{u} \in \overset{o}{H}{}^{np+n}_\partial(\mathfrak{U}_\mathbf{K} \mathbf{L}_2)$, minimizing the penalty functional

$$J(\eta, u) = \mu \sum_{q=0}^{n} \int_0^\tau \| \overset{o}{\eta}{}^{(q)} - \overset{o}{\tilde{\eta}}{}^{(q)} \|^2_{\mathfrak{G}_\mathbf{K} \mathbf{L}_2} dt + \nu \sum_{q=0}^{np+n} \int_0^\tau \langle N_q \overset{o}{u}{}^{(q)}, \overset{o}{u}{}^{(q)} \rangle_{\mathfrak{U}_\mathbf{K} \mathbf{L}_2} dt. \tag{17}$$

Here $\mu, \nu > 0$, $\mu + \nu = 1$, $N_q \in \mathcal{L}(\mathfrak{U}_\mathbf{K} \mathbf{L}_2)$, $q = 0, 1, \ldots, np+n$, are self-adjoint positively defined operators, and $\tilde{\eta}(t)$ is the target state of the system.

Theorem 6 *Let the operator B be (A,p)-bounded, $p \in \{0\} \cup \mathbb{N}$. Then for arbitrary $w \in H^{np+n}(\mathfrak{G}_\mathbf{K} \mathbf{L}_2)$ there exists a unique optimal control to solutions of problem (4), (5).*

Proof By Theorem 5, for arbitrary $w \in H^{np+n}(\mathfrak{G}_\mathbf{K} \mathbf{L}_2)$, $\xi_m \in \mathfrak{V}_\mathbf{K} \mathbf{L}_2$, and $u \in H^{np+n}(\mathfrak{U}_\mathbf{K} \mathbf{L}_2)$ there exists a unique strong solution $\eta \in H^n(\mathfrak{G}_\mathbf{K} \mathbf{L}_2)$ to problem (4), (5), given by

$$\eta(t) = (A_1 + A_2)(w + Cu)(t) + k(t), \quad (18)$$

where the operators A_1, A_2 and the vector function k are defined in Lemma 3.

Fix $w \in H^{np+n}(\mathfrak{G_K L}_2)$ and $\xi_m \in \mathfrak{V_K L}_2$, and consider function (18) as a mapping $D: u \mapsto \eta(u)$. The mapping $D: H^{np+n}(\mathfrak{U_K L}_2) \to H^n(\mathfrak{G_K L}_2)$ is continuous. Therefore, the penalty functional depends only on u: $J(\eta, u) = J(u)$.

We write out the functional (17, p. 13 of the Russian translation) in the form

$$J(u) = \mu \|\eta(t, u) - \tilde{\eta}\|^2_{H^n(\mathfrak{G_K L}_2)} + \nu[v, u],$$

where $v^{(q)}(t) = N_q u^{(q)}(t)$, $q = 0, \ldots, np + n$. Hence it follows that

$$J(u) = \pi(u, u) - 2\lambda(u) + \mu \|\tilde{\eta} - \eta(t, 0)\|^2_{H^n(\mathfrak{G_K L}_2)},$$

where

$$\pi(u, u) = \mu \|\eta(t, u) - \eta(t, 0)\|^2_{H^n(\mathfrak{G_K L}_2)} + \nu[v, u]$$

is a bilinear continuous coercive form on $H^{p+n}(\mathfrak{U_K L}_2)$ and

$$\lambda(u) = \mu \langle \tilde{\eta} - \eta(t, 0), \eta(t, u) - \eta(t, 0) \rangle_{H^n(\mathfrak{G_K L}_2)}$$

is a linear continuous form on $H^{np+n}(\mathfrak{U_K L}_2)$. Therefore, the assumptions of theorem in [18, p. 13, Theorem 1.1], are satisfied. The proof of the theorem is complete.

References

1. Al'shin, A.B., Korpusov, M.O., Sveshnikov, A.G.: Blow-up in Nonlinear Sobolev Type Equations. Walter de Gruyter and Co., Berlin (2011)
2. Demidenko, G.V., Uspenskii, S.V.: Partial Differential Equations and Systems Not Solvable with Respect to the Highest Order Derivative. Basel, Hong Kong, Marcel Dekker, Inc, N.Y. (2003)
3. Favini, A., Yagi, A.: Degenerate Differential Equations in Banach Spaces. Basel, Hong Kong, Marcel Dekker, Inc, N.Y. (1999)
4. Showalter, R.E.: Hilbert Space Methods for Partial Differential Equations. Pitman, London, San Francisco, Melbourne (1977)
5. Sviridyuk, G.A., Fedorov, V.E.: Linear Sobolev Type Equations and Degenerate Semigroups of Operators. VSP, Utrecht, Boston, Koln, Tokyo (2003)
6. Zamyshlyaeva, A.A.: The higher-order Sobolev-type models. Bull. South Ural State Univ. Ser.: Math. Model. Program. Comput. Softw. **7**(2), 5–28 (2014). https://doi.org/10.14529/mmp140201 (in Russian)
7. Gliklikh, YuE: Global and Stochastic Analysis with Applications to Mathematical Physics. Dordrecht, Heidelberg, N.Y., Springer, London (2011)
8. Zagrebina, S.A., Soldatova, E.A.: The linear Sobolev-type equations with relatively p-bounded operators and additive white noise. Bull. South Ural State Univ. Ser.: Math. Model. Program. Comput. Softw. **6**(1), 20–34 (2013)

9. Kovács, M., Larsson, S.: Introduction to stochastic partial differential equations. In: Proceedings of "New Directions in the Mathematical and Computer Sciences". National Universities Commission, Abuja, Nigeria, October 8–12, 2007. Publications of the ICMCS, vol. 4, pp. 159–232 (2008)
10. Favini, A., Sviridyuk, G.A., Zamyshlyaeva, A.A.: One class of Sobolev type equations of higher order with additive "White Noise". Commun. Pure Appl. Anal. **15**(1), 185–196 (2016)
11. Shestakov, A.L., Sviridyuk, G.A.: On a new conception of white noise. Obozrenie Prikladnoy i Promyshlennoy Matematiki, Moscow, **19**(2), 287–288 (2012)
12. Shestakov, A.L., Sviridyuk, G.A.: On the measurement of the "white noise". Bull. South Ural State Univ. Ser.: Math. Model. Program. Comput. Softw. **27**(286), 99–108 (2012)
13. Gliklikh, Yu.E.: Investigation of Leontieff type equations with white noise protect by the methods of mean derivatives of stochastic processes. Bull. South Ural State Univ. Ser.: Math. Model. Program. Comput. Softw. **27**(286), 24–34 (2012)
14. Gliklikh, YuE: Mean derivatives of stochastic processes and their applications. SMI VSC RAS, Vladikavkaz (2016). (in Russian)
15. Sviridyuk, G.A., Manakova, N.A.: Dynamic models of Sobolev type with the Showalter-Sidorov condition and additive "Noises". Bull. South Ural State Univ. Ser.: Math. Model. Program. Comput. Softw. **7**(1), 90–103 (2014). (in Russian)
16. Favini, A., Sviridyuk, G.A., Manakova, N.A.: Linear Sobolev type equations with relatively p-sectorial operators in space of " Noises". Abstract Appl. Analy. Article ID: 69741, 8 (2015). https://doi.org/10.1155/2015/697410
17. Zamyshlyaeva, A.A.: Stochastic incomplete linear Sobolev type high-ordered equations with additive white noise. Bull. South Ural State Univ. Bull. South Ural State Univ. Ser.: Math. Model. Program. Comput. Softw. **40**(299), 73–82 (2012). (in Russian)
18. Lions, Zh-L: Optimal'noe upravlenie sistemami, opisyvaemymi uravneniyami s chastnymi proizvodnymi. Mir, Optimal Control of Systems Described by Equations with Partial Derivatives. Moscow (1972). (in Russian)

Printed by Books on Demand, Germany